Ecological Studies

Analysis and Synthesis

Edited by

J. Jacobs, München · O. L. Lange, Würzburg
J. S. Olson, Oak Ridge · W. Wieser, Innsbruck

Volume 7

Mediterranean Type Ecosystems

Origin and Structure

Edited by

Francesco di Castri and Harold A. Mooney

With 88 Figures

Springer-Verlag New York · Heidelberg · Berlin 1973

ISBN 0-387-06106-1 Springer-Verlag New York · Heidelberg · Berlin
ISBN 3-540-06106-1 Springer-Verlag Berlin · Heidelberg · New York
ISBN 0-412-68420-9 Chapman & Hall Limited London

Typesetting: Union Druckerei GmbH Stuttgart. Printing and bookbinding: Georg Appl, Wemding

Preface

No other disjunct pieces of land present such striking similarities as the widely separated regions with a mediterranean[1] type of climate, that is, the territories fringing the Mediterranean Sea, California, Central Chile and the southernmost strips of South Africa and Australia. Similarities are not confined to climatic trends, but are also reflected in the physiognomy of the vegetation, in land use patterns and frequently in the general appearance of the landscape. The very close similarities in agricultural practices and sometimes also in rural settlements are dependent on the climatic and edaphic analogies, as well as on a certain commonality in cultural history. This is certainly true for the Mediterranean Sea basin which in many ways represents a sort of ecological-cultural unit; this is also valid for California and Chile, which were both settled by Spaniards and which showed periods of vigorous commercial and cultural interchanges as during the California gold rush. One other general feature is the massive interchange of cultivated and weed species of plants that has occurred between the five areas of the world that have a mediterranean-type climate, with the Mediterranean basin region itself as a major source.

In spite of their limited territorial extension, probably no other parts of the world have played a more fundamental rôle in the history of mankind. Phoenician, Etruscan, Hellenic, Jewish, Roman, Christian and Arab civilizations, among others, have shaped many of man's present attitudes, including his position and perception vis-à-vis nature. Also the mediterranean-climate areas spotted in other continents had and are having an important rôle in the cultural and economic development of the adjacent regions. On the other hand, few areas have been more radically transformed by man, to an extent that recognition of what would have been the "natural" ecosystem becomes a very difficult task. A profound understanding of historical factors is therefore essential for explaining the structure and functioning of today's mediterranean ecosystems.

Furthermore, because of the rolling topography of these regions, of the occurrence of many different local climates, of their transitional position in regard to recent and ancient geological events, and of the different degrees of intervention by man, mediterranean ecosystems and mediterranean landscapes are mainly characterized by structural heterogeneity. In each given area, there is interpenetration of different plant formations (from woody to herbaceous and desertic), juxtaposition of many biogeographical elements of different origin, and interplay of different land uses.

Due to the scientific interest both from a basic and an applied point of view of the aspects evoked above, and also because of the presence in these regions of a number of research teams, many studies have been carried out in these five areas of the world in the framework of the International Biological Programme (IBP). An integrated research programme focussing on Chilean and Californian ecosystems, and supported since 1970 by

1 To avoid misunderstandings concerning the meaning of the word *mediterranean*, in this volume Mediterranean is written with a capital letter when the geographical region of the Mediterranean Sea basin is referred to, while mediterranean is written with a small letter when reference is made to mediterranean type features, such as mediterranean type climate or mediterranean type vegetation.

the National Science Foundation of the USA, is taking advantage of the peculiarity of these two disjunct regions – that is, regions of analogous patterns of climate and topography but with native plants and animals of different biogeographical origin – in an attempt to answer a fundamental biological question: "whether very similar physical environments, acting on phylogenetically dissimilar organisms in different parts of the world, will produce structurally and functionally similar ecosystems".

One of the greatest barriers to understanding most ecosystems is the lack of repeatability. Hypotheses based on detailed studies of a particular ecosystem cannot be put in a context of generality because there are often no standards of comparison, that is no precisely similar ecosystems built up from different starting points in which to test the principles involved. This question has major practical implications for the management of natural resources. In effect, in order to assess the transfer of technologies, it is essential to know to what extent information acquired from studying one particular ecosystem is applicable to another ecosystem of the same type but situated in a different geographic location. Mediterranean climate regions, widely separated in geographic terms but with evident ecological similarities, provide the best basis for this kind of comparative study. Since the IBP has logically concentrated on intensive site studies of the functioning of ecosystems, it is necessary to ensure the validity of the generalizations that will emerge, through promoting more extensive comparable research in which particular emphasis is given to the evolutionary aspects of ecosystems. This is a basic requirement for predicting the future evolution of populations and ecosystems as human pressures on these biotic units increase.

In the framework of the integrated research programme on the origin and structure of ecosystems mentioned above, a symposium was organized by the Universidad Austral de Chile, Valdivia. Supported mainly by the National Science Foundation of the USA, this symposium was held in Valdivia, Chile, in late March 1971. One of the purposes of this meeting was to make a preliminary critical overview of the information available on mediterranean climate ecosystems in the light of the approach adopted in the research programme, that is, with special emphasis on evolutionary and structural aspects. Contributors to this symposium were selected according to this approach, having in mind a fairly precise outline of the publication which would result from the symposium. As presented at the meeting, most papers took the form of preliminary drafts serving as a basis for fruitful discussions on the different themes. In effect, this discussion helped in the further elaboration of these contributions and in avoiding undue overlap between the different chapters, as can be seen by the frequency of cross references in this book. Most of the manuscripts, including new bibliography, were finalized by late 1971-early 1972, when the true editorial work started.

It is not claimed that this volume presents a complete review of the different approaches to the study of mediterranean ecosystems; nor is it the intention to treat in depth a single approach or to be comprehensive for one or more of the mediterranean regions. The principal aim is to show that different approaches can and should coexist in an interdisciplinary exercise, and that different levels of sophistication in ecological research can interplay to reflect the nature of the problem, the present development of different scientific fields, and the state of knowledge in different regions.

It is true that, because of the very nature of the Valdivia symposium and of the co-operative research undertaken in California and Chile, this book put a greater emphasis on the mediterranean regions of the New World, while including for comparison selected chapters dealing with other areas. But, on the other hand, more works of synthesis have previously been done in Europe, at least from the aspect of mediterranean vegetation.

The text of this volume consists of twenty two chapters, grouped into seven sections. Following a general section introducing the concept of convergence in ecosystems, one section deals with the physical geography of mediterranean areas – climate, landforms and soils – and two sections consider the structure and some functional responses of vegetation and litter-soil subsystems. The following two sections are devoted to evolutionary and biogeographical aspects of the biotic components of these ecosystems, both plants and animals. A final section reviews different kinds and phases of human activities in mediterranean ecosystems. Short introductions to the seven sections provide a summary of the chapters included in each section and attempt to show the main interrelationships between them as well as possible links with other sections. In each section we have tried to include not only general statements of problems, but also reports of more detailed research, or a case study reflecting the situation in one of the five mediterranean regions, or examples from selected taxonomic groups. In this way, diverse approximations are given, ranging from generality toward different approaches to specificity.

Paris and Stanford, September 1973 F. DI CASTRI and H. A. MOONEY

Contents

Section V: Plant Biogeography

Section VI: Animal Biogeography and Ecological Niche

Section VII: Human Activities Affecting Mediterranean Ecosystems

Contributors

ASCHMANN, HOMER, Department of Geography, University of California, Riverside, CA, USA

AXELROD, DANIEL I., Department of Geology, University of California, Davis, CA, USA

BRADBURY, DAVID E., Department of Geography, University of California, Los Angeles, CA, USA

DI CASTRI, FRANCESCO, Instituto de Ecología, Universidad Austral de Chile, Valdivia, Chile. Present address: Section of Ecology, UNESCO, Paris, France

CODY, MARTIN L., Department of Zoology, University of California, Los Angeles, CA, USA

DAN, JOEL, Volcani Institute of Agriculture, Rehovot, Israel

DIFEO, JR., DAN R., The Cell Research Institute and Department of Botany, University of Texas, Austin, TX, USA

HURTUBIA, JAIME, Instituto de Ecología, Universidad Austral de Chile, Valdivia, Chile

JOHNSON, ALBERT W., Department of Biology, San Diego State College, San Diego, CA, USA

KUMMEROW, JOCHEN, Laboratorio de Botánica, Universidad Católica de Chile, Santiago, Chile

LOSSAINT, PAUL, Laboratoire d'Eco-Pédologie, Centre d'Etudes Phytosociologiques et Ecologiques, CEPE, CNRS, Montpellier, France

MABRY, TOM J., The Cell Research Institute and Department of Botany, University of Texas, Austin, TX, USA

MOONEY, HAROLD A., Department of Biological Sciences, Stanford University, Stanford, CA, USA

NAVEH, ZEV, Faculty of Agricultural Engineering, Technion-Israel Institute of Technology, Haifa, Israel

PARSONS, DAVID J., Department of Biological Sciences, Stanford University, Stanford, CA, USA

PASKOFF, ROLAND P., Centre d'Etudes Géographiques, Université de Tunis, Tunis, Tunisia

RAVEN, PETER H., Missouri Botanical Garden and Washington University, St. Louis, MO, USA

SAGE, RICHARD D., Department of Zoology, University of Texas, Austin, TX, USA

SÁIZ, FRANCISCO, Departamento de Biología, Universidad Católica de Valparaíso, Valparaíso, Chile

SCHAEFER, ROGER, Laboratoire d'Ecologie Végétale, Université de Paris-Sud, Orsay, France

SPECHT, RAYMOND L., Department of Botany, University of Queensland, St. Lucia, Brisbane, Queensland, Australia

THROWER, NORMAN J. W., Department of Geography, University of California, Los Angeles, CA, USA

VITALI-DI CASTRI, VALERIA, Muséum National d'Histoire Naturelle, Laboratoire de Zoologie (Arthropodes), Paris, France

ZINKE, PAUL J., School of Forestry, University of California, Berkeley, CA, USA

Section I: Convergence in Ecosystems

An historical view of the concept of the evolution of ecosystems with a particular emphasis on the phenomenon of convergence is provided by JOHNSON in the following chapter. He documents the antiquity of the observations on convergence of ecosystems and follows the development of thought in this area.

It is clear that although there has long been recognition of the existence of convergent vegetation types in homologous climates of the world, the basis for this observation is quite superficial. Although physiognomic vegetation analogs can be cited, in no case do we know the detailed aspects of convergence in any single ecosystem type. That is, do convergent systems which have comparable vegetation structural attributes, also have equally comparable structural (and presumably functional) consumer and decomposer types although they may have little or no taxonomic affinity? More specifically will there be, for example, the same approximate number of mammal species in the two systems which are dividing the available resources in the same manner?

The International Biological Program on the Origin of Ecosystems is attempting to provide this depth of detail for two convergent ecosystems – the mediterranean scrub of California and Chile and the desert scrub of Arizona and Argentina.

The evidence is accumulating that for any given climate there is an optimal solar energy capture system which is manifested by plants which has, among other things, certain leaf types and amounts held at prescribed angles for given durations. Homologous climates have vegetation which are generally analogous in these features.

The physical configuration of the energy capture system, or vegetation structural unit, in turn determines to a large degree the microclimate of the area which places constraints on the spatial distribution of consumers.

The output of energy to consumers in the form of leaves, flowers, fruits, etc. is regulated to a large degree by climate, but also by the competitive strategies between the producers themselves (e.g., the timing of canopy development) as well as by the activities of the consumers which act as selective forces (e.g., the asynchrony of flowering by different plants which may relate to competition for pollinators).

Thus, although the spatial and temporal distribution of ecosystem resources may be determined to a large degree by strictly climatic considerations, co-evolutionary forces may also play a large role. This may mean that convergence within the systems may be less noticeable at higher trophic levels. Another evolutionary force which may lead to dissimilarity in analogous systems is the lineage constraint of the colonizers to the system. However, if the colonizers come from the same general source region, and if past climatic histories have been similar, these effects may be minimal.

Hopefully, within the next few years, we will have more evidence as to whether equal environment produces equal structural – functional units independent of phylogeny.

The introduction to this section was prepared by HAROLD A. MOONEY.

1

Historical View of the Concept of Ecosystem Convergence

ALBERT W. JOHNSON

Introduction

Plants (or animals) living in separate but climatically similar areas often resemble each other to some extent. Sometimes this is explained by common ancestry; where organisms have only remote ancestral connections, however, their similarities are more often attributed to the effects of similar environmental conditions on the evolution of each group. Traditionally, similarity through common descent, homology, has been called parallel evolution; similarity based on analogous structures is referred to as convergent evolution. Because it is often difficult to distinguish between the two, in practice the term convergence refers loosely to those cases where no obvious close relationship exists between the organisms showing convergent characteristics.

The extremes of these two situations illustrate this point. The plants of the circumpolar arctic tundra look alike and are rather closely related from place to place; although some of the northern land areas are isolated from each other today, they have been connected intermittently and opportunities have existed for the exchange of species. On the other hand, the deserts of the world are relatively distant from one another, have had considerably different histories and are composed of similar appearing but quite unrelated plants and animals. Of course, many intermediate situations exist.

Convergent evolution gives rise to organisms that are ecologically equivalent. ALLEE *et al.* (1949) review evidence that illustrates the pervasiveness of convergence in animals. Similarities in methods of food gathering, body support, patterns of fat deposition, behavior and coloration among distantly related animal groups are common. Convergence among plants is often seen in growth form, morphological details, and other features of the life history and has been demonstrated in physiological processes as well e.g. MOONEY and DUNN (1970 a).

In a strict sense convergence applies to structures and functions, not to organisms. It is important to note that the similarities that arise through convergence do not obscure the fundamental distinctions between the groups. DARWIN (1859) for example, pointed out the "utter dissimilarity" among the faunas and floras of the areas of mediterranean climate in the southern hemisphere despite the extreme similarity in all their conditions. He contrasted this with the "incomparably" closer relationship of organisms living in geographically contiguous but climatically different areas. An ecologist confronted with the same situation would probably react almost oppositely, because he would focus his attention on structure and form and their relationship to habitat. Thus, convergence as a phenomenon, is of considerable interest to the ecologist because it gives him a chance to test his ideas about adaptation in similar though widely separate environments.

DARWIN admitted that cases of "analogous variation" occurred as a result of species being "exposed to similar influences". He preferred this explanation to one requiring "separate yet closely related acts of creation". Although DARWIN himself does not seem to have stressed the point, convergence is usually interpreted as strong evidence for the "efficacy of selection and for its adaptive orientation of evolution" (SIMPSON, 1953). Since it is improbable that convergent characteristics would arise separately by chance, the way in which organisms resemble each other are probably adaptive. Following FISHER (1930) most evolutionary biologists are inclined to support the view that "no character is likely to remain immune from selection for very long", so in all likelihood the ways in which organisms showing convergent characters differ are also adaptive.

BATES (1863) considered mimicry to be different from convergence, primarily because he assumed, probably correctly, that a predator common to the mimicked species and its copy is essential. Also, both must live in the same area. Neither is necessary in the examples of convergence discussed thus far. Since natural selection is clearly responsible for both situations it is probably better to regard mimicry as a special case of convergence.

Total agreement has not always existed on the nature of and causes for convergence. BERG (1926) writing at a time when disenchantment with Darwinian evolution was high, believed that the fact of convergence was the strongest kind of evidence *against* selection, since "it is scarcely credible that such (chance) variations should arise accidentally in even one species; but still more incredible would be its occurrence in different species having no common ancestor". BERG's essentially Lamarckian theory of nomogenesis (an "unfolding of pre-existing rudiments") to explain such similarities generally discounted the entire concept of evolution by natural selection.

Evolution by the slow accumulation of small differences has always had its sceptics. In genetics, R. B. GOLDSCHMIDT (1940) viewed this kind of evolution as being entirely too slow to account for known changes through geologic time and proposed the idea that macromutations might play a major role in evolution. Ecologists often forget that the father of American dynamic ecology, F. E. CLEMENTS, proposed essentially Lamarckian processes to account for some of his later work on supposed species transmutations. Even today, doubt is expressed that natural selection is adequate to explain convergence. WENT (1971) for example, is of the opinion that the very large number of convergences that occur among unrelated organisms could not have arisen by natural selection. He thinks that some of them might be "just locally (non-adaptive) available characters" whose widespread occurrence in an area can be explained by the one-time non-sexual transfer (by insects, for example) of chromosome segments between unrelated plants.

Because there is little or no basis for accepting views such as this, we assume, here, that neoDarwinian concepts of evolution are sufficient to account for the production of organic adaptations and that convergence provides especially convincing evidence for their support.

Convergence in Higher Units

Classification of the earth's vegetation into communities of similar physiognomy, life-form or physiological characteristics (GRISEBACH, 1872; DRUDE, 1890; SCHIMPER, 1903) was based on tacit recognition of convergence as was biological community classification based on climatic comparisons, e.g. HOLDRIDGE, 1947). Ecological generalizations such as RAUNKIER's Biological Spectrum, BERGMANN's Principle, and ALLEN's Rule support the idea that the climate has widespread and similar effects on whole groups of organisms.

SCHIMPER, in particular, came close to providing a modern explanation for plant distribution on the basis of the interactions between plants and the environment. While his subjective interpretations of the adaptive significance of the structural features of plants is not always in agreement with more recent work, they constitute a beginning for understanding the general classes of adaptations significant in different climates. With respect to the areas of mediterranean climate, for example, he observed "the vegetation bears essentially the same stamp, in spite of deep seated differences in composition of the flora; it is dominated by sclerophyllous plants, and always, although to a subordinate extent, by tuberous and bulbous plants".

The mediterranean climatic areas have attracted most attention in studies of vegetation convergence, probably because they are all reasonably accessible, are located near educational and research centers and are in the greatest danger of exploitation and destruction. Recent studies by NAVEH (1967), SPECHT (1969 a, 1969 b), MOONEY and DUNN (1970 a, 1970 b) and MOONEY *et al.* (1970) compare structural and physiological features of the plant communities of California and Israel, California, Australia and France, and California and Chile, respectively. These specific studies are not summarized here, but they show that plants of these vegetation types are more than superficially alike and in some cases their similarities extend to basic patterns of growth, morphology and physiology.

Convergence in Ecosystems

Abundant evidence already exists to show that certain biological and physical features of some ecosystems are much alike. Comparative data, not only on structural and functional properties, but on rates of processes such as energy transfer, are becoming available from the International Biological Program and other ecosystem studies. Very little information is available, however, that compares the details of structure and function of ecosystems in the same type of climate. Thus, this discussion concentrates on problems associated with measuring and analyzing convergence rather than attempting a synthesis of what is known from isolated studies that have little relevance to one another. We are still in a relatively early and descriptive phase of investigating an evolutionary phenomenon that until now has been examined primarily at the species population level.

The fundamental question of interest in comparative ecosystem studies is one that addresses our ability to generalize about ecosystems. If one knows intimately the details of composition, structure and function in one ecosystem can one predict similar properties in another ecosystem of the same type but whose species composition may be almost totally different? Or is each ecosystem a unique collection of organisms which, despite superficial similarities is so different from all others as to defy generalization?

Thus far, ecosystem studies have emphasized structure, trophodynamics, nutrient cycling, succession, and diversity and stability. MARGALEF (1968) and ODUM (1969) propose "strategies" common to developmental processes in all ecosystems. Some of these processes, e.g. succession, have been studied in great detail and in many different ecosystems and communities. Here, it would seem, generalization is possible. In other cases, such as diversity, some doubt exists that the concept is even meaningful biologically (HURLBURT, 1971), but however one regards it, generalization would be premature.

Some assumptions are necessary before any of the questions raised above can be answered. First, ecosystems evolve by additions and subtractions of species or by changes in the properties of those of which they are comprised. There is little or no basis for recognizing the ecosystem as the unit upon which natural selection works, although its properties are altered by evolution. At this stage in our understanding ecosystem convergence probably should be defined in terms of the tendencies for certain features of organisms belonging to two systems to resemble each other. Presumably, these evolve separately from one another but under the aegis of similar selection agents.

Secondly, ecosystems converge only to the extent possible given the evolutionary potential of the organisms concerned. For example, forest vegetation, common to all high altitude ecosystems of North America, is absent from similar climates in South America. The indigenous forest vegetation of South America seems to lack the potential for occupying these kinds of places.

Third, the evolution of the dominant vegetation is probably the most fundamental aspect of ecosystem history. Climate exercises primary influence on plants and determines (through evolution) their physiognomic and physiological characteristics. Vegetation, then viewed as a resource, also poses constraints on what is possible in animal evolution.

HUTCHINSON (1959) emphasizes the importance of plant structural and chemical diversity in contributing to animal diversity. One should expect, all other things being equal, that the greater the plant diversity with respect to architecture and composition, the greater the different kinds of animals there could be.

As indicated above, however, vagaries of migrations and other historical accidents have provided ecosystems with different collections of evolutionary "pioneers". Some kinds of plants and animals common in one ecosystem may be completely absent in its climatic homolog. There is no assurance that their functions will be assumed by other organisms, nor can one assume that a single species in one ecosystem will evolve to utilize all of the resources several species use at the same trophic level in another. Thus, comparative analysis of the number of kinds of animals occupying the same trophic level in two otherwise similar ecosystems may show large differences. It seems likely, as has been emphasized by HAIRSTON, SMITH and SLOBODKIN (1960), that in most terrestrial ecosystems only negligible amounts of photosynthate remain unused on an annual basis. If it is consumed on the herbivore level in one ecosystem there seems to be no fundamental reason why it could not be consumed primarily by the decomposers in another one. In fact, the increasing species richness of ecosystems through evolution has probably resulted in less and less of the energy fixed by the primary producers being available below the herbivore level. Clearly some upper limit on species richness must be characteristic of any ecosystem type. And it must be determined by that point at which a further subdivision of ecosystem resources would provide insufficient energy for reproduction of another species. We have no evidence that any ecosystem has ever approached this upper limit.

The point is that until recently ecologists have not had the opportunity to gather the basic physical and biological data to understand even single ecosystems. Studying pairs of ecosystems where one can have some confidence in the similarity of the physical and historical backgrounds of the two systems should help settle some of these unresolved questions raised above. It is possible that our interpretation of events in single ecosystems is clouded by subjectivity. When two systems, composed of quite unrelated organisms, are structured and are functioning in similar ways, our explanations seem more probable. The differences between them also are more susceptible to analysis. The answers to all of these questions relate to variation, selection and adaptation. Hopefully, a comparison of two ecosystems that have developed against similar physical backgrounds will provide some basis for determining whether similar evolutionary pathways have been used in each.

References

ALLEE, W. C., EMERSON, A. E., PARK, O., PARK, T., SCHMIDT, K. P.: Principles of Animal Ecology. Philadelphia and London: W. B. Saunders Co. 1949.

BATES, H. W.: The Naturalist on the River Amazon. London: Murray 1863.

BERG, E. S.: Nomogenesis or evolution determined by law. London: 1926.

DARWIN, C. R.: The Origin of Species by Means of Natural Selection or the Preservation of Favored Races in the Struggle for Life. London: Murray 1859.

DRUDE, O.: Handbuch der Pflanzengeographie. Stuttgart: J. Engelhorn 1890.

FISHER, R. A.: The Genetical Theory of Natural Selection. Oxford: Clarendon Press 1930.

GRISEBACH, A. H. R.: Die Vegetation der Erde nach ihrer klimatischen Anordnung. Leipzig: W. Engelmann 1872.

GOLDSCHMIDT, R.: The Material Basis of Evolution. New Haven: Yale University Press 1940.

HAIRSTON, N. G., SMITH, F. E., SLOBODKIN, L. B.: Community structure, population control and competition. Amer. Nat. **94,** 421–425 (1960).

HOLDRIDGE, L. R.: Determination of world plant formations from simple climatic data. Science **105,** 267–368 (1947).

HURLBURT, S. H.: The non-concept of species diversity: a critique and alternative parameters. Ecology **52,** 577–586 (1971).

HUTCHINSON, G. E.: Homage to Santa Rosalia or Why are there so many kinds of animals. Amer. Nat. **92,** 145–159 (1959).

MARGALEF, R.: Perspectives in Ecological Theory. Chicago: University of Chicago Press 1968.

MOONEY, H. A., DUNN, E. L.: Photosynthetic systems of mediterranean-climate shrubs and trees of California and Chile. Amer. Nat. **104,** 447–453 (1970 a).

MOONEY, H. A., DUNN, E. L.: Convergent evolution in mediterranean-climate evergreen sclerophyll shrubs. Evol. **24,** 292–303 (1970 b).

MOONEY, H. A., DUNN, E. L., SHROPSHIRE, F., SONG, L.: Vegetation comparisons between the mediterranean climatic areas of California and Chile. Flora **159,** 480–496 (1970).

NAVEH, Z.: Mediterranean ecosystems and vegetation types in California and Israel. Ecology **48,** 445–459 (1967).

ODUM, E. P.: The strategy of ecosystem development. Science **164,** 262–270 (1969).

SCHIMPER, A. F. W.: Plant geography upon a physiological basis. (Transl. by W. R. FISHER). Oxford: Clarendon Press 1903.

SIMPSON, G. G.: The Major Features of Evolution. New York: Columbia University Press 1953.

SPECHT, R. L.: A comparison of the sclerophyllous vegetation characteristic of mediterranean type climates in France, California, and southern Australia. I Structure, morphology, and succession. Aust. J. Bot. **17,** 277–292 (1969 a).

SPECHT, R. L.: A comparison of the sclerophyllous vegetation characteristic of mediterranean type climate in France, California, and southern Australia. II. Dry matter energy, and nutrient accumulation. Aust. J. Bot. **17,** 293–308 (1969 b).

WENT, F. W.: Parallel evolution. Taxon **20,** 197–226 (1971).

Section II: Physical Geography of Lands with Mediterranean Climates

Abundant data are available concerning the physical geographies of each of the several areas in the world that experience mediterranean climates. In general, however, they are couched in different descriptive or theoretical matrices and are not directly comparable. The chapters presented here attempt to express points of view or tentative methodologies for examining the non-biotic framework or environment in which an ecosystem functions. To the extent that such methodologies can be applied consistently to different areas, intercontinental comparisons of ecosystems can rest on a firm base of physical understanding.

In describing one or more physical features of factors as they occur in distant but presumably similar environments the student is led to take one of two approaches. He may point up how the two areas resemble each other and differ from the remainder of the world. Or he may assume a basic similarity and focus on the more or less subtle variations, differences that invariably exist among distantly located places on the earth's surface.

Climate, the environmental feature that defines the areas of interest to this volume, is considered to some extent in each of the chapters in this section, with two giving it their prime attention. ASCHMANN's approach is to define the true mediterranean climate quite narrowly, identifying remarkably tiny areas but still showing a distribution on all the inhabited continents. He recognizes another, more extensive set of semi-mediterranean climates on their borders. Even the narrowly defined true mediterranean climate regions, however, prove to have distinctive regional characters on each of the several continents, with Chile's being notably cool, Australia's notably warm and the Southern Hemisphere areas showing far less intraregional variability than those in the Northern Hemisphere. DI CASTRI is less restrictive in his definition, requiring only some concentration of rainfall in winter with at least one month of drought in summer and the absence of continental winter cold. Thus he considers areas in North America and Chile that are occupied by deserts or dense west coast marine forests. His immediate comparison of data for the most analogous station pairs on the two continents shows that remarkably good matches can be found, and has uncovered a singular, and as yet inexplicable, consistent difference. The beginnings of the arid and humid periods often occur about a month earlier in their respective seasons in Chile than in California. The ecological implications of this situation are not clear.

The chapters by THROWER and PASKOFF are primarily concerned with topography and landforms, the former focussing on the specific characteristics of each of the regions on their several continents. As products of depositional and tectonic histories throughout geologic time, during which continental displacements and major variations in world climatic patterns have occurred, the incidence of parallel situations, though not universal, is surprizingly great. It is, however,hard to regard as other than fortuitous. PASKOFF, on the other hand, is concerned with the general presence of a distinctive set of landform features, of both erosional and depositional origin, that are characteristic of regions of mediterranean climate. He finds them a product of specific, recurring climatic conditions. The features, however, are more enduring than the climatic patterns, and many now present can be

---

The introduction to this section was prepared by HOMER ASCHMANN.

explained only in terms of repeated systematic displacements of climatic belts in Pleistocene and Upper Tertiary times. Further, valley cutting or filling is extremely sensitive to changes in the vegetative cover, and this can be altered by man as well as by climatic change.

Soil conditions, which both affect and are affected by lithology, topography, climate, and vegetative cover, constitute the most complex of the features of the physical environment. The present soils are products of ancient geologic events, old and modern climatic conditions and changes, and modifications in the vegetative cover for whatever cause. The character of the latter, however, is significantly influenced by the soil. ZINKE's chapter deals empirically with specific soil samples along transects through considerable distance and elevation changes in California and Italy and Greece. He finds remarkably good analogies when he is able to match climate, slope, exposure, lithology, and vegetation. Perhaps his most significant conclusion is that in all the regions he studied soils vary sharply over short distances, but in regular fashion. Meaningful statements can be made in terms of catenas, not in terms of regional or zonal soils.

It is not imagined that the chapters presented here provide the data of physical geography needed for a comparative study of ecological systems in any of the localities concerned. Rather they point to the kinds of data that are relevant and make suggestions for their organization and analysis.

1

Distribution and Peculiarity of Mediterranean Ecosystems

HOMER ASCHMANN

Parameters of the Mediterranean Climate

The lands of mediterranean scrub or chaparral climate and the ecosystems that have developed in them must be defined in climatic terms. If we attempt to focus on the characteristic core of a mediterranean scrub or chaparral climate three terms stand out, two involving precipitation and one temperature. The most distinctive term involves the concentration of rainfall in the winter half year, November through April in the northern hemisphere and May through October in the southern. Although at a large number of stations, especially in California and Chile, 80 or even 90% of the precipitation occurs in winter, so large a proportion rarely obtains around the Mediterranean Basin itself. The value of at least sixty five percent of the year's precipitation occurring in the winter half year seems, on the basis of examining a considerable number of station records, to form a satisfactory boundary. Winter rainfall, because of lower evaporation, is more effective in sustaining plant growth than is warm season precipitation; nonetheless, in this climatic region all but favorably located phreatophytic vegetation is subject to drought stress in summer.

The total amount of precipitation must be sufficient to support a continuous vegetative cover on all but the most rocky sites. On the other hand it should be insufficient to support a mesophytic arboreal growth except where phreatophytes have access to groundwater. The problem of establishing satisfactory objectively determinable boundaries at both the dry and the wet ends of a region of mediterranean climate is singularly complex. Moisture effectiveness is the desired datum, and the following semi-independent elements contribute to it: total annual precipitation, seasonal concentration of precipitation that recognizes the greater effectiveness of rainfall in winter, annual temperature and seasonal temperatures, and relative humidity, particularly during the dry season. The definition used here is that devised by BAILEY (1958), and it considers the precipitation effectiveness as a function of the ratio of precipitation divided by temperature corrected for seasonality of precipitation [1]. It proves to be the case that introducing an average annual temperature

---

1 The equation may be expressed:

EP = $0.178\,(P/1.045^{T+x})$
P = precipitation in centimeters
T = temperature in C°
x = a correction factor for seasonal concentration of precipitation
EP = effective precipitation; the mediterranean climate falls between values of 3.4 and 8.7. Sample values are Los Angeles 3.65, Santiago 4.2, Concepcion, Chile 14.7 (BAILEY, 1958 and personal communication).

range term obviates the need for computing monthly precipitation effectiveness values. In terms of moisture then the mediterranean scrub climate is BAILEY's sub-humid and the wetter half of the semi-arid zone. Approximate annual precipitation values for the dry boundary are about 275 mm for cool coastal stations and 350 mm for warm interior stations that meet the winter rainfall concentration requirement of a mediterranean climate. The value of 900 mm works fairly well for the humid boundary.

The mediterranean climate must have a winter, which can readily be defined as a month with an average temperature below 15° C. As climatic regions on the earth's surface which satisfy the two aforementioned precipitation criteria actually occur no mid-latitude station approaches so high a winter temperature value. Summer-dry tropical areas such as Northern Ceylon are much warmer in winter. Winters, however, are mild. Although frost may occur in almost all areas of mediterranean climate it is relatively rare and normally not severe. As a phenomenon that directly affects plant life the proportion of the time that temperatures are below 0° C. is a good index of the severity of winters, and we can accept as a boundary that the hours per year at which the temperature at weather stations height falls below freezing (0° C.) should not exceed 3 percent of the total. As BAILEY (1966) has demonstrated in areas of mediterranean climate, this percentage value calculated from the annual range of mean monthly temperatures agrees remarkably well with the actual hourly temperature data from those stations at which such data are available (see Fig. 3).

The Köppen climatic classification with its *Cs* or olive climate defining the mediterranean climatic region (KÖPPEN and GEIGER, 1936) is well known. An examination of station data will show that the mediterranean climate as defined here is somewhat more restricted in its distribution. In particular, northerly and interior areas in North America and Eurasia such as the Oregon and Washington coast and the Columbia Plateaux prove to be too wet and too cold, respectively, supporting the impressions of the average naive observer. The significant presence of a broadleafed, deciduous forest or a dense coniferous one is incongruous to the mediterranean landscape. KÖPPEN's definition of the cold boundary of his *C* climate as −3° C. for the average temperature of the coldest month would allow at least 6 percent of the hours in a year to be below freezing. Even the ACKERMAN (1941) variant popular in the United States, which bounds the *C* climate at 0° C. for the coldest month would allow more than 4 percent of the hours in the year to be below freezing. These are absolutely minimal values. Fig. 3 shows statistically more probable ones.

Except for Australia, where the topography is characterized by moderate relief, the mediterranean lands are singularly rugged, with mountain ranges disposed sometimes parallel to and sometimes normal to the paths of atmospheric circulation. As a result climate can vary drastically over short horizontal distances. A transect of 25 kilometers can go from a valley too dry to be mediterranean, through middle slopes that are, into high areas too wet and cold to qualify, and reversing the sequence on the far side of the mountain. A world map of mediterranean climates then must exaggerate their area by failing to show local districts with different climates, compensated only partly by even smaller patches of mediterranean climate, that lie outside the principal areas. Considerably less than one percent of the earth's land surface qualifies as having a mediterranean climate.

Elevation, distance from the sea, and exposure, as contrasted with shelter by a mountain barrier, make for systematically and more or less predictably arranged variability within the mediterranean climate itself. Increasing elevation correlates positively and strongly with increasing precipitation. Stations exposed to the open ocean have cool summers, sometimes strikingly cool, and nearly frost free winters, while interior stations exhibit greater annual temperature ranges.

Distribution of Mediterranean Climates and Sub-Mediterranean Climates

The worldwide distribution of this singular climatic type shows a pattern of notable regularity, displaying a direct relationship to the general circulation of the world's atmosphere and its seasonal displacements. Areas of mediterranean climate are found between latitudes 32° and 40° north and south of the equator on the west coasts of continents. On the north side of the Mediterranean Sea they extend into somewhat higher latitudes and into somewhat lower latitudes in Western Australia. Equatorward the climate becomes desertic, and poleward precipitation values increase and rainfall is less concentrated in the winter season. If a major mountain range parallels the western continental coast the area of mediterranean climate is likely to extend only to its middle, seaward facing slopes, sometimes quite a short distance. In no case does this climate extend far into the interior of a continent because winter cold, drought, or summer rain serve to alter the category. The summer drought is associated with the presence of the subtropical high pressure belt at these latitudes, emphasized on west coasts by stable equatorward moving air from persistent oceanic anticyclonic cells. In winter the high pressure belt and the anticyclonic cells are displaced toward the equator and mid-latitude cyclones penetrate the mediterranean zone. The duration of the season in which this occurs and the frequency of such cyclonic penetration increases polewards. The relative mildness of the winters results from the fact that polar air masses associated with cyclonic storms almost always come from the west, hence from oceanic sources.

At a small scale on Figs. 1 and 2 I have endeavored to map the areas in the world that have truly mediterranean climates as defined above. The rarity of this climatic type as well as its notably wide distribution, occurring in each of the six inhabited continents, is immediately apparent. There is also a set of sub-mediterranean climates, those which qualify on two of the three criteria noted above. Since there is both an upper and a lower limit in terms of moisture availability there are four such sub-mediterranean possibilities: places with too much winter frost, places too dry, places too wet, and those in which precipitation is too evenly distributed through the year. Because of the disposition and extent of the several continents at the critical latitudes, and the sharp barrier effects of major topographic features such as the Andes not all the sub-mediterranean climates are represented on all continents.

In a few places, such as Anatolia, two or even three of the terms that define the boundary of the mediterranean climate lie in a zone only a few miles wide. In that example the country becomes too dry, too cold, and receives too much summer rain at almost the same place. As might be expected such sudden complete changes in climatic patterns are associated with mountain ranges lying athwart an important path of atmospheric flow. The northern littoral of the Mediterranean Sea in general is notably characterized by such drastic shifts.

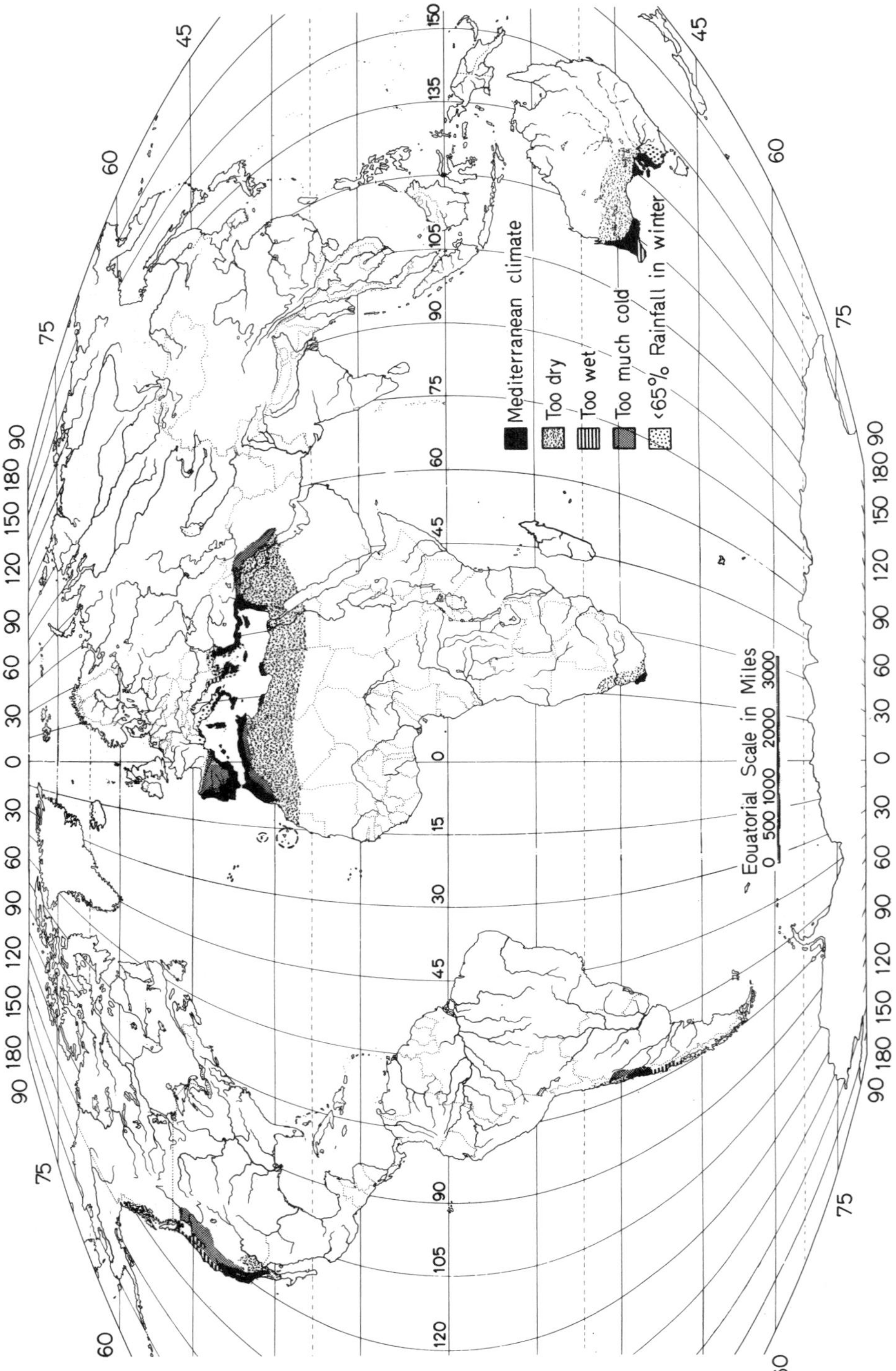

Fig. 1. World map of the areas with mediterranean climates

As will be seen on the map the sub-mediterranean climate that is most extensive is the one which meets the temperature and precipitation seasonality requirements but is just too dry. Wherever there is a continental expanse without major topographic barriers equatorward and inland from the mediterranean climatic zone the subtropical high pressure system dominates in fall and spring as well as in summer to the point that the climate becomes effectively desertic. Poleward from the mediterranean zone, the cyclonic storms of winter are more frequent, begin earlier in fall, and end later in spring. Although the summer drought may be relatively pronounced, especially in North America and Chile, its duration is so brief and the soil so soaked from the previous season that the arboreal vegetation does not suffer much drought stress. Broadleafed deciduous trees or dense evergreen forests that occur there do not fit the mediterranean pattern.

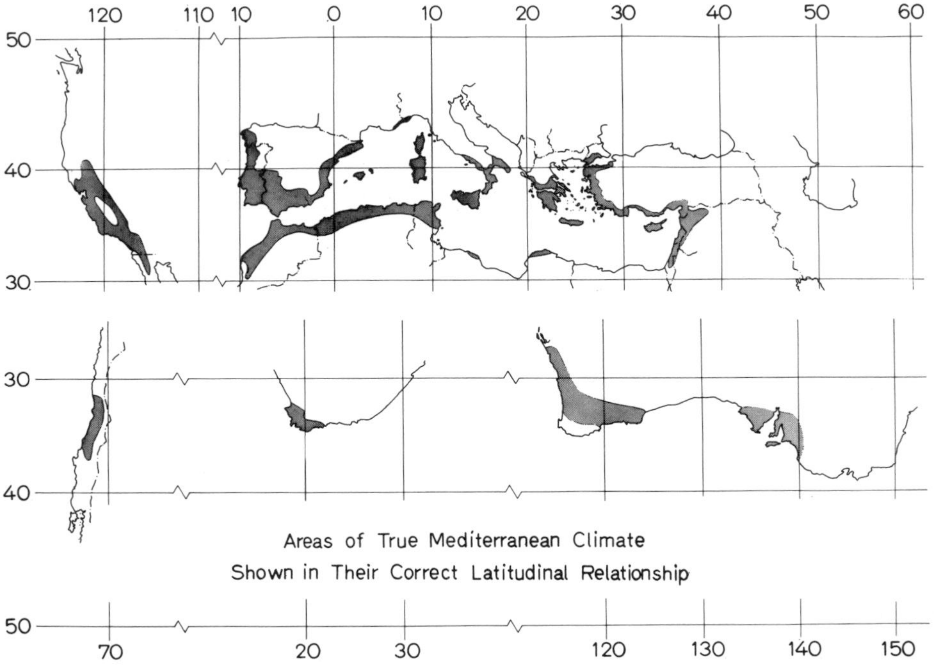

Fig. 2. Areas of true mediterranean climate shown in their correct latitudinal relationship

The areas with excessive cold are associated with continental climatic patterns and considerable elevation. North American plateaux and the Iberian *meseta* are the most extensive regions but mountainous areas in the zone of mediterranean climate also are not properly of it. Some are shown on the map, others like the mountains of Southern California or the Sierra Nevada of Southern Spain are too small to show. The districts that receive proportionately too much summer rain but have the other mediterranean climatic attributes are distinctly European and Australian. They tend to occur on the high latitude side of the mediterranean zone, but this is not an adequate explanation for their existence. It would seem to depend on a combination of the arrangements of land and sea, along with topographic barriers, that induces cyclonic storms fo follow more equatorward paths for a major fraction of the year.

The Köppen system of climatic classification subdivides its mediterranean climate into hot summer *(Csa)* and cool summer *(Csb)* phases. Since in the classification used here Köppen's *Csb* stations with coolness caused by elevation would have too much winter cold to qualify as mediterranean, the distinction is between cool marine and hot interior sites. Only in California does the topography permit both climatic types to express themselves fully. In Central Chile by the time one is far enough from the sea to escape the main marine influence he is at an elevation great enough to effectively cool the summer nights. Conversely, the Mediterranean Sea has little cooling effect in summer and stations all around its littoral are hot. Although the distinction between hot and cool summers is very striking in terms of human comfort its effect on plants is modest. Many though not all of the same chaparral species that grow on the cool south facing slopes of the Santa Monicas also occur on the hot similarly exposed slopes of the San Bernardinos. It would appear that, if moisture availability is equivalent, winter cold has a more determinant effect on plant distributions.

Peculiar Characteristics of the Individual Regions

Although the widely distributed mediterranean climates are defined quite narrowly in this analysis, and because of variations in exposure and elevation each areal unit contains almost the complete range along one or more parameters, each region has a climatic character distinct from the others. Topography plays a part in this distinctiveness, but differences in the orientations of continental coastlines, surface temperatures in the several mid-latitude oceans, and standing wave-like features in the world's general atmospheric flow serve to give each continent's mediterranean region or regions its character.

Along the parameter of winter concentration of rainfall and extreme summer drought California may be said to have the most mediterranean climate in the world. At almost all stations more than 85 percent of the precipitation falls in the winter half year. Because of diversified topography, however, all the other variance allowable with the mediterranean climatic type occurs repeatedly across very short distances. Coastal stations have extremely cool summers and almost completely frost free winters. As little as twenty miles inland a sheltered valley can have summers that are extremely hot with scores of days recording more than 40° C., even at elevations above 500 meters, and sharp though brief winter frosts. As is typical of the mediterranean lands precipitation correlates positively with elevation and latitude but in extreme degree. An isohyetal map can almost be substituted for a hypsometric one. Because cold weather is almost invariably associated with clear skies, frosts occur only in the late night and early morning hours. Stations at moderate elevations, over 1000 meters, can experience occasional extremely severe frosts (−12° C.) although less than 3 percent of the hours in a year are below freezing. Ice days, in which the temperature does not rise above freezing, are rare to non-existent.

The tiny area of true mediterranean climate in Chile experiences extremely rapid latitudinal climatic transitions. The narrow strip, about 100 kilometers wide, between the Andes and the sea passes from an area too dry to be mediterranean to one that is too wet in less than five degrees of latitude. At its center, at Valparaiso and Santiago, the summer drought is extremely intense with more than 90 percent of the precipitation occurring in the winter half year, but the concentration fades rapidly to the south and to some degree to the north as well. The proximity of all stations to the coast and the extremely cold water offshore combine with the many breaks in the Chilean Coast Ranges to assure a strong marine influence right up to the slopes of the Andes. Consequently summers are relatively cool and there is almost no frost except at considerable elevation. Such station data as are available support the impression created by the barrenness of the middle slopes of the Andes at these latitudes that these mountains receive very little more precipitation than the lower lands to their west, a sharp contrast with the Sierra Nevada in California and not one that is readily explicable.

The lands that lie around the Mediterranean Sea include more than half of the total area of mediterranean climate in the world. In one respect they have a pronounced unity. All low elevation stations are hot in summer. Only outside the basin, in Northern Portugal and Galicia and along the southern coast of Morocco does the cool summer variant occur and then only mildly. In terms of the intensity of the summer drought, however, there are two distinct regions, an eastern and a western one, the latter including Italy and Tunisia. June, July, and August are fairly dry around the western basin, but they definitely are not rainless. Late spring and September and October are notably rainy, with October often being the wettest month of the year. Typically the winter half year gets between 65 and 70 percent of the annual precipitation. The midsummer moisture deficit is intensified by the hot summers, but plants are not called on to endure six dry months as they are likely to be in California or Chile. The French Riviera is in the highest latitude ($>43°$ N.) where a true mediterranean climate occurs, and the whole western basin lies on the high latitude side in terms of worldwide locations for this climatic type. In addition major east-west trending topographic features parallel a steep barometric gradient and seem to have the effect of diverting some mid-latitude cyclones into the Mediterranean basin at any time of the year.

Except for the Aegean area north of Athens, areas of mediterranean climate around the eastern basin show a sharp summer drought, which, together with the hot summers and moderate frost probability in winter, makes the climate of even coastal stations resemble that of interior valleys in California. As in California, all around the Mediterranean Sea there is a pronounced increase in precipitation with increased elevation. Many Köppen maps show extensive areas of mediterranean climate in the uplands of northern and western Iran and Russian Turkestan, well east of the Mediterranean Sea[2]. By the definitions used here these areas are too cold or too dry or both to have a true mediterranean climate.

One of the most curious features of the worldwide distribution of mediterranean climates is the contrast between the extensive area in Western Australia and the very limited one in South Africa despite the fact that both areas are in the same latitude and equally exposed to a broad open ocean. Restricted to a narrow zone at the southwest tip of the continent, the South African area of mediterranean climate largely involves a strong marine influence, fairly cool summers, and only a little frost danger. A series of folded ranges parallels the coast. The most exposed of them receive a substantial precipi-

---

2 Examples are the Köppen-Geiger wall map published by JUSTUS PERTHES, Darmstadt, and GOODE'S World Atlas, 1970 edition, pp. 12–13. JAMES (1951, pp. 592–593) has one of few American textbooks that presents a Köppen map that is plotted according to the formal definitions and does not alter the boundaries to accord with the author's perception of the landscape.

tation increment, and the sheltered valleys behind are drier and have quite warm summers and somewhat greater frost danger, resembling to near-coast valleys of Southern California (San Fernando and Elsinore). Before the plateau or the Karoo country to the east are reached rainfall is too low and too evenly distributed through the year to qualify the climate as mediterranean.

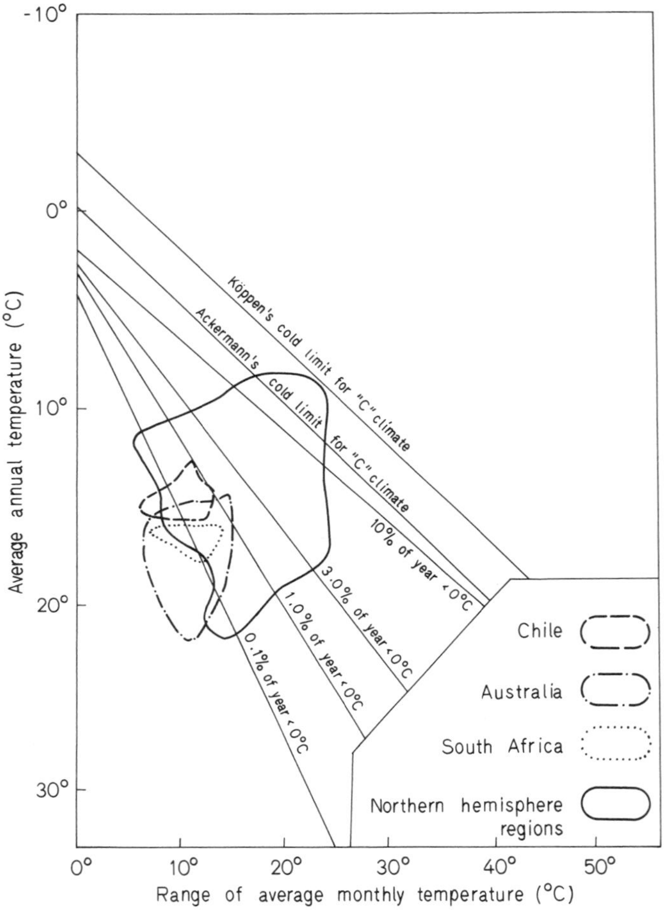

Fig. 3. Thermal characteristics of mediterranean regions as defined by KÖPPEN (*Cs*)

On the Western Australian coast a stormier and rainier environment prevails despite the absence of significant relief to enhance precipitation values. There is enough rain to establish a full mediterranean climate north of Latitude 28° S., several degrees closer to the equator than in any other part of the world. The southwestern tip of the continent, beyond Latitude 34° S., receives so much rain and has so limited a dry season that it cannot be considered as having a mediterranean climate at all despite being at the central latitude for that climate on all other continents. The instrumental record confirms the vegetational impression as this area is forested with gigantic eucalyptus trees. The probable cause of Western Australia's relatively heavy rainfall is the absence of a strong north-setting current with cold upwelling waters near shore, a feature present off Chile, California, and South Africa. For the same reason the marine cooling influence on summer temperatures is limited, and throughout the region of mediterranean climate summers are warm. The inflow of marine air in winter, unobstructed by terrain, keeps the frost danger at a minimal level. In the north rainfall values fall off quite rapidly as one moves inland, but the south coast has a mediterranean climate eastward to where it swings north to Latitude 32° S.

On the east side of the Great Australian Bight the coastline swings southward, getting again into the paths of the mid-latitude cyclones. There is a concentration of rainfall in winter, but the cyclonic paths in this part of the Southern Hemisphere experience relatively little seasonal displacement, and rainstorms can occur at any season. As one proceeds eastward and inland the winter concentration quickly falls to a level too low to qualify for a mediterranean climate. Summers are warm and frosts are uncommon and mild as there are minimal topographic barriers to an inflow of marine air in winter.

Utilizing a portion of a nomogram prepared by HARRY P. BAILEY[3] (Fig. 3) it is possible to compare graphically the several regions of the world with mediterranean climates. The areas outlined are from a representative sampling of stations within the regions mapped as *Cs* by KÖPPEN (JAMES, 1951)[4]. It will be noted that a considerable fraction of the North American and Eurasian stations experience too much frost to qualify as mediterranean by the definitions used here. Other features of note are the consistently small annual temperature ranges in all the Southern Hemisphere stations and the extreme coolness and mildness found in Chile in contrast to the warmth and mildness in Australia.

References

ACKERMAN, E. A.: The Köppen Classification of Climates in North America. Geographical Review **31**, 105–111 (1941).

AXELROD, D. I.: A Method for Determining the Altitudes of Tertiary Floras. The Paleobotanist **14**, 144–171 (1966).

BAILEY, H. P.: A Simple Moisture Index Based upon a Primary Law of Evaporation. Geografiska Annaler **40**, 196–215 (1958).

BAILEY, H. P.: The Mean Annual Range and Standard Deviation as Measures of Dispersion of Temperature around the Annual Mean. Geografiska Annaler **48 A**, 183–194 (1966).

JAMES, P. E.: A Geography of Man. Boston: Ginn. 1951.

KÖPPEN, W., GEIGER, R.: Handbuch der Klimatologie, Band 1, Teil C, C 42–43. Berlin: Gebrüder Bornträger 1936.

3 Professor BAILEY's complex complete nomogram which permits the comparison of climatic stations on many temperature parameters has not been fully published. A considerable number of its applications are illustrated in AXELROD (1966).

4 See Note 2.

2

Climatographical Comparisons between Chile and the Western Coast of North America

FRANCESCO DI CASTRI

Introduction

The measure of the mediterranean bioclimate, and its relations with the physiognomy of vegetation and with agricultural potentials, is a classical theme of research – a theme which has been given special attention by scientists of the Mediterranean regions of the Old World. It is not the purpose here to make a complete bibliographic review in this respect, but mention should at least be made of the contributions of BAGNOULS and GAUSSEN (1953), BOYKO (1962), EMBERGER (1955a, 1955b, 1958, 1959, 1962), GAUSSEN (1955), GIACOBBE (1958, 1959, 1962, 1964) and SAUVAGE (1963). The main results obtained in these and related studies have been summarized in two series of maps (UNESCO-FAO, 1963, 1970), complemented by explanatory notes and a fairly comprehensive bibliography.

For this kind of bioclimatic classification, a great number of climatic indices and different climatic diagrams have been proposed. These methods are, of course, simply conventional and their correspondence with vegetation is frequently unsatisfactory, particularly at a small scale. They provide, however, a useful basis for a first rough approximation, which is of particular value in regions where more precise information on the vegetation is lacking.

In Chile there have been many attempts at elaborating a climatic classification of the country, a classification which would also be meaningful from a biogeographic point of view. Particular mention should be made of the maps of Chile elaborated by FUENZALIDA (1950), LAUER (1960), OBERDORFER (1960) and SCHMITHÜSEN (1956), the latter two mainly based on vegetational characteristics. Through study of the applicability of most climatic indices to Chilean conditions, an attempt has been made to identify the major elements for partitioning Chile into bioclimatic zones (DI CASTRI and HAJEK, 1961). The resulting bioclimatic map (DI CASTRI, 1968) describes fifteen bioclimatic regions in Chile, six of which have a mediterranean type climate; from North to South, these regions are the perarid, the arid, the semi-arid, the subhumid, the humid and the perhumid types. This classification is mainly based on the principles of EMBERGER, with a number of modifications inspired by local peculiarities of vegetation and by consideration of the phenology of soil animal communities. The mediterranean climate can be broadly defined as a climate with concentration of rainfall in winter and with a period of drought lasting from a minimum of one month (in summer) to a maximum of 12 months; in this latter case the winter rainfall is not sufficient to overcome the conditions of aridity. If a restrictive approach is adopted, only the semi-arid and subhumid types can be considered as belonging to a true mediterranean climate ("eumediterranean"); the arid and humid types represent transitional conditions that can be defined as submediterranean climates, and the perarid and perhumid categories are extreme types which exist in regions where the dominant plant formations are deserts and fairly dense forests, respectively. These formations contain, however, a number of biogeographical elements of mediterranean type.

Basis and Aims of This Comparison

Because of the biogeographical interest of disjunct areas with marked climatic and topographic similarities, and also because of the importance of the transfer of agricultural technologies, there have been many comparative studies of the Chilean and Californian environments. Among the most recent, and relevant, studies is that of KOHLER (1966), who showed that the distribution of the psammophyte *Ambrosia chamissonis* was remarkably similar in the coastal fringe of Chile and western North America. NAZAR, HAJEK and DI CASTRI (1966), during a search for climatic analogues to the fifteen bioclimatic regions of Chile in other parts of the world, payed special attention to the mediterranean zones of western North America. NAZAR (1969) used hytergraphs in an essentially agricultural and zootechnical approach. Finally MOONEY *et al.* (1970), as climatic background to their general comparison on vegetation of California and Chile, described two sequences of climadiagrams from stations located only in the coastal sector.

A major objective of this study was to take advantage of the possibility of collecting climatic data to a much greater extent than in previous research, and to present these data in the simplest graphic way, as easily accessible background information for all participants in the integrated project on mediterranean scrub ecosystems undertaken within the framework of the International Biological Programme. In total, use was made of reliable climatic data from 116 stations in Chile, 120 in Baja California (Mexico), 351 in California, 121 in Oregon, 162 in Washington State and 148 in British Columbia (Canada).

Preliminary synthesis of all this climatic information showed that a number of conclusions could be obtained through graphical comparison of the climate of localities in the western fringes of North and South America. It is not the aim here to provide a comprehensive climatic description of these two regions, but rather to help identify possible critical points to be considered in research on the structure and functioning of mediterranean ecosystems, or in future studies on the biogeographical distribution of mediterranean elements.

The type of diagram used in this comparison is fundamentally that proposed by GAUSSEN (1955) and subsequently modified by WALTER (1955) and developed in the series of atlas of WALTER and LIETH (1960–1967). It provides probably the easiest way of representing the climate graphically, and is also the most familiar format for nonclimatologists. It could be argued that this formula does not include consideration of the variability of climatic data; inter-annual variability of precipitation is probably the most critical factor in mediterranean areas. However, to date there are too few data on this subject to permit intercontinental comparison, especially as regards stations in Chile. This is in spite of some preliminary research in Chile by GASTÓ (1966).

The climatic diagram of Santiago (Fig. 1) is given as an example. The climadiagram of WALTER has been simplified. Only the two essential elements, precipitation (solid line) and temperature (dashed line), have been maintained. In addition to the standard measurement of the length of the dry and the humid periods (given by the points of crossing of the thermic and the pluviometric curves), the surface areas of the humid parts (shaded) and the dry part (dotted) were also measured. The measurement of these surface areas, which was done with planimeters, is no doubt an arbitrary method, but it has facilitated comparisons in a really significant way.

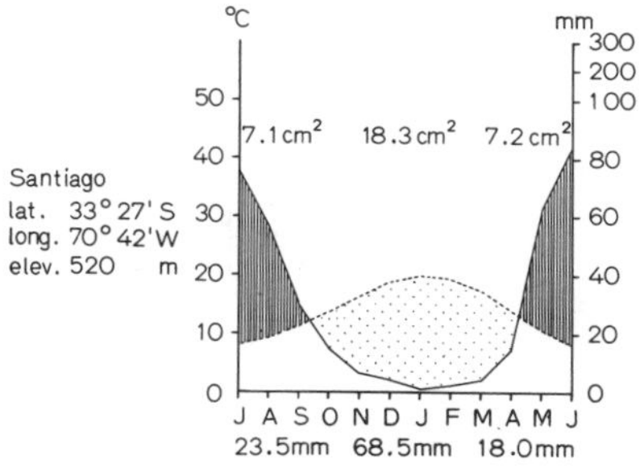

Fig. 1. Climatic diagram of Santiago, Chile. The order of months is from July to June, while in the stations of the northern hemisphere it is from January to December. Surface areas and lengths on the abscissa are given for the arid (dotted) and the humid periods (shaded)

For many of the 116 stations in Chile and the 902 stations in western North America, this graphic representation was repeated for different chronological periods. Thus, more than 1,500 diagrams were drawn, codifying all data on length (an arbitrary estimation of the duration) and surface area (an arbitrary estimation of the intensity) of the dry and humid periods.

Latitudinal Comparison between Chile and Western North America

For a latitudinal comparison, distinction was made between coastal, interior and mountain stations in Chile and western North America. Intercontinental pairs of stations having as similar a latitude as possible, were identified. Since there are much fewer meteorological stations in Chile than in North America, a Chilean locality was first selected. Its latitudinal and physiographic analogue in western North America was then identified. Because of the irregular distribution of the Chilean stations, it was not possible to maintain the same territorial distance between the pairs of diagrams in the latitudinal sequences.

Coastal Sector. The 10 pairs of localities, selected as above, are shown in Fig. 2. The main conclusions from these diagrams, which are largely self-explanatory, are as follows:

a) The mediterranean climate, considered here in the non-restrictive sense, is found at lower latitudes in Chile than in North America. In addition, in the sequences from arid towards humid conditions, the existence of a humid season appears at lower latitudes in Chile (see La Serena) than in North America.

b) The mediterranean climate ends at higher latitudes in North America than in Chile. The latitudinal equivalent in North America of first Chilean locality (Puerto Dominguez)

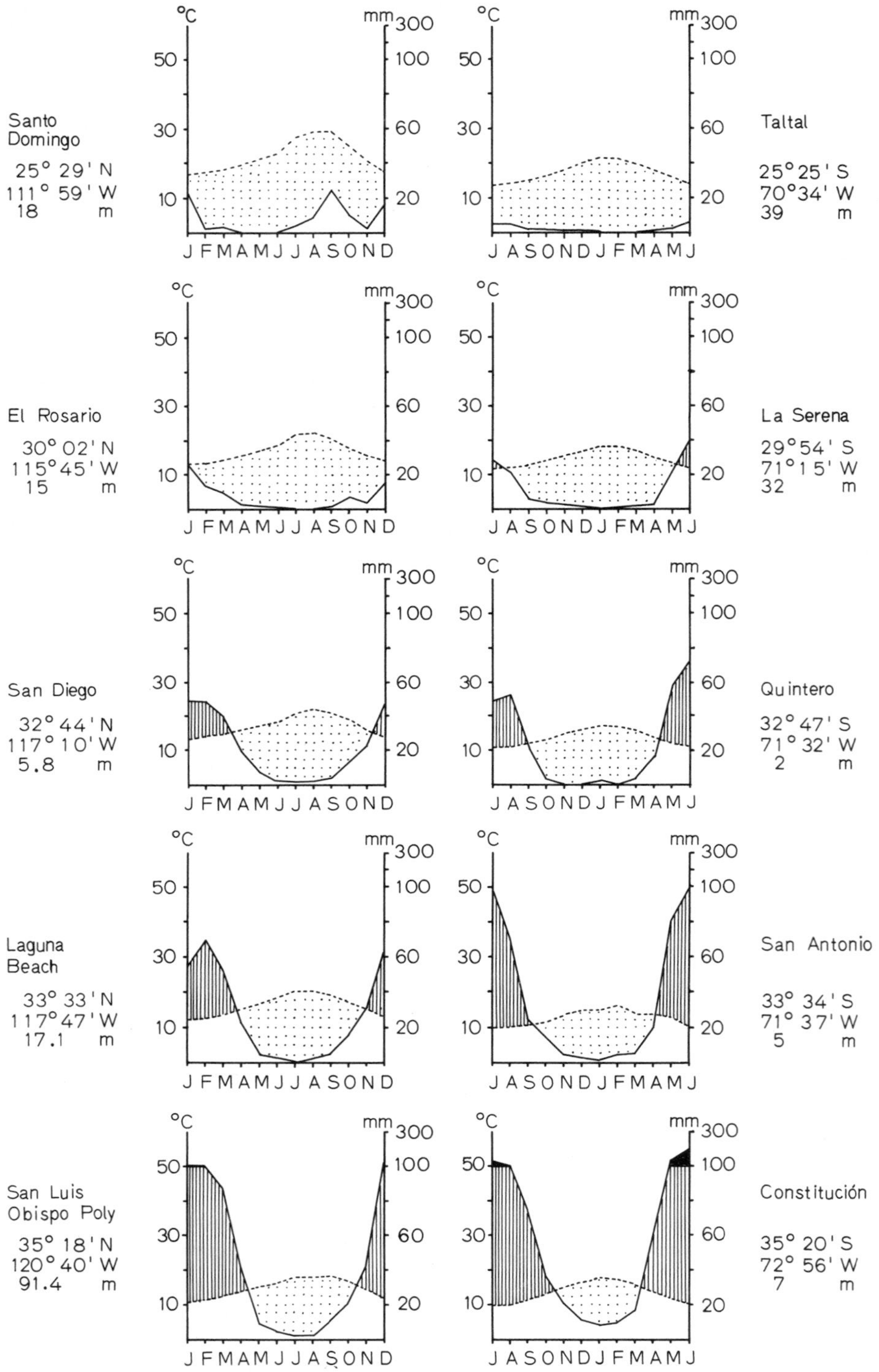

Fig. 2. Pairs of stations from the coastal sector in western North America

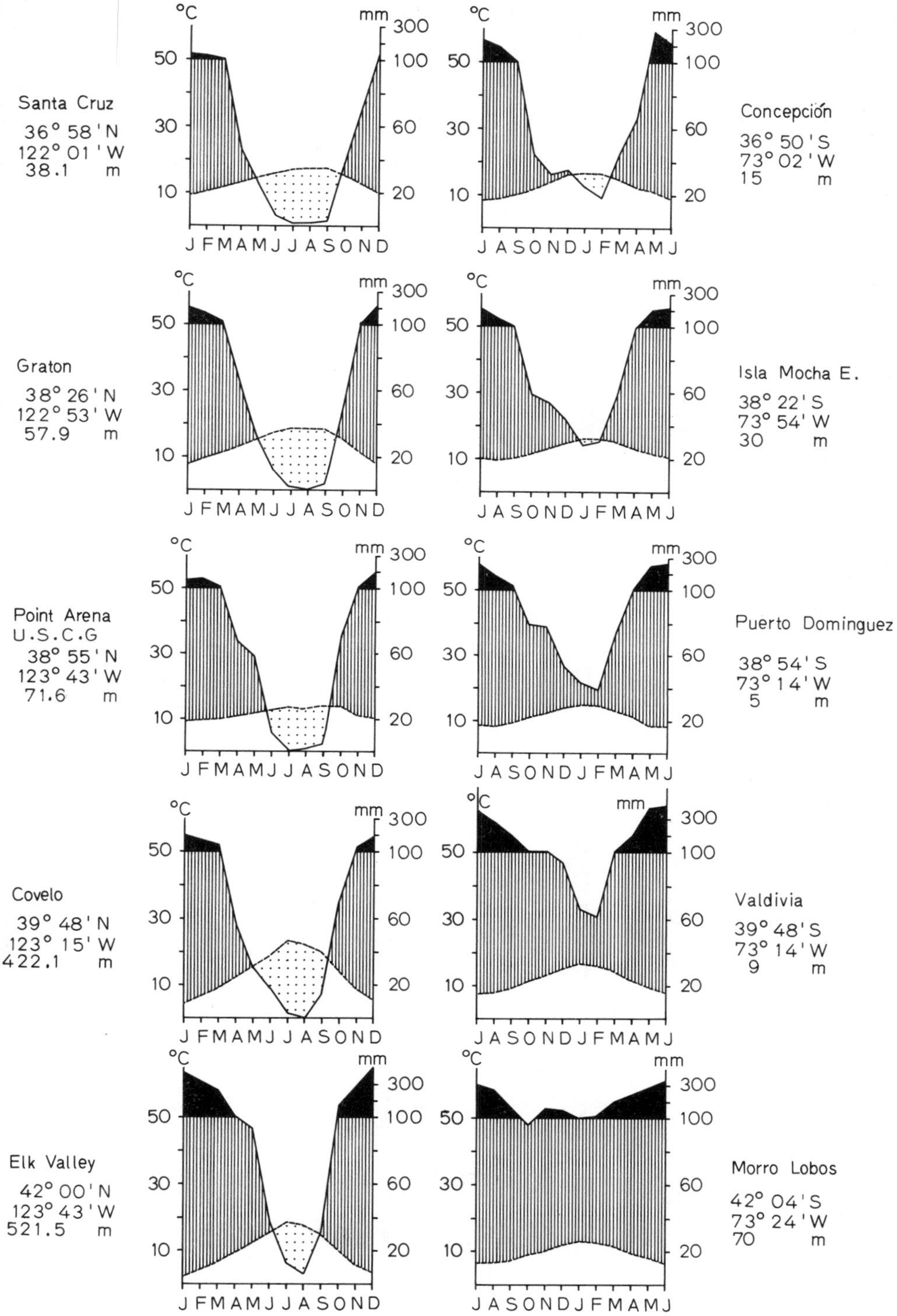

(left sides) and Chile (right sides) located at a comparable latitude

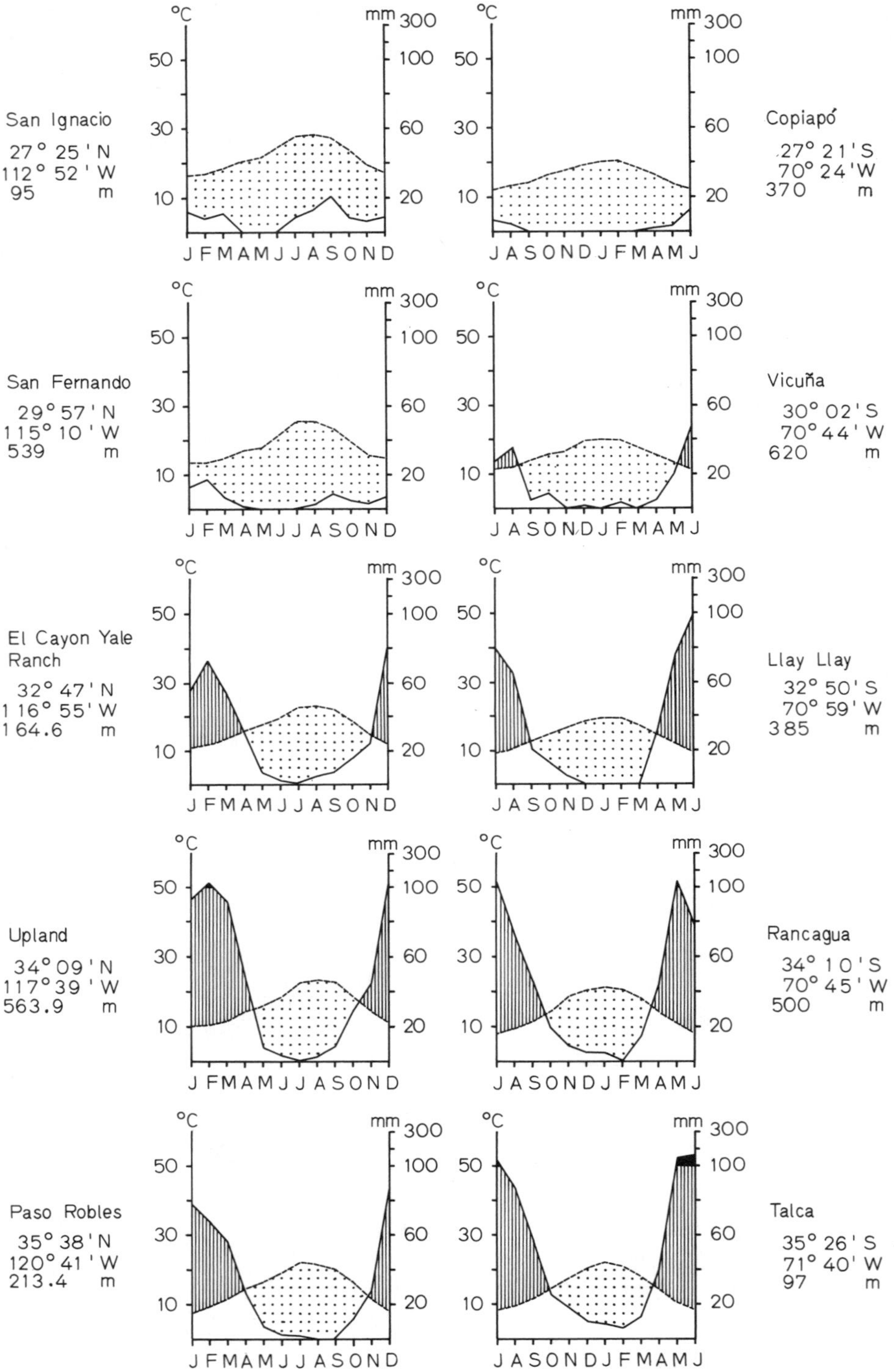

Fig. 3. Pairs of stations from the interior sector in western North America

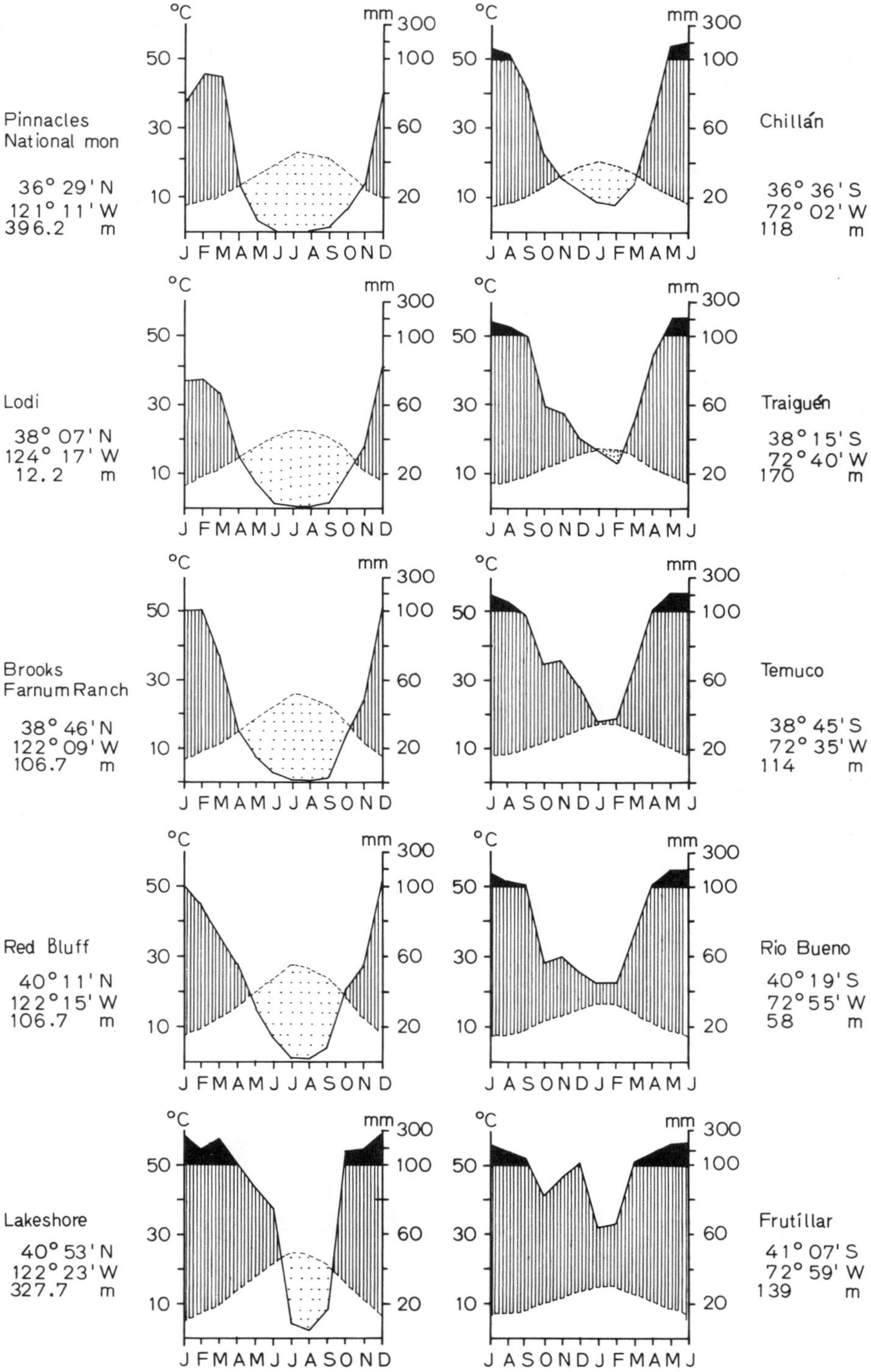

(left sides) and Chile (right sides) located at a comparable latitude

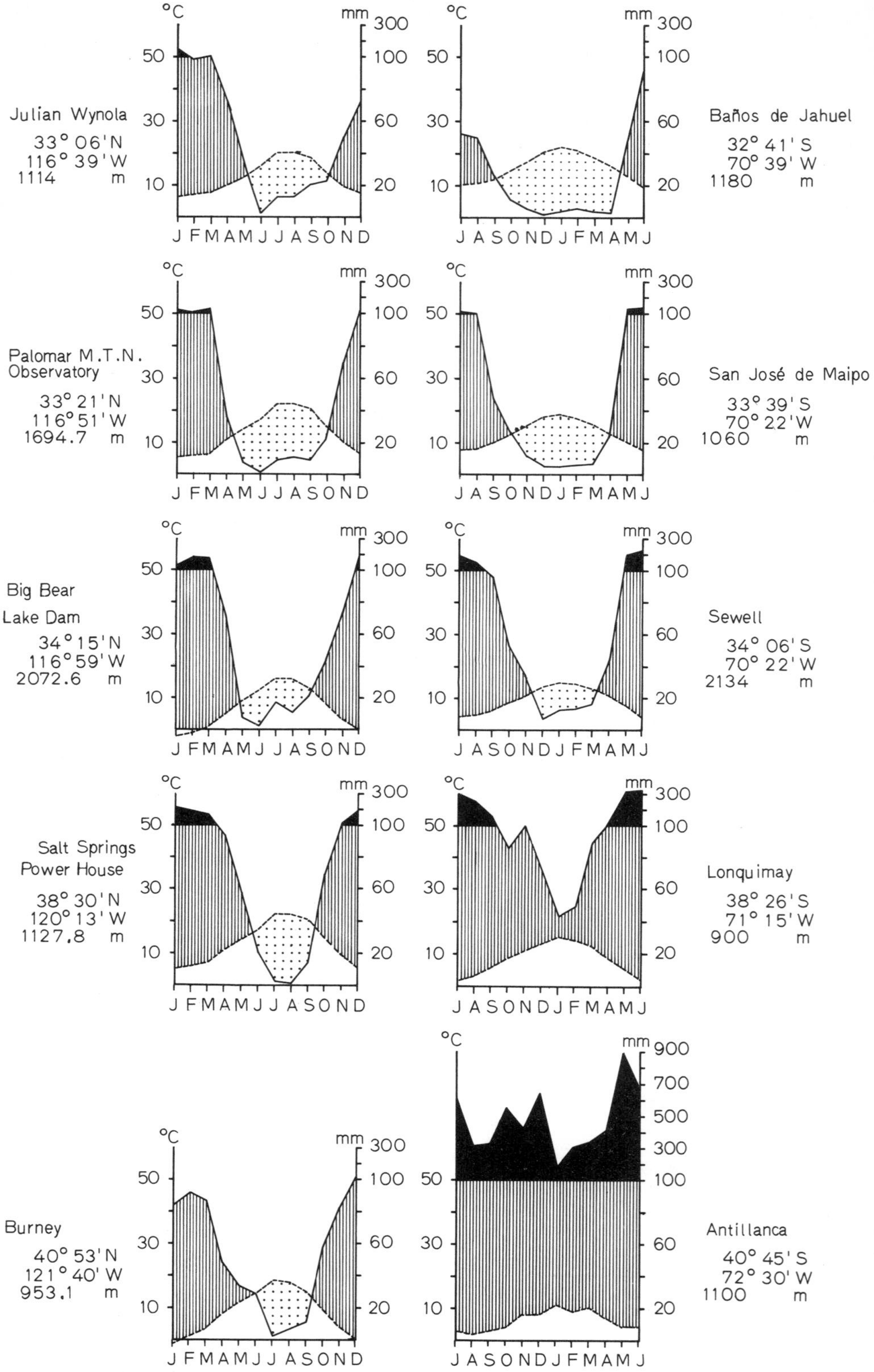

Fig. 4. Pairs of montane stations in western North America (left side) and Chile (right side)

where there is no arid period in summer is Point Arena. Though average conditions only are being considered, it is salient to note that at Point Arena there would be a 4-month dry season.

c) Thus there appears to be a climatic shifting, in a latitudinal sense, between the Pacific coasts of North and South America.

d) The Chilean stations are consistently less arid than those located at the same latitude in North America, with the exception of the most arid locations (see Taltal); this fact is due to the occurrence in Baja California of summer precipitation, itself referable to tropical regime influences. Furthermore, less high temperatures are reached in summer at the Chilean localities compared with their North American analogues.

Interior Sector. Comparison is more difficult in this case, due to the fact that the mediterranean climate is strongly influenced by topographical effects. The 10 pairs of selected stations are presented in Fig. 3 and the main conclusions are as follows:

a) To a great extent the same latitudinal shift discussed above is repeated, the mediterranean climate in Chile occurring at comparatively low latitudes but extending less far into the high latitudes.

b) Reference to the first North American station in this sequence (San Ignacio in Baja California) shows that the tropical influences on summer precipitation are greater than in the coastal sector. Detailed information on precipitation regimes in Baja California is given by HASTINGS and TURNER (1965).

c) In general, temperatures are higher in summer and lower in winter in western North America when compared to analogous situations in Chile. The climate in Chile is not characterised by continental trends, due to the narrowness of the land mass between coast and mountains and because of the influence of the Humboldt current in climatic stabilization.

Mountain Sector. Comparisons for this sector are of a crude nature, because of the scarcity of meteorological stations in the Chilean Andes, because of the difficulty in finding comparable stations located both at the same latitude and altitude, and also because of the atypical character of many montane climates.

Furthermore, a fundamental question in this respect remains unanswered. To what extend can a mountain ecosystem located in a region of mediterranean climate really be considered as "mediterranean" from a bioclimatic and biological viewpoint? Certainly, the period of greatest biological activity in this kind of ecosystem is not usually in winter or in equinoctial seasons, as in the corresponding lowlands, but in summer. This problem has been evoked by DI CASTRI and HAJEK (1967, 1969) and various opinions on the existence of a truly mediterranean mountain vegetation have been discussed by CORTI (1958).

Allowing for these methodological and conceptual constraints, five comparable pairs of diagrams are presented in Fig. 4. In general, the mountain climate in western North America has, from a thermic point of view, a more continental aspect than in Chile. At the higher latitudes in western North America, it appears to be less rainy, at least from the evidence given in the last pairs of diagrams. However, when the stations located in the most typically mediterranean areas (latidudes 33°–34°) are considered, summer precipitation is significantly greater in California than in Central Chile. CALISTRI (1962), working in Tuscany (Italy), reported a highly significant correlation between the amount of precipitation in the warmer semester (May–October) and the thickness of the annual rings in trees, thus illustrating the importance of the summer rainfall in mediterranean climates for tree growth. Nevertheless, it is unlikely that this factor provides a complete explanation

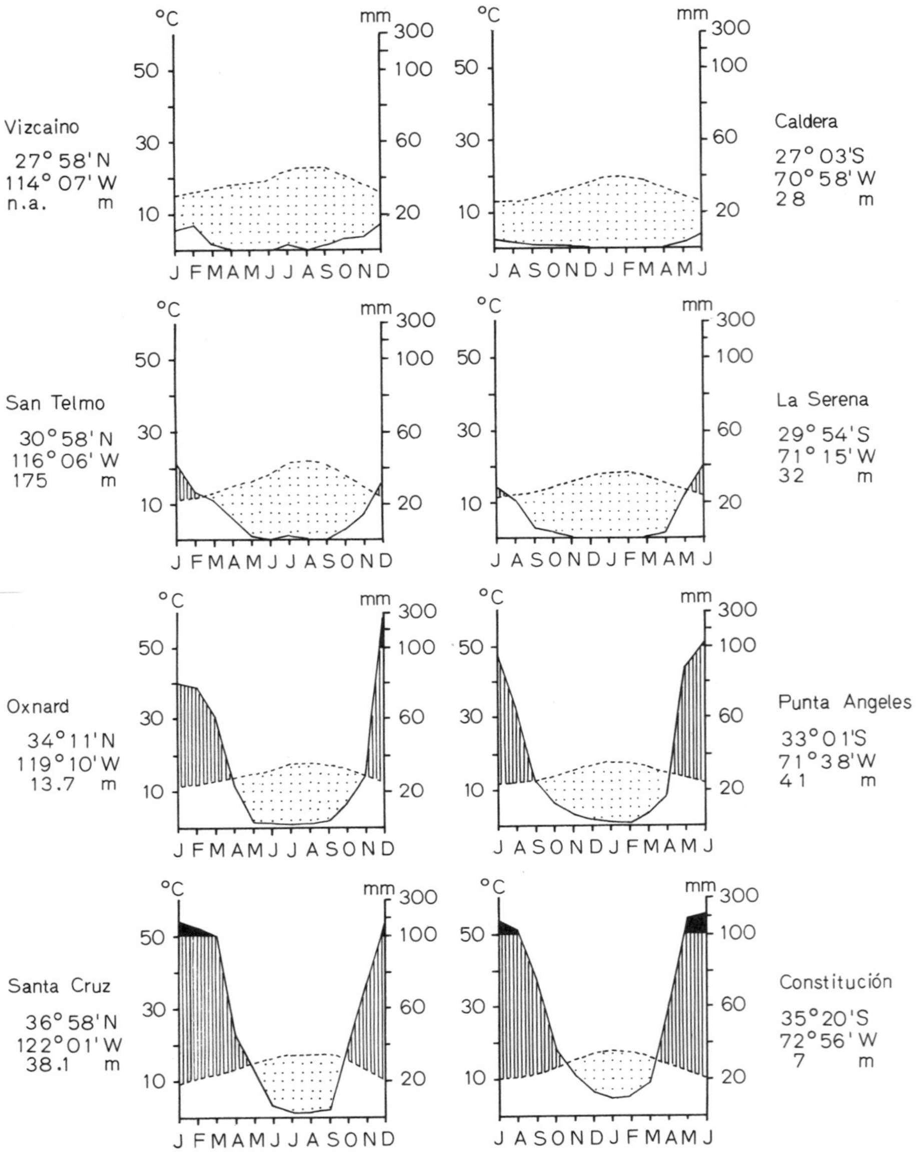

Fig. 5. Pairs of "homoclimatic" stations from the coastal sector in western North America

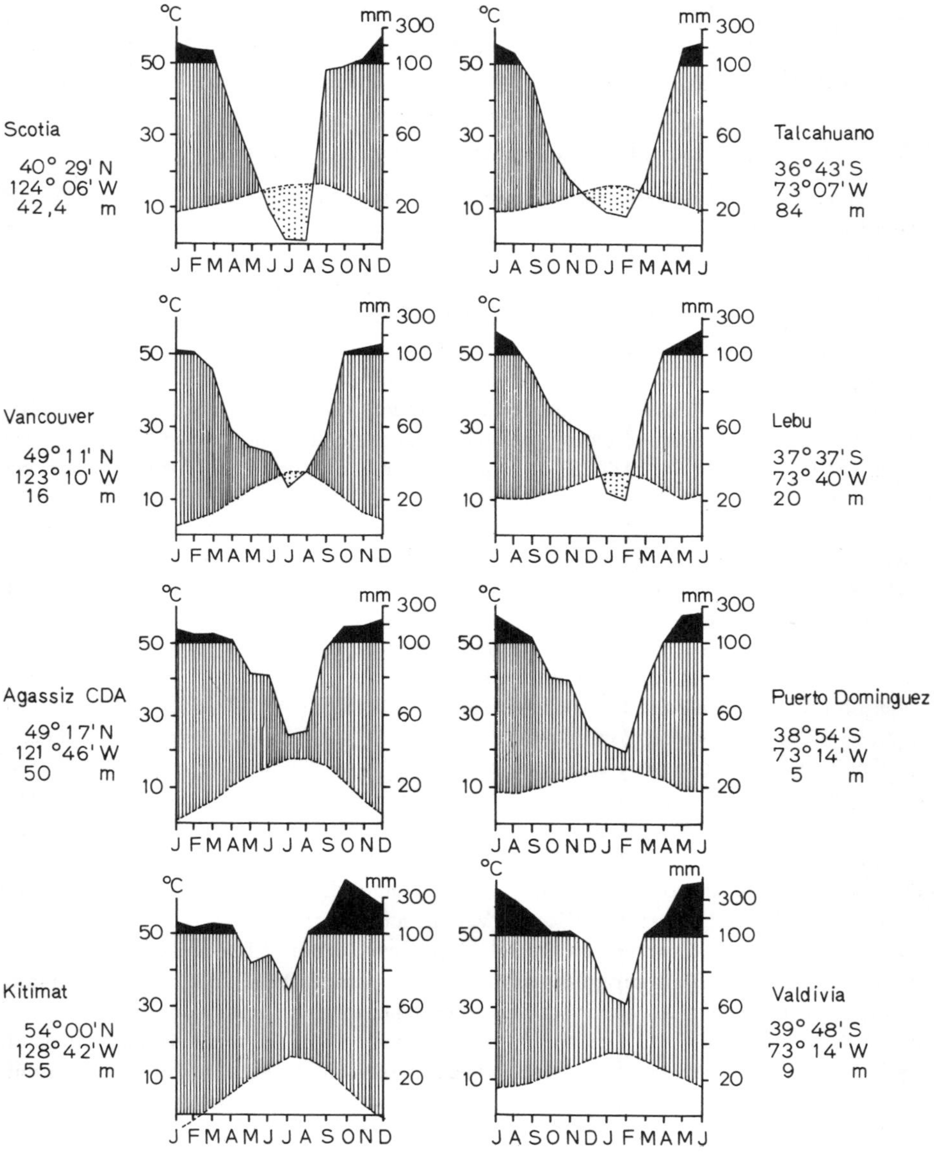

(left sides) and Chile (right sides), showing the latitudinal differences. See text for further explanation for the major dissimilarity between the vegetation of California and Chile, namely the lack of a montane forest in the Central Andes compared with the rich montane forests of California (MOONEY *et al.*, 1970).

Latitudinal Extension of the Mediterranean Climates

Although the mediterranean climate occurs at lower latitudes in Chile than in North America, the total latitudinal extension of the mediterranean areas is greater in North America. This is particularly true when the broad definition of mediterranean climate is adopted, but also, to a lesser extent, when the restrictive definition is taken.

The arbitrary nature of these boundaries necessitates that the figures given here must be considered as indicative only. They are based on preliminary interpretation of the sets of diagrams, taking due account of the works listed in the references to this chapter.

According to the broad definition (thus including the perarid, arid, semi-arid, subhumid, humid and perhumid types), the mediterranean climate in Chile embraces an area from 25° 30'–26° S to 38° 30'–39° S, about 13 degrees of latitude. In North America, the extension of the same climatic types is from 27° 30'–28° N to 49° N, more than 21 degrees of latitude. However, it must be noted that in Chile there exist disjunct areas with mediterranean climate far to the south of the limits given here; for instance the climate is of a mediterranean type in the interior part of the province of Osorno, in restricted areas of Chiloé Island and even of the Chilean Patagonia (Chile Chico, 46 ° 36' S).

When a more classical definition of the mediterranean climate is accepted (this would comprise the arid, semi-arid, subhumid and humid types detailed above), the comparable extension would be as follows: in Chile, from 29° 30' S to 37° 30' S, about 8 degrees of latitude; in North America, from 30 ° 30' N to 44 °–45 ° N, about 14 degrees of latitude.

Finally, taking the most restrictive definition which corresponds approximately to the semi-arid and subhumid ("eumediterranean") climates given above, the mediterranean climate extends in Chile from 31°–31° 30' S to 36° 30' S (5 degrees of latitude) and in North America from 32 °–32 ° 30' N to 39 °–40 ° N (about 7–8 degrees). According to the different interpretations, the latitudinal extension of the areas having a mediterranean climate in Chile and in North America can vary by a factor of nearly three.

Towards the humid parts of the latitudinal sequence in both Chile and North America, the mediterranean climate progressively merges into temperate or cold oceanic climates. At the dry extreme of the sequence on the other hand, there are significant differences between these two regions. In fact, the northern boundaries of the mediterranean areas of Chile give way to a very extensive desert where there is an absolute lack of precipitation; areas with tropical summer regime of precipitation are situated far away in the northern Andes of Chile and at altitudes over 2,500–3,000 m (in general, at 4,000–5,000 m).

Mention has already been made of the more frequent occurrence of precipitation in summer in California, in part indicative of the influence of a tropical regime; this is particularly evident in the southern, non-mediterranean part of Baja California. Here, there is not the territorial discontinuity between mediterranean and tropical regimes of precipitation. In addition, in the deserts surrounding the mediterranean areas of California, rainfall is irregularly distributed in almost all the months (but with a very high variability from one year to another). This precipitation is not sufficient to counter the conditions of dryness and maximum temperatures are much higher than in the Atacama Desert of Chile. In general, however, the desertic conditions are less extreme than in Chile.

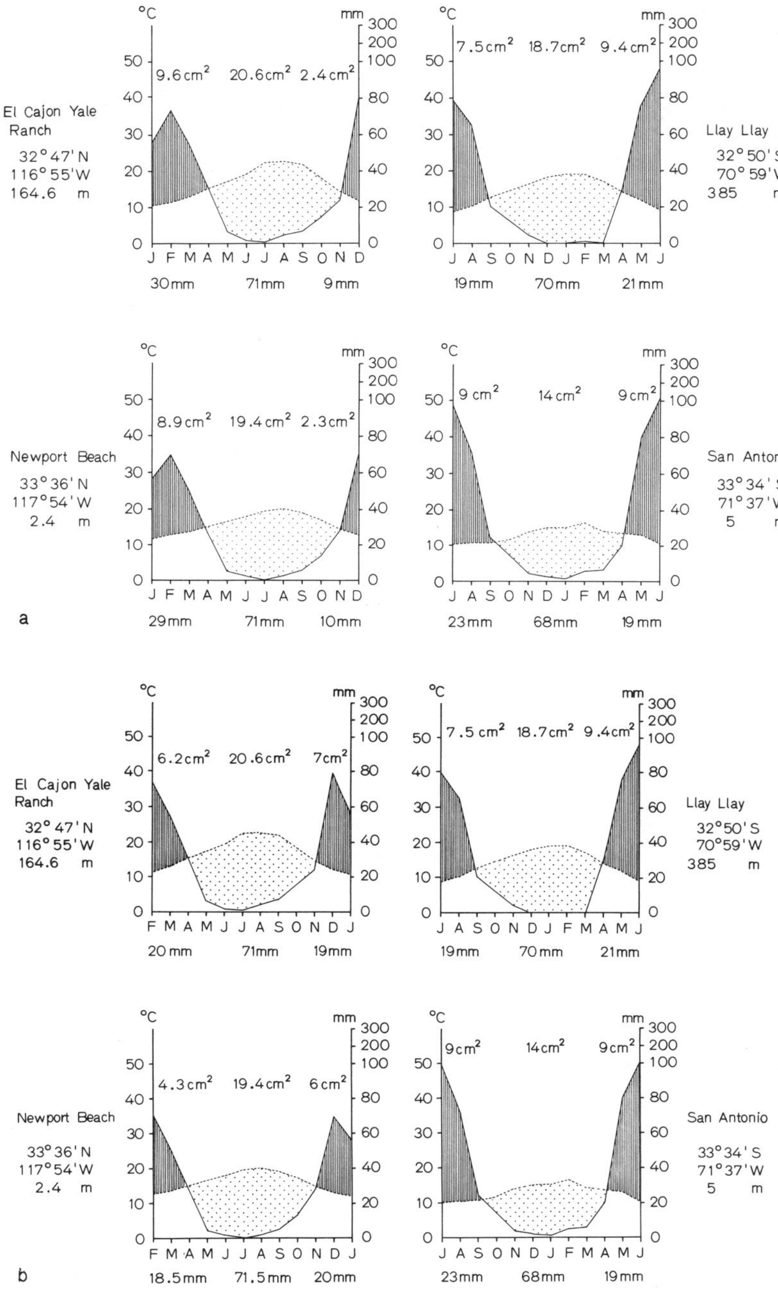

Fig. 6. Comparison of two pairs of coastal (Newport Beach and San Antonio) and interior (El Cajon Yale Ranch and Llay Llay) stations from California (left side) and Central Chile (right side). The usual sequence of months is maintained in Fig. 6a, while the sequence for California stations in Fig. 6b starts with February. See text for further explanation

Latitudinal Shifting of Climate between Chile and Western North America

In order to make the latitudinal dissimilarity more evident, the criteria adopted in the previous comparisons were changed. Homoclimatic stations in both hemispheres were firstly identified, irrespective of latitude. The latitudinal difference between them was subsequently measured. Fig. 5 shows 8 pairs of coastal localities selected according to this approach.

The amount of this "shifting" is about one degree of latitude in the areas with arid or "true" mediterranean climates (up to about the 36° parallel), but with increase of precipitation in the humid part of these latitudinal gradients, this gap of one degree progressively widens and over 10–14 degrees of difference may occur. It must be stressed, however, that homoclimate is a very relative term which must be used with caution. Here "homoclimate" is taken to refer mainly to the precipitation regime and to the length of the period of aridity, and not to the thermic conditions.

Seasonal Shifting of Climate between Chile and Western North America

Close analysis of Figs. 2–5 reveals a certain dissimilarity in the form of the North American climadiagrams especially when compared with those for Chile. Thus, in the Chilean diagrams the central dry part (dotted) is sandwiched between two humid parts (shaded) which have the same length on the abscissa and the same surface area. In the North American diagrams, on the other hand, the humid period at the left side of the diagram is almost invariably larger than that on the right side.

In order to better illustrate this phenomenon, Fig. 6a presents two pairs of diagrams from stations in California and Chile located at approximately the same latitude in the interior sector (El Cajon Yale Ranch and Llay Llay) and in the coastal strip (Newport Beach and San Antonio). The order of the months is, as in all the previous figures, from January to December for the North American localities and from July to June (the classical seasonal correspondence) for the Chilean stations. To facilitate comparison, figures are given on the lengths and surface areas of the different parts of the diagrams. It can be observed that the length in months of the dry period is very similar in the Californian and the Chilean stations. However, the dry period occurs, in relative terms, one month later in California compared to Chile. That is to say, the dry period starts in April in California and in September in Chile (in calendar terms, September in the southern hemisphere corresponds to March in the northern hemisphere). Similarly, the humid period begins in November in California and in April in Chile. The causes of this phenomenon, as well as its possible implications in phenological processes, are as yet unexplained. The diagrams of the same localities present a greater similarity when the months for the Californian diagrams are artificially ordered from February to January as in Fig. 6b.

References

BAGNOULS, F., GAUSSEN, H.: Saison sèche et indice xérothermique. Bull. Soc. His. nat. Toulouse. **88**, 193–239 (1953).

BOYKO, H.: Old and new principles of phytobiological climatic classification. 113–127. In: Biometeorology, S. W. TROMP (Ed.). Oxford: Pergamon Press 1962.

CALISTRI, I.: Studio dell'influenza dell'andamento pluviometrico sull'ampiezza degli anelli legnosi in *Abies Alba* MILL. Italia Forestale e Montana. **17**, 1–16 (1962).

CASTRI, F. DI: Esquisse écologique du Chili. 7–52. In: Biologie de l'Amérique Australe, C. DELAMARE DEBOUTTEVILLE and E. RAPOPORT (Eds.), Vol. 4. Paris: C.N.R.S. 1968.

CASTRI, F. DI, HAJEK, E. R.: Indices pluviotérmicos como base para una clasificación del país en zonas bioclimáticas. Bol. IV Conv. Méd. Vet. Santiago. 19–23 (1961).

CASTRI, F. DI, HAJEK, E. R.: Bioclimatic favorable period in the Chilean zone of mediterranean trend. 307–308. In: Biometeorology, S. W. TROMP and W. H. WEIHE (Eds.), Vol. 3. Amsterdam: Swets and Zeitlinger 1967.

CASTRI, F. DI, HAJEK, E. R.: Biological response to the climate of the Chilean Andes. 111–112. In: Biometeorology, S. W. TROMP and W. H. WEIHE (Eds.), Vol. 4. Amsterdam: Swets and Zeitlinger 1969.

CORTI, R.: Esiste una vegetazione mediterranea montana? Ann. Acc. Ital. Sc. Forest. **7**, 61–86 (1958).

EMBERGER, L.: Projet d'une classification biogéographique des climats. Ann. biol. **31**, 249–255 (1955 a).

EMBERGER, L.: Une classification biogéographique des climats. Recueil Trav. Lab. Botanique, Géologie et Zoologie Fac. Sci. Montpellier, série bot., fasc. 7, 3–43 (1955 b).

EMBERGER, L.: Afrique du Nord et Australie Méditerranéenne. 141–147. In: Climatologie et Microclimatologie, Actes du Colloque de Canberra. Recherches sur la Zone Aride 11. Paris: UNESCO 1958.

EMBERGER, L.: La place de l'Australie méditerranéenne dans l'ensemble des pays méditerraneéens du Vieux Monde (Remarques sur le climat méditerranéen de l'Australie). 259–273. In: Biogeography and Ecology in Australia, Monographiae Biologicae, Vol. 8. Den Haag: Uitgeverij Dr. W. Junk 1959.

EMBERGER, L.: Comment comprendre le territoire phytogéographique méditerranéen français et la position "systématique" de celui-ci. Naturalia Monspeliensia, Série Bot., fasc. 14, 47–54 (1962).

FUENZALIDA, H.: Clima. 188–257. In: Geografía Económica de Chile, CORFO, Tomo 1. Santiago: Edit. Universitaria 1950.

GASTÓ, J. M.: Variación de las precipitaciones anuales en Chile. Fac. Agronomía, Univ. Chile, Bol. Tec. Nº 24, 4–20 (1966).

GAUSSEN, H.: Expression des milieux par des formules écologiques. Leur représentation cartographique. Ann. Biol. **31**, 257–269 (1955).

GIACOBBE, A.: Ricerche ecologiche sull'aridità nei paesi del Mediterraneo occidentale. Webbia **14**, 1–79 (1958).

GIACOBBE, A.: Nuove ricerche ecologiche sull' aridità nei paesi del Mediterraneo occidentale. Webbia **15**, 311–345 (1959).

GIACOBBE, A.: Problemi di bioclimatologia mediterranea. Italia Forestale e Montana. **17**, 3–15 (1962).

GIACOBBE, A.: La misura del bioclima mediterraneo. Ann. Acc. Ital. Sc. Forest. **13**, 37–69 (1964).

HASTINGS, J. R., TURNER, R. M.: Seasonal precipitation regimes in Baja California, Mexico. Geografiska Annaler **47**, 204–223 (1965).

KOHLER, A.: *Ambrosia chamissonis* (Less.) GREENE, ein Neophyt der chilenischen Pazifikküste. Ber. dtsch. bot. Ges. **79**, 313–323 (1966).

LAUER, W.: Klimadiagramme. Gedanken und Bemerkungen über die Verwendung von Klimadiagrammen für die Typisierung und den Vergleich von Klimaten. Erdkunde **14**, 232–242 (1960).

MOONEY, H. A., DUNN, E. L., SHROPSHIRE, F., SONG, L.: Vegetation comparisons between the mediterranean climatic areas of California and Chile. Flora **159**, 480–496 (1970).

NAZAR, J.: Analogías bioclimáticas entre Chile y California. Zooiatría **9–10**, 62–84 (1969).

NAZAR, J., HAJEK, E. R., DI CASTRI, F.: Determinación para Chile de algunas analogías bioclimáticas mundiales. Bol. Prod. anim. (Chile) **4**, 103–173 (1966).

OBERDORFER, E.: Pflanzensoziologische Studien in Chile. Flora et Vegetatio Mundi. Band II. Weinheim: J. Cramer 1960.

SAUVAGE, C.: Etages bioclimatiques. Atlas du Maroc. Notices explicatives. Rabat: Comité National de Géographie du Maroc 1963.

SCHMITHÜSEN, J.: Die räumliche Ordnung der chilenischen Vegetation. Bonner Geogr. Abh. Heft 17, 1–86 (1956).

UNESCO-FAO: Bioclimatic map of the mediterranean zone. Explanatory notes. Arid Zone Research 21. Paris: UNESCO 1963.

UNESCO-FAO: Carte de la végétation de la région méditerranéenne. Notice explicative. Recherches sur la Zone Aride 30. Paris: UNESCO 1970.
WALTER, H.: Die Klimadiagramme als Mittel zur Beurteilung der Klimaverhältnisse für ökologische Vegetationskundliche und landwirtschaftliche Zwecke. Ber. dtsch. bot. Ges. **68**, 331–344 (1955).
WALTER, H., LIETH, H.: Klimadiagramm Weltatlas. Jena: VEB Gustav Fischer Verlag 1960–1967.

3

The Physiography of the Mediterranean Lands with Special Emphasis on California and Chile

NORMAN J. W. THROWER and DAVID E. BRADBURY

On the basis of climatic similarity a few small areas on the surface of the earth are classified as mediterranean lands. Elsewhere in this volume the climate of these areas is discussed (see H. ASCHMANN and F. DI CASTRI). Therefore it is only necessary for us here to refer to those aspects of climate which affect other physical phenomena with which we are particularly concerned. In brief these climatic characteristics include a warm to hot and emphatically dry, high sun season with high evaporation rates. Contrasting with this is a mild and wet low sun period when the precipitation is more effective. Areas with these moisture and temperature characteristics are found in a number of separated, but expected

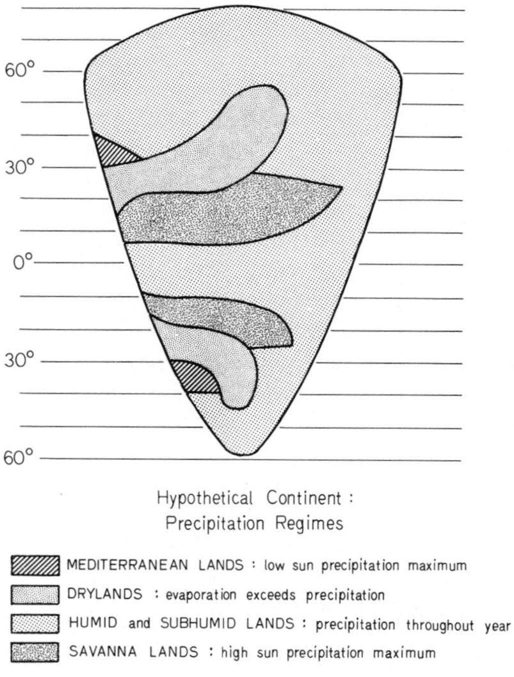

Fig. 1. Precipitation regimes of a hypothetical continent

locations on the globe. Mediterranean core areas are typically centered at about 35° north and south of the equator on the west sides of continents. On a hypothetical continent they would extend through about 10° of latitude being bounded on their equatorward margins by drylands, on their polar margins by cool humid areas and on the east by humid or subhumid lands with greater seasonal temperature contrasts (Fig. 1). In the real world, because of different coastal configurations and other factors, the extent of mediterranean lands varies widely (see H. ASCHMANN).

Location and Extent of Mediterranean Areas [1]

The type region borders the nearly enclosed Mediterranean Sea and embraces parts of three continents – Europe, Africa, and Asia. The maximum latitudinal spread of this area is from about 30° N latitude in Morocco or Israel to approximately 45° N in the Po Delta in Italy. The general east-west trend of the relief allows a deep inland penetration of mediterranean climate from Portugal to Iraq, a distance of some 4,000 kilometers (10° W to 50° E longitude). Approximately three-quarters of the coasts of the enclosed sea experience mediterranean climate, the remaining quarter in Egypt, Libya, and part of Tunisia being too dry to be so classified. Altogether well over half of the total area of mediterranean climate on the earth occurs in the type region. At the southern extremity of Africa is a much smaller mediterranean area in the Cape Province. This region extends from about 32° to 35° S latitude being limited on the north by dryland conditions and on the south by the termination of the continent. Similarly limited is the mediterranean area of West Australia. This locality is separated from another small area of mediterranean climate in South Australia by dryer lands. Australian mediterranean areas extend, in total, from about 32° S latitude in West Australia to 38° S latitude in South Australia. One might speculate on the size of the mediterranean regions in Australia and South Africa if the coastal configuration were different in both cases. The absence of land in the appropriate latitudes is not a limiting factor to the size of mediterranean areas on the west coasts of the Americas but they are constricted to the east by mountains. There are two such regions – much of California in North America and the central part of Chile in South America. The California mediterranean area extends from about 41° to 31° N latitude. Between 41° and 32° S latitude in Chile is an area remarkably like the California mediterranean region.

Structure, terrain, drainage, and soils characteristics of mediterranean areas are stressed in this chapter (Fig. 2). It is intended to provide an overview of the physiography of the mediterranean lands and not to concentrate on particular areas which will be attempted in subsequent reports.

For further background information on this subject, reference should be made to BIROT (1966), BUNTING (1965), HOUSTON (1967), KING (1967), and MCGILL (1958). Some valid generalizations can be made about certain aspects of the physical geography of the mediterranean lands, but this will be deferred until after each of the areas noted above has been examined individually.

1 Definition of mediterranean areas has been the subject of much debate (see H. ASCHMANN's introduction above). Latitudinal and longitudinal limits used in the present chapter are not meant to define, but rather to include the major extent of generally accepted mediterranean areas of the earth.

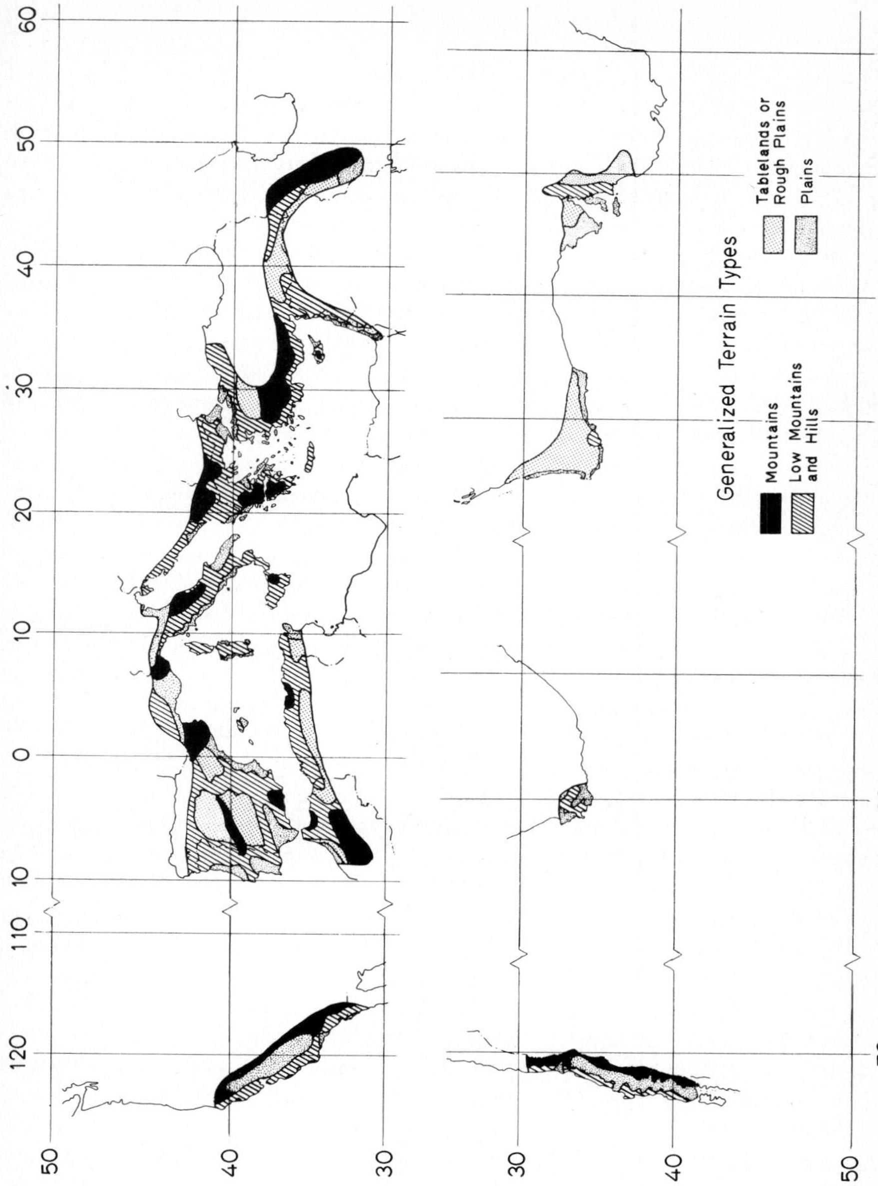

Fig. 2. Generalized terrain types in mediterranean lands

The Mediterranean Type Region

During the Mesozoic era the region was part of a large geosyncline – Tethys Sea. This depression, which was bounded on the north by the Laurasian continent and on the south by Gondwanaland, received thousands of meters of sediments from the erosion of these areas. In response to isostatic movements resulting from the relative approach of Africa and Europe, sediments and other associated materials began to rise out of the water to reach a maximum elevation in mid-Tertiary times. This orogeny was marked by profound faulting and folding, producing the characteristic nappes and other highly contorted structures. Combined with volcanism, this activity led to the development of the Alpine mountain systems including the Atlas, Sierra Nevada, Pyrenees, Appenines, Dinarides, Taurus, and Anatolian chains. Contemporaneous sub-crustal movement resulted in the formation of resistant regional blocks or cratons, such as the Iberian massif. Subsequent faulting in the Pliocene and Pleistocene with associated subsidence and uplift, led to the present basic configuration of the Mediterranean area. Depending upon climate, erosional forces of varying intensity have been at work on these features from their earliest appearance to the present time. Although Pleistocene continental glaciation did not directly affect the Mediterranean lowlands, intensified mountain glaciation during this epoch is important in understanding contemporary physiography. In addition to the erosional effect of glaciers on the mountains themselves, outwash streams deposited sheets of sands and gravels in neighboring valleys. These deposits in turn provided additional material with which to nourish the coastlands. As this material was introduced to the shore zone by regional drainage arteries it was re-worked by waves and currents to form many of the detailed deposition features which presently are found around the Mediterranean Sea.

Characteristically the present terrain of the Mediterranean consists of relatively high fold mountains (up to 3,000 meters high in the mediterranean climatic zone proper but over 4,500 in adjacent areas) rising close to the shore, with narrow coastal plains or small, discontinuous, isolated lowlands. An exception to this generalization is the southern Levant which is a rifted shield area with horsts and grabens.

The Mediterranean shore zone exhibits a striking alternation of complex stream eroded mountains and hills and alluvial plains. Moderate to steep slopes with thin soils and much bare rock are commonplace. Karst features predominate along the coasts of Yugoslavia and Apulia (the heel of Italy) while recent volcanism is especially important in and around the Tyrrhenian Sea off southern Italy. Sea cliffs are especially noteworthy along the coasts of central Portugal, Apulia, and western Greece. The Po and Rhone and some smaller rivers have well-developed and expanding delta plains at their mouths and dune crested plains are scattered throughout a number of localities. Other common features are marine and riverine terraces indicative of past climatic and marine oscillations and of tectonic disturbances. These events have given rise to the characteristic valley profile of a well-defined channel cut into a broad, smooth valley floor. Although the Mediterranean has numerous short intermittent streams a number of large perennial rivers, which have their sources in more humid lands, reach the shores of the nearly enclosed sea.

Erosional features in different parts of the Mediterranean suggest its location as a climatic borderland. Thus in the more humid areas, such as Liguria, V-shaped valleys predominate indicative of linear erosion. In semi-arid regions, including the Maghreb of North Africa, lateral erosion is characteristic producing more vertically sided channels.

Alluvial fans and pediments are found especially in the dryer areas, and some of the present erosional features, such as the granitic forms at Montserrat in Spain, inselbergs in Portugal, and karstic relief in Yugoslavia appear to reflect earlier, warmer and more humid regimes. Present summer drought does not favor rapid chemical or organic weathering. Mechanical weathering is especially evident in the higher mountains and in the relict glacial features of erosion, e.g. arêtes. Periglaciation has been a potent force in the Pyrenees.

Just as certain erosion surfaces give evidence of having been developed under earlier and different climatic conditions, so it is inferred that some of the present soils were formed under climatic conditions different from those now obtaining in the area. Accordingly the Mediterranean today presents a mosaic of old and newer soils. A distinctive soil of rather wide distribution on the lowlands is the *terra rossa.* Strictly speaking this term is applied to soils developed from limestone and, as the name implies, they are reddish in color. These clay-rich soils are highly variable in depth depending on location and tend to contain much rock debris. Many podzolic soils of the Mediterranean which resemble the *terra rossa* have been classified with these soils on the basis of color, but rubrification may result from physiochemical action on rocks other than limestone. Richer in organic content than these reddish soils but of more restricted occurrence, are the brown forest (mull) soils considered by some workers to be intrazonal soils which were developed under coniferous or deciduous cover in the Mediterranean. Podzols are found on the mountains with higher precipitation, while alluvial soils are important in river valleys throughout the area. Intrazonal soils, reflecting local controls, and azonal soils (mainly lithosols) are very widespread. Over large areas the removal of vegetation has led to accelerated erosion and modification of the surface materials.

Further information on this subject can be found also in HOFFMAN (1961) and VITA-FINZI (1969).

The Mediterranean Tip of South Africa

Three major geological formations dominate the mediterranean area of southern Africa – the Archaean or Primitive System, the Paleozoic or Cape System and deposits of Tertiary to Quaternary age. These formations are arranged in more or less concentric form with respect to the coast with the oldest occupying the innermost position. The Archaean platform consists of a basement complex of old granites, gneisses, and various metamorphosed sediments with some scattered post-Archaean eruptive rocks. The areally more important Paleozoic Cape System extends in a great arc covering most of the region and extending to the coast in the southwest. The Cape System is a massive series of predominantly anticlinal ridges composed mainly of sandstones and shales. The attitude of these rocks varies widely from nearly horizontal at Table Mountain to strongly folded in the northwest. Tertiary and Quaternary deposits mantle much of the west and southern coastlands though, as indicated earlier, they are interrupted by the upstanding Cape System in the Cape peninsula. The Tertiary deposits are largely Mio-Pliocene calcareous formations while the Quaternary material is largely blown sand, notably that flanking the Cape Flats east of Cape Town. Sea level changes during the Quarternary are evidenced by raised beaches and terraces.

Thus, in physiographic terms, the mediterranean Cape area consists of a tabular interior bordered by folded mountains with coastal forelands on the west and south. The folded belt is a series of ranges more or less parallel to the coast trending, therefore, north-south in the west and east-west in the south. Extreme elevations are in the neighborhood of 2,300 meters. The coastal foreland is topographically an irregular plain divided into two main segments. On the west the Malmesbury Plain is formed of shales and slates covered with sand deposits and with granitic outcrops locally. The southern part of the area is similar in its topography. A series of intermediate streams originating on the plateau cut across the grain of the structures to reach the sea on the south and to a lesser extent on the west coast.

Soils in this region tend to show a close relationship with parent materials especially where steep slopes and resultant rapid weathering and erosion occur. In the Cape folded area, for example, the dominant type is a grey sandy soil poorly developed in the mountains with local outcrops of bare rock. This is bordered on the south by a zone of coarse-textured clay loams. At the coastal margins aeolian sand and sandy soils predominate. Understandably the rocks have weathered to better and deeper soils in the sub-humid south than in the dryer northwest. Viewed from the coast, the mediterranean tip of southern Africa alternates between alluvial plains, hills, and mountains including Table Mountain (1,087 meters) which rises impressively behind Cape Town.

Detailed information on this region is given by KING (1963), TALBOT and TALBOT (1960) and WELLINGTON (1955).

West and South Australian Mediterranean Areas

The two Australian areas exhibit the least relief of any of the mediterranean areas of the world. In its lithology mediterranean West Australia superficially resembles South Africa in that older rocks inland are bordered on the coast by those laid down in more recent geological times. Most of the area is part of a pre-Cambrian shield or basement complex of Archaean granitic rocks. In some localities, notably in the southeast interior, sandstones predominate while at the southwest tip, Mesozoic rocks are bordered at the coast by the pre-Cambrian series. Elsewhere at the coast Quaternary sedimentary formations predominate. The highest relief features are a few eroded knobs reaching an extreme elevation, near the southeast coast of the area, of about 1,200 meters. A number of small streams, notably the Blackwood River, reach the coast especially at the southwest corner of the area. Coastal lagoons and marshes are also present.

The South Australian mediterranean region is separated from the West Australian area, just discussed, by over a 1,000 kilometers of steppe climate. A rifted shield of the same general type as the southern Levant forms the greater part of the South Australian mediterranean area. The core of this "Shatter Belt" as it is known locally, forms the north-south trending Flinders and Mount Lofty Ranges consisting of pre-Cambrian and Cambrian rocks, including thick quartzites. Elsewhere there are widely distributed Quaternary deposits with small outcrops of Tertiary sediments. The main features of the area – the structurally-controlled uplands, the depressions with intermittent lakes, the grossly indented coastland with its gulf, peninsula and insular form, owe their origin to Tertiary block faulting. The Flinders Range with extreme elevation of approximately 1,300 meters in a few localities, separates two different topographic surfaces. To the west are the low Gawler Ranges consisting of broken hill country with average peaks of 500 meters terminating in Eyre Peninsula. On the east is the mouth and flood plain of Australia's most important river system – the Murray-Darling. This river experiences great seasonal variations in flow. Its drainage area overlies an important artesian basin. Other exterior drainage features in the area are minor and intermittent in character.

For background information on these Australian areas, see Commonwealth Scientific and Industrial Research Organization (1960), FRENZEL, BAUER and PLUMB (1959), JENNINGS and MABBUTT (1967) and NUTTONSON (1958).

The South African and Australian mediterranean areas, which we have discussed, though somewhat similar to each other in their physiography differ in important ways from the mediterranean regions of the Americas. In turn the California and Chilean mediterranean areas are remarkably alike in their terrain. Both have coastal mountains rising steeply from the sea with small interrupted, coastal plains with few good natural harbors. Inland from the coastal ranges, in both cases, is a central trough or plain. The central valley in both instances is bordered on the east by high mountains with few low passes. Naturally there are differences as well as similarities in these two areas and these will be considered in the following sections.

The Mediterranean Area of California

California can be divided into a number of physiographic provinces of which we are concerned mainly with: the Sierra Nevada, the Great Valley, and the Coast, Transverse, and Peninsula Ranges (Fig. 3). Of the Sierra, only the lower western elevations can be considered within the mediterranean climatic area. As in the case of the European Alps, the Sierra Nevada occupies the site of an ancient geosynclinal system. While the sediments were being elevated in the Jurassic Period the terrain underwent profound disturbance and deformation resulting in folded and broken structures at the surface while subsurface rocks were greatly metamorphosed. Extensive, deep-seated igneous activity occurred and with the cooling of the magma, granitization took place leading to the creation of massive batholiths. Widespread erosion obtaining over a very long period led to exposure of the granite batholith. The Sierra was further elevated during late Tertiary and Quaternary times and took on its present asymmetrical form, with a long western slope and short, steep eastern face resulting from block faulting. During the Pleistocene the range was sufficiently elevated to support extensive glaciers which have left strong erosional evidence in such places as Yosemite Valley. The northern part of the Sierra Nevada exhibits numerous volcanic landforms more so than to the south, where volcanic activity is largely confined to the zone immediately east of the mountain front. Peaks in the Sierra, in the extra-mediterranean area, reach extreme elevations of about 4,500 meters with passes averaging 2,500 meters.

The Great Valley is a structural depression now some 600 kilometers in length and averaging 70 kilometers in width. In the past, before the Coast Ranges defined its western limits, the Valley was much more extensive than at present and was intermittently submerged. The building of the Coast Ranges in Plio-Pleistocene times reduced the size of the Valley to approximately its present extent. The erosion of the surrounding mountains, particularly the Sierra Nevada, has deposited sediments to great depth masking the lower slopes of the Sierra Nevada and, to a lesser extent, the Coast Ranges. This recent deposition is underlain by clastic sedimentary rocks of Cretaceous and younger age.

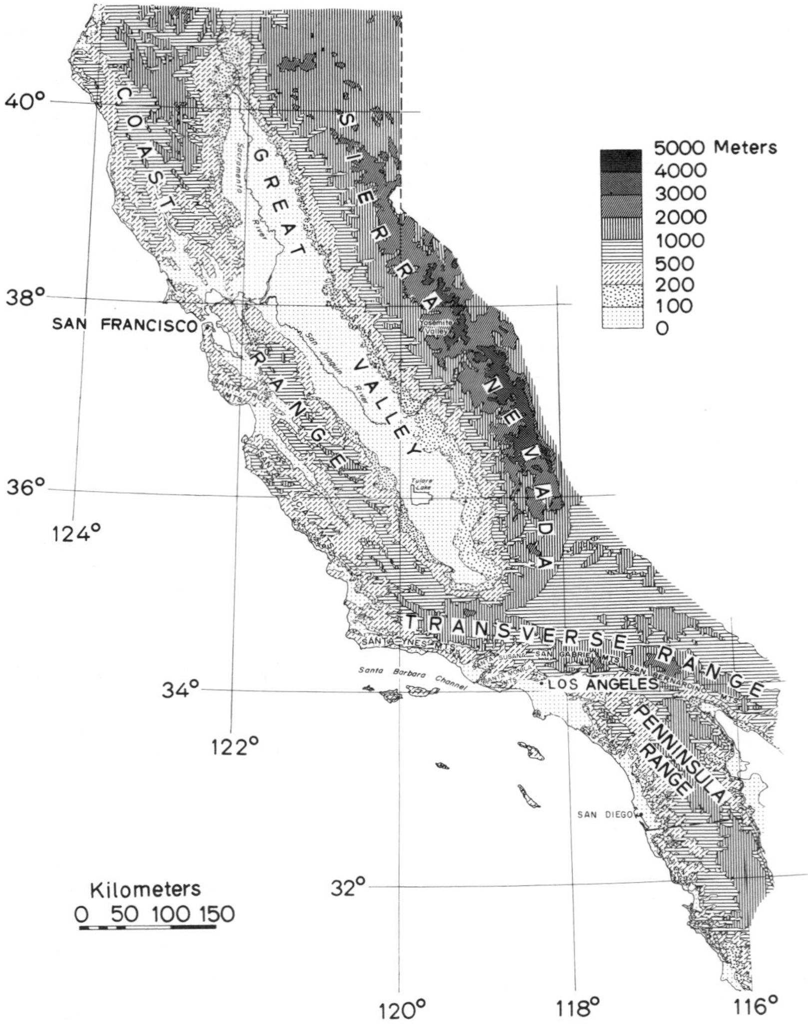

Fig. 3. California relief

The Coast Ranges, covering an area comparable in size to the Great Valley are composed of late Mesozoic and younger formations. Their evolution is roughly analogous to that of the Sierra Nevada but on a much smaller scale. The topography gives strong evidence of folding and faulting and Pleistocene deformation has produced the present gross morphology. Intermittent uplift since the mid-Pleistocene is evidenced by marine and river terraces. The San Andreas Fault, extending almost the entire length of the Coast Ranges and beyond in the south through the Transverse Ranges, is the most remarkable of a series of faults which continue to dislocate the coastal and interior landforms. At a maximum the Coast Ranges attain elevations of about 2,000 meters but are commonly much lower.

The Transverse Ranges are unusual in the overall north-south trending topographic pattern of the Americas for their east-west structural orientation. Although geologically complex the western section of these mountains more closely resembles the Coast Ranges, and the eastern, the Sierra Nevada in their relief, structure, and lithology. In the eastern section the imposing San Gabriel and San Bernardino Mountains are composed of Mesozoic granitics and metamorphosed sediments and rise to some 3,000 meters. The somewhat lower western ranges, the Santa Ynes and Santa Monica, are characterized by beds of Tertiary sediments including shales and siltstones. Partially submerged portions of these mountains form a series of rugged islands separated from the mainland by the Santa Barbara Channel. Again, deformation since mid-Pleistocene times has led to the broad features of present day relief, and marine terraces are found up to 400 meters above sea level.

South of the Transverse Ranges coastal California and northern Baja California are dominated by the Peninsula Ranges. Structurally these mountains are a massive elevated fault block with an extensive western slope bounded by a very steep eastern fault scarp. The bulk of this physiographic province consists of plutonic rocks of the southern California batholith. Marine and non-marine Tertiary deposits flank the uplands to the west. At the extreme coastal margin Quaternary marine terrace deposits occur. Elevations of nearly 3,000 meters are attained in the inland portion of the Peninsular Ranges which extend as a series of tabular surfaces declining in elevation as they approach the coast. A series of faults, including the Elsinore, traverse the province, and crustal movements associated with these give the area its present character of alternating basins and uplands.

Fluctuating sea levels have produced raised wave cut benches. Stream courses cut into these features, produce bold cliffed headlands to give considerable relief to the shore zone. Elsewhere the coast is marked by beach bars, marshes and lagoons.

Extending northward along the shore these coastal forms are repeated but with some local differences. The Los Angeles lowland, the largest alluvial plain on the coast, is bounded on the north by the seaward extensions of the Santa Monica Mountains, the coastward expression of the Transverse Ranges.

Other mountains fronting the sea, in the Coast Ranges, are the Santa Ynez, Santa Lucia and Santa Cruz Ranges. These are punctuated by coastal alluvial plains of varying size often bordered on their seaward margins by dune sheets. The great inlet of San Francisco Bay can be generally characterized as having a coastline of complex stream eroded hills controlled by structure, with alluvial deposits at the mouths of rivers.

The San Joaquin and Sacramento Rivers reach the sea at San Francisco Bay. These rivers, which derive their water from many tributaries in the Sierra Nevada, drain the southern and northern sections of the Great Valley respectively. Understandably a greater and more regular hydrologic regime characterizes the northern Sacramento River Valley than the more southerly San Joaquin drainage area. Apart from these two master streams and some interior drainage, rivers of widely fluctuating seasonal flow reach the Pacific from the Coast Ranges through short but often steep courses offering further evidence of recent regional uplift. The southern portion of the San Joaquin Valley has great alluvial fans and interior drainage to Tulare Lake, now much modified by man's work.

California soils are extremely complex locally but a few useful generalizations can be made (see also P. ZINKE). Much of the rugged topography of the Sierra Nevada, the Transverse, and the Peninsula Ranges, is characterized by shallow, stony, coarse-textured lithosols. Elsewhere undifferentiated residual soils occur on the hilly upland flanks of the mountain masses and vary locally according to climate, parent material, vegetative cover, and time.

The greatest extent of transported (alluvial) lowland soil occurs in the Central Valley. Other significant deposits are found in the Santa Clara, Salinas, and Santa Maria Valleys, the Oxnard Plain, and the Los Angeles Basin. In the Central Valley three major alluvial soil types are recognized – terrace, valley, and valley basin varieties. Terrace soils on the higher Valley slopes are commonly the most extensively weathered of the three types. Valley soils characterize the alluvial fans and flood plains; they are well-drained and most valuable for agriculture. Valley basin soils occupy the lowest portion of the Valley and are often heavy–textured, poorly drained, and of only moderate agricultural value.

Other terrace soils are associated with coastal benches, areas where the sea once stood at higher levels than at present. As in other mediterranean areas these are often older soils formed on terrace sites under different climatic regimes than those now obtaining in the area.

CARTER (1957), HINDS (1952), LANTIS, STEINER and KARINEN (1963) and STORIE and WEIR (1953) give further information on this subject.

The Mediterranean Area of Chile

Three major physiographic provinces comprise the mediterranean area of Chile – the Coast Ranges, the Central (or Longitudinal) Valley, and the Andes (Fig. 4). As in California, of the inner upland realm only the lower part can be considered within the mediterranean climatic zone. The orientation and arrangement of these landform regions of Chile is similar to their California counterparts.

The Andes owe their evolution to several tectonic stages with the earliest movements beginning in the late Cretaceous following previous periods of deposition. In the early Oligocene a second period of orogeny was marked by widespread volcanic activity. Further strong movement took place in the Miocene with igneous intrusions, now exposed in the western foothills of the Andes. Following this, widespread erosion reduced the elevations of the cordillera considerably. The modern Andes owe their form to profound Pleistocene uplifting, involving folding and faulting of Mesozoic sediments. These processes continue, as evidenced by present-day seismic, as well as volcanic, activity. The resulting ranges are much dissected by deep valleys and modified by glacial action. The result is that the Andean cordillera of central Chile is higher (with elevations up to 7,000 meters) and narrower, than the Sierra Nevada in California. South of Santiago volcanos are superimposed on the Andes similar to the condition which obtains in extreme northern California. Major volcanic features of Chile appear to be coincident with crestal rift valleys or grabens. The lithology of the Chilean Andes reflects their genesis, being formed of metamorphosed sediments underlain by igneous batholithic rocks and surmounted along zones of

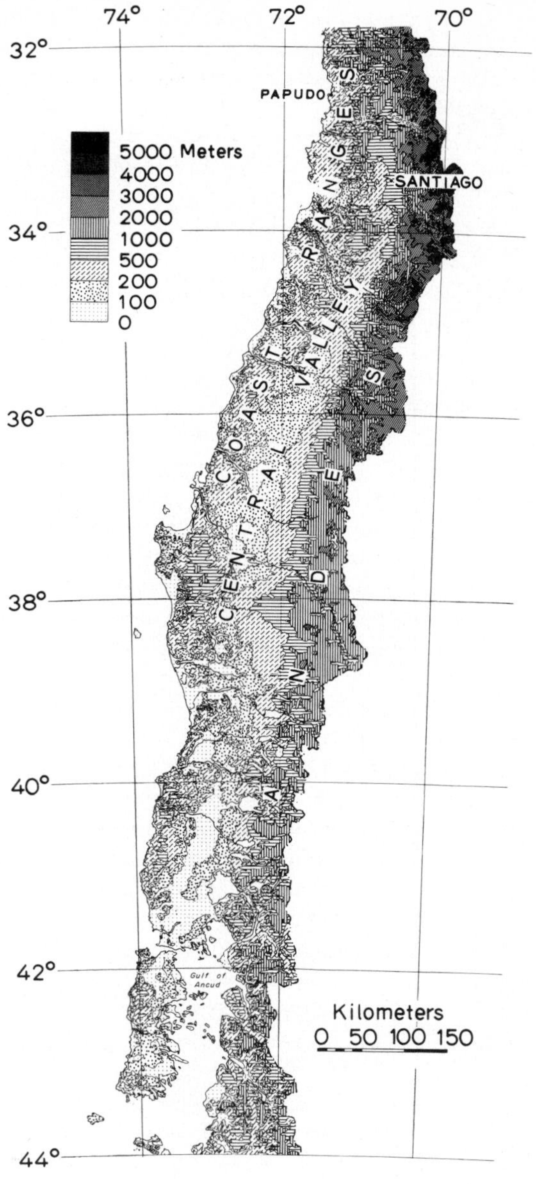

Fig. 4. Central Chile relief

weakness by volcanic formations. The average elevation of the Andes, backing mediterranean Chile, declines from north to south with passes becoming lower and more numerous to the south. In this more humid area, beyond the mediterranean proper, a series of lakes interfinger the mountains.

The Central Valley of Chile, like the Great Valley of California, is a structural trough filled to great depths with sediments derived from erosion of the surrounding mountains. In Chile spurs of the Andes break the continuity of the Central Valley to form a series of nearly separated basins in contrast to the simpler morphology of the Great Valley of California. Interestingly though, at the arid extremities of both California and Chile, respectively, mountain ranges isolate a lowland extending to the coast. Thus the Aconcagua Valley is cut off from the remainder of the Central Valley by prominent spurs of the Andes creating a comparable situation to the Transverse Ranges which separate the Great Valley of California from the Los Angeles Basin. North of the Aconcagua Valley spurs of the Andes reach so far westward as to preclude the extension of lowlands in this direction. The Central Valley proper extends some 900 kilometers from Santiago to the Gulf of Ancud, in humid Chile, where it is finally submerged. Virtual discontinuity into a series of basins of Chile's Central Valley is not the only difference between this province and its California counterpart. Thus the Great Valley of California is more nearly horizontal and below 100 meters. The Central Valley of Chile has a regional slope from some 700 meters in the north to sea level in the south. However, in both the Chilean and California Central Valleys the eastern sections, flanking the larger mountain masses, tend to be higher because of greater alluvial deposition than on the eastern side where they abut the lower Coast Ranges. Thus in the latitude of Santiago the eastern edge is about 700 meters, while the western side is some 300 meters above sea level.

In their drainage patterns the two American mediterranean areas present striking contrasts. Whereas in California two master streams and their tributaries drain almost the entire Great Valley, in the case of Chile a series of rivers, rising in the Andes, cut across the shorter axis of the Central Valley and break through the Coast Ranges to reach the sea. As in all mediterranean areas seasonal fluctuation of rivers, varies from torrent conditions in the low sun period to minimal flow in the high sun season, with a braided pattern in the valley. Smaller streams, locally called *esteros,* dry up completely in the high sun season. Understandably valley fill shows gradations from boulder-sized materials at the margins of the uplands to fine alluvial sediments in the lower portions of the valleys.

The Coast Ranges consist basically of a Paleo-crystalline cordillera folded in the Tertiary. This upland is composed of complex rocks including metamorphosed pre-Cambrian outcrops and Cretaceous granites overlain by Tertiary and more recent deposits. The arrangement of these rocks has suggested to some workers a geanticlinal condition. In places the Coast Ranges rise abruptly from the Central Valley; on the seaward margins they are bordered only locally by a coastal plain. It is not always possible to differentiate between the coastal plain and the Coast Ranges, because the contact zone is masked by alluvium and is not evident even in the drainage patterns. The maximum peaks of the Coast Ranges rise to over 2,000 meters on their eastern margins in the Front Range. West or seaward of this range discontinuous remnants of the crystalline massif are more rounded and lower in elevation. Rivers flowing generally at right angles to the coast cut deep gorges, locally called *quebradas,* in order to reach the sea. This gives to the western margins a somewhat tabular or plateaulike appearance, more similar to the lower parts of the Peninsular Ranges than to the Coast Ranges of California. Near the coastal margins marine terraces, partly cut in bedrock, are found up to elevations of 400 meters. They decline from this height in the north to only a few meters in the south. Elsewhere lowlands occur at the mouths of rivers which have continued downcutting as the coast has generally been emerging. Viewed from the sea the coastal zone presents an alternation of small alluvial plains at river mouths, complex stream eroded plains with dunes locally (as in the vicinity of Papudo), and plateaux and hills, cliffed at the coast in some localities.

Mediterranean soils in Chile show the same complexity which we have noticed in California and other areas, but again some generalizations can be made. Thus the sub-humid central area exhibits leached red and brown soils. On the arid margins more chestnut brown-like soils predominate while increasing podsolization occurs south of central Chile reflecting broad climatic patterns.

In the Chilean Andes as in other mediterranean upland areas soils strongly reflect the parent material and relief. For example in the central Andes lithosols predominate at higher elevations while brown forest soils characterize the piedmont. A thick mantle of relatively infertile andesols is found in the southern Andes known in Chile as Trumao soils. The deep graded deposits of alluvium in the Central Valley have great concentrations of humus, silt and clay in their more fertile, lower portions. In limited areas of the Valley soil textures are so fine as to inhibit drainage. These localities are occupied by shallow lakes in the low sun season and become dry wastes in the summer, similar to the former Tulare Lake in California's Great Valley before recent modification by man. Related to the crystalline rock and Tertiary sediments of the Coast Ranges are red brown mediterranean soils. In certain small areas near the coast reddish-brown lateritic soils have developed on old volcanic material. Although gulley and sheet erosion are common problems, as in many other mediterranean areas, the reddish soils in Chile appear to be somewhat resistant to removal.

Further information on the physiography of Chile can be found in FITTKAU *et al.* (1969), JAMES (1942), MC BRIDE (1936), SAUER (1950) and SMOLE (1963).

Generalizations and Conclusions

Mediterranean areas are defined in terms of climatic similarity. Although their structural origins may be diverse they possess a similarity of landform process and character (see R. PASKOFF). Three areas i. e. most of the Mediterranean type region (except the southern Levant), California, and Chile are marked by relatively young orogenic systems exhibiting high, sharp, folded and faulted mountains and hills rising close to the coast. These uplands, in the three areas, gained their present gross morphology through violent upthrusting in late Tertiary and early Quaternary periods. Tectonic instability and volcanic activity are associated with these linear mountain chains to the present time (Fig. 5). Glaciation, even at maximum extent, was very localized within these areas and probably only modified pre-existing valleys cut before the Pleistocene.

The southern Levant and South Australian mediterranean areas are rifted shields and therefore differ in their genesis and present form from the three areas mentioned previously; the topographic expressions of such features, grabens and horsts, are more subdued. The remaining two mediterranean areas, South Africa and West Australia, are character-

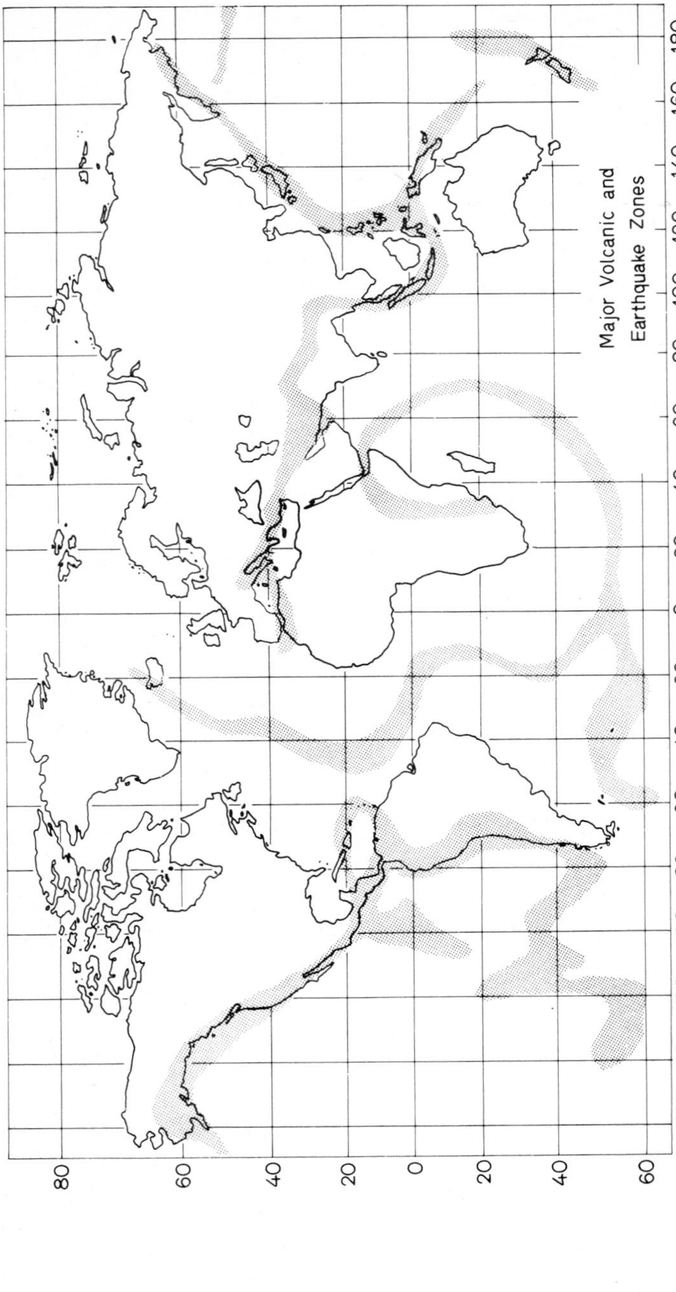

Fig. 5. Major volcanic and earthquake zones of the Earth

ized by ancient basement complexes. The South African example is by far the more topographically diverse of the two but they both have more mature, subdued landforms than California, Chile, and the Mediterranean, outside of the southern Levant. However, in all mediterranean areas coastal lowlands are more or less discontinuous interrupted by areas of greater relief reaching the shore. Extensive plains inland occur only in California and Chile, if we exclude Italy's Po Valley, the greater part of which is not truly within the mediterranean climatic area.

Rivers commonly exhibit strong seasonal variations except for the few large exotic streams, e.g. Rhone, which drain to the Mediterranean Sea. Typically rivers follow short, steep courses to the sea, in some cases, as in Chile, cutting across the grain of the topography. The two master streams draining the Great Valley of California do not, of course, fit this generalization except at their confluence. Denudation is rapid under prevailing mediterranean conditions. This, in combination with recent emergence on the coasts, leads to strongly dissected scenery. Gully and sheet erosion are common problems caused by relief, high intensity but infrequent precipitation exaggerated by the removal of vegetation by man (see PASKOFF, and NAVEH and DAN). Alternating down-cutting and aggradation have tended to produce characteristic valley profiles. These consist, typically, of a broad, smooth valley floor cut by a well defined channel. Such valleys are often bordered by riverine terraces, the remnants of past valley cut and fill. Another common feature of mediterranean lands is the marine abrasion surface sometimes found at extraordinary elevations. These are also products of the combined effect of eustatic and tectonic disturbances.

Soils in the mediterranean areas have some common characteristics (see also P. ZINKE); thus upland soils are often thin and show strong relationship to parent material from which they were formed. At lower elevations alluvium is deposited to great depths in some areas as in the central valleys of California and Chile. Many areas exhibit paleosols giving evident of past climatic conditions different from those obtaining today (see D. AXELROD).

References

BIROT, P.: General Physical Geography. New York: John Wiley & Sons 1966.

BUNTING, B. T.: The Geography of Soil. Chicago: Aldine Publishing Company 1965.

CARTER, G. F.: Pleistocene Man at San Diego. Baltimore: Johns Hopkins University Press 1957.

Commonwealth Scientific and Industrial Research Organization, Australia, with Melbourne University Press: The Australian Environment. London and New York: The Cambridge University Press 1960.

FITTKAU, E. J., ILLIES, J., KLINGE, H., SCHWABE, G., SIOLI, H. (eds.): Biogeography and Ecology in South America, Vols. I and II, Monographiae Biologicae, Vol. 19. The Hague: Dr. W. JUNK N. V. 1969.

FRENZEL, K., BAUER, A., PLUMB, T. (eds.): Atlas of Australian Resources. Canberra: Department of National Development 1959.

HINDS, N. E. A.: Evolution of the California Landscape. San Francisco: Department of Natural Resources, Division of Mines, Bulletin 158, 1952.

HOFFMAN, G. W. (ed.): A Geography of Europe. New York: Ronald Press Company 1961.

HOUSTON, J. M.: The Western Mediterranean World. New York: Praeger 1967.

JAMES, P. E.: Latin America. New York: The Odyssey Press 1942.

JENNINGS, J. N., MABBUTT, J. A. (eds.): Landform Studies from Australia and New Guinea. London: Cambridge University Press 1967.

KING, L. C.: South African Scenery. London: Oliver & Boyd Ltd. 1963.

KING, L. C.: Morphology of the Earth: A Study and Synthesis of World Scenery. London: Oliver & Boyd Ltd. 1967.

LANTIS, D. W., STEINER, R., KARINEN, A. E.: California: Land of Contrasts. California: Wadsworth Publishing Company 1963.

MC BRIDE, G. M.: Chile: Land and Society. New York: American Geographical Society, Research Series No. 19, 1936.

MC GILL, J. T.: Coastal Landforms of the World. Col. Map. Scale 1 : 25,000,000. New York: American Geographical Society 1958.

NUTTONSON, M. Y.: The Physical Environment and Agriculture of Australia with Special Reference to its Winter Rainfall Regions and to Climatic and Latitudinal Areas Analogous to Israel. Washington: American Institute of Crop Ecology 1958.

SAUER, C. O.: Geography of South America. 319–344. In: J. H. STEWARD (ed.), Handbook of South American Indians, Vol. 6. Smithsonian Institution, Bureau of American Ethnology, Bulletin No. 143. Washington: U.S. Government Printing Office 1950.

SMOLE, W. J.: Owner-Cultivatorship in Middle Chile. Department of Geographical Research No. 89. Chicago: University of Chicago Press 1963.

STORIE, E. R., WEIR, W. W.: Generalized Soil Map of California. Manual 6, with map. California: University of California Agricultural Extension Service 1953.

TALBOT, A. M., TALBOT, W. J.: Atlas of the Union of South Africa. Pretoria: The Government Printer 1960.

VITA-FINZI, C.: The Mediterranean Valleys: Geological Changes in Historical Times. Cambridge:The University of Cambridge Press 1969.

WELLINGTON, J. H.: Southern Africa: A Geographical Study, Vol. I. Cambridge: The University of Cambridge Press 1955.

4

Geomorphological Processes and Characteristic Landforms in the Mediterranean Regions of the World

ROLAND P. PASKOFF

The landforms of the mediterranean regions of the world constitute a distinct group. This distinctiveness is primarily due to the geographical position of these regions, which has largely determined the peculiar present morphoclimatic conditions and the marked climatic changes of the recent geological past. In addition to these natural elements, human intervention has disturbed the fragile geomorphic equilibrium of mediterranean regions at different times in the past.

Morphoclimatic Distinctiveness of Mediterranean Regions

Climatic Data of Interest from a Geomorphic Point of View

Precipitation. In mediterranean regions, precipitation is characterized by: 1) an appreciable annual average rainfall, which generally ranges between 300 mm (Santiago, Chile: 360 mm) and 1,000 mm, although sometimes more (Concepción, Chile: 1,300 mm); the average annual rainfall of a mediterranean station may be higher than that recorded at a station located in humid temperate areas (for instance, Nice 862 mm and Paris only 585 mm); 2) a marked interannual irregularity; the extreme maxima and minima recorded at Athens (average amount: 375 mm) are 830 mm for the wettest year and 125 mm for the driest one. In 1924, only 66 mm fell in Santiago, but the rainfall in 1900 was 820 mm (the month of July alone received 353 mm). The concept of average rainfall is not of great significance in mediterranean regions; at Marseille (average rainfall: 572 mm) the most frequent annual amounts are about 480 mm and 640 mm; 3) a marked seasonal concentration; the dry season which corresponds to summer lasts between two (Concepción) and seven months (Santiago); 4) violent downpours; the number of days with precipitation is relatively low: 40 to 80 per year against 150 to 200 in temperate oceanic areas. Thus it is easy to understand the serious floods, very often catastrophic, which occur from time to time in mediterranean regions; 313 mm in 1 hour 30 minutes on 20 March 1888 at Moligt-les-Bains (Eastern Pyrenees); 210 mm in 3 hours 50 minutes (150 mm in 2 hours) at Marseille on 1 October 1892. Contrary to general opinion, mediterranean downpours may last for a considerable period. Some 1,495 mm fell in three days at Santa Cristina of Aspromonte (Italy) between 16 and 18 October 1951.

The Thermal Regime. Winter temperatures are mild. The average temperature for the coldest month ranges between 7° and 13° C according to geographical position. Low temperatures are exceptional (absolute minimum at Santiago: –4,6° C). Frost and snow are rare, at least on the plains. On the other hand, summers are hot. The average temperature for a month in mid-summer (July or January, according to hemisphere) generally ranges between 14° and 25° C, although it may sometimes be greater. The absolute maximum recorded at Montpellier is 45° C. The daily thermal oscillation, especially in winter, may be large. At Santiago, differences of about 20° C between day and night temperatures are frequent in June and July.

These climatic features explain why, from a geomorphic point of view, mediterranean regions have greater affinities to subtropical arid areas than to humid temperate zones. This resemblance is revealed in the vegetation, which is evergreen but sclerophyllous. Its stunted aspect and its discontinuity are referable to its metabolic activity. The processes of erosion, with the exception of the wind which plays a negligible geomorphic role in mediterranean regions, are also reminiscent of those acting in desert regions.

Morphogenetic Processes

Weathering of Rocks. There is relatively little chemical weathering of rocks in mediterranean regions. This is due to the drought which prevails during most of the year and also to the fact that, even during the wet season, a good deal of precipitation is in the form of violent downpours; run-off rates are high and infiltration rates relatively low. The phenomenon of hydrolysis, without being wholly negligible, is relatively reduced and clay formation is light. Nevertheless, under forest cover, biochemical action may play an active role in soil formation. Mechanical weathering is also relatively unimportant. Shattering through frost is exceptional, except in mountainous areas. The influence of temperature variations alone seems minor. Generally speaking, physical-chemical processes, such as hydration or salt cracking along coasts, are comparatively important and are probably the cause of the granular disintegration of granitic rocks (Fig. 1).

Fig. 1. An aspect of weathering in mediterranean regions: granular disintegration; cavity called "taffoni" in an acid volcanic rock near Guercif, northeastern Morocco

Run-off. The run-off coefficient (ratio between precipitation which runs off and incident precipitation) is always high in mediterranean regions. Several factors account for this phenomenon; marked concentration of rainfall in winter which, due to the mild temperatures, does not accumulate as snow; violence of downpours; steepness of slopes; discontinuity of vegetation; and thinness of soils. These features explain why rates of infiltration and retention are relatively low.

The concentrated run-off of surface waters in mediterranean regions exhibits several unique hydrological characteristics. Mediterranean streams have marked annual and inter-annual flow variations and, above all, sudden and violent floods. This last characteristic is crucially important from a geomorphic viewpoint, because floods can cause substantial changes in fluvial plains, primarily through lateral erosion. Because they occur suddenly and with high velocity, they pick up heavy loads from the bottoms of the river beds. Banks are very susceptible to erosion because of the high viscosity and turbulence of the flood waters.

At the arid margins of mediterranean climates at least, distinctive types of run-off may occur. These are known by the name of rill or sheet wash. They result from a particular combination of pluviometric conditions (concentration of precipitation and violence of downpour), biogeographical conditions (discontinuity of vegetation cover) and topographic conditions (smooth and gentle slopes). This kind of "blanket" run-off must also be taken into account in attempts to explain, and make inventories of the landforms of mediterranean regions.

Landforms

Three main kinds of landforms are characteristic of mediterranean lands: bare and steep slopes, broad flood plains, torrential forms.

Bare and Steep Slopes. These represent a conspicuous feature of mediterranean landscapes even when mountainous volumes are reduced (Fig. 2). These rocky walls, without vegetation or soils, where structural influences (lithology, tectonics) are clearly visible, owe their remarkable persistence to the poor development of rock weathering processes.

Broad Flood Plains. Flood plains in mediterranean regions are very broad compared with the narrowness of river channels. This geomorphic characteristic is attributable to the large variations in stream flow and to the high rates of lateral erosion resulting from serious flooding.

Torrential Forms. These probably constitute the most striking geomorphic aspect of mediterranean landscapes. Southern Italy, the hills of the Algerian Tell, the coastal range of central Chile, all show a typically dissected topography of deeply entrenched, ephemeral streams and related coarse detritic accumulations (alluvial fans). Torrentiality is mainly due to the climatic conditions mentioned above, but its development also has topographic, structural and biogeographic requirements. Around the Mediterranean Sea, as in California, the powerfull tectonic movements which occurred at the end of the Tertiary Era, and extended into the Quaternary, created pronounced topographic gradients. Rock outcrops in these areas are often soft either by nature (clays, marls ...) or by weathering (decayed granites). Plant cover, moreover, does not provide a very great protection.

Detailed studies of these torrential forms indicate two facts which have to be taken into account in consideration of mediterranean landforms. Firstly, these torrential forms are generally constituted within similar, but older and larger, forms. Very often the present-day alluvial fans are small in comparison with older detritic accumulations, which may be still bulky in spite of partial destruction. Secondly, contemporary torrential forms are well

Fig. 2. Steep and bare slope, Jebel Mahsseur (1,354 m), near Oujda, northeastern Morocco; remnant of a syncline in a folded Jurassic formation

developed mainly in places where there has been strong human intervention. Man's modification and destruction of mediterranean vegetation has undoubtedly favored torrential dissection.

The Paleoclimatic Inheritance

Most of the morphoclimatic zones of the world have been marked several times by the strong climatic changes that occurred during the Quaternary. Mediterranean regions, because of their transitional position between the tropical and temperate zones, have been particularly prone to these changes. During the glacial periods in higher latitudes, mediterranean regions have been affected by the appreciable decrease of temperature, and, above all, by the considerable increase of precipitation. A pluvial regime ruled over mediterranean regions during these times, similar in some aspects to the present climate of mid-latitude lands on the western sides of continents. Paleobotanical data on the shift of temperate forest ecosystems toward lower latitudes during periods of glaciation support this supposition. Because present-day climatic conditions have existed for less than 10,000 years, it is necessary to take into account the paleoclimatic conditions when considering the mediterranean landscapes of today.

Fluvial Terraces

Large mediterranean rivers very often display step-like alluvial benches which indicate an alternation of filling and down-cutting stages. If terraces in the lowermost part of the stream courses may be explained by glacio-eustatic variations of sea-level during Quaternary time, further upstream they are a direct consequence of climatic changes. The following sequence is proposed for their genesis (Fig. 3). During the pluvial period, precipitation was sufficient to maintain a dense vegetation cover on slopes when these were not very

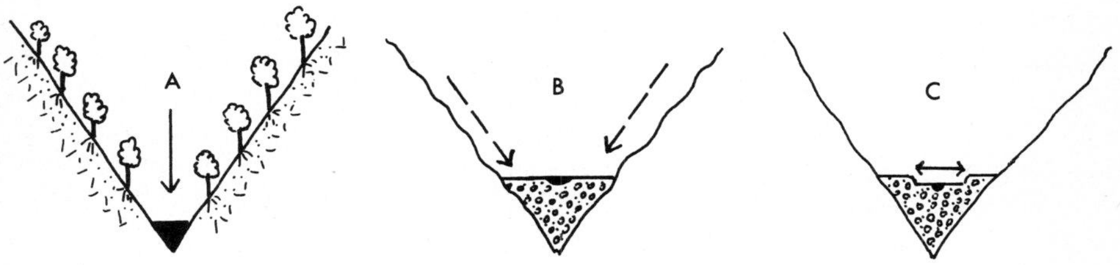

Fig. 3. Genesis of fluvial terraces

A. Pluvial period: down-cutting; vegetation covering slopes; plentiful and permanent water supply; small amount of load supplied to river.

B. Transition between pluvial and interpluvial period: filling up; decreasing water supply and excessive amount of load supplied to river.

C. Interpluvial period: light down-cutting and active lateral erosion; weak load supplied to river and poor water flow; serious floods from time to time

steep. Therefore, the slope evolution was mainly pedological; little coarse debris moved down, and rivers, because their flow was plentiful and permanent, were able to make cuttings. The transition between the pluvial and interpluvial periods was characterized by a degradation of the vegetation. Debris was transported down the slopes and an excessive amount of load supplied to rivers. At the same time the water flow became insufficient to carry the load downstream, which was thus deposited. Finally during the interpluvial period itself – the present situation – rivers are poorly supplied with clastic detritus because slopes are not strongly attacked by weathering. They cut down slightly, but primarily undermine the most recent alluvial filling through lateral erosion during floods.

Pediments or "glacis d'érosion"[1]

Toward the arid borders of mediterranean regions, at the foot of mountains or hills, gently sloping rock-cut surfaces (1° to 5°) are found which are covered by a thin veneer of angular detritus. They often merge downstream into fluvial terraces. French scientists working in North Africa have commonly used the term *glacis d'érosion* to describe these features. At these latitudes such forms are no longer active, since they are dissected

1 Some French geographers have recommended the use of the term "*glacis d'érosion*" for surfaces which truncate soft sedimentary rocks and the term "pediment" when they cut granitic outcrops, because the conditions of genesis would not be exactly similar in both cases (Birot and Dresch, 1966). American scientists, however, do not differentiate between these two phenomena and use the single term "pediment".

by ephemeral streams. Step-like remnants of older *glacis d'érosion* suggest an alternation of periods favorable to their genesis – probably three or four times during the Quaternary – and periods marked by a down-cutting of the hydrographic network.

Scientists are still divided in their opinions on how and when these *glacis d'érosion* have been formed. Nevertheless it appears that their genesis took place in a physical environment characterized by a dry bioclimatic trend, a rather open vegetation cover, heavy downpours at wide intervals, and relatively active rock weathering which produced mainly fine and medium size detritus which could be easily transported. These conditions might have occurred during Pleistocene pluvial periods in what are today transitional areas between mediterranean and subtropical arid climates. But is the main geomorphic process of planation the unchanneled run-off of surface waters (rill wash, sheet wash) or the lateral erosion of concentrated run-off? The debate remains open. Even though most geologists and geographers agree that periods of down-cutting are caused by climatic and not by tectonic factors, the conditions which are required for stream entrenchment are still debatable. Whatever these divergences of opinion there is general acceptance that, in mediterranean regions, these pediments and/or *glacis d'érosion* represent paleoforms.

Relict Torrential Forms

Reference has already been made to the old detrital accumulations deposited in the past by streams at the foot of mountains. Sometimes, despite their dismantlement, different alluvial fans, one built up into another, are clearly visible (Fig. 4). Clastic

Fig. 4. Alluvial fan dating back to the last pluvial period (cultivations and dwellings on its surface indicate that it is no longer active today) – built up within the remnants of a bigger fan formed during the penultimate pluvial period; northcentral Chile, Claro River Valley, Monte Grande (Lat. 30° S)

components are more or less weathered, according to their age. These observations suggest the occurence of several pluvial periods, separated by less rainy ones (PASKOFF, 1970). During pluvial periods, where slopes were sufficiently steep and elevations high enough to prevent a dense vegetation cover, torrential actions were much more active than today, particularly when frost also contributed to rock shattering.

Paleoclimatic inheritance also dates back beyond Quaternary time. It has often been pointed out that in mediterranean regions outcrops of granitic rocks correspond to depressed areas in comparison with sedimentary and volcanic formations which remain in elevated topographic positions. This phenomenon is explained by the weathering of crystalline rocks, which take on the geomorphic behavior of soft rocks (Fig. 5). Such strong

Fig. 5. Deeply weathered Upper Cretaceous granitic rocks; preformed rounded boulders limited by orthogonal and curved fractures; most of this intense alteration probably dates back to the Neogene period; central Chile, Cuesta de la Dormida (Lat. 33° S)

weathering is not a present-day phenomenon. It results from the tropical climates which prevailed over mediterranean regions during the Middle to Upper Tertiary Era. These climates, rather hot and above all rainy, with a summer precipitation peak (see AXELROD'S chapter), were very favorable to the strong chemical weathering of rocks. Sometimes, for instance in central Chile, internal alterations by hydrothermal actions augmented this external weathering.

Morphogenetic Effects of Human Intervention

The consequences of human intervention on the morphogenesis of mediterranean regions are quite important. Because of the fragile equilibrium between the vegetation and the present climatic conditions, transformation to arable land is almost always irreversible. Despite its discontinuity, the natural vegetation has an important protective role which is lost on transformation. There is an increase in torrential flow. Accelerated rates of erosion are favored by the seasonal contrasts of the mediterranean climate. This erosion breaks up soils, brings detrital materials to the surface and finally attacks the bedrock itself when it is soft (clays, marls, sands, etc.).

Therefore, deforestation to obtain grazing and arable lands (see chapter by NAVEH and DAN) induces imbalance in the precarious morphoclimatic equilibrium of mediterranean regions. Some cultivation systems are more destructive than others. Traditional ancient techniques of mediterranean agriculture limited the catastrophic effects of accelerated erosion by the construction of terraces on slopes. This contrasts with the disastrous consequences of certain modern agricultural methods introduced into relatively dry areas of mediterranean regions at the end of the last century. This was the case, for example, in the dry farming practice – a mechanized and speculative wheat monoculture on land under long fallow – in eastern Algeria.

The destruction of the vegetation cover explains why a concentration of the water flow in surface channels soon takes the place of rill wash. The summer drought favors the occurrence of desiccation cracks in the soil. With the winter rains, the first rills appear, which rapidly turn into gullies. These gullies are rapidly deepened and extended. When the gully network is dense, the resulting dissected landscape, often called "badlands" or "calanchi", constitutes a rugged topography characterized by short steep slopes and unstable narrow interfluves. The consequences of this erosion pattern are found in the sediments which fill water reservoirs and in the sterile sand and pebble deposits which cover cultivated soil after heavy rains.

Toward the humid borders of mediterranean regions, anthropogenic morphogenesis may manifest itself on clay slopes which are commonly affected by landslides. The cause of these masses of rocks and soil moving downslope is usually complex, involving slumping and flowage. This form of slope slippage arises mainly from the seasonal climatic contrast when the stabilizing vegetation no longer exists. During the hot and dry summer, the clays crack, thus facilitating the infiltration of water at the beginning of the rainy season. Saturation of the rocky mass is made possible and solifluction phenomena follow.

To summarize this short chapter on mediterranean landforms, three important points must be stressed:

1) the distinctiveness of the present-day processes of erosion which show affinities with geomorphic processes acting in arid lands;

2) the importance of the paleoclimatic inheritance, referable to the transitional latitudinal position of mediterranean climates;

3) the vulnerability and fragility of the natural equilibrium which is continuously threatened by human intervention.

Acknowledgements

The author is grateful to Dr. NORMAN THROWER and Mr. IAN MILMINE for critically reading the manuscript of this chapter.

References

BIROT, P., DRESCH, J.: Pédiments et glacis dans l'Ouest des Etats-Unis. Annales de Géographie, Paris, N. 411, 513–552 (1966).

PASKOFF, R.: Recherches géomorphologiques dans le Chili semi-aride. Bordeaux: Biscaye Frères 1970.

5

Analogies between the Soil and Vegetation Types of Italy, Greece, and California

PAUL J. ZINKE

Introduction

It has been a fairly common observation that there are similarities between the vegetation and the soils of Mediterranean countries and California. However, this has usually been limited to the gross characteristics; for example, noting that there is chaparral or that red soils are prevalent in both areas. Plant species introductions from the Mediterranean countries to California indicate that the California environment of climate and soil is within the tolerance of many typical Mediterranean plant species such as olive *(Olea europaea)*, Aleppo pine *(Pinus halepensis)*, oleander *(Nerium oleander)*, cork oak *(Quercus suber)*, and various species of *Cistus*. The world-wide plant geography study of SCHIMPER (1903) called attention to the similarities between the vegetation of the Mediterranean countries and California, but he also pointed out that not much was known about California.

Since that time there has been gained a considerable knowledge in California concerning the vegetation and, in the past fifteen years, the soils of the wildland areas of California. A soil-vegetation survey has covered more than 15 million acres (more than 6 million ha) of the California landscape, and the vegetation type maps that were the forerunner of the soil-vegetation survey (WIESLANDER, 1935) also covered millions of acres of California. It was noticed in this work that typical groupings of vegetation species into types occurred, and that these could be described in terms of vegetation structure. This led to an aerial photo technique of describing vegetation types (JENSEN, 1947) by structure, which then formed the basis for subsequent vegetation mapping work. Generally it was found in the relatively undisturbed vegetation of the wildland areas of California that characteristic groupings of plant species and associated vegetation structure occurred in response to different soil types, in turn dependent upon geology, with modification by the past landscape history, the topography, and the climate. The result is a mosaic of soil-vegetation types.

The study reported here was planned to investigate some of the analogies that might exist between the soil-vegetation types of Italy and Greece and those of California, with observations of the soil and vegetation similar to those made in the California Soil-Vegetation Survey.

Objectives and Methods of the Study

The objectives of this study were to observe soil-vegetation types in Italy and Greece from the standpoint of determining what similarities they had to those in California. Observations were to be made at selected locations in Italy and Greece that would give a range of conditions similar to the range of climatic, geologic, and altitudinal conditions which exist in California areas having relatively undisturbed vegetation. Studies made in connection with the California Soil-Vegetation Survey were then used in making the analogies with comparable areas in Italy and Greece. The vegetation comparison was made on the basis of structure with notation of the dominant vegetation species and their abundance; the soils comparison was made on the basis of field characteristics of soil profiles.

Field work was begun in Italy with the observation of sampling plots on two elevation sequences, one on basalt rock and one on granite rock located in Sardegna. A similar sequence on sedimentary rocks in Sicily was observed, and one sequence was studied on granitic rocks in southern Italy (Calabria). A sequence of observations was made on marl and limestone rocks in central Italy (Abruzzi). Sequences of observation points were visited in Greece on granitic rocks (west from Florina toward the Albanian border, and north from Serrai toward the Bulgarian border), on metamorphic rocks (Mt. Olympos, and near Sparta), on limestone rocks (near Sparta), and on a landscape sequence involving sedimentary rocks into which serpentine and peridotite had been intruded (in the northern Pindus Mountains near Metsovon). Observations were completed with a sequence of points on sedimentary rocks in northern Italy (Tuscany).

Systematic Relationships among California Forest Soils

The extensive mapping which has been carried out in California in the Soil-Vegetation Survey has provided observations not only for a systematic description of soil series in areas of natural vegetation but also a factual basis for relating and grouping these soils.

Field men mapping in the Soil-Vegetation Survey have observed predictable relationships between the type of soil profile and certain variables in the landscape, particularly the type of parent material or the geology of the area, the relative age and the topography of the land surface, and the large-scale differences in elevation. Associated with the interaction of these landscape variables are soils having various degrees of development.

That soils can be grouped according to degree of development on a given parent material is obviously not new to the soils literature. GERHARDT (1900) noted that he could classify sand dunes, from white to yellow to reddish, in relation to increasing age of dune. SHAW (1928) established a grouping of soils into families ranging from an immature soil to a mature soil on each parent material. STORIE and WEIR (1953) have developed these concepts into a guide to soil series in California.

Soil series described during the Soil-Vegetation Survey can be related as sequences of increasing degree of development on each rock type. There will be some variation in the developmental sequence on a given parent rock, depending upon present or past climatic differences. Also, in some areas an entire sequence may not be present due to excessive soil erosion, to colluvial deposition, or to past stripping of the soils by glaciation, thus resulting in a lack of developed soils (ZINKE and COLWELL, 1965).

Examples of soil development sequences can be found on soil-vegetation maps published for areas dominated by each of the main parent rock types. The broad categories of rock types in California are basic igneous, acid igneous (with associated metamorphic rocks), and sedimentary rocks. A transect across a landscape dominated by any one of these rock types will often show variations in properties of soil profiles which can be arranged in sequence of degree of development. A developmental sequence on schist rock, for example, is shown on the Soil-Vegetation Survey map of the Hoopa Quadrangle (DELAPP and SKOLMEN, 1961).

Sequences of decreasing soil development with increasing elevation over wide ranges of elevation change are usually the rule in California mountain forests. In the Blocksburg Quadrangle (COLWELL, GEE and MEYER, 1955) on a long slope of increasing elevation, a sequence of soil profiles of lessening degree of development with increasing elevation occurs on graywacke sandstone.

These observations in California have indicated two general relationships: 1) Each parent rock type will usually have a sequence of soil profiles of increasing degree of development related either to climatic differences or topographic variation. 2) Similar sequences occur on each of the parent rock types with variation reflecting differences in the physical and chemical properties of the rocks. These conclusions were then used as a basis for assessing the soils in Greece and Italy as to their similarity to California.

Soil Properties

Certain soil properties of the soil series in each of these sequences on the various rock types show a fairly consistent relationship to degree of development of the soil.

Color. The most obvious change apparent in the field with increasing soil development was in the soil color. Thus, in most sequences the color progressively changes from grayish-brown to reddish-brown with increasing soil development. This color change is more noticeable in the lower horizons, and less in the A horizons, apparently because of organic matter from forest litter that was incorporated in the soil. The general relationship for most rock types in the forest areas of California is that a sequence of soil series with increasing redness of the soil can be found. Presumably this parallels soil development. Mature red soils are a characteristic of many California areas. There are notable exceptions to this color sequence on a minor portion of the areas mapped that are related to initial strong color in the parent material, or to soil development under conditions of poor drainage. The question then posed for the analogies to be tested in Greece and Italy is whether this holds for the soils in these countries.

Gravel and Stone Content. The field mappers of the Soil-Vegetation Survey also noticed that the degree of stoniness of the soil changed with the degree of development of the profile on a given rock type. Data generally indicate a progressive decrease in the percent by weight of the greater than 2 mm fraction in bulk soil samples (stones greater than 5 cm discarded) with increasing development of the soil. This change is most apparent at a depth of 30 in (76 cm) in the soil. Differences related to parent material are evident. The soils derived from granitic rocks had fewer coarse fragments because initial weathering produced grains the size of the mineral crystals in the rock. The other parent rock tended to weather initially into larger fragments. The soils derived from basic igneous rocks are more variable in rock content and as a result show a less consistent trend of diminishing rock content with increasing development. This is due partly to the large size of the rocks in the field profiles, many of which were beyond the size limits of the samples taken. A general relationship apparent in most of these sequences is that the content of coarse fragments decreases with increasing development of these forest soils. A similar situation is to be expected in the Italian and Greek soils if they are analogous to those of California.

Clay Content. Again from the observation of many profiles, the mappers found clay content of the soil profile another obvious criterion of increasing degree of development of the soil on any of the parent rocks. The *percent clay* (less than 2 microns) *in the fine earth fraction* of the soil increases with soil development from low amounts of 6% in the immature soils. The soils derived from granitic rocks had the lowest clay content, and those derived from sandstone (graywacke) had the highest. The clay content became greater with increasing development, especially in the subsoil, reaching 36% in soils from granitic rocks and more than 60% in soils derived from graywacke sandstone. The general relationship, an old one in soil science, is that on the developmental sequences associated with each of the main rock types in California forest areas, clay content increases with increasing maturity of soil. An important observation is that clay content at comparable stages of soil development will differ on different parent rocks. Presumably this should hold for analogous sequences of change in soil development with elevation in Italy and Greece.

These observations indicate that it is possible to relate the numerous soil classification units currently being used in the classification of California forest lands according to degree of development from a given parent rock type and that likewise could be applied to areas similar to California. The usual sequence in its simplest form progresses from soils with slight development, through moderate development to well-developed soils. In the terminology of STORIE and WEIR (1953) this involves a sequence from lithosols to gray-brown podzols to red-yellow podzolic soils. According to that of KUBIENA (1953) the sequence would begin with ranker soils, progressing through brown earths with increasing degree of leaching to end with red loams. The soils in Italy and Greece have been examined for analogous sequences.

Vegetation Types in California

In order to make comparisons between the vegetation at selected sites in Italy and Greece with the vegetation of California, it was felt necessary to either develop or utilize an existing method of categorizing the vegetation types in California, that in its broader generalities could then be applied to the vegetation types of Italy and Greece. A regional zoning of vegetation could be fitted into the early life zone formulations of MERRIAM (1898) which, although supposedly based on climatic variables, are more usually defined in terms of vegetation zones (BAKER, 1934). A classification of the vegetation of California deriving the information principally from aerial photographs has been described by JENSEN

(1947). It provides a technique which can be applied to such areas as Italy and Greece, as well as other lands around the Mediterranean and Black Seas. More recently, MUNZ and KECK (1959) have categorized California plant communities. These classifications simplify the making of analogies between the vegetations of the Mediterranean area and California. There are nearly 5,000 species of plants comprising the vegetation of California (MUNZ and KECK, 1959) and 3,446 species in the vegetation of Italy according to BARONI (1963); but by the technique of grouping these species into vegetation types a system for making the analogies between the California and the Mediterranean vegetation can be made.

Life Zones and California Vegetation

MERRIAM (1898) defined life zones for the United States which have some utility if used wisely in application to California conditions. The application of these zones to categorize California vegetation types was made by JEPSON (1923). They can be summarized as follows:

Zone	*Typical form and type of vegetation*
Lower Sonoran	Desert vegetation
Upper Sonoran	Grassland and Oak woodland, Juniper woodlands, Chaparral, and Digger pine
Transition (arid)	White fir, incense cedar, sugar pine; mixed conifer forests of the Sierra Nevada, the south coast ranges, and the southern Cascade Mtns. Some hardwoods such as *Quercus kelloggii* and *Cornus nuttallii*
Transition (humid)	The coastal forests of northern California, and the interbedded brush types and grasslands. Typical species are redwood, Douglas fir
Canadian	High elevation conifer forests characterized by such species as *Abies magnifica, Pinus monticola,* and *Pinus contorta*
Hudsonian	Open conifer woodlands of small bushy trees such as *Tsuga mertensiana* (hemlock), *Juniperus occidentalis* (western juniper), and *Juniperus communis.*
Arctic Alpine	Open fields of herbs, sedges, and grasses with low dwarf shrubs of *Erica* (heather), willow, etc. Characteristic genera are *Carex, Erica, Salix, Juniperus.*

A similar life zone scheme exists for Italy, and comparisons and analogies to this will be made later.

Plant Communities in California

MUNZ and KECK (1959) in their flora of California have defined major vegetation types and plant communities in California as follows:

Vegetation type	*Plant community*
I. Strand	1. Coastal strand
II. Salt Marsh	2. Coastal Salt Marsh
III. Freshwater Marsh	3. Freshwater Marsh
IV. Scrub (brush)	4. Northern Coastal scrub
	5. Coastal sage scrub
	6. Sagebrush scrub
	7. Shadscale scrub
	8. Creosote bush scrub
	9. Alkali sink
V. Coniferous forest	10. North Coast coniferous forest
	11. Closed cone pine forest
	12. Redwood forest
	13. Douglas fir forest
	14. Yellow pine forest
	15. Red fir forest
	16. Lodgepole pine forest
	17. Subalpine forest
	18. Bristle cone pine forest
VI. Mixed Evergreen Forest	19. Mixed evergreen forest
VII. Woodland Savanna	20. Northern oak woodland
	21. Southern oak woodland
	22. Foothill woodland
VIII. Chaparral	23. Chaparral
IX. Grassland	24. Coastal prairie
	25. Valley grassland
X. Alpine fell fields	26. Alpine fell fields
XI. Desert Woodlands	27. Northern juniper woodland
	28. Pinyon-juniper woodland
	29. Joshua tree woodland

Lists of typical species of vegetation for these plant communities have been given by MUNZ and KECK (1959).

A Technique of Classifying Vegetation and Land Cover

A technique of classifying vegetation and land cover based upon elements of the vegetation and land cover which are readily identifiable from aerial photographs has been presented by JENSEN (1947), and is being used as a basis for widespread mapping of Soils and Vegetation in California (California Department of Natural Resources, 1958). The advantage of this classification is that it can be applied without seeing the area, except on aerial photographs. It depends upon the recognition of readily identifiable structures in the vegetation seen on the aerial photographs. The delineation made is based upon various structural classes of the vegetation or the land cover. These are as follows:

C	Conifers, commonly so called commercial or large coniferous trees
K	Scrubby conifers. Coniferous trees of small size or stature
H	Hardwoods. Broad leaved trees. Further subdivided into old or large hardwoods H_o, and young hardwoods H_y
S	Chaparral. Shrubs of the tall, dense, heavily branched type such as manzanitas, scrub oaks, and chamise
T	Sage. Soft shrubs lower in stature, usually grayish to light colored in tone and characteristic of desert, or coastal sand areas
F	Bushy herbs, mainly ferns such as *Pteris aquilina*
G	Grasses. Grasses, sedges and other associated herbaceous vegetation
M	Marsh
B	Bare ground
R	Rock
A	Cultivated
U	Urban Industrial

For the purposes of classifying the vegetation and land cover at the various observation points visited in Greece and Italy, the following adaptation of the JENSEN classification was made.

C	Conifers
H	Hardwoods with H_o and H_y
S	Chaparral and Sage (for example, including *Artemisia* spp. as well as shrubby oaks)
G	Grasses, Herbs and Fern cover
M	Marsh
B	Bare ground
R	Rock
A	Cultivated

These would be elements of vegetation cover as they would be viewed from aerial photographs, but likewise observed on the ground. A further stratification is obtained by identifying a total woody vegetation density. This density classification is expressed in such terms that it is also identifiable on aerial photographs, the density being the crown cover density of the woody vegetation. The following legend is used for density classes (JENSEN, 1947):

1) Dense. Stands in which the crowns of the vegetation elements cover more than 80% of the ground area.
2) Semidense. Stands in which the crowns of the vegetation elements being considered cover 50–80% of the ground space.
3) Open. Stands in which the crowns of the vegetation element being considered cover from 20–50% of the ground space.
4) Very Open. Stands in which the crowns of the vegetation element being considered cover from 5–20% of the ground space.
5) Unstocked. Areas having less than 5% of ground space covered by crowns of the vegetation element being considered.

Examples of aerial photo interpretations using these types of legends are presented by JENSEN (1947). For the study areas in Italy and Greece, a classification was given to the vegetation which included the crown density of the woody vegetation including conifer and hardwood trees, and shrubs, this being noted in the numerator of a fraction. The vegetation elements present were then grouped in order of abundance in the denominator of the same fraction. The densities and orders of abundance are those which would appear on the basis of ground cover as viewed from the air. A symbol reading 5/G for a type indicates less than 5% woody vegetation cover, and a cover which is composed mainly of herbaceous vegetation such as grasses, etc. When coniferous trees are present, an age class has been given to the conifers, designated by Y or O or a mixture depending upon whether the trees are greater than 150 years old (O), or less (Y). Three density figures are used to characterize the vegetation type with conifers. These are: density of all conifers greater than approximately 1 in (2.5 cm) in diameter, density of conifer trees of all sizes, and finally the density for all woody vegetation. Thus a typical cover class symbol describing a dense young stand of coniferous trees is Y111/C, signifying that the vegetation type is coniferous forest less than 150 years old, with greater than 80% coniferous trees, and greater than 80% total woody vegetation canopy density. This allows a comparison to be made directly with Timber Stand Density Maps which are published for most of the wildland areas of California (California Department of Natural Resources, 1958).

The Vegetation of Italy and Adjacent Mediterranean Areas

Life Zones and Italian Vegetation

A similar life zone classification to that previously described for California has been developed for Italy by using a classification developed by MAYR for Europe and modifying it for the Italian dry summer conditions (DE PHILIPPIS, 1937). This climatic classification is as follows:

Zone	*Temperature*		
	Average annual	Coldest months mean	Minimum mean
A) *Lauretum*			
1. Type 1 uniform precipitation, hot	15–23°	>7°	> −4°
2. Type 2 with summer drought	14–18°	>5°	> −7°
3. Type 3 with summer rain	12–17°	>3°	> −9°
B) *Castanetum*			
1. Warm subzone	10–15°	>0°	> −12°
Type 1 without summer drought			
Type 2 with summer drought			
2. Cold subzone	10–15°	> −1°	> −15°
Type 1 precipitation more than 700 mm			
Type 2 precipitation less than 700 mm			
C) *Fagetum*			
1. Warm subzone	7–12°	> −2°	> −20°
2. Cold subzone	6–12°	> −4°	> −25°
D) *Picetum*			
1. Warm subzone	3–6°	> −6°	> −30°
2. Cold subzone	3–6°	< −6°	< − 30°
E) *Alpinetum*	< 2°	< −20°	< − 40°

An analogy can be made between these climatic types and those of California described earlier, as follows:

California Life Zone (MERRIAM, 1898)	*Italian Life Zone* (DE PHILIPPIS, 1937)
Upper Sonoran	*Lauretum* warm subzone
Transition	*Lauretum* cold subzone and *Castanetum*
Canadian	*Fagetum*
Hudsonian	*Picetum*
Arctic Alpine	*Alpinetum*

Missing from the Italian life zones and also those of Greece would be the Lower Sonoran representing the deserts of California. However if one were to go further east or south in the Mediterranean region one can find life zones which would encompass the desert areas of California. For example, GINDEL (1964) has referred to phytogeographic zones in Israel to include the Mediterranean Maqui zone which coincides with the Upper Sonoran in California and then defines the following dryer zones:

Irano-Turanian	Semi-Desert	200mm–350mm precipitation
Saharo-Sindian	Desert Zone	25mm–200mm precipitation

Both of these would fall within the Lower Sonoran Zone in California, the first being within the Mojave and the Valley Sonoran of JEPSON (1923), and the second within the Colorado Desert Sonoran of JEPSON (1923).

These life zones, of course, progress from the warmer at the lower elevations to the cooler at higher elevations.

Italian Plant Communities and Some California Analogies

The major plant formations of Italy have been described by GIACOMINI (1958). They are listed below with the California plant community analogy as described by MUNZ and KECK (1959), and presented earlier:

Italian Vegetation and Plant Community	*Analogous California Plant Community*
I. Montane Forests	
1. Oak-chestnut woodland *(Quercus-Castanea)*	Northern oak woodland.
2. Beech forest *(Fagus)*	Partially Northern oak woodland; however, the closest species analogy is the tanoak of northwestern California.
3. Fir forest *(Abies)*	Red fir forest or North Coast coniferous forest, and to some extent the Douglas fir forest.
4. Mountain pine forest *(Pinus nigra)*	The Yellow pine forest, Lodgepole pine forest.
II. Evergreen Forests	
A. Evergreen oak forests	
1. Cork oak forest *(Quercus suber)*	Southern oak woodland (mainly those with *Quercus engelmannii*).
2. Holly leaf oak forest *(Quercus ilex)*	Foothill woodland (mainly those with species such as *Quercus wislizenii*, and *Quercus chrysolepis*).
3. Scrub oak forest *(Quercus cerris, pubescens)*	Foothill woodland, i.e. *Quercus douglasii*.
B. Littoral pine forests	
1. Domestic pine forest *(Pinus pinea)*	Closed cone pine forest (especially the coastal Torrey pine forest of southern California).

Italian Vegetation and Plant Community	*Analogous California Plant Community*
2. Aleppo pine forest *(Pinus halepensis)*	Closed cone pine forest (particularly the Monterey pine forest of central California).
3. Cypress forest *(Cupressus sempervirens)*	Either the northern portion of the closed cone pine forest or the upper portions of the Pinyon Juniper woodland.
C. Olive and Carob forests *(Olea europea* and *Ceratonia siliqua)*	Southern oak woodland.
III. Macchia (Brushfields)	
A. Holly leaf oak macchia *(Quercus ilex)*	Chaparral (especially with *Quercus dumosa).*
B. Macchia of *Erica* and *Arbutus* *(Erica scoparia – Arbutus unedo)*	Chaparral (especially with chamise and manzanita).
C. Cistus macchia *(Cistus salviifolius)*	Chaparral (especially with species of *Salvia).*
D. Olive macchia *(Olea europaea)*	Chaparral.
E. Dwarf palm macchia *(Chamaerops humilis)*	None in California.
F. Broom macchia *(Cytisus scoparius)*	Chaparral with *Pickeringia montana* or *Ceanothus* spp.
G. Oleander macchia *(Nerium oleander)*	Chaparral with *Rhus laurina.*
IV. Gariga	Various types of scrub as defined. Northern Coast scrub, Coastal sage scrub, but excluding the desert and alkali soil scrubs, such as sagebrush, shadscale, creosote bush and alkali sink which were not observed in Italy and Greece.
V. Degraded Macchia and Gariga	Degraded very low, sparse chaparral or scrub.
VI. Steppes and Prairies	
A. High altitude pasture (above timber line) *(Carex curvula)*	Alpine fell fields.
B. Mediterranean steppes (low altitude grasslands) *(Festuca – Bromus)*	Coastal prairie and valley grassland.
VII. Littoral sand and salt flat vegetation	Coastal strand and coastal salt marsh.

Observations on Soil-Vegetation Types of Italy and Greece

The planning of a program of observations of soil vegetation conditions in Italy and Greece, with the objective of establishing the analogies with the soil and vegetation of California, involved a review of the literature on the subjects, the determination of map locations which would predict the field locations which would be most representative for sampling and description, and then the carrying out of an extensive field program.

In making the review, it soon became apparent that there would be similar sequences of soil property change with increasing altitudes on a given parent material as has been reported earlier for California. In Italy, COMEL (1939) noted that the properties of soils derived from limestone changed progressively with increasing elevation. In mounting from low to high elevations across the karst north of Trieste, one progressed through the following soil sequence:

Elevation (m)	100	300	1000	1500
Soil type	terra rossa without humus	terra rossa with humus	terra gialla	humus rendzina
Precipitation (mm)	1000	1200	2000	2500

This sequence on limestone is somewhat analogous to those from California in which, with increase in elevation, the soil progressed from a red soil through a brown soil to a dark humus rich soil. The general effect is a decrease in the amount of the clay rich B horizon of the soil and an increase in the amount of humus rich A horizon of the soil with increasing elevation, and with the accompanying increase in precipitation and decrease in mean annual temperature.

RODE (1962) has reported a similar example from the Caucasus Mountains in southern Russia. There are changes in soils from low to high elevations in this area of mediterranean type climate in which the soil is a mediterranean red earth at the lower elevations and progresses through soils of lesser development to a humus rich ranker type soil at high elevations.

MANCINI (1960) in his soils map of Italy implies such sequences. Wherever there is a change in elevation over a wide range, one can trace changes in soil types, usually in the sequences mentioned previously.

Thus the literature supports the hypothesis that elevation sequences of sampling sites on the same geologic rock type facilitates the establishment of presence or absence of analogies between the areas sampled in Italy and Greece and those of California.

A key element in locating the sampling sites thus became the geology of the landscape, and a sufficient altitude difference to allow the broadest possible range in soil development on the rock type.

Since in California the major rock types forming soils are basic igneous rocks (basalt), acid igneous rocks (granite), sedimentary rocks (sandstones), and metamorphic rocks (schists or serpentines), it was desirable to sample on these types of rocks in Italy and Greece. Also, the extensive areas of calcareous rocks such as limestone and marl in these countries necessitated sampling on them, although areas of these rock types, extensive enough to give full developmental sequences of soils, are absent in California. After reference to the geologic and topographic maps of Italy and Greece, the locations for sampling sequences were determined.

The field work of establishing soil and vegetation sampling points on these elevation sequences was begun in February 1964 and completed in August 1964. On each sampling sequence, except for the limestone sequence on Mt. Terminillo, locations were established which represented the range of soil development present on the rock type. This involved a sampling point located at low elevation and low rainfall, followed by others at intermediate elevations and finally one at the uppermost elevation (and highest rainfall) available on that parent material. At each location observations were made on vegetation characteristics and soil properties. Soil samples were taken at each observation point. Following sieving-out of coarse materials, these samples were shipped to Berkeley for subsequent analysis.

Summaries of the sequences of soils and vegetation at various sampling locations are given in Tables 1–7. ZINKE (1965) gives extended descriptions of these sequences, as well as presenting information on other sampling transects in Italy and Greece.

Table 1. Vegetation-soil sequence on basic igneous rocks (basalt and andesite) in Sardegna, Italy, between Oristano and Macomer

Landscape	Elevation (m) 0–200	200–500	500–800	800+
Vegetation				
Type	Open woodland, herbs, shrubs, hardwoods *(Olea, Pistacia, Myrtus, Pinus)*	Open woodland, herbs, shrubby sprouts, hardwoods *(Quercus pubescens, Q. suber)*	Open woodland, herbs, fern, shrubby hardwoods *(Q. pubescens, Q. suber, Q. ilex)*	Open woodland, fern, hardwoods *(Ilex aquifolium, Q. ilex, Q. pubescens)*
Soil:				
Classification	Mediterranean red earth	Terra bruna	Terra bruna leached	Ranker
Subsoil:				
Color	Reddish brown	Reddish brown	Dark grayish brown	Dark grayish brown
pH	6.5	6.5	5.5	4.5
Texture	Clay	Clay	Clay loam	Loose loam

Table 2. Vegetation-soil sequence on acid igneous rock (granite) in Sardegna, Italy, between Lago Coghinas and Monte Limbara

Landscape	Elevation (m) 300	500	1000
Vegetation			
Type	Shrub hardwood	Shrub hardwood	Shrub
Soil:			
Classification	Mediterranean red earth	Terra bruna or reddish brown lateritic	Ranker
Subsoil:			
Color	Yellowish red	Pale brown	Brown
pH	5.0	5.0	5.5
Texture	Loamy clay	Gravelly clay	Loamy sand

Table 3. Vegetation-soil sequence on acid igneous rock (granite) in Calabria, Italy, between Rosarno and Serra San Bruno

Landscape	Elevation (m) 75	650	950	1300
Vegetation:				
Cover class	$\frac{3}{\text{GS Hy}}$	$\frac{5}{\text{G}}$	$\frac{\text{Y 221}}{\text{C Hy}}$	$\frac{1}{\text{Hy}}$
Type	Olive macchia	Grassland cleared from scrub oak forest	Fir forest	Beech forest
Soil:				
Classification	Mediterranean red earth (KUBIENA)	Mediterranean red earth (Terra bruna)	Terra bruna lessivée Podzolic brown earth	Terra bruna lessivée Podzolic brown earth (KUBIENA)
Subsoil:				
Color	Dark red brown	Yellowish red	Brown	Brown
Texture	Clay	Gritty clay	Gritty clay loam	Gritty sandy loam

Table 4. Vegetation-soil sequence on calcareous rock (marl) in Abruzzi, Italy, between Sangro River Valley and La Maiella Mtn.

Landscape	Elevation (m) 200	875	1500	1900
Vegetation:				
Cover class	$\frac{1}{\text{SHy}}$	$\frac{5}{\text{G}}$	$\frac{1}{\text{Hy}}$	$\frac{5}{\text{G}}$
Type	Hollyleaf oak macchia	Grassland cleared from scrub oak forest	Beech forest	High altitude pasture
Soil:				
Classification	Sierozem with A/Ca/C	Earthy terra fusca (KUBIENA) Terra Gialla (COMEL)	Mull rendzina (KUBIENA)	Alpine pitch rendzina (KUBIENA)
Subsoil:				
Color	Light grey	Brown	Very dark brown	Dark brown
pH	8.2	8.0	7.5	5.0
Texture	Stony clay	Clay	Loam	Clay loam

Table 5. Vegetation-soil sequence on calcareous rock (limestone) in Umbria, Italy, Mt. Terminillo

Landscape	Elevation (m) 600	1500	1800	1950
Vegetation:				
Cover class	$\frac{1}{Hy}$	$\frac{5}{G}$ or $\frac{1}{Hy}$	$\frac{5}{G}$	$\frac{4}{GS}$
Type	Oak-Chestnut woodland	Beech forest or high altitude pasture	High altitude pasture	High altitude pasture
Soil:				
Classification	Terra Rossa CH	Mull rendzina CH	Mull rendzina CL	Mull-like rendzina GM
Subsoil:				
Color	Red	Pale brown	Dark brown	Dark brown over very pale brown
pH	6.1	6.5	5.0–4.5	6.8
Texture	Clay	Stony clay	Clay loam	Stony sandy loam

Table 6. Vegetation-soil sequence on sedimentary rock (sandstone) in Tuscany, Italy, from south of Florence to Pratomagno (near Vallombrosa)

Landscape	Elevation (m) 300	900	1460
Vegetation:			
Cover Class	$\frac{Y441}{SHyC}$	$\frac{Y111}{C}$	$\frac{5}{G}$ or $\frac{4}{GS}$
Type	Scrub oak forest[a]	Fir forest	High mountain pasture
Soil:			
Classification	Terra bruna lessivée	Gray brown podzolic	Alpine humus (JENNY)
Subsoil:			
Color	Reddish yellow	Very pale brown	Very pale brown
pH	6.9	5.0	5.0
Texture	Clay loam	Loam	Loam

[a] Degraded to macchia and planted with *Pinus pinea*

Table 7. Vegetation-soil sequence on acid igneous rocks (granite) in northern Greece, Macedonia and Thrace from Serrai to Lailia and Florina to crest of mnts. to west

Landscape	Elevation (m) 770	1000	1000	1600	1800
Vegetation:					
Cover class	5/G	5/G	1/Hy	YO322/CG	3/GS
Type	Scrub oak woodland cleared to grass	Scrub oak woodland cleared to grass	Beech forest (*Quercus robur* var. *sessiliflora, Fagus silvatica*)	*Pinus sylvestris* forest, misc. grasses and herbs	High mountain pasture with *Juniperus communis*
Soil:					
Classification	Mediterranean red earth	Brown forest soil	Gray brown podzolic	Gray brown podzolic	Ranker
Subsoil:					
Color	Yellowish red	Yellowish brown	Brown	Dark grayish brown	Dark grayish brown-light yellow brown
pH	5.5	5.5	5.5	4.5	5.5
Texture	Clay loam-clay	Stony loam	Sandy loam	Gritty loam	Stony sandy loam

The Mediterranean Soil-Vegetation Catena

During the field work that was carried out in this project, it became obvious that there were certain generalities that were true in nearly all of the sequences of soil and vegetation examined. One of these generalities was the consistency of the change in soil properties and vegetation with increasing elevation; the other was the existence of a local mosaic of soil and vegetation types.

The similarities between most of the sequences of change of soil vegetation, whether in Italy, Greece or California, are presented in Table 8. The usual sequence of change is from a mediterranean red earth at low elevations to a dark lithosol (ranker, rendzina or andosoil) at higher elevations, with brown soils of various types (brown forest soil, gray-brown podzol, leached brown soil) in between.

The main exception to this sequence occurred in the soils sequence on marl rock. This leads one to expect that where the parent material weathers readily, and is rich in clay and such bases as calcium or possibly magnesium, another type of sequence typified by that on marl rock will be found. The main difference between this sequence and the more usual one is that in place of the mediterranean red earth at the low elevations (or a terra rossa), a xerorendzina or black earth would be found. This probably accounts for the difference in the types of soil development on the tertiary deposits, in which those tertiary deposits which were clay rich and calcium rich developed dark rendzina-like soils, often called terre nere (PRINCIPI, 1943), or mediterranean black earths. The vegetation sequences associated with these soil sequences were also consistent in beginning with grassland at lower elevations, almost always on the black earths, and frequently on the red earths, and progressing successively through macchia, oak woodland, beech and coniferous forest to high elevation pasture of grasses and *Carex*. Thus sequences of soil vegetation types in terms of such broad categories of classification as World Soil Groups or Plant Communities exist that are typical of the areas examined. These usual categories of classification, however, would require boundaries to be placed on the landscape whenever the limits of definition of one of the groups had been reached in the sequence.

Table 8. Soil-Vegetation Catenas observed with elevation change on most of the rock types of Italy, Greece, and California

Elevation[a] (m)	Soil	Vegetation
High 1600+	Ranker soil (GM or OL), dark grayish brown A horizon, A/C profile, pH 4.5–5.5. Stony, loam-sandy loam. Lithosol-minimal soil profile development. Inceptisols	High mountain pasture – *Carex* and grassland with some *Juniperus communis*
Mod. high 1000–1600	Terra bruna lisciviata or Gray brown podzolic (Cl, MH), brown-dark brown A horizon, B horizon present. Loam texture, pH 5.5–6.5. Minimal soil profile development. Alfisols	Coniferous forest with *Abies* or *Pinus*. Hardwood forest, beech in Europe
Low 500–1000	Terra bruna – Brown forest soil (CH–MH–CL), brown-reddish brown. Loam–clay loam, B horizon present. Medial soil profile development	Hardwood forest – Lower beech forests, Scrub oak forest Closed cone pine forests – *Pinus pinea, P. pinaster, P. attenuata*
Lowest 0–500	Mediterranean red earths (CH), reddish brown – red. Clay loams – clay. pH may be up to 7.5. Maximal soil profile development with clay accumulation. Ultisols	Oak woodlands – *Quercus ilex, Quercus chrysolepis, Quercus suber, Quercus engelmannii* Macchia or chaparral – *Quercus coccifera, Quercus dumosa, Erica arborea, Adenostoma fasciculatum*
	Xerorendzina or Serozem (CH) on carbonate rich, clay rich easily weathered rocks. Dark brown A horizon, nearly white C horizon, with $CaCO_3$ deposition in C horizon. pH up to 8.5. Maximal soil profile development. Aridosol-calcorthid	Grassland. Annual grasses and herbs

[a] Elevations noted for Italy and Greece

Each elevation sequence is seen, however, to amount to a gradual change in properties of the soil or of the vegetation as one ascends a uniform slope of broad elevation change. Since this was such a typical occurrence, it seems that the literature of soil science should encompass a definition and description of such a sequence in which a predictable change in soil properties was consistently related to a topographic change. MILNE (1935) deals with just such a problem. He felt the need for a classification category for a situation in which there is a regular repetition of a certain sequence of soil properties in association with a certain topography, and in which uniformity of parent material might be of subsidiary interest. For this situation, MILNE proposed the word *Catena* to indicate "a grouping

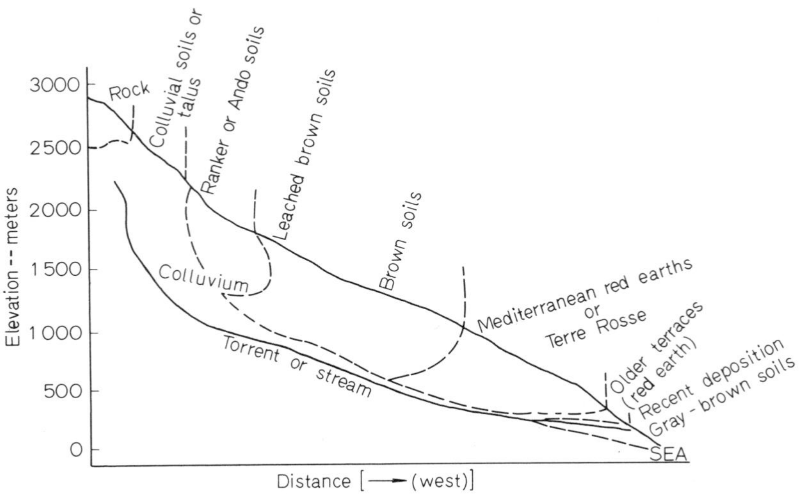

Fig. 1. A typical sequence or catena of soils found in this study in Greece, Italy, and California

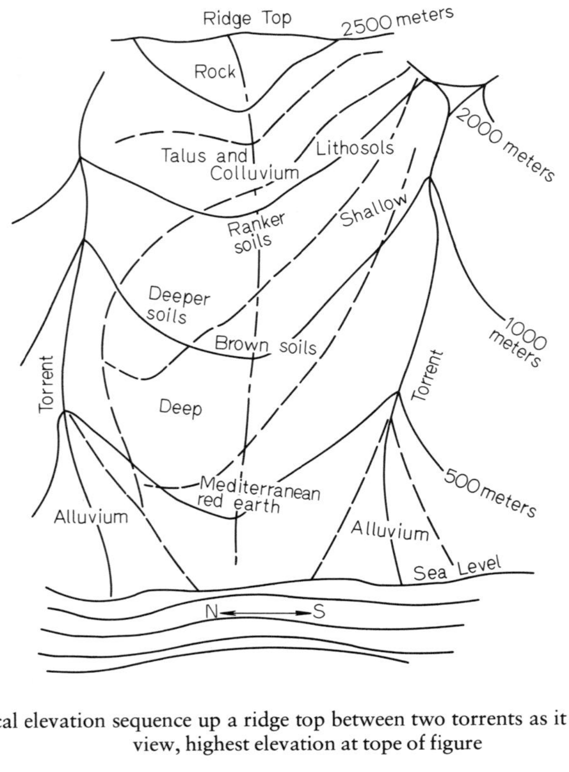

Fig. 2. The typical elevation sequence up a ridge top between two torrents as it would appear in plan view, highest elevation at tope of figure

of soils which, while they fall wide apart in a natural system of classification on account of fundamental genetic and morphological differences, are yet linked in their occurrence by conditions of topography and are repeated in the same relationship to each other wherever the same conditions are met with." He advocated the use of such a composite mapping unit in preference to the suppression of all but one of the associated soil types on grounds of their being of arrested development, or intrazonal. He was referring particularly to a basin-ridge system of small dimensions in which the soil vegetation landscape continuously repeated itself. The problem had arisen, in using such units as great world soil groups, as to how to define the zonal soil for the area. He wished to base his classification upon the soils that were present in terms of the evidence that they themselves yield, not by a supposition of what would happen in the production of a zonal soil. He elaborated that in certain forms of the catena, only portions of the relationship with the topography may be present. The broader aspects of the catena would include also climatic zones that were related to topography, and thus catenas on a grand scale would be found on such mountains as Mt. Kenya in the tropics, or on Mt. Kilimanjaro, where conditions range from alpine to semi-arid within a score of miles. Thus the catena can express relationships of soils in directions that run across the main phyla of genetic and morphological classification.

The conclusion of the present study is that the various sequences of soil property change that have been described are in effect descriptions of typical mediterranean catenas (see Figs. 1 and 2), and that there are two main ones. The most prevalent is based on a red soil at the lower elevations; the other, less widely distributed (except for the lowest member), is based on a black soil at the lower elevations. It is apparent that in Italy and Greece the elevation limits for the various types of soil in a catena are lowered with increasing north latitude or with north slopes, so that the red soil lower member is attenuated and the others are lowered in elevation. The same catenas are present in Italy, Greece, and California; and presumably the other countries bordering on the Mediterranean and Black Seas; and probably in other areas of the world with a dry summer climate and temperate moist winter, and having the requisite topographic change.

References

Baker, F. S.: Theory and practice of silviculture. New York: McGraw-Hill 1934.

Baroni, E.: Guida Botanica d'Italia. Bologna: Cappelli 1963.

California Department of Natural Resources: Soil-vegetation surveys in California. Sacramento: California Department of Natural Resources 1958.

Colwell, W. L., Gee, W., Meyer, J.: Soil-vegetation map SE 1/4 Blocksburg Quadrangle, Humboldt County, California. San Francisco: U.S. Forest Service 1955.

Comel, A.: Ricerche chimiche-pedologiche sulle "terra rossa" dei Monti di Medea Gorizia. Boll. Soc. Geol. Ital. **58**, 6–14 (1939).

Delapp, J., Skolmen, R.: Soil-vegetation map NW 1/4 Hoopa Quadrangle, Humboldt County, California. San Francisco: U.S. Forest Service 1961.

De Philippis, A.: Classificazione ed indici del clima in rapporto alla vegetazione forestale italiana. N. Giorn. Bot. Ital. **44**, (1937).

Gerhardt, P.: Handbuch des deutschen Dunenbaues. Berlin: Paul Parey 1900.

Giacomini, V.: La Flora. Serie Conosci l'Italia, **2**. Roma: Touring Club Italiano 1958.

Gindel, I.: Seasonal fluctuations in soil moisture under the canopy of xerophytes and in open areas. Commonwealth Forestry Review **43**, 219–234 (1964).

Jensen, H. A.: A system for classifying vegetation in California. Calif. Fish Game **33**, 199–266 (1947).
Jepson, W. L.: A manual of the flowering plants of California. Berkeley: University of California Assoc. Students Store 1923.
Kubiena, W. L.: The soils of Europe. London: George Allen and Unwin 1953.
Mancini, F.: Carta dei Suoli d'Italia. Roma: Istituto di Tecnica e Propaganda Agraria 1960.
Merriam, C. H.: Life zones and crop zones of the United States. Bull. U.S. Bur. biol. Surv. **10** (1898).
Milne, G.: Some suggested units of classification and mapping, particularly for East African soil. Soil Research **4**, 183–198 (1935).
Munz, P. A., Keck, D. D.: A California Flora. Berkeley: University of California Press 1959.
Principi, P.: I Terreni d'Italia. Genova: Dante Alighieri 1943.
Rode, A.: Soil Science. Trans. from Russian by Israel Program for Scientific Translations. Washington: Nat. Sci. Found. 1962.
Schimper, A. F. W.: Plant geography upon a physiological basis. Oxford: Clarendon Press 1903.
Shaw, C. F.: Profile development and the relationship of soils in California. Proc. I Int. Congr. Soil Sci. **4**, 291–397 (1928).
Storie, R. E., Weir, W. W.: Soil series of California. Berkeley: University of California Book Store 1953.
Wieslander, A. E.: A vegetation type map of California. Madroño **3**, 140–144 (1935).
Zinke, P. J.: Mediterranean analogs of California soil vegetation types. Berkeley: University of California Wildland Research Center 1965.
Zinke, P. J., Colwell, W. L.: Some general relationships among California forest soils. 353. In: Forest Soil Relationships in North America. Proc. Second N. American Forest Soils Conference. Corvallis: Oregon State University Press 1965.

Section III: Vegetation in Mediterranean Climate Regions

The mediterranean climatic areas of the world offer an opportunity to test the hypothesis that similarity of environment will produce similarities in structure and function of vegetation. Physiognomic comparisons of these areas show a predominance of evergreen sclerophyllous shrubs and trees in these regions, indicating the validity of this hypothesis. However, few studies have documented the pervasiveness of convergence in terms of vegetation structural detail or functional attributes of the plant components.

There exists an unusually complete account of the floristics, biogeography, and function and patterning of the vegetation of the Mediterranean Basin (see, for example, BRAUN BLANQUET, 1936; RIKILI, 1948; ZOHARY, 1962; ECKARDT, 1967). In this volume, LOSSAINT describes the long-term studies on the mineral and water cycles of the French *maquis* and *garrigue*, and ZINKE indicates the relationships of regional soil types to vegetation patterning in Italy and Greece, as well as California.

In spite of this detailed documentation of the specifics of taxonomy, structure, and function of these systems, there is still controversy over the exact degree of modification by man of the communities which are seen today in the Mediterranean Basin. Elsewhere in this volume, NAVEH and DAN discuss the long-term and overwhelming impact of man on the landscapes of Israel. They conclude that all of today's vegetation types in the Mediterranean Basin are variants of differing degrees of degradation or regeneration from those which evolved in equilibrium with the mediterranean-climatic type.

Thus, the utility of this detailed background information to answer questions of how biological systems have evolved to respond to the limitations of mediterranean climates is limited, at least at the present state of our knowledge. The impact of man in modifying these systems probably overwhelms the effects of the physical environment and past evolutionary history on molding their structural and functional attributes. Instead, it may be more productive initially, at least, to focus on areas with a shorter history of man's influence, such as California, Chile, and Australia, even though the overall data base is far more scanty.

According to information presented by MOONEY and PARSONS and SPECHT, in this section, these areas share vegetations which have a similar life form composition. Further, they have vegetations which are comparable in their adaptation to fire and in the low nutrient soils they occupy. Disturbance to these ecosystems has brought the very same weedy invaders. However, there appear to be important differences among these communities, particularly the nature of their herbaceous components and in the growth rhythm of the dominants.

More detailed studies are needed to assess the basis for these similarities and differences. Do these regions differ in some important environmental parameter or, as SPECHT suggests, has past evolutionary history played a large role in establishing these divergencies?

A preliminary chemical view of convergence by MABRY and DIFEO indicates that similar results to those discussed at the vegetation-structural level will be found. That is, there will be important similarities in the secondary chemistry of the plant components

---

The introduction to this section was prepared by HAROLD A. MOONEY of the disjunct areas – such as a rarity of alkaloids and the abundance of saponins. However, there will be a great number of differences which will be related to evolutionary history rather than current selective events.

Finally, KUMMEROW, in a discussion of the leaf anatomy of sclerophylls of mediterranean climatic regions, indicates that morphological plasticity is one of their most distinctive features.

In totality, it is rather remarkable that despite the great influence of past evolutionary history, the large potential for plastic responses to the environment, and the significant impact of man on these systems, there are still so many highly convergent features in the vegetations of mediterranean climatic regions of the world. It would thus appear that selective forces are similar and stringent in these regions, and they produce a limited number of evolutionary outcomes.

References

BRAUN-BLANQUET, J.: La Chênaie d' yeuse méditerranéenne *(Quercus ilicis)*. Mem. Soc. Nimes **5**, 1–150 (1936).

ECKARDT, F.: Mécanisme de la production primaire des écosystèmes terrestres sous climat Méditerranéen. Oecol. Plant. **2**, 367–393 (1967).

RIKILI, M.: Das Pflanzenkleid der Mittelmeerländer. Berne: Hans Huber 1948.

ZOHARY, M.: Plant Life of Palestine. New York: Ronald Press 1962.

1

Structure and Function of the California Chaparral – an Example from San Dimas

HAROLD A. MOONEY and DAVID J. PARSONS

Introduction

Within the mediterranean climatic region of California there are diverse types of woody plant communities, which include evergreen forests and woodlands, evergreen scrubs (chaparral) in addition to a drought deciduous semi-arid scrub (coastal sage). These types cover extensive areas of California (Fig. 1). Certain ecological aspects of these communities have been discussed recently and compared with homologous types in other parts of the world (NAVEH, 1967; MOONEY and DUNN, 1970; MOONEY *et al.*, 1970; SPECHT, 1969a, b). These studies give a general view of many aspects of the environmental relationships of these ecosystems throughout their extent.

Of all the mediterranean climatic ecosystems in California, perhaps the most is known about the ecology of a small area in southern California near Los Angeles. This area, the San Dimas Experimental Forest, has been a focal point of research by the U.S. Forest Service for almost 40 years. Since the objective of the present volume is to synthesize our current knowledge on the origin and structure of mediterranean climate ecosystems, it is appropriate to summarize certain of the San Dimas studies in an attempt to obtain a model of how a specific example of a California mediterranean climate ecosystem is structured. This framework can then be used for detailed comparisons between other mediterranean climate ecosystems in California as well as elsewhere in the world. If we are to understand the basis for the apparent convergence in the structure of the mediterranean climate ecosystems we must identify in full detail those aspects of these environments which are common and those which are unique to one or more of them.

San Dimas Experimental Forest – Location and Physical Characteristics

The San Dimas Experimental Forest is located in the San Gabriel Mountains of southern California approximately 45 km east of Los Angeles (Fig. 2). The Forest occupies 6,885 ha of land including elevations from 458 to 1,678 meters (HILL, 1963). The area is dissected into a number of drainages which vary in size from less than one-quarter of a square km to nearly 50 sq. km. The topography is generally quite steep with the average slope of the land about 68% and with the angle of some drainage channels over 30%. Only 7% of the slopes are less than 40%, whereas nearly half of all of the slopes of the Forest are extremely steep, having angles greater than 70% (BENTLEY, 1961) (Fig. 3).

The San Gabriel Mountains, which attain elevations up to 3,350 m, stand out in sharp relief from the coastal plain to the south and the Mojave Desert to the north. The bedrock of these mountains is predominantly granitic or pre-Cretaceous or pre-Cambrian crystalline metamorphics (STOREY, 1948). The San Gabriel Mountains are a fault block which

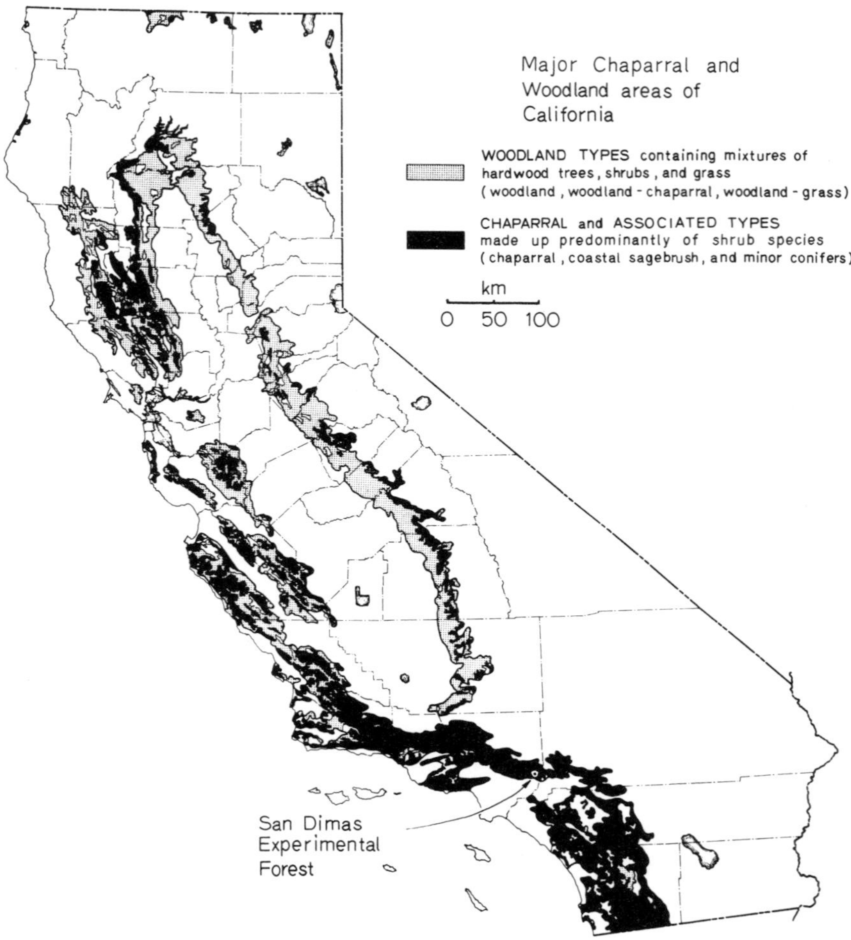

Fig. 1. Distribution of chaparral and woodland types in California. (From WIESLANDER and GLEASON, 1954)

has been uplifted during recurrent mountain building periods during the past 90 million years (STOREY, 1948). The mountains at present are deeply dissected with numerous V-shaped valleys and sharp ridges. The whole mountain mass has been intensively faulted and fractured. The faulting has determined in part the pattern of canyon formation. Importantly to the hydrology of the region, the extreme fracturing and subsequent deep weathering has resulted in possible water penetration to considerable depths into the bedrock.

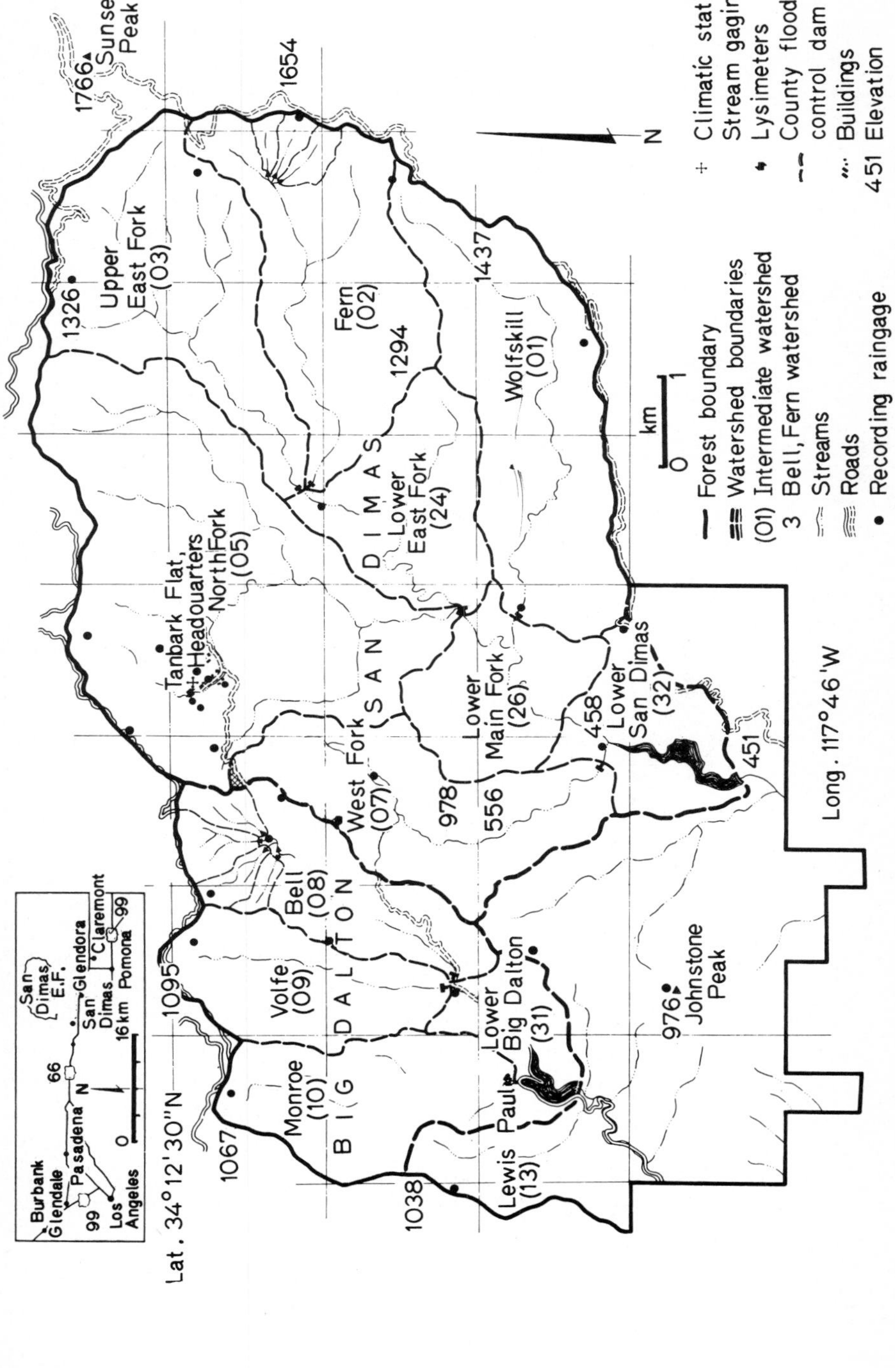

Fig. 2. Location and features of the San Dimas Experimental Forest. Elevations given are in meters. (From HILL, 1963)

Within the Experimental Forest most of the soils are derived from a coarse-grained igneous-metamorphic complex of rocks of which diorites, schists, gneisses, granites and granodiorites are the most common. Areas of andesite and basalt are also present (CRAWFORD, 1962). The soils of the Experimental Forest have been recently mapped. The soils are predominantly very shallow (BENTLEY, 1961). Seventy-two percent of the area is covered with soils less than 0.6 m deep. The most prevalent soil type of the San Dimas Experimental Forest (SDEF) has been designated series "A" by CRAWFORD (1962). This soil type

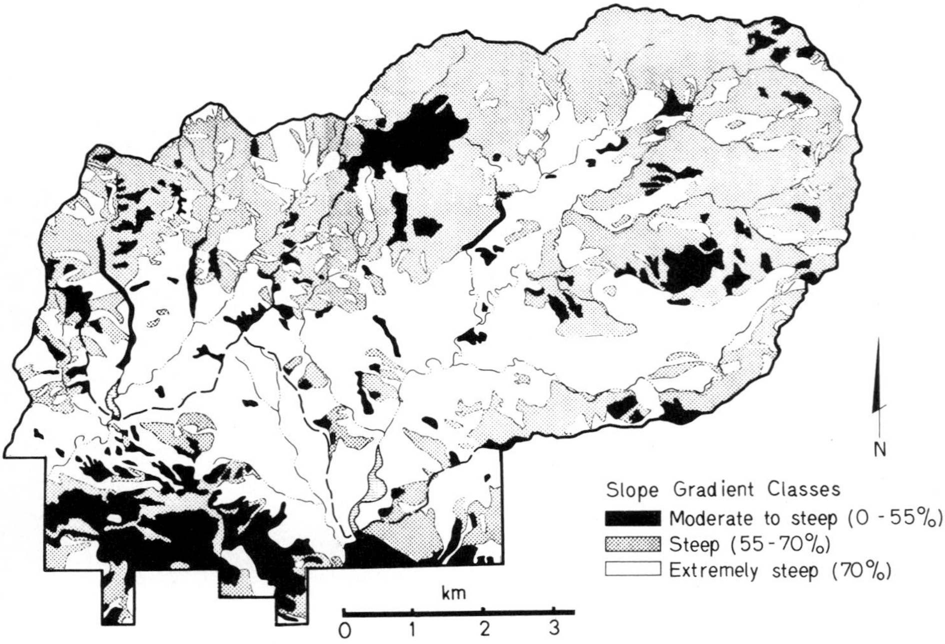

Fig. 3. The distribution of slope classes in the San Dimas Experimental Forest. (From BENTLEY, 1961)

occurred in 90% of 200 random sample points within the Forest studied by RICE, CORBETT and BAILEY (1969). According to CRAWFORD (1962) these soils are shallow, coarse textured, and well drained. They occur on steep slopes at elevations below 1500 m. Apparently this soil is common throughout the mountains of southern California. The characteristics of the type are given in Table 1. As noted by HORTON and KRAEBEL (1955) the high erosion rates on the steep slopes prevent the development of soil horizons.

Six other soil types occur within the SDEF but only to a limited extent. They include those derived from parent materials with high iron and quartz content; from calcareous tufa; from andesite. At elevations above 1500 m a soil type has developed which is slightly acid near the surface to strongly acid near the parent material. These soils, which are developed under relatively low temperatures and high precipitation attain depths of greater than a meter and a half on gentle slopes.

Table 1. Characteristics of the series "A" soils[a]

Excessively-drained, shallow, and coarse-textured, the "A" soils are derived from diorites, granodiorites, schists, and gneiss. These soils are typical for the entire San Dimas area up to about 1,375 m elevation. The "A" soils are common on steep slopes, especially those over 40%.

The "A" soils range from neutral at the surface to slightly acid at about 60 cm below. These soils are usually gravelly and stony throughout the profile. The boundary between the soil and the underlying decomposed parent material is very gradual; the soil is often indistinguishable from the parent material.

These soils have weak, angular blocky structure. Their consistency is loose when dry, and loose to very friable when moist. Depths are mostly less than 61 cm, measured vertically.

Soil profile.

1. 9–12 cm	Pale brown gravelly, sandy loam; weak to moderate, medium, angular, blocky structure; soft when dry to very friable when moist; very slightly acid to neutral; 2.5–15 cm thick; gradual, wavy lower boundary.
2.14–28 cm	Pale brown, gravelly, loamy sand; weak to moderate, fine, angular, blocky structure; loose when dry to very friable when moist; very slightly acid; 13–25 cm thick; gradual, irregular lower boundary.
3. 29–41 cm	Light yellowish brown sandy gravel; weak, fine, angular, blocky to single-grain structure; loose when dry to loose when moist; slightly acid; 13–25 cm thick; diffuse lower boundary.
4. 42 cm	Heavily weathered igneous and metamorphic rocks, mostly gneiss, schists, diorites, and granodiorites; variations of rock type occurring within short distances, sometimes within a meter.

Range in characteristics. – The surface color (dry) ranges from pale brown to brown. The subsoil color (dry) ranges from light yellowish brown to yellowish brown. Subsoil texture ranges from a gravelly, sandy loan to a gravelly sand. The soils are typically gravelly, and become increasingly rocky with depth. Typically more than 10% of the surface is covered by rocks over 8 cm in diameter. Soil depth is variable within short distances. Shallow and very shallow depth phases predominate. Distinct soil horizons may often be lacking.

Topography. – Slopes vary from steep to extremely steep. Very steep slopes are by far the most common. The "A" soils generally occur below 1,375 meters elevation.

Drainage and permeability. – The soils are excessively drained. The underlying decomposed rock can absorb large amounts of water.

Vegetation. – Chaparral. *Rhus ovata, Quercus dumosa, Quercus wislizenii, Adenostoma fasciculatum, Yucca whipplei, Eriogonum* sp., *Salvia mellifera, Eriodictyon trichocalyx.* and *Heteromeles arbutifolia* are the most common shrubs.

[a] Data and text from CRAWFORD (1962).

Climate

The climate of the SDEF has been studied in some detail (BURNS, 1952; HAMILTON, 1944, 1951, 1954; HAMILTON and REIMANN, 1958; QASHU and ZINKE, 1964; REIMANN, 1959 a, b; REIMANN and HAMILTON, 1959). An unusually comprehensive network of rain gauges was utilized in the Forest for a period to determine the influence of the mountainous terrain on the sampling of rainfall. This network included over 300 gauges distributed along contours of 300 m elevational intervals (HAMILTON, 1954).

In addition to the precipitation network, air temperature observation stations were maintained for varying periods at elevations of 457, 518, 853, 1,310, 1,554, and 1,585 m (REIMANN, 1959 a). Evaporation measurements were also made at a few of these elevations.

The most intensive and lengthy record has been made at the Tanbark Flat Station which is centered in the Forest at an elevation of 815 m (Fig. 2). The climate of this station is characteristic in its general features of the entire Forest and will be discussed first.

Table 2. Climatic characteristics of the Tanbark Flat Climatic Station, San Dimas Experimental Forest (elevation 850 m, latitude 34° 12′, longitude 117° 46′)

	Rainfall mm[a]	% of total	Evaporation mm[b]	% of total
Jan	132.1	19.7	58.2	3.6
Feb	139.7	20.8	61.0	3.8
Mar	104.1	15.5	86.4	5.3
April	63.5	9.5	104.1	6.4
May	15.2	2.3	142.2	8.7
June	2.5	0.4	177.8	10.9
July	T	0	246.4	15.2
Aug	2.5	0.4	243.8	15.0
Sept	7.6	1.1	208.3	12.8
Oct.	27.9	4.2	142.2	8.7
Nov	53.3	8.0	94.0	5.8
Dec	121.9	18.2	61.0	3.8
Annual Average	670.3		1625.4	

[a] 32-year average, data from HILL, 1963.
[b] 25-year average, data from HILL, 1963.

Air temperatures, °C[c]					Soil temperatures, °C[d] Mean	Maximum day length, hrs.
Absolute		Mean				
Max	Min	Max	Min	Mean		
28.9	−7.8	14.8	2.6	8.0	12.4	10.3
28.6	−5.6	15.1	2.8	8.4	10.9	11.2
28.9	−4.4	16.1	3.3	9.3	11.0	12.3
33.1	−3.3	18.8	5.3	11.8	12.4	13.3
37.8	−1.9	21.4	7.1	14.0	13.9	14.1
39.4	0.6	25.1	9.4	17.2	14.9	14.3
40.0	3.9	30.5	13.8	22.0	17.1	14.2
41.7	3.3	30.7	14.0	22.2	18.9	13.5
42.5	3.1	29.7	13.3	21.0	19.9	12.5
38.9	−3.6	23.7	9.1	15.8	19.9	11.5
31.9	−2.2	19.8	5.9	12.5	17.4	10.5
29.2	−5.8	16.6	4.0	9.9	14.6	10.0
42.5	−7.8	21.9	7.6	14.3	15.3	

[c] 25-year period, 1933–58, data from REIMANN, 1959 a.
[d] Soil temperature data, for a depth of 1.7 m under a vegetation cover of *Quercus dumosa*, from QASHU and ZINKE, 1964, for an eight-year period, from Jan. 1952 to Dec. 1959.

The climate of the SDEF is a mediterranean type with summer drought and a mild winter (Table 2). The bulk of the precipitation, which is generally entirely rain, falls in the four months from December through March, when the coolest temperatures of the year prevail. Most of this rain falls in less than 20 storms per year (HAMILTON, 1951). These storms may be intensive. In over a ten-year period it was found that 23% of the precipitation fell in 3% of the storms. Approximately half of the annual rain can be accounted for in less than 10% of the storms (Table 3).

The intensity of some of the winter storms is indicated by the fact that, during a 25-year period from 1933 to 1958 with a total of 460 storms (precipitation greater than 2.54 mm normally followed by 24 hours without rainfall), ten produced precipitation in excess of 127 mm in 24 hours, and two greater than 254 mm (REIMANN and HAMILTON, 1959). A concentration of most of the yearly rainfall in a relatively brief period, and often at a high intensity, is a feature of this climate which is of great importance in the operation of the southern California mediterranean climate ecosystem.

Table 3. Storm characteristics at the Tanbark Flat Station of the San Dimas Experimental Forest[a]

Storm size class, mm	Percentage of storms	Percentage of rainfall
0 – 6.4	30	2
6.5– 12.8	15	3
12.9– 25.5	20	10
25.6– 50.9	15	15
51.0– 76.3	7	12
76.4–101.7	3	7
101.8–127.1	3	9
127.2–152.5	2	10
152.6–177.8	2	9
over 177.8	3	23

[a] Data for the period 1933–1947. From HAMILTON, 1951.

During the summer months there may be occasional thunder showers which bring small amounts of precipitation. However, the summers are often completely dry as well as hot. Evaporation rates are high (Table 2).

In the winter, mean maxima temperatures average about 15° C. Mean minima in the winter are above freezing, although freezing days occur.

The variation of temperatures with elevation in the SDEF is indicated for stations from 450 to 1,550 m (Table 4). These elevations encompass the full altitudinal limits of evergreen scrub.

Mean temperatures generally decrease with an increase in elevation throughout most of the year. In the summer, however, penetration of maritime air to the lower slopes of the mountains evidently results in inversion conditions. During the months of June, July and August, mean temperatures of the 1,326 m station exceed those of the lower elevation stations. No doubt, this maritime influence is of great significance for organisms, since it occurs during the height of the drought. Inversion conditions, also bring high levels of "smog" to the area.

Average annual precipitation varies from 660 to over 965 mm within the SDEF (STOREY, 1948). The higher amounts of precipitation are found generally at the highest elevations. The Tanbark Flat area, however, receives less rain than what would be expected for the elevation because the air ascending up the major canyons bypasses this area somewhat.

Table 4. Mean monthly temperature (° C) and annual precipitation (mm) at four elevations within the San Dimas Experimental Forest[a]

	Stations			
	San Dimas Canyon	Tanbark Flat	San Gabriel Divide	Fern Canyon
Elevation, m	457	853	1326	1554
Month				
January	10.2	8.6	9.9	6.4
February	10.6	8.3	6.8	6.1
March	11.9	10.1	9.6	8.4
April	14.2	11.9	9.4	10.9
May	16.8	14.7	14.2	15.0
June	18.7	17.4	19.6	18.9
July	22.7	22.1	26.3	23.3
August	22.7	22.7	24.4	23.3
September	21.4	20.4	21.9	20.3
October	17.1	15.7	13.9	14.3
November	14.3	12.5	14.6	11.3
December	11.7	10.1	8.1	8.6
Annual	16.1	14.6	14.9	13.9
Mean Annual Precipitation	752.1	879.9	978.9	1052.6

[a] Temperature data for a 10-year period from 1933 to 1943, REIMANN, 1959 a.
Precipitation values for 1933 to 1944 from REIMANN and HAMILTON, 1959 and SDEF files.

Table 5. Fire History of the SDEF[a]

Year	Total hectares burned
1896	1635
1911	60
1919	4589
1932	66
1938	206
1947	52
1953	241
1960	6705
Total area of SDEF = 6885 ha.	

[a] Data from files of the SDEF.

Table 6. Summary of the San Dimas Experimental Forest flora[a]

Total number of families	83
Total number of genera	288
Total number of species	517
Number of introduced species	75

[a] These data are derived from a checklist prepared by J. S. HORTON, of the U.S. Forest Service.

Periodic violent winds (Santa Anas) sweep through the lowlands of southern California, mostly in fall and winter. These winds, which can reach gale intensities, bring very dry, and usually warm, air and clear conditions. These winds often are important in the explosive spread of chaparral fires.

Fire

In southern California, as in any mediterranean climatic area, fire is not only a natural component of the environment, but also perhaps one of the most important evolutionary forces. There have been detailed fire records for the SDEF since the late 1800's. During three-quarters of a century there have been eight fires in the area, two of which destroyed more than half the watershed (Table 5).

The most recent, and largest, conflagration was started by lightning on a hot summer day, July 20th, 1960. Within a day-and-a-half nearly five thousand ha had been consumed (HOPKINS, BENTLEY and RICE, 1961), and by the end of the week the vegetation on virtually the entire SDEF had been destroyed.

There is some question, which will be discussed later, whether fires in this area have changed in frequency, intensity, and timing due to the recent influence of man.

The Flora

There are over 500 species of plants in the SDEF belonging to 288 genera and 83 families (Table 6). Nearly 15% of this flora is introduced. The Compositae is the single most important family, with 50 representative genera. The Gramineae, Cruciferae, Scrophulariaceae, and Umbelliferae are all well represented, with 25, 13, 11, and 11 genera respectively. The genera which include 6 or more species are all herbaceous (*Lotus*, 11; *Phacelia*, 11; *Bromus*, 9; *Mimulus*, 9; *Gnaphalium*, 7; *Lupinus*, 7; *Eriogonum*, 6; *Juncus*, 6; *Trifolium*, 6) with exception of the arboreal genus *Quercus*, which has 6 representatives.

The flora consists predominantly of therophytes and hemicryptophytes (Table 7). The herbs, which constitute the great bulk of the flora, are mostly forbs. The shrubs are mostly evergreen, as are approximately half of the trees. The deciduous trees are mainly found either in riparian or high elevation habitats.

Periodicity. The greatest bulk of the flora is in bloom during the spring months of April, May and June (Table 8). The trees flower in abundance earliest and for the shortest period of time. The shrubs and herbs, although flowering most abundantly in spring and early summer, have representatives blooming throughout the entire year.

Flowering activity is, of course, not entirely an index of growth activity. Some species flower prior to their growth cycle, whereas others flower either during or subsequent to their growth phase.

Table 7. Growth and life forms of the flora of the San Dimas Experimental Forest[a]

Growth Forms

	Form		*Number*	*Percent*
Trees	Broadleaf deciduous		12	
	broadleaf evergreen		6	
	needleleaf evergreen		5	
		Total	23	= 4% flora
Shrubs	broadleaf deciduous		14	
	broadleaf evergreen		21	
	narrowleaf dicot		12	
	suffrutescent		20	
	pinnate leguminous		1	
	woody vine or climbing semi-shrub		6	
	stem-succulent		1	
	.parasitic		1	
	spinose		1	
	monocot rosette-shrub		1	
		Total	78	= 15% flora
Herbs	herbaceous perennial forb		154	
	herbaceous annual forb		186	
	perennial grass		37	
	annual grass		26	
	fern		13	
		Total	416	= 81% flora

Life Forms

	Number of species	*Percent of flora*
Phanerophytes	80	15
Chamaephytes	24	5
Hemicryptophytes	172	33
Geophytes	34	7
Therophytes	207	40
Hydrophytes	2	

[a] These data are derived from a checklist prepared by J. S. HORTON of the U.S. Forest Service.

Table 8. Percentage of given growth forms flowering by month[a]

Growth Form	Jan	Feb	Mar	Apr	May	June	July	Aug	Sept	Oct	Nov	Dec
Trees	9	18	45	77	64	23	4	4	4	0	0	4
Shrubs	8	12	35	52	60	53	36	27	19	18	8	5
Annual Herbs	4	11	41	72	89	76	46	30	22	13	5	2
Perennial Herbs	5	11	25	43	67	71	62	45	31	14	4	3
Geophytes	3	9	26	44	62	59	38	21	6	3	0	0
Percentage of Total Flora	5.0	11.2	33.9	57.6	74.0	67.1	47.3	32.6	22.3	12.6	4.5	2.7

[a] These data are derived from the flowering times of the species as indicated in MUNZ and KECK, 1959. The flowering data are for California in general and not specifically for San Dimas.

The Fauna

The vertebrates found within the confines of the SDEF have been enumerated by WRIGHT and HORTON (1951, 1953). Of a total of 201 species of vertebrates there are 35 kinds of mammals, 143 types of birds, and 23 species of amphibians and reptiles. Of these, there are 7 species of large carnivorous mammals, including wildcats, foxes, coyotes, lions, skunks, and racoons. Particularly characteristic and important mammals include the California mule deer *(Odocoileus hemionus)* which is common throughout the Forest, and the wood rat *(Neotoma fuscipes)*. The ecology of the wood rat on the SDEF has been studied in some detail by HORTON and WRIGHT (1944). These abundant and large rodents build big, above-ground nests of branches and twigs which contain living as well as food-storage areas. At the high elevations especially, they store large amounts of food material, including oak acorns. They have been noted to harvest the entire acorn crop of one species of oak during one year of observation.

There is no statistical information available on the arthropods of the SDEF.

Vegetation

The vegetation of the SDEF is dominated by an evergreen scrub (chaparral) (Fig. 4). Numerous chaparral types are often distinguished on the basis of the principal dominants. The two most commonly recognized types are chamise chaparral which is dominated by *Adenostoma fasciculatum*, and which usually occurs on the driest sites, and mixed chaparral which is composed of broad-leaved evergreen scrubs such as *Ceanothus* spp., *Arctostaphylos* spp., *Quercus dumosa*, and *Cercocarpus betuloides*.

The driest sites on the SDEF, and those which have unstable soil, are dominated by a second major vegetation type, an open sub-shrub type called the coastal sage or soft chaparral. The most important components of this type in the SDEF are *Salvia apiana*, and *Eriogonum fasciculatum*.

There are also a number of woodland types present at the SDEF. A winter deciduous riparian woodland composed of *Alnus rhombifolia, Acer macrophyllum*, and *Platanus racemosa* (often in combination with the evergreen trees, *Quercus agrifolia* and *Umbellularia californica)* is found along the major water courses. These habitats, in addition to being moist, are probably cold, due to nightly cold air accumulation.

On somewhat drier habitats than riparian, there is a woodland of evergreen oaks, including *Quercus agrifolia, Q. chrysolepis*, and *Q. wislizenii*. In certain limited areas within this woodland there may be scattered small stands of big cone spruce *(Pseudotsuga macrocarpa)*.

One could align the above communities along an aridity gradient from dry to wet in the following manner: coastal sage, chamise chaparral, mixed chaparral, oak woodland, and finally riparian woodland. In terms of principal life forms of the dominants, this series extends from drought-deciduous shrubs, evergreen needle-leaf shrubs, broad-leaf evergreen shrubs, broad-leaf evergreen trees, to winter deciduous trees (Figs. 5 and 6).

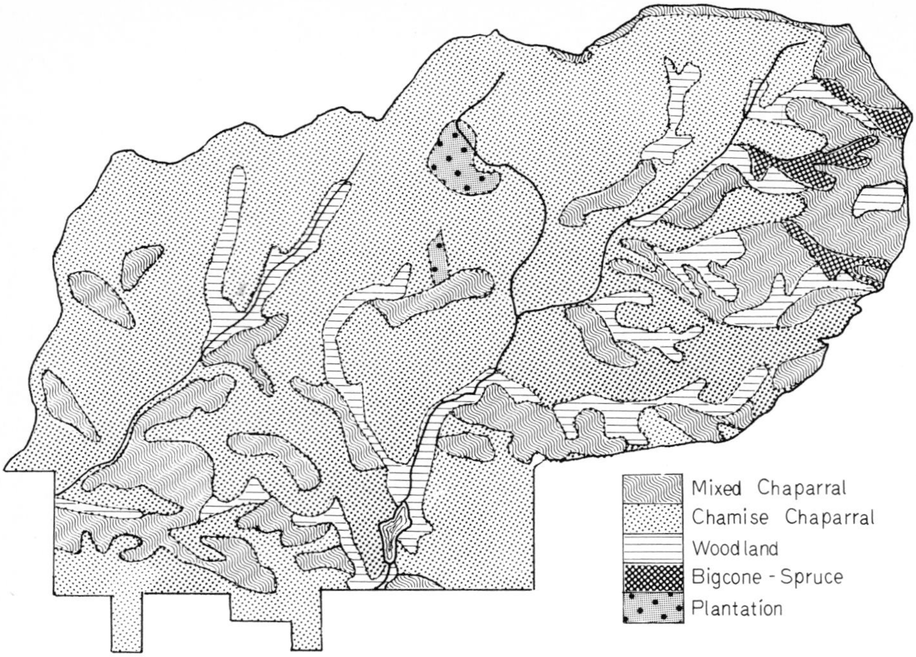

Fig. 4. Vegetation map of the San Dimas Experimental Forest. (From USDA, 1942)

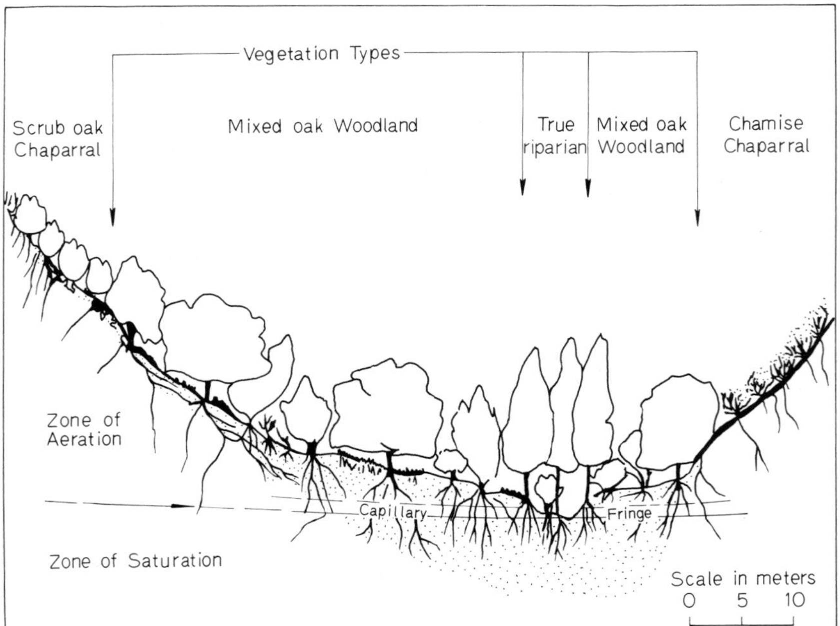

Fig. 5. Vegetation profile through Monroe Canyon indicating ground water. The chamise chaparral faces south and the scrub oak, north. (From ROWE, 1963)

Fire Succession

Vegetation Succession. There have been a number of studies on the changes in vegetation structure of the chaparral subsequent to fire in southern California. HANES (1971) has recently given a comprehensive account of chaparral succession throughout the coastal and interior regions of the San Gabriel and San Bernardino Mountains. Studies of southern California chaparral fire succession including areas of the SDEF have been made by HANES and JONES (1967), HORTON and KRAEBEL (1955), PATRIC and HANES (1964), PLUMB (1961, 1963), and SPECHT (1969 b).

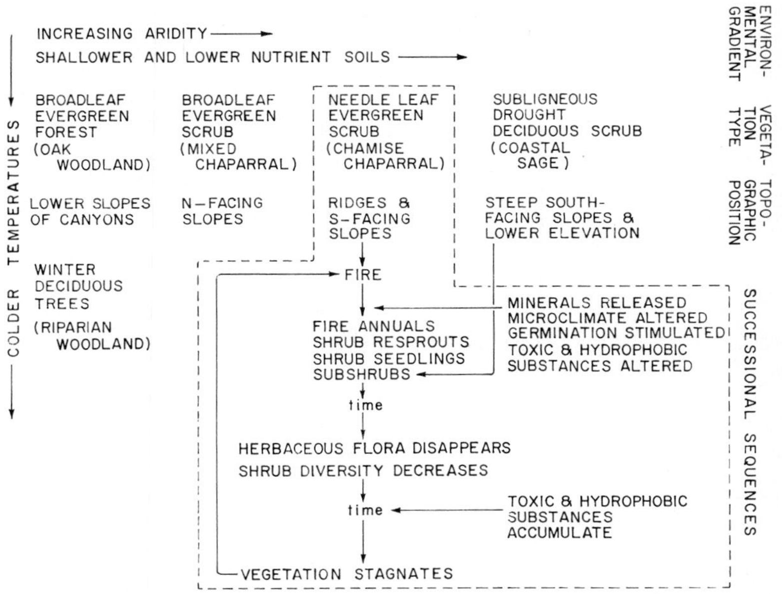

Fig. 6. The vegetation of Monroe Canyon as related to environmental gradients. The fire successional cycle is given for the chamise chaparral only

A general sequence of events can be outlined as follows: the fire is most likely to occur during the drought period in late summer or fall when the vegetation is very dry. If the fire is associated with a Santa Ana wind, as it often is, it can rapidly become a holocaust consuming vast areas in a short period.

In addition to ashing most of the vegetation and litter, thus releasing tied-up minerals, the fire has other profound effects on the soil characteristics. There is normally a distinctive "non-wettable" layer in the soils supporting chaparral. This layer, composed of soil particles coated by hydrophobic substances leached from the shrubs or their litter, is usually associated with the upper part of the soil profile just below the litter. These substances build up through time in the unburned chaparral, particularly under chamise (DE BANO, 1969, DE BANO *et al.*, 1967).

The high temperatures which normally accompany chaparral fires (1100°C above the soil surface, 650° C at the surface, 180–290° C five cm below the surface) cause distillation of these hydrophobic substances which then condense on lower soil layers. This process results in a shallow layer of wettable soil overlying a non-wettable layer. Such a condition, particularly on steep slopes, can result in severe surface erosion (DE BANO, 1969).

There are also numerous phytotoxic compounds which accumulate in the soil of mature chaparral. In addition to preventing germination (MCPHERSON and MULLER, 1969), these substances could interfere with litter decomposition as well as soil nitrification (HANES, 1971). They are evidently destroyed by fire. Thus fire considerably alters the chemical nature of the upper layers of the soil as well as destroying the organisms above the soil and probably those in the upper few centimeters.

After the fire, but before the first winter rains, many shrub species may begin sprouting from lignotubers, and trees from adventitious stem buds. Riparian species may sprout almost immediately. Shrubs of the slopes in the SDEF have been noted to sprout within ten days of the fire and to grow as much as 25 cm within 30 days (PLUMB, 1961). Since this occurs during the drought period, these plants are presumably utilizing water reserves from the soil depths or from deep rock crevices. In dry sites, shrubs may delay resprouting, and during particularly dry periods, sprouts may be delayed as long as two years (PLUMB, 1963).

Not all phanerophytes have the capacity to resprout after fire. HANES (1971) noted that, of 59 dominants characteristic of the coastal regions, approximately 50% are fire-sprouters. Of those species which do fire-sprout, there is a differential capacity among individuals. For example, PLUMB (1961) surveyed the sprouting behavior of 20 dominant species in the SDEF after the 1960 fire. All the observed individuals of *Quercus agrifolia* and *Rhus laurina* resprouted, compared to only slightly over 50% of *Adenostoma fasciculatum*. Of those sprouting species greater than 50% of the individuals resprouted.

The sequence of events following the first rains after the fire depends greatly on the intensity and duration of the storms. COLMAN (1953) has described chaparral burns which were followed by little erosion or flooding because of less than normal precipitation. In sharp contrast, a November fire in the San Gabriel Mountains covering an area of 18 ha was followed by a severe storm in January, during which 456,000 cu m of soil and debris were washed into the lower valley. HORTON and KRAEBEL (1955) measured soil loss from their study plots after fire to depths as great as 7.5 cm during a five-year period. High-intensity rains subsequent to the fire may account for a great deal of this erosion, which also moves buried seeds downslope with the soil. In general, erosion rates exceed normal for at least 8 years after a fire, and stream flow rates are higher than average for at least 20 years. Winter rains also stimulate germination, and both sprouting and non-sprouting shrub species may produce abundant seedlings.

One of the distinctive features of chaparral succession is the great abundance of annual plants in the first years following fires and their virtual absence from older stands. Some species, termed "fire-type" annuals, can be extremely rare in the vegetation for periods longer than 50 years. After a fire, they may form masses of cover, only to disappear again after a few years (HORTON and KRAEBEL, 1955) (Fig. 7). Some fire-type species also occur in cleared areas. In addition to the fire-types, there are annual herbs occurring naturally on rock outcrops and openings within the chaparral which increase their numbers for a period after a fire, though more slowly than the fire-type species. Finally there are weedy

Fig. 7. Fire succession in Fern Canyon of the San Dimas Experimental Forest. The top picture was taken after a fall fire. The middle picture shows the same site in the following spring, and the bottom picture, four years later. (From HORTON and KRAEBEL, 1955)

annuals generally restricted to roadside and trails which may invade recently burned areas. All these annuals, regardless of origin, are virtually eliminated from mature chaparral.

This annual flora associated with burned areas is of further interest because it may be peculiar to California chaparral; SWEENEY has studied the fire-type annual in detail in California (1956, 1967) and is of the opinion that this distinctive flora lacks homologous types in the Mediterranean region and Australia (personal communication).

Although the vegetation may appear very different the first couple of years after a fire than it did prior to it, it soon regains its former physiognomy and many of the same dominant plants. Many of the shrubs which will form the canopy of the mature community are present the very first year as resprouts or seedlings.

Table 9. Percent cover and plant heights in typical stands of southern California coastal chaparral of different ages[a]

Species	Age class, years								Fire sprouts
	2–8		9–21		22–40		41–96		
	% cover	shrub ht, m	% cover	shrub ht, m	% cover	shrub ht, m	% cover	shrub ht, m	
Adenostoma fasciculatum	25.1	1.0	28.3	1.2	49.4	1.5	41.6	1.6	Yes
Lotus scoparius	13.1	0.7	0.2	0.6	0.4	0.5	0.0	0.7	No
Quercus dumosa	10.5	1.6	18.6	1.7	12.5	2.4	12.5	2.9	Yes
Ceanothus crassifolius	10.4	1.1	4.7	1.2	7.8	2.0	15.1	2.4	No
Salvia mellifera	9.9	1.0	6.1	1.0	4.2	1.1	8.7	1.1	No
Eriogonum fasciculatum	5.7	0.8	2.9	0.7	2.2	0.6	2.2	0.8	No
Arctostaphylos glandulosa	2.4	0.9	6.5	0.9	2.8	2.2	3.3	1.5	Yes
Cercocarpus betuloides	1.9	1.4	1.4	2.4	0.6	2.3	1.2	2.8	Yes
Ceanothus leucodermis	1.8	1.3	7.1	1.4	4.2	2.1	0.4	2.2	Yes
Yucca whipplei	0.4	0.8	0.0	0.8	0.6	0.8	0.6	0.9	Yes
Ceanothus greggii	0.2	0.7	2.7	1.2	1.6	1.5	0.7	1.2	No
Arctostaphylos glauca	0.2	1.1	0.1	1.2	3.0	1.5	2.6	2.0	No
Mean for all shrubs	79		97		107		126		

[a] Data from HANES, 1971

HANES (1971) has described the changes in plant cover and stature in stands of coastal chaparral of various ages (Table 9). *Adenostoma fasciculatum*, the principle dominant, attains its greatest cover between 22 and 40 years after fire. Height increase continues, at a reduced rate, as the stand matures further. *Lotus scoparius*, a suffrutescent plant, is important only in the early successional stages. *Salvia mellifera*, another subligneous shrub is also an early successional species, but often becomes important in later stages as a successional replacement for woody shrubs which die after 50 years or so, such as certain of the *Ceanothus* species. The canopy generally becomes closed after about 20 years, and, as shrubs overlap, cover greatly exceeds 100% within 50 to 100 years. Stands of such extreme age are rare however since fire frequency is generally of much shorter time intervals.

HANES (1971) has stated that the successional sequences are quite different on north- and south-facing slopes. On south-facing slopes at elevations below 1,000 m in the coastal regions a vegetation develops which is dominated by *Adenostoma fasciculatum*, and in which coastal sage subligneous scrub elements play an important early successional role. Most of the coastal sage species arise from seedlings, as does the chaparral shrub, *Ceano-*

thus crassifolius. After 40 years most of the subshrubs have been eliminated, and the stand is dominated by tall *Adenostoma fasciculatum* with no understory. After 60 years the vegetation has stagnated, and there is little annual growth. Diversity decreases, and only a few dominant species remain. Herbs are absent. Stands of this age are very rare, the

Table 10. Major vegetation types of Monroe Canyon

Type[a]	Percentage of watershed[a]	Principal dominants	Growth form[b]
1. Sage and barren areas	11.9	*Salvia apiana*	Ssf
		Eriogonum fasciculatum	Ssf
		Selaginella bigelovii	Hp
2. Chamise chaparral	47.3	*Adenostoma fasciculatum*	Sn
		Ceanothus crassifolius	Sbe
		Arctostaphylos glauca	Sbe
3. Mixed chaparral	3.0	(In order of importance)	
		Ceanothus crassifolius	Sbe
		Heteromeles arbutifolia	Sbe
		Arctostaphylos glandulosa	Sbe
		Garrya veatchii	Sbe
		Rhus ovata	Sbe
		Prunus ilicifolia	Sbe
		Rhamnus crocea var. *ilicifolia*	Sbe
		Rhus laurina	Sbe
		Arctostaphylos glauca	Sbe
4. Scrub oak chaparral	24.3	*Quercus dumosa*	Sbe
		Q. wislizenii var. *frutescens*	Sbe
		Ceanothus oliganthus	Sbe
		Cercocarpus betuloides	Sbe
5. Woodland sage	5.2	*Quercus agrifolia*	Tbe
		Q. chrysolepsis	Tbe
		Salvia apiana	Ssf
6. Big cone spruce	0.6	*Pseudotsuga macrocarpa*	Tn
7. Riparian woodland	7.7	*Alnus rhombifolia*	Tbd
		Acer macrophyllum	Tbd
		Umbellularia californica	Tbe
		Platanus racemosa	Tbd
		Baccharis viminea	Sbe

[a] Unpublished data from U.S. Forest Service, J. S. HORTON.

[b] Ssf – suffrutescent shrub
Hp – perennial herb
Sn – narrowleaf dicot shrub
Sbe – broadleaf evergreen shrub
Tbe – broadleaf evergreen tree
Tr – needleleaf tree
Tbd – broadleaf deciduous tree

usual vegetation noted in southern California being at an earlier successional stage. Apparently, fire is necessary to rejuvenate these old stands (the successional sequence of this type is illustrated in Fig. 6).

HANES (1971) describes succession on the north-facing slopes as faster than that on the south-facing slopes. Coastal sage subshrubs do not play an important role in early succession as they do on the south slopes. Resprouts, particularly of *Quercus dumosa,* and seedlings of various shrub species form a cover which after 30 years no longer changes substantially with time, although short-lived species such as *Ceanothus oliganthus* may die out in 20 to 30 years. The canopy of this vegetation becomes closed with age, but, because of shade pruning, the understory becomes quite open.

Monroe Canyon – General Features

Because of the wealth of information for one small drainage within the SDEF, Monroe Canyon, it will be used as a basis for a description of certain functioning aspects of the California mediterranean climate ecosystem. Data from other studies within the SDEF will be used when appropriate.

Monroe Canyon lies at the western edge of the SDEF (Fig. 2). It consists of approximately 355 ha of surface area and includes altitudes of approximately 525 m to about 1,085 m. Most of the canyon was burned in 1896 and again in 1919. There were no further fires in the canyon until the fire of 1960 which destroyed the vegetation of the entire area.

Monroe Canyon has been intensively studied for a variety of purposes over the years. ROWE and COLMAN (1951) made a detailed study of the disposition of rainfall in the canyon. PATRIC and HANES (1964) and HANES and JONES (1967) investigated post-fire chaparral succession and HORTON (unpublished data) has compiled a vegetation map of the area and CRAWFORD (1962) a soils map.

Monroe Canyon has also served as a study site for vegetation management. Just prior to the 1960 fire, 15 ha of riparian woodland were removed in an attempt to increase stream flow (SINCLAIR, 1960). Post-clearing and post-fire stream flow were reported by CROUSE (1961) and ROWE (1963). Further vegetation manipulations were made after the 1960 fire. The area was seeded with grasses and certain portions were sprayed with herbicides to prevent shrub regrowth (HILL, 1963).

The pre-1960 burn vegetation of Monroe Canyon was classified and mapped (HORTON, unpublished data). Almost one-half of the land surface was covered with chamise chaparral (Table 10). The more diverse mixed chaparral occupied only small areas at the head of the watershed. The fairly abundant shrub oak chaparral was found generally on north-facing slopes and sage on the steepest south-facing slopes. The woodland types which occupied only a small part of a total land area were generally restricted to the valley bottoms.

Water Balance

ROWE and COLMAN (1951), in a classic study of the disposition of rainfall on a watershed, installed a series of rain gauges and soil moisture monitoring stations throughout Monroe Canyon. From these installations they determined the input, as well as the storage and use of precipitation within the canyon. Nearby studies in the chaparral gave estimates of the interception of rain by the vegetation, as well as surface runoff. They concluded the latter was negligible in Monroe. A gauging station at the mouth of the canyon yielded an index of loss of water from the system.

The soils of the moisture plots were characterized and, on the basis of a large survey, were found to be typical of soils of the entire watershed (Table 11). The amount of water held in the soil mantle at the various plots under the given tensions was calculated from the depth of the profile, the density of the soil, and the percentage of rock content.

Table 11. Mean characteristics of moisture sampling plots in Monroe Canyon[a]

Soil depth m	Slope %	Rock content %	Mechanical analysis Sand %	Silt %	Clay %	Field capacity mm	Wilting point mm	Available water mm
Northerly Hillsides								
1.15	83.3	10.0	79.8	15.3	4.8	230.1	80.5	149.3
Southerly Hillsides								
1.20	73.1	12.5	81.1	14.8	4.1	190.5	88.9	109.1

[a] Data from ROWE and COLMAN, 1951.

The soils of the slopes of Monroe Canyon are all of the "A" series (CRAWFORD, 1962). They are residual and fairly homogenous in their physical characteristics. The steep slopes have generally shallow soils which show active dry creep (ROWE and COLMAN, 1951).

ROWE and COLMAN determined that the debris-filled fractured rock under the soil mantle could hold no more than 4 mm of available water per meter of root depth. According to their findings, a chaparral shrub with a 6-meter-deep root system would have only about 25 mm of water available to it from the rock mantle. More recent studies however have shown that the granitic parent material itself can hold significant amounts of moisture —amounting to 95 mm per meter depth in contrast to the 4 mm value given by ROWE and COLMAN (RICE, personal communication).

Detailed studies were carried out during the two-year periods of 1943–44 and 1944–45. Both of these years had virtually identical total amounts of rainfall, although during the second year the rainfall period commenced a month later.

In 1943–44 the total soil profiles reached field capacity in mid-December after an average total of 178 mm of rain had been received. Percolation of water occurred through the middle of March, at which time evapo-transpiration losses started drawing on stored soil moisture. The dates on which the "wilting point" was reached varied considerably between sites, depending on the microsite characteristics. These dates extended between June 20th, at the earliest, until October 18th, at the latest. Such vast differences are no doubt of importance in the patterning of the vegetation.

On the basis of additional calculations, which included a consideration of dry season evapo-transpiration from the riparian zone, ROWE and HAMILTON were able to calculate a water budget for the total watershed (Table 12).

Since they could not account for all of the losses of water by interception, evapo-transpiration, and stream flow, they concluded that considerable amounts of water were being lost from the watershed to underground flow. The extensively faulted nature of the bedrock which leads to these losses makes these watersheds particularly difficult to study for total accounting of mineral balances.

Studies by ROWE and COLMAN (1951), on plots of chaparral elsewhere in the SDEF during wet and dry years showed how different the lengths of time are at which the entire

Table 12. Disposition of annual rainfall in Monroe Canyon

	mm		
	1940–41[a]	1943–44[b]	1950–51[a]
Rainfall received	1321	798	305
Interception loss	127	69	51
Evapo-transpiration loss	356	282	254
Streamflow yield	279	117	trace
Ground water yield	559	330	trace

[a] Data from HILL, 1963, for the wettest and driest years observed.
[b] Data from ROWE and COLMAN, 1951.

profile can be below the wilting point. In a relatively dry year of 427 mm of precipitation, sites with a 1 and 1.2 meter-deep soil were below the wilting point for as long as 4 months, whereas, in a wet year (1214 mm), in one site, for as little as 10 days.

Carbon Balance

SPECHT (1969b) has recently provided data on biomass accumulation in chamise chaparral in the SDEF. He compared biomass, plant density, caloric, mineral, and water content of stands of various ages. Data for one of his stands are given in Table 13. This stand occurred at 830 meters elevation on a south-facing slope. The area had not burned for 37 years. This stand was typical of much of the vegetation on Monroe Canyon prior to the 1960 burn.

The vegetation had approximately equal numbers of *Adenostoma* and *Ceanothus* per hectare; however, there was a greater amount of living biomass of *Adenostoma*. As is typical of older stands of chamise chaparral, there were many shrubs of *Ceanothus* which had died. The suffrutescent shrub which is also present, *Salvia mellifera,* could in part represent temporal successional replacements to the *Ceanothus*. The total above-ground biomass of the stand was nearly 50,000 kg per ha. No estimates are available for below-ground biomass, but it could be as much as twice this amount (HELLMERS *et al.*, 1955). Mean annual above-ground production is estimated at 1000 kg per ha. This estimate is based on the weight differences between stands of known ages. SPECHT calculates this vegetation fixes in its above-ground parts only .04% of the total incoming radiation. When calculated on the percent of incoming radiation fixed during the growing season only, this amounts to one-tenth of one percent. As will be shown, however, these plants are fixing energy during the entire year rather than just during the spring growing season, so that the efficiency would be nearer the lower value.

DUNN (1970) studied the seasonal patterns of carbon fixation by a number of chaparral species at SDEF during 1966 and 1967 emcompassing two summer drought periods. He found that these plants photosynthesized throughout the entire year (Fig. 8), although at reduced amounts during the drought period. He calculated that, during six months of winter and spring, these plants fixed about 6.4 g of carbon dioxide per gram dry weight of leaves. During a dry six months of summer and fall, they fixed 5.3 g, and during a wet summer and fall, 7.8 g of carbon dioxide per gram dry weight. These values are net gains, since they account for respiration losses.

Table 13. Characteristics of a 37-year-old stand of chamise chaparral[a]

Species	Biomass dry wt kg/ha	Density per/ha	Mean dry wt per plant, kg
Adenostoma fasciculatum	12,084	2,941	4.1
Ceanothus crassifolius	9,241	2,644	3.5
Salvia mellifera	5,931	2,397	2.5
Total living	27,256	7,982	
Standing dead	21,836		
Total living plus dead	49,092		

For the Stand	
a) Water content in standing live and dead, kg/ha	15,491
b) Percentage of growing season radiation fixed in tops	0.10
b) Caloric content, kcal/g dry wt	4,742
d) Total caloric content of tops cal/ha	129×10^9
e) Annual increment of tops kg/ha	1,000
f) Annual increment of energy of tops, cal/ha	50×10^8
g) Total solar radiation received per year cal/ha	153×10^{11}
h) Percentage of total radiation fixed in tops	.036
i) Total solar radiation received during growing season (March-June) cal/ha	526×10^{10}
j) Mineral content of tops[b], kg ha	
N	145
K	105
Ca	85
Na	45
Mg	35
P	10
Total	425

[a] Data from SPECHT, 1969 b

[b] Estimated to nearest 5 kg from curves provided by SPECHT, 1969 b.

One of the important points is that, although growth is periodic and mainly during the most favorable period of the year, carbon fixation is continuous.

On the basis of DUNN's measurements, and published values of shrub dimensions, annual production can be calculated. The dry weight of living shrubs on the hectare of land at San Dimas which had not burned for 37 years was 27,255 kg, most of which was chamise (SPECHT, 1969b). The average percentage of leaves to stem material for chamise is 13.2% (the Leaf Area Index averages 4.1) (COUNTRYMAN and PHILPOT, 1971). Thus, assuming the whole stand was chamise, the amount of dry leaf material would be 3,597 kg. During the wet year each gram of leaf material would fix 14 g of carbon dioxide (DUNN, 1970). Thus, a stand would fix approximately 50,000 kg of CO_2. Assuming that one gram of CO_2 equals .614 g of dry matter, total yearly production of dry matter would be approximately 30,000 kg per year. This would be net production without consideration of respiration losses of roots and non-leafy stems, root production (which could conceivably be as much as two-thirds of that amount), leaf and reproductive structure losses (KITTREDGE, 1955, has measured annual litter fall in this vegetation type at San Dimas as 2,800 kg per ha), nor losses of any stem and leaf leachates or volatiles. Obviously, more

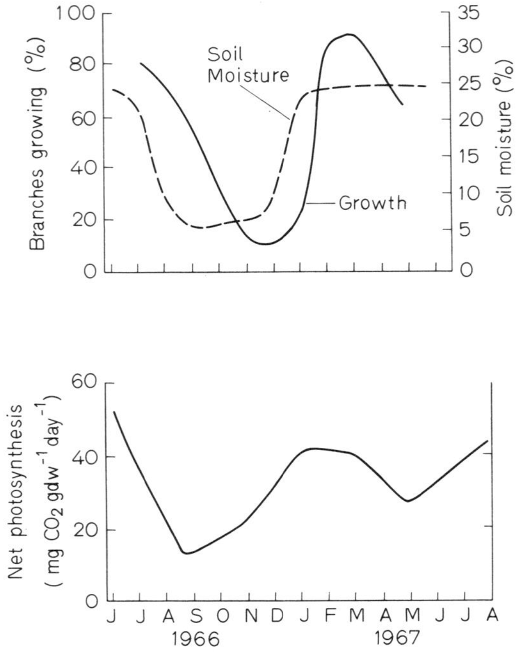

Fig. 8. Seasonal course of soil moisture and the growth and net photosynthesis of *Adenostoma fasciculatum* (chamise) at the San Dimas Experimental Forest. (Data from DUNN, 1970)

measurements are needed to complete this analysis and to reconcile the harvest and gas analysis estimates. It is particularly important that reliable estimates of root weights be obtained.

Mineral Balance

The soils of the San Gabriel Mountains are nutrient deficient, particularly in nitrogen and phosphorus, even for native shrubs (HELLMERS *et al.*, 1955). It is thus not surprising that one of the early successional shrubs (*Ceanothus* ssp.) in the chaparral is a nitrogen fixer (HELLMERS and KELLEHER, 1959).

A great proportion of the nutrients returned to the soil in the annual litter fall (Table 14), as well as that accumulated, is lost through erosion. Erosion rates in the San Gabriel Mountains, in areas of full vegetation cover, can average as high as 8,500 kg per ha (SINCLAIR, 1954).

In a study of erosion in a number of sites in the San Gabriel Mountains (ANDERSON, COLEMAN and ZINKE, 1959) it was found that erosion could be greater in the dry season, when the soil had low cohesion, than in the rainy period. A large proportion of the material eroded was organic matter (Table 15).

Table 14. Total nitrogen and phosphorus content in plants and soils at San Dimas[a]

	kg/ha	
	Nitrogen	Phosphorus
Accumulated in tops	145	12
Annual uptake	6	0.8
Annual in litter loss[b]	~ 2.8	~ 0.2
Accumulated in soil		
0–2.5 cm	1290	278
2.6–15 cm	1940	1330
Total in soil	3230	1608

[a] Data from SPECHT, 1969b for a 37-year-old stand, and assuming bulk density of 1.5 gm/cc.
[b] Annual litter fall for chamise-ceanothus chaparral is 2800 kg/ha, KITTREDGE, 1955. Assume litter has 1% N and 0,1% P.

Table 15. Slope erosion in the San Gabriel Mountains[a]

	Location			
	Silver Springs North	Silver Springs South	Falls	Upper Brown
Slope aspect	NE	SW	NE	SW
Slope angle, %	55	60	60	55
Vegetation cover, %	85	65	95	95
Substrate	Diorite	Diorite	Diorite	Diorite
Erosion, kg/ha				
Dry season	220	220	484	924
Wet season	220	440	308	660
Annual	440	660	792	1584
Percentage				
Rock > 3,7 cm	0	0	2	3
Organic matter	24	17	29	36

[a] Data from ANDERSON, COLEMAN, and ZINKE, 1959, for a 5-year study of an area long unburned.

The steep slopes of the San Gabriel Mountains are vulnerable to soil slippage. The south-facing sage-covered slopes have higher numbers of slips than the mesic slopes which have deeper rooted chamise and broad-leaved chaparral shrubs (RICE, CORBETT and BAYLEY, 1964). In areas which have been unburned for five years soil slippage amounted to 32,000 kg par ha.

Removal of vegetation by fire greatly accelerates erosion on these mountain slopes. In a typical area of the San Gabriel Mountains it was found that pre-burn erosion rates were 7,980 kg per ha (ROWE, STOREY and HAMILTON, 1951). The first year after fire this amount increased to 230,000 kg per ha. The second and third years after fire the erosion rates were reduced to 52,700 and 26,600 kg per ha, respectively.

Santa Ana winds may be an important agent in eroding dry soils after fire.

Thus, loss of minerals through soil erosion is great, even under a full vegetation cover, during the dry as well as the wet season. Fire can result in large mineral losses from the system.

Man and the Chaparral

The California chaparral ecosystem has not been intensively utilized by man to any great extent. Since this type generally occurs on non-agricultural soils, it has not been extensively cleared for cultivation in the past; however, clearing and conversion to grassland is becoming increasingly common (BENTLEY, 1967). Chaparral plants have not been utilized directly to any degree for commercial purposes. There is, however, interest, mostly relatively recent, in management of the chaparral for its game, particularly deer and quail. Since this type covers the mountains surrounding the most populous areas of the State, there has been a long-term interest in its management in order to increase water yield in this arid region, as well as to protect the watershed from erosion. Interest in chaparral manipulation for increased water yield in southern California has lessened in recent years however because it is uneconomical. National interest in the problems of this region stem from the turn of the century when large forest reserves were established in southern California (SINCLAIR, HAMILTON and WAITE, 1958). These reserves were subsequently designated national forests. Early studies on stream flow and erosion stimulated the establishment of the San Dimas Experimental Forest in 1933 as a center for watershed research. During this same year there was a large fire in the San Gabriel Mountains which was followed by a severe storm. Subsequent flood damage devastated large areas, killing 30 people (COLMAN, 1953). This particular event was important in stimulating a vigorous policy of fire protection and flood control for the southern California watersheds. Millions of dollars are spent annually in fire detection and suppression, as well as in the construction of debris basins, reservoirs and channels. The fire control policy, as practiced, has been considered to be of questionable merit. HANES (1971) has recently noted that "A fire exclusion policy does not prevent fire, it only forestalls fire. In chaparral stands where fire has been excluded for decades, the threat of fire is greater."

Controlled burning is being utilized in the chaparral to a greater extent in certain areas of southern California. However, its practice is still subject to considerable controversy (DODGE, 1970; HELLMERS, 1969).

The past research at the SDEF has focused on understanding how watersheds function. However, studies have also been directed toward developing land management techniques for the chaparral (HILL, 1963). For example, considerable research has been directed toward learning of the effect on water budgets of the conversion of chaparral to grassland as well as the removal of riparian vegetation. The reported benefits of the conversion of chaparral to grassland – increasing water yield (particularly on deeper soils), increasing grassland range, improving fire control, and improving game habitat – has stimulated extensive conversion programs in southern California (BENTLEY, 1967). Mechanical clearing, burning and chemical control are used in conversions. These programs are not successful on all sites – particularly where slopes are greater than 55% (the majority of the slopes in the SDEF exceed this amount). Where conversion is not possible the chaparral vegetation is often broken into smaller units by fuel-break systems. These breaks are usually 60–120 m wide (BENTLEY and WHITE, 1961).

There have also been considerable efforts to manage the succession of chaparral subsequent to fire. As already noted, burned watersheds can lead to disastrous floods. At the SDEF, after the fire of 1960, an extensive program of watershed treatment was initiated which included broadcast sowing of exotic annual and perennial grasses, as well as contour planting and contour trenching. Aerial broadcast of annual grass seeds is one of the commonest treatments utilized subsequent to fire in southern California. Its effectiveness in erosion control varies considerably, dependent upon subsequent weather conditions (CORBETT and GREEN, 1965).

As the population has grown in southern California there has been increasing utilization of hillside chaparral lands for home building sites. These homes are often embedded in a matrix of this vegetation and both are periodically consumed in fires. Recent fires have devastated hundreds of homes.

Man, in recent times, at least, has not come to any balance with the chaparral ecosystem. The system is fire-dependent, and yet man has tended to exclude fire. As the population grows there will be even more vigorous attempts to manipulate and convert the system into one which is less hostile to the activities of man.

A Comparison of the Chile and California Sclerophyll Scrub Ecosystem

MOONEY *et al.* (1970, 1972) have recently compared the vegetations of the mediterranean climatic regions of California and Chile. They have pointed to the great similarity in the climate and vegetation structure between these regions. It was noted (MOONEY *et al.*, 1972) that there are, however, certain differences in the structure of the vegetation between homoclimates in California and Chile which may be related to land-use history rather than differences in adaptive modes. As noted above, man has not depended on the California chaparral ecosystem for subsistence. In contrast, in Chile man has long utilized the native chaparral-type vegetation (matorral) for fuel and for feed for livestock. The apparent result of this use is an opening of the vegetation, particularly at the more xeric parts of this range. Since fuel does not accumulate, when fires occur they are not so catastrophic.

Thus, in Chile man has gradually been changing the nature of the chaparral to a more xerophytic type. In California he is attempting to make the change quickly from its more or less native state to a form more compatible with his activities.

Summary

1) An analysis is made of the structure and function of the San Dimas Experimental Forest (SDEF), a 6,885 ha area of land near Los Angeles, California. A graphical summary of its environmental peculiarities, ecosystem features, and management problems are given in Fig. 9.

2) This region, which encompasses elevations from 458 to 1,678 m, is clothed with a vegetation which is predominantly an evergreen scrub (the chaparral) such as characterizes much of California. In drier habitats than those which support chaparral there is a subligneous drought-deciduous scrub (coastal sage), and on moister sites an evergreen forest. Winter-deciduous trees predominate in riparian areas.

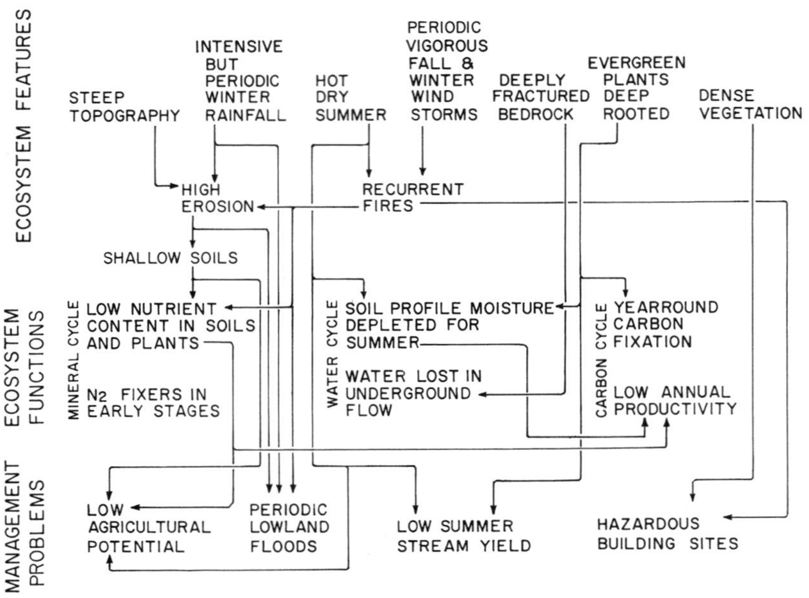

Fig. 9. A summary of the peculiarities of the southern California chaparral environment and its effect on ecosystem function and management

3) The climate of the area is a mediterranean type. Precipitation falls predominantly in the cool winters and as rain which can fall in large amounts for brief periods. Summer is generally hot and dry with high evaporative potential, although morning fog may moderate this somewhat. Violent winds (the Santa Anas) characterize the region, particularly in fall and winter.

4) The soils of the SDEF are shallow and nutrient-deficient. Erosion rates are high. The underlying bedrock is highly fractured and thus serves as a small reservoir for deeply rooted shrubs as well as a pathway for water loss from watersheds.

5) There are approximately 517 species of vascular plants on the SDEF, of which 75 are introduced. Eighty percent of the flora are herbs, of which half are annuals. Flowering activity is concentrated in spring. Seventy-five percent of all the species may flower during May.

6) Two hundred and one species of vertebrates are found in the SDEF, three-quarters of which are birds. There are seven large carnivores.

7) In an average year, about 8% of the incoming precipitation is intercepted by the vegetation and evaporated, and 15% flows into streams. Over 40% of the rainfall pene-

trates the fractured geological substrate and flows out as ground water. More than a third of the precipitation is lost through evapo-transpiration. Soils normally reach field capacity around December and – 15 bars between early summer and late fall, depending on site and season.

8) The evergreen dominants can fix carbon throughout the year; however, the amount fixed in summer depends greatly on the amount of rainfall received the previous winter. Growth of these shrubs is, however, primarily restricted to spring. Yearly dry weight gain of tops of chaparral shrubs of an intermediate age is 1,000 kg per ha. Root production is unknown.

9) Minerals are readily lost from the steep slopes through erosion which occurs nearly equally during both the dry and wet seasons and at a greatly accelerated rate after fire. Nitrogen-fixing species are important in the chaparral vegetation.

10) Vegetation succession subsequent to fire differs somewhat, depending on slope exposure. In general, however, about 50% of the shrub species will resprout after fire, even prior to winter rains. Winter rain will stimulate germination of those shrub species which are fire sprouters, as well as the non-sprouters. A distinctive "fire type" annual flora will also germinate following the rains. The vegetation will become closed again within a decade and the herbs will disappear. Toxic materials and water-repellant substances will accumulate in the soil, particularly in chamise chaparral, and the system may stagnate after about 50 years. Fire will rejuvenate the system by releasing minerals from the vegetation, detoxifying the soil, and vaporizing the hydrophobic soil materials. The rejuvenated system has a higher productivity, higher diversity, and a greater carrying capacity for consumer organisms.

11) The chaparral ecosystem has been of little direct utility to man; however, he has great interest in its management. Attempts of conversion of this type to grassland for grazing purposes is not successful on the steep slopes characteristic of the study region. Because of the high potential for flood damage subsequent to fires, a great effort is directed toward fire prevention. Fuel-break systems have been utilized to assist this program. The fire exclusion practice results in the production of mature stands of chaparral of high fuel content, which, when they do burn, can be catastrophic in scope. The use of controlled burning in this region is still controversial.

Additional management efforts in the chaparral ecosystem have been directed toward the "repair" of burned watersheds by artificial seeding and in increasing watershed yield by removal of riparian vegetation.

12) The SDEF chaparral ecosystem is characteristic, in many regards, of much of the coastal regions of southern California, although the steep slopes, high erosion rates, and depauperate soils are particularly prevalent in the San Gabriel Mountains.

13) The California chaparral is homologous to its Chilean counterpart in many regards, although it apparently differs in the important feature of being less utilized by man and hence of a less xerophytic nature.

Acknowledgements

We express appreciation to the National Science Foundation for support of this study. This is a contribution to the U.S. International Biological Program, Origin and Structure of Ecosystems Project. We are greatly indebted to Mr. JEROME S. HORTON of the U.S. Forest Service for providing important unpublished data regarding the San Dimas Experimental

Forest. Mr. RAYMOND RICE, in charge of the Forest, and his staff helped us in many ways for which we are grateful. Further assistance was provided by the staff of the U.S. Forest Service, Pacific Southwest Experimental Station.

References

ANDERSON, H. W., COLEMAN, G. B., ZINKE, P. J.: Summer slides and winter scour-dry-wet erosion in southern California montains. U.S. Forest Serv. Pacific Southwest Forest and Range Expt. Sta. Tech. Paper 36. 12 p. (1959).

BENTLEY, J. R.: Fitting brush conversion to San Gabriel watersheds. U.S. Forest Service Pacific Southwest Forest and Range Expt. Sta. Misc. Paper 61. 8 p. (1961).

BENTLEY, J. R.: Conversion of chaparral areas to grassland: Techniques used in California. U.S.D.A. Agriculture Handbook No. 328. 35 p. (1967).

BENTLEY, J. R., WHITE, V. E.: The fuel-break system for the San Dimas Experimental Forest. U.S. Forest Service Pacific Southwest Forest and Range Expt. Sta. Misc. Paper 63. 9 p. (1961).

BURNS, J. I.: Small scale topographic effects on precipitation distribution in San Dimas Experimental Forest, California. Unpublished Civil Engineering Thesis. Stanford Univ. 21 p. (1952).

COLMAN, E. A.: Fire and water in southern California mountains. U.S. Forest Service, Calif. Forest and Range Expt. Sta. Misc. Paper 3. 7 p. (1953).

CORBETT, E. S., GREEN, L. R.: Emergency revegetation to rehabilitate burned watersheds in southern California. U.S. Forest Service, Pacific Southwest Forest and Range Expt. Sta. Res. Paper PSW-22. 14 p (1965).

COUNTRYMAN, C., PHILPOT, C.: Chamise as a wild life fuel. U.S. Forest Service Pacific Southwest Forest and Range Expt. Sta., Res. Paper PSW-66. (1970).

CRAWFORD, J. M., Jr.: Soils of the San Dimas Experimental Forest. U.S. Forest Service Pacific Southwest Forest and Range Expt. Sta. Misc. Paper 76. 20 p. (1962).

CROUSE, R. P.: First-year effects of land treatment on dry season streamflow after a fire in southern California. U.S. Forest Service Pacific Forest and Range Expt. Sta. Res. Paper 191. 5 p. (1961).

DE BANO, L. F.: Water repellant soils: a worldwide concern in management of soil and vegetation. Agri. Sci. Rev. 7, 11–18 (1969).

DE BANO, L. F., OSBORN, J. F., KRAMMES, J. F., LETEY, J., Jr.: Soil wettability and wetting agents ... our current knowledge of the problem. U.S. Forest Service Pacific Southwest Forest and Range Expt. Sta. Res. Paper PSW-43. 13 p (1967).

DODGE, M.: Fire control of chaparral. Science **168**, 420 (1970).

DUNN, E. L.: Seasonal patterns of carbon dioxide metabolism in evergreen sclerophylls in California and Chile. Unpublished Ph. D. thesis. Univ. of Calif. (1970).

HAMILTON, E. L.: Rainfall-measurement as influenced by storm-characteristics in southern California mountains. Amer. Geophys. Union Trans. **1944**, 502–518 (1944).

HAMILTON, E. L.: The climate of southern California. In: Some aspects of watershed management in southern California. Staff Div. of Forest Influences Research. U.S. Forest Service Calif. Forest and Range Expt. Sta. Misc. Paper 1. 29 p. (1951).

HAMILTON, E. L.: Rainfall sampling on rugged terrain. USDA Tech. Bull. 1096. 41 p. (1954).

HAMILTON, E. L., REIMANN, L. F.: Simplified method of sampling rainfall on the San Dimas Experimental Forest. U.S. Forest Service Calif. Forest and Range Expt. Sta. Tech. Paper 26. 8 p. (1958).

HANES, T. L.: Succession after fire in the chaparral of southern California. Ecol. Monogr. **41**, 27–52 (1971).

HANES, T. L., JONES, H. W.: Postfire chaparral succession in southern California. Ecology **48**, 259–264 (1967).

HELLMERS, H.: Fight fire with fire. Science **166**, 945–946 (1969).

HELLMERS, H., BONNER, J., KELLEHER, J. M.: Soil fertility: a watershed management problem in the San Gabriel mountains of southern California. Soil Sci. **80**, 189–197 (1955).

HELLMERS, H., KELLEHER, J. M.: *Ceanothus leucodermis* and soil nitrogen in southern California mountains. Forest Sci. **5**, 275–278 (1959).

HILL, L. W.: The San Dimas Experimental Forest. U.S. Forest Service Pacific Southwest Forest and Range Expt. Sta. 25 p. (1963).

HOPKINS, W., BENTLEY, J., RICE, R.: Research and a land management model for southern California watersheds. U.S. Forest Service Pacific Southwest Forest and Range Expt. Sta. Misc. Paper 56. 12 p. (1961).

HORTON, J. S., KRAEBEL, C. J.: Development of vegetation after fire in the chamise chaparral of southern California. Ecology **36**, 244–262 (1955).

HORTON, J. S., WRIGHT, J. T.: The wood rat as an ecological factor in southern California watersheds. Ecology **25**, 341–351 (1944).

KITTREDGE, J.: Litter and forest floor of the chaparral in parts of the San Dimas Experimental Forest, California. Hilgardia **23**, 563–596 (1955).

MCPHERSON, J. K., MÜLLER, C. H.: Allelopathic effects of *Adenostoma fasciculatum*, "chamise", in the California chaparral. Ecol. Monog. **39**, 177–198 (1969).

MOONEY, H. A., DUNN, E. L.: Convergent evolution of mediterranean-climate evergreen sclerophyll shrubs. Evolution **24**, 292–303 (1970).

MOONEY, H. A., DUNN, E. L., SHROPSHIRE, F., SONG, L.: Vegetation comparisons between the mediterranean climatic areas of California and Chile. Flora **159**, 480–496 (1970).

MOONEY, H. A., DUNN, E. L., SHROPSHIRE, F., SONG, L.: Land-use history of California and Chile as related to the structure of the sclerophyll scrub vegetations. Madroño **21**, 305–319 (1972).

MUNZ, P. A., KECK, D. D.: A California Flora. Berkeley: University of California Press 1959.

NAVEH, Z.: Mediterranean ecosystems and vegetation types in California and Israel. Ecology **48**, 445–459 (1967).

PATRIC, J. H., HANES, T. L.: Chaparral succession in a San Gabriel mountain area in California.Ecology **45**, 353–360 (1964).

PLUMB, T. R.: Sprouting of chaparral by December after a wildfire in July. U.S. Forest Service Pacific Southwest Forest and Range Expt. Sta. Tech. Paper 57. 12 p. (1961).

PLUMB, T. R.: Delayed sprouting of scrub oak after a fire. U.S. Forest Service Res. Note PSW-1. 4 p. (1963).

QASHU, H., ZINKE, P.: The influence of vegetation on soil thermal regime at the San Dimas Lysimeters. Soil Sci. Soc. Proc. **1964**, 703–706 (1964).

REIMANN, L. F.: Mountain temperatures. Data from the San Dimas Experimental Forest. U.S. Forest Service Pacific Southwest Forest Expt. Sta. Misc. Paper 36 (1959 a).

REIMANN, L. F.: Mountain evaporation data from the San Dimas Experimental Forest. U.S. Forest Service Pacific Southwest Forest and Range Expt. Sta. Misc. Paper 35. 14 p. (1959 b).

REIMANN, L. F., HAMILTON, E. L.: Four hundred sixty storms-data from the San Dimas Experimental Forest. U.S. Forest Service Pacific Southwest Forest and Range Expt. Sta. Misc. Paper 37. 101 p. (1959).

RICE, R. M., CORBETT, E. S., BAILEY, R. G.: Soil slips related to vegetation, topography, and soil in southern California. Water Resources Res. **5**, 647–659 (1969).

ROWE, P. B.: Streamflow increases after removing woodland-riparian vegetation from a southern California watershed. J. Forestry **61**, 365–370 (1963).

ROWE, P. B., COLMAN, R. A.: Disposition of rainfall in two mountain areas of California. USDA. Forest Service Tech. Bull. 1018. 84 p. (1951).

ROWE, P. B., STOREY, H., HAMILTON, E.: Some results of hydrologic research. In: Some aspects of watershed management in southern California. U.S. Forest Service California Forest and Range Expt. Sta. Misc. Paper 1. p 19–29 (1951).

SINCLAIR, J. D.: Erosion in the San Gabriel Mountains of California. Amer. Geophys. Union Trans. **35**, 264–268 (1954).

SINCLAIR, J. D.: Watershed management research in southern California's brush covered mountains. J. Forestry **58**, 266–268 (1960).

SINCLAIR, J. D., HAMILTON, E. L., WAITE, M. N.: A guide to the San Dimas Experimental Forest. U.S. Forest Service California Forest and Range Expt. Sta. Misc. Paper 11. 8 p. (1958).

SPECHT, R. L.: A comparison of the sclerophyllous vegetation characteristic of mediterranean type climates in France, California, and southern Australia. I. Structure, morphology, and succession. Austr. J. Bot. **17**, 277–292 (1969 a).

SPECHT, R. L.: A comparison of the sclerophyllous vegetation characteristic of mediterranean type climates in France, California, and southern Australia. II. Dry matter, energy, and nutrient accumulation. Austr. J. Bot. **17**, 293–308 (1969 b).

Storey, H. C.: Geology of the San Gabriel Mountains, California, and its relation to water distribution. U.S. Forest Service California Forest and Range Expt. Sta. 19 p. (1948).
Sweeney, J. R.: Responses of vegetation to fire. Univ. California Pub. Bot. **28**, 143–250 (1956).
Sweeney, J. R.: Ecology of some "fire type" vegetations in northern California. p. 111–125. In: Proc. Seventh Ann. Tall Timbers Fire Ecology Conf. Tallahassee, Florida: Tall Timbers Res. Sta. 1967.
USDA Forest Service: Vegetation types for the Cucomonga and Pomona Quadrangles (1942).
Wieslander, A. E., Gleason, C. H.: Major brushland areas of the Coast Ranges and the Sierra-Cascade foothills in California. U.S. Forest Service California Forest and Range Expt. Sta. Misc. Paper 15. 9 p. (1954).
Wright, J. T., Horton, J. S.: Checklist of the vertebrate fauna of San Dimas Experimental Forest. California Forest and Range Expt. Sta. Misc. Paper 7. 15 p. (1951).
Wright, J. T., Horton, J. S.: Supplement to checklist of the vertebrate fauna of San Dimas Experimental Forest. California Forest and Range Expt. Sta. Misc. Paper 13. 4 p. (1953).

2

Structure and Functional Response of Ecosystems in the Mediterranean Climate of Australia

RAYMOND L. SPECHT

Climate

The mediterranean-type climate of Australia, with wet winter periods alternating with dry summer periods, is confined to the south-west corner of Western Australia, the southern portion of South Australia, and the western half of Victoria. To the north it grades into the semi-arid zone of the interior of Australia; in the east, it merges with areas receiving increasing amounts of summer rainfall. Within this zone, annual precipitation varies from 250 to 1,500 mm with 5-15% of this precipitation falling during the summer months, December to February. This small incidence of summer rainfall appears to have enabled certain growth patterns atypical of most mediterranean regions of the world to persist in the southern Australian vegetation.

The general climatic conditions of the area are summarised in Table 1.

Table 1. Climatic range of plant communities of the mediterranean region of southern Australia

Climatic factor	January	July	Year
Solar radiation[a] (cal. cm^{-2} day^{-1})	350–700	100–300	220–500
Mean maximum temperature (° C)	20–33	10–20	17–25
Mean minimum temperature (° C)	9–19	2–12	6–15
Precipitation (mm)	6–40	18–270	230–1,300
Pan evaporation (mm)[b]	120–260	30–70	880–1,770

[a] Estimated by the formula developed by BLACK (1956).

[b] Estimated by the formula developed by FITZPATRICK (1963).

Plant Formations

The structure of the plant formations dominating the landscape in the mediterranean area of southern Australia varies considerably according to the water balance in the area (Table 2). Typically, with decrease in annual precipitation, the vegetation grades from an open-forest to a woodland to an open-scrub formation, all dominated by species of *Eucalyptus* with evergreen, sclerophyllous leaves (mesophyll in size). The understorey – as well as the dominant species of *Eucalyptus* – is markedly influenced by the fertility of the soil on which it occurs. Fertile soils tend to support an herbaceous understorey of perennial, tussock grasses (e.g. *Themeda, Danthonia, Stipa*) and geophytes. Infertile soils, especially low in phosphorus and nitrogen, tend to support a dense assemblage of sclerophyllous

Table 2. Structural characteristics of vegetation in the mediterranean region of southern Australia

Formation[a]	Dominant species Life-form[b]	Height (m)	Foliage projective cover %	Understorey on relatively fertile soils	Understorey on infertile soils
Open-forest	evergreen trees	5–30	30–70	Perennial grasses and herbs	Sclerophyllous shrubs
Woodland	evergreen trees	5–30	10–30	Perennial grasses and herbs	Sclerophyllous shrubs
Open-scrub	evergreen shrubs	2–8	30–70	Perennial grasses, herbs and chenopods	Sclerophyllous shrubs
Open-heath	evergreen shrubs	0–2	30–70	–	Sclerophyllous shrubs
Tussock grassland	perennial grasses & herbs	0–0.5	30–70	Perennial grasses and herbs	–

[a] Nomenclature follows Specht (1970)

[b] A *tree* is defined as a woody plant more than 5 m tall, usually with a single stem. A *shrub* is defined as a woody plant less than 8 m tall, usually with several stems arising at or near the base.

shrubs (0.5-2.0 m tall, and with leaves ranging in size from leptophyll to microphyll) with many species codominant (Specht and Rayson, 1957a). Some common genera are as follows:

Casuarinaceae –	*Casuarina*
Epacridaceae –	*Epacris, Astroloma, Acrotriche, Leucopogon*
Mimosaceae –	*Acacia*
Myrtaceae –	*Leptospermum, Melaleuca, Calytrix, Baeckea*
Papilionaceae –	*Daviesia, Pultenaea, Dillwynia, Phyllota*
Proteaceae –	*Banksia, Hakea, Grevillea*
Xanthorrhoeaceae –	*Xanthorrhoea*

On deeply-cracking, heavy, fertile soils, such as on the basaltic areas of western Victoria, *Eucalyptus* trees are absent and the herbaceous layer persists as a tussock grassland formation (Patton, 1936; Groves, 1965b).

Eucalyptus trees are also absent from water-logged, infertile soils in areas with more than 650 mm annual precipitation and from well-drained, deep infertile sands (more than 1.5 m deep) in 350-600 mm annual precipitation. Here the understorey of sclerophyllous shrubs (0.5–2.0 m tall) persists as an open-heath formation (Groves, 1965a; Groves and Specht, 1965; Patton, 1933; Specht and Rayson, 1957a).

The Effect of Calcium Carbonate

Infertile soils, as defined above, occur over the whole climatic range of the mediterranean region of Australia. Calcium carbonate is invariably low or absent in these soils.

Calcareous soils (terra rossa, rendzinas, red brown earths, brown solonised soils) are common among the more fertile soils. Calcium carbonate may be absent from the profile, be scattered in particle form throughout the profile or be concreted into nodules and strata in the subsoil. The pH of the soil thus varies from acid to markedly alkaline. The dominant species of *Eucalyptus* appear to be subtly affected by the range of calcium carbonate and pH (PARSONS, 1968 a, 1968 b, 1968 c, 1969; PARSONS and SPECHT, 1967). The understorey remains herbaceous on all soils, with, probably, clinal sifting in component species – although this has not been investigated.

The Effect of Fire

The formations are all subject to periodic bushfires (BEADLE, 1940; GILL and ASHTON, 1968; GROVES and SPECHT, 1965; JONES, GROVES and SPECHT, 1969; SPECHT, 1966, 1969a, 1969b; SPECHT, RAYSON and JACKMAN, 1958; VINES, 1968). The *Eucalyptus* trees of the area are all fire-resistant; even following a severe crown fire, epicormic shoots rapidly sprout all over the tree. The slender stems of the mallee-shrub species of *Eucalyptus*, in the open-scrub formation, may be destroyed during a fire but the plants regenerate rapidly from underground ligno-tubers. Many of the sclerophyllous shrubs of the understorey on infertile soils show the same capacity to regenerate from ligno-tubers after a fire (SPECHT and RAYSON, 1957b). Some, however, are killed but regenerate from seeds released from capsules and follicles by the heat of the fire.

Fire simply razes the dry aerial portions of perennial grasses, herbs and geophytes in the herbaceous understorey on fertile soils; the perennating buds just below the surface of the soil survive to sprout on the next favourable rain (GROVES, 1965 b).

The Effect of Grazing

Native Australian fauna, such as the kangaroo and wallaby, apparently exerted minimal grazing pressure on the plant communities of Australia – even on those communities with herbaceous understories.

The introduction of sheep, cattle and horses during the Nineteenth Century, when Europeans settled in Australia, rapidly altered the ecosystem on fertile soils. Native perennial grasses tended to disappear under intensive grazing pressure and their places were taken by annual species of grasses and herbs (*Hordeum, Avena, Bromus, Lolium, Asphodelus, Homeria, Trifolium, Medicago, Melilotus, Oxalis, Echium, Rumex, Plantago, Hypochoeris, Cryptostemma,* etc.) introduced into Australia from other mediterranean climatic regions, especially from around the Mediterranean Sea and from the Cape Province of South Africa (SPECHT, 1969 a).

The plant formations on infertile soils proved less profitable for the grazing of domestic animals. "Scrub-grazing" was undertaken at low-density stocking rates especially on the young shoots regenerating after a bushfire – which increased in frequency with European occupation of the land. Very few introduced plants invaded these sclerophyllous understories – and then only along roadsides and clearings where fertility had increased.

Growth Rhythm of Dominant Species

Growth of the dominant species in the chaparral vegetation of southern California and in the maquis-garrigue vegetation in southern France is inhibited by two periods of stress – the extreme drought during the summer months and the low temperatures and frost during the winter months (SPECHT, 1969 a, 1969 b). Some growth is initiated during autumn but ceases during winter; most growth occurs when the temperatures rise during the spring months. Introduced mediterranean annuals continue to show the same growth rhythm in Australia.

In the mediterranean region of southern Australia, only a few native species – largely annual geophytes and a few shrubs such as *Leptospermum* – show a spring growth rhythm (JONES, 1968 a, 1968 b). The dominant trees and shrubs, *Eucalyptus, Banksia, Casuarina* – and even the perennial grasses, *Themeda* and *Danthonia* – tend to begin growth in late spring – early summer and to continue growth well into summer. It appears that the mean daily temperature has to rise above 60–65° F (16–18° C) before growth occurs – a threshold more characteristic of tropical species (FITZPATRICK and NIX, 1970; GROVES, 1965 a; SPECHT and RAYSON, 1957). Growth continues spasmodically throughout summer whenever soil moisture is available (Fig. 1) – either as soil storage from the winter-spring period or from erratic summer rains (HOLMES, 1960; MARTIN and SPECHT, 1962; SPECHT, 1957 a, 1957 b; SPECHT and JONES, 1971).

Origin of the Australian "Mediterranean" Flora

The growth rhythm of the "mediterranean" flora of southern Australia is markedly out of phase with the current climate (SPECHT and RAYSON, 1957 a). It appears from scattered palaeontological records and from fossil soil and rock deposits that much of southern Australia experienced a much more humid, and probably warmer, climate during the Miocene to Pleistocene Periods when the current flora, with other formations, became widespread across the landscape. This flora extended from the tropical regions – and genera characteristic of the flora are still to be found in lowland heaths (pandangs) in Borneo and Malaysia, in tropical Cape York in Queensland, and in isolated, but extensive, patches of infertile soil along the whole length of the eastern coast of Australia.

The "mediterranean-climate" flora of southern Australia thus appears to be a relic flora of tropical origin, stranded in southern Australia as the climate became more arid during the Pleistocene-Recent Periods. Fortunately, erratic summer rains, supplemented by soil moisture storage maximized by reduced evapotranspiration rates, enabled the evergreen, perennial shrubs and trees to persist in a climate approaching that of a mediterranean region with a dry summer period.

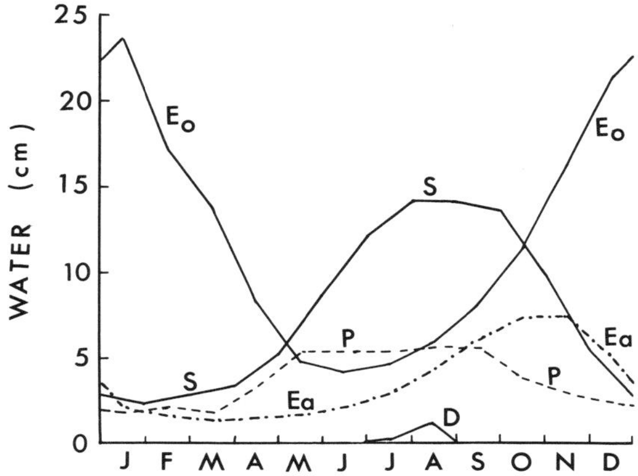

Fig. 1. Mean monthly values of precipitation (P), pan evaporation (E_o), actual evapotranspiration (E_a), total soil moisture storage (S), and drainage (D) estimated by using the equation,

$$E_a/E_o = 0.0433\,(P + S) - 0.101,$$

for 10–14 years old heath growing on 2 m of sand near Dark Island Soak, South Australia. The basal value of 2.34 cm of soil moisture, above pF 4.2 is assumed to be unavailable to the plant community. Growth of the dominant species of the heath vegetation occurs from December to March. Litter fall occurs throughout the year with a maximum fall occurring when soil storage becomes limiting towards the end of summer (February-March). Litter decomposition is most rapid during spring (September-October). Most species in the heath tend to flower during spring (September to November), but the dominant species do not flower until late summer-early autumn or even mid-winter

The herbaceous flora on the fertile soils also shows the late spring-summer growth rhythm but this growth becomes restricted as higher rates of evapotranspiration more rapidly reduce soil moisture storage. The aerial portions of the perennial grasses and herbs then tend to shrivel and die. The herbaceous flora thus more closely approximates to that of a typical mediterranean climate, but not quite – as revealed by its replacement by introduced mediterranean species.

Evolution towards evergreen, perennial shrubs and trees with growth rhythms more adjusted to the mediterranean-type climate may either require more time or may be inhibited by the poor nutrient status of the environment.

Effect of Fertilizer

The sclerophyllous understorey, characteristic of infertile soils, responds markedly to the addition of phosphorus fertilizer; growth of many dominant species increases two to three times (Specht, 1963; Specht and Groves, 1966). In certain cases phosphorus-toxicity may result, but this is a rare occurrence. An additional response may be obtained in some species when nitrogen fertilizer is applied with phosphorus. Response to other nutrients has, so far, not been demonstrated.

Spaces between the shrubs of the fertilized community are invaded by introduced mediterranean annuals.

The leaf area index of the community increases with fertilizer; the young summer-growing shoots appear 1–2 weeks earlier; and the degree of mesophytism of the young shoots appears to persist for a longer period than normal (BEADLE, 1953, 1954, 1962, 1966; SPECHT, 1963; SPECHT and GROVES, 1966; SPECHT and JONES, 1971). In drought years this growth leads to minimal water supply and death of about 50 per cent of dominant shrubs (SPECHT, 1963; SPECHT and JONES, 1971).

The community tends to open out following death of these sclerophyllous, shrubby species and is invaded by herbaceous plants (SPECHT, 1963).

Although the fertilizer increases the growth of the sclerophyllous shrubs, it upsets the water balance of the community during periodic drought years and death of species showing a strong summer-growth rhythm results. Selection of species with a more typical mediterranean climate spring-growth rhythm appears to have been induced by the addition of the phosphorus fertilizer.

The sclerophyllous species grow quite successfully – at least for a number of years – when planted on more fertile soils. Under normal uncultivated conditions, the seedlings have to establish themselves in competition with faster-growing herbaceous grasses and herbs and are smothered.

Phosphorus Conservation

The remarkable annual growth (780–1,900 kg per hectare – SPECHT, 1963; SPECHT, RAYSON and JACKMAN, 1958) of sclerophyllous shrubs growing on soils very low in phosphorus has interested a number of Australian ecologists. The element must be very efficiently conserved within the ecosystem to enable the community to fix so much solar energy.

Phosphorus found in senescing organs appears to be more efficiently recirculated than in "normal" plants (SPECHT and GROVES, 1966). Some, however, is lost from the plant in the litter.

A fine mat of roots, either proteoid or mycorrhizal, penetrates the decomposing litter containing phosphorus lost from the plant (JEFFREY, 1967; PURNELL, 1960). In fact, the proteoid roots, with markedly increased surface area, appear to be induced by phosphorus-deficiency in many genera of the family Proteaceae and are not found on plants growing in well-fertilized soil.

Another conservation mechanism appears to be the presence of polyphosphate-forming and – hydrolysing enzymes in the roots (and in micro-organisms in the roots) of most sclerophyllous plants (JEFFREY, 1964, 1967, 1968). Orthophosphate, released from decomposing litter in spring, is stored in the roots as long-chain polyphosphate; subsequently, in late spring-early summer when growth is initiated, the polyphosphate appears to be hydrolysed back to orthophosphate and transported to the growing shoot apices.

Such phosphorus-conservation mechanisms (and possibly others) enable the sclerophyllous flora to persist on phosphorus-deficient soils.

Summary

The ecosystems found in the mediterranean climatic region of southern Australia have a summer-growth rhythm which appears to be a relic of a tropical flora which is still adjusting – by evolution and invasion – to the present climatic conditions.

Evolution within the flora has been markedly influenced by soil fertility. Soils low in plant nutrients, especially phosphorus, are common in the area and support an understorey of sclerophyllous shrubs. Certain phosphorus-conservation mechanisms, which have been discovered by this flora, may have contributed to the survival of tropical aspects of the flora.

On more fertile soils, an herbaceous understorey is widespread. The dominant trees and shrubs show evolutionary sifting with decrease in available soil water and increase in calcium carbonate content of the soil.

Fire appears to have played a major role in the evolution of the flora.

The grazing pressure of native animals has been minimal on the flora.

References

BEADLE, N. C. W.: Soil temperature during forest fires and their effect on the survival of vegetation. J. Ecol. **28**, 180–192 (1940).

BEADLE, N. C. W.: The edaphic factor in plant ecology with a special note on soil phosphates. Ecology **34**, 426–428 (1953).

BEADLE, N. C. W.: Soil phosphate and the delimitation of plant communities in eastern Australia. Ecology **35**, 370–375 (1954).

BEADLE, N. C. W.: Soil phosphate and the delimitation of plant communities in eastern Australia II. Ecology **43**, 281–288 (1962).

BEADLE, N. C. W.: Soil phosphate and its role in molding segments of the Australian flora and vegetation with special reference to xeromorphology and sclerophylly. Ecology **47**, 991–1007 (1966).

BLACK, J. N.: The distribution of solar radiation over the earth's surface. Arch. Met. Geoph. Biokl. B. **7**, 165–189 (1956).

FITZPATRICK, E. A.: Estimates of pan evaporation from mean maximum temperature and vapour pressure. J. Appl. Met. **2**, 780–792 (1963).

FITZPATRICK, E. A., NIX, H. A.: The climatic factor in Australian grassland ecology. In: Australian Grasslands. (R. M. MOORE, Ed.). Canberra: Aust. Nat. Univ. Press 1970.

GILL, A. M., ASHTON, D. H.: The role of bark type in relative tolerance to fire of three central Victorian eucalypts. Aust. J. Bot. **16**, 491–498 (1968).

GROVES, R. H.: Growth of heath vegetation. II. The seasonal growth of a heath on ground-water podzol at Wilson's Promontory, Victoria. Aust. J. Bot. **13**, 281–289 (1965 a).

GROVES, R. H.: Growth of *Themeda australis* tussock grassland at St. Albans, Victoria. Aust. J. Bot. **13**, 291–302 (1965 b).

GROVES, R. H., SPECHT, R. L.: Growth of heath vegetation I. Annual growth curves of two heath ecosystems in Australia. Aust. J. Bot. **13**, 261–280 (1965).

HOLMES, J. W.: Water balance and the water-table in deep sandy soil of the Upper South-East, South Australia. Aust. J. Agric. Res. **11**, 970–988 (1960).

JEFFREY, D. W.: The formation of polyphosphate in *Banksia ornata*, an Australian heath plant. Aust. J. Biol. Sci. **17**, 845–854 (1964).

JEFFREY, D. W.: Phosphate nutrition of Australian heath plants. I. The importance of proteoid roots in *Banksia* (Proteaceae). Aust. J. Bot. **15**, 403–412 (1967).

JEFFREY, D. W.: Phosphate nutrition of Australian heath plants. II. The formation of polyphosphate by five heath species. Aust. J. Bot. **16**, 603–613 (1968).

JONES, R.: The leaf area of an Australian heathland with reference to seasonal changes and the contribution of individual species. Aust. J. Bot. **16**, 579–588 (1968 a).

JONES, R.: Estimating productivity and apparent photosynthesis from differences in consecutive measurements of total living parts of an Australian heathland. Aust. J. Bot. **16**, 589–602 (1968 b).

JONES, R., GROVES, R. H., SPECHT, R. L.: Growth of heath vegetation. III. Growth curves for heaths in southern Australia: A reassessment. Aust. J. Bot. **17**, 309–314 (1969).

MARTIN, H. A., SPECHT, R. L.: Are mesic communities less drought resistant? A study on moisture relationships in dry sclerophyll forest at Inglewood, South Australia. Aust. J. Bot. **10**, 106–118 (1962).

PARSONS, R. F.: Ecological aspects of the growth and mineral nutrition of three mallee species of *Eucalyptus*. Oecol. Plant. **3**, 121–136 (1968 a).

PARSONS, R. F.: Effects of waterlogging and salinity on growth and distribution of three mallee species of *Eucalyptus*. Aust. J. Bot. **16**, 101–108 (1968 b).

PARSONS, R. F.: The significance of growth-rate comparisons for plant ecology. Amer. Nat. **102**, 595–597 (1968 c).

PARSONS, R. F.: Physiological and ecological tolerances of *Eucalyptus incrassata* and *E. socialis* to edaphic factors. Ecology **50**, 386–390 (1969).

PARSONS, R. F., SPECHT, R. L.: Lime chlorosis and other factors affecting the distribution of *Eucalyptus* on coastal sands in southern Australia. Aust. J. Bot. **15**, 95–105 (1967).

PATTON, R. T.: Ecological studies in Victoria. The Cheltenham flora. Proc. Roy. Soc. Vict. **45**, 205–218 (1933).

PATTON, R. T.: Ecological studies in Victoria. Part IV. Basalt Plains association. Proc. Roy. Soc. Vict. **48**, 172–191 (1936).

PURNELL, H. M.: Studies of the family Proteaceae. I. Anatomy and morphology of the roots of some Victorian species. Aust. J. Bot. **8**, 38–50 (1960).

SPECHT, R. L.: Dark Island heath (Ninety-Mile Plain, South Australia). IV. Soil moisture patterns produced by rainfall interception and stem-flow. Aust. J. Bot. **5**, 137–150 (1957 a).

SPECHT, R. L.: Dark Island heath (Ninety-Mile Plain, South Australia). V. The water relationships in heath vegetation and pastures on the Makin Sand. Aust. J. Bot. **5**, 151–172 (1957 b).

SPECHT, R. L.: Dark Island heath (Ninety-Mile Plain, South Australia). VII. The effect of fertilizers on composition and growth, 1950–60. Aust. J. Bot. **11**, 67–94 (1963).

SPECHT, R. L.: The growth and distribution of mallee-broombush (*Eucalyptus incrassata – Melaleuca uncinata* association) and heath vegetation near Dark Island Soak, Ninety-Mile Plain, South Australia. Aust. J. Bot. **14**, 361–371 (1966).

SPECHT, R. L.: A comparison of the sclerophyllous vegetation characteristic of Mediterranean type climates in France, California, and southern Australia. I. Structure, morphology, and succession. Aust. J. Bot. **17**, 277–292 (1969 a).

SPECHT, R. L.: A comparison of the sclerophyllous vegetation characteristic of Mediterranean type climates in France, California, and southern Australia. II. Dry matter, energy, and nutrient accumulation. Aust. J. Bot. **17**, 293–308 (1969 b).

SPECHT, R. L.: Vegetation. 44–67. In: The Australian Environment. Fourth Edition (G. W. LEEPER, Ed.). Melbourne: C.S.I.R.O. – Melbourne University Press 1970.

SPECHT, R. L., GROVES, R. H.: A comparison of the phosphorus nutrition of Australian heath plants and introduced economic plants. Aust. J. Bot. **14**, 201–221 (1966).

SPECHT, R. L., JONES, R.: A comparison of the water-use by heath vegetation at Frankston, Victoria, and Dark Island Soak, South Australia. Aust. J. Bot. **19**, 311–326 (1971).

SPECHT, R. L., RAYSON, P.: Dark Island heath (Ninety-Mile Plain, South Australia). I. Definition of the ecosystem. Aust. J. Bot. **5**, 52–85 (1957 a).

SPECHT, R. L., RAYSON, P.: Dark Island heath (Ninety-Mile Plain, South Australia). III. The root systems. Aust. J. Bot. **5**, 103–114 (1957 b).

SPECHT, R. L., RAYSON, P., JACKMAN, M. E.: Dark Island heath (Ninety-Mile Plain, South Australia). VI. Pyric succession: Changes in composition, coverage, dry weight, and mineral nutrient status. Aust. J. Bot. **6**, 59–88 (1958).

VINES, R. G.: Heat transfer through bark, and the resistance of trees to fire. Aust. J. Bot. **16**, 499–514 (1968).

3

The Role of the Secondary Plant Chemistry in the Evolution of the Mediterranean Scrub Vegetation

TOM J. MABRY and DAN R. DIFEO, Jr.

Introduction

As far as is known every plant species produces some unique secondary compounds, often in remarkably large quantities. Although some of these alkaloids, terpenes, phenolics, etc. regulate vital biochemical processes at the cellular level, others mediate the interaction of the plant with other organisms and with other elements of the ecosystem. Some of the secondary chemical constituents appear to be relic compounds in that they are produced today in a wide variety of genetically related species by a genome which was present in the ancestral stock; others may be present in only a very few populations which are restricted to a unique ecosystem and appear to be of more recent origin and thus represent one type of ecochemistry.

WHITTAKER and FEENY (1971) have emphasized that chemical agents are of major significance in the adaptation of species and organization of communities; indeed the expression allelochemics is now used to denote all types of interactions between species which involve chemicals by which organisms of one species affect the growth, health, behavior, or population biology of organisms of another species.

Some of these secondary compounds are known to be hormones which mediate a variety of developmental responses, others are metabolically active in that they undergo further transformations and turnover during the life cycle of the plant. For example, BARZ and GRISEBACH (1970) have observed that in some instance flavonoids are not metabolically inactive end products but are subject to further turnover and degradation; indeed the halflife of some of these substances appears to be only a matter of hours. Other workers such as LOOMIS (1967) have used pulse labeling techniques to detect rapid turnover of volatile mono- and sesqui- terpenes in plants belonging to the mint family; the turnover rates, which are only a matter of hours in some instances, were found to vary with the stage of development of the plant.

Current studies of the ecochemistry of the dominant vegetation in the similar mediterranean climatic regions of Chile and California are designed to distinguish the relic chemistry in the mediterranean scrub vegetation and to determine which elements in the Californian and Chilean ecosystems provided the pressures for the selection of the compounds of recent origin. Microchemical gradients will be determined over microecological gradients in restricted study sites; while macrochemical patterns will be obtained for populations of a given taxon in widely separated and ecologically quite distinct regions.

The present chapter presents examples of the way certain types of both relic and recent secondary compounds are distributed in plant populations and the experimental approaches employed for determining this chemistry.

Secondary Plant Chemistry: Examples of Relic and Recent Ecochemistry

We and others have successfully employed flavonoids, terpenes, pigments and alkaloids for a variety of evolutionary problems in the plant kingdom; some of the approaches are illustrated here by the application of terpenes[1] and betalains to phylogeny. As mentioned in the introduction some of the chemistry appears to be relic in that it represents an ancestral stock; other substances appear to be of more recent origin and many reflect selection pressures readily discernable in today's ecosystems.

Most of the examples noted here emphasize the application of secondary plant chemistry for recognizing genetically-related taxa; therefore, the data, for the most part, have not been correlated with either the climatic and physical features of the environment or the predators and pollinators associated with the taxa. Thus, the significance of the "mediterranean scrub vegetation project" is that it should permit the complete integration of the secondary chemical data, both recent and relic, with all elements of the ecosystem.

A. Relic Chemistry: The Sesquiterpene Lactone Chemistry of the Genus Parthenium (Compositae)

The present example was selected in order to illustrate some of the problems we can expect to encounter in our efforts to correlate plant chemistry with other elements of a current ecosystem.

In our investigation of the sesquiterpene lactone chemistry of 17 taxa of *Parthenium* (RODRIGUEZ, YOSHIOKA and MABRY, 1971), we found that members of all four morphologically-distinct subgenera are characterized by their ability to elaborate a series of C_{14} and C_{15} oxygenated sesquiterpene lactones (Fig. 1). A given plant will normally produce 2–5 of these substances and the total concentration of these compounds will often be as much as 2–5 % of the dry weight of the leaves. They are the major secondary compounds elaborated by these species, and their synthesis represents a major energy commitment in these plants. Moreover, since these compounds have not been detected elsewhere in the plant kingdom, they appear to be reliable genetic markers for the *Parthenium* species.

An intriguing example of what we choose to call "relic chemistry" concerns the sesquiterpene lactone chemistry of such *Parthenium* species as *P. alpinum* var. *tetraneuris*, *P. fructicosum* var. *trilobatum* and *P. integrifolium*; in these three species, all of which are not only morphologically quite distinct but occur in climatically, geologically, and organismically different ecosystems, we find a secondary-compound genome representing an ancestral stock still to be active in each species.

1 Many other classes of secondary compounds as well as such primary substances as proteins and nucleic acids have been used equally well for problems similar to those we expect to encounter in our studies of the secondary chemistry of the mediterranean scrub vegetation (see, for example: MABRY and MEARS, 1970; MARKHAM, MABRY and SWIFT, 1970).

Species	Distribution	Sesquiterpene lactone chemistry
P. alpinum var. *tetraneuris* (small clumps of plants only a few cm in diameter and height)	Occurs at elevations of over 5,000 ft (1520 m) on gypsum and limestone outcroppings in south-central Colorado. Prabably reproduces apomictically	tetraneurin-C plus -A and -D
P. fructicosum var. *trilobatum* (large, vigorous plants, often over 2 m in height)	Occurs in tropical regions south of Monterrey, Mexico	tetraneurin-C plus -B and -D
P. integrifolium (plants are intermediate in both height and foliage)	Occurs in the disturbed plains of Illinois and the north-central United States	tetraneurin-C plus -E

The three taxa are so distinct morphological that one of them, *P. alpinum* var. *tetraneuris*, was in earlier times often aligned along with other members of the section Bolophytum into a separate genus (MEARS, 1970). Of particular interest, however, is the observation that the three taxa have adapted to three distinct ecosystems, but nevertheless continue to produce large quantities of either identical or very similar sesquiterpene lactones. Although the matter is only now being investigated, it is assumed that the plant-plant and plant-animal interactions associated with each of the three species are quite different in the three ecosystems. Therefore, it appears that the taxa may be using their "relic chemistry" representing an ancestral genome to meet rather different problems associated with their survival in distinct ecosystems.

B. Relic Chemistry: Betalains in the Order Centrospermae

Perhaps the most remarkable example of relic secondary chemistry concerns the occurrence of betalains in ten plant families belonging to the order Centrospermae (see MABRY, 1964, 1966, and 1970a). Either the same or slightly modified compounds (based on the general betalain formula I) occurs in species of these ten families and, notably, some of the species can be found in essentially every conceivable ecosystem utilized by higher plants on the earth. Possibly, these plants continue to form a particular secondary compound (produced by ancestral genes) because it is still important to their survival regardless of the other elements present in the ecosystems. These unique red and yellow betalain-alkaloids are restricted to ten plant families which, at one time or another, were positioned in the order Centrospermae for morphological and anatomical reasons. However the first time that all of the ten betalain-containing families, and only these ten families, were placed in the Centrospermae was in 1963 (MABRY *et al.*, 1963), shortly after it was recognized (MABRY *et al.*, 1962) that the betalains were structurally unrelated to the more common anthocyanins [cyanin (II), the red rose pigment is typical]. Now many systematists acknowledge that the betalain-containing families are evolutionarily related and that the betalains do indeed represent a phylogenetically-important class of relic compounds.

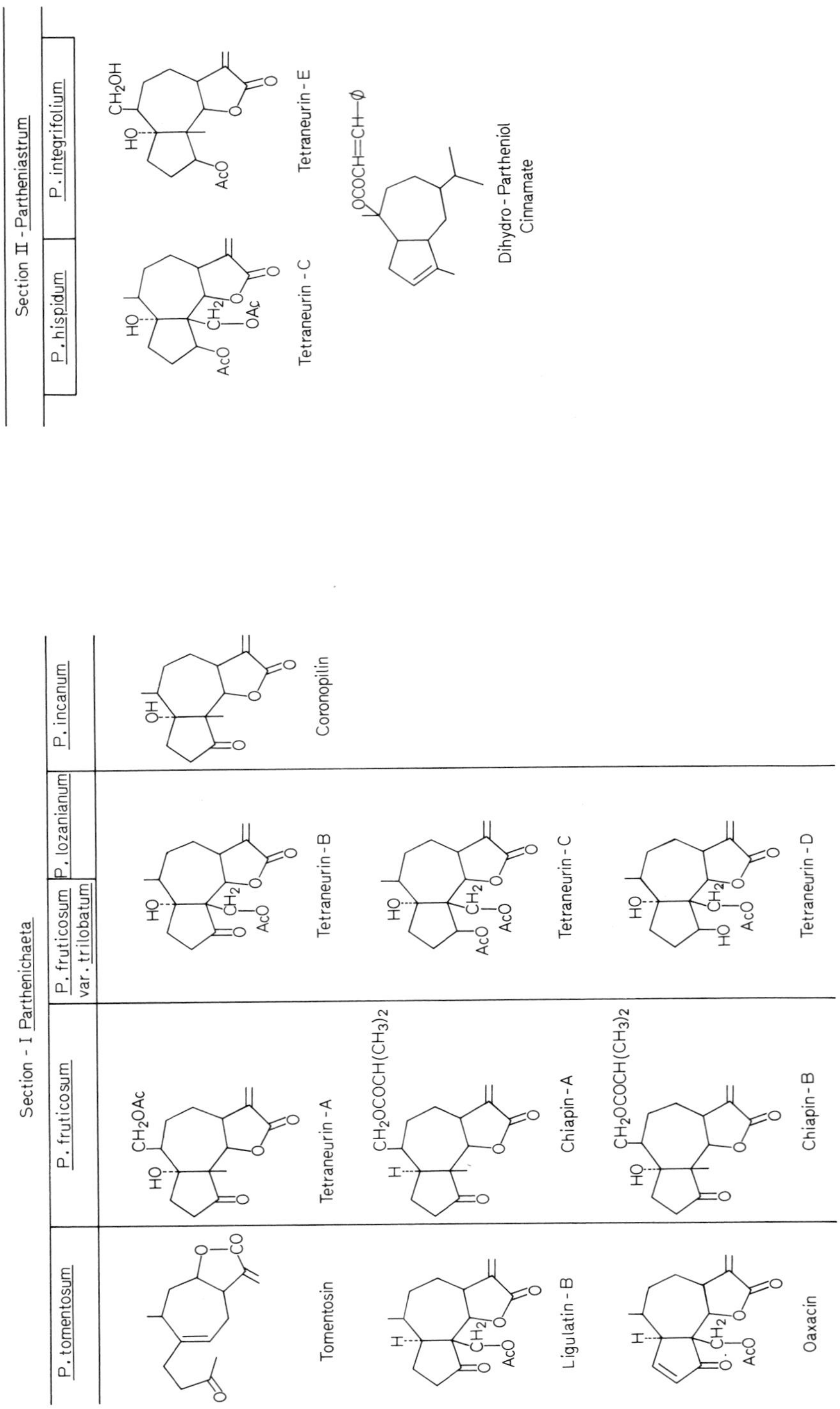
Section - I Parthenichaeta
P. tomentosum
P. fruticosum
P. fruticosum var. trilobatum
P. lozanianum
P. incanum
Tomentosin
Ligulatin - B
Oaxacin
Tetraneurin - A
Chiapin - A
Chiapin - B
Tetraneurin - B
Tetraneurin - C
Tetraneurin - D
Coronopilin
Section II - Partheniastrum
P. hispidum
P. integrifolium
Tetraneurin - C
Tetraneurin - E
Dihydro - Partheniol Cinnamate

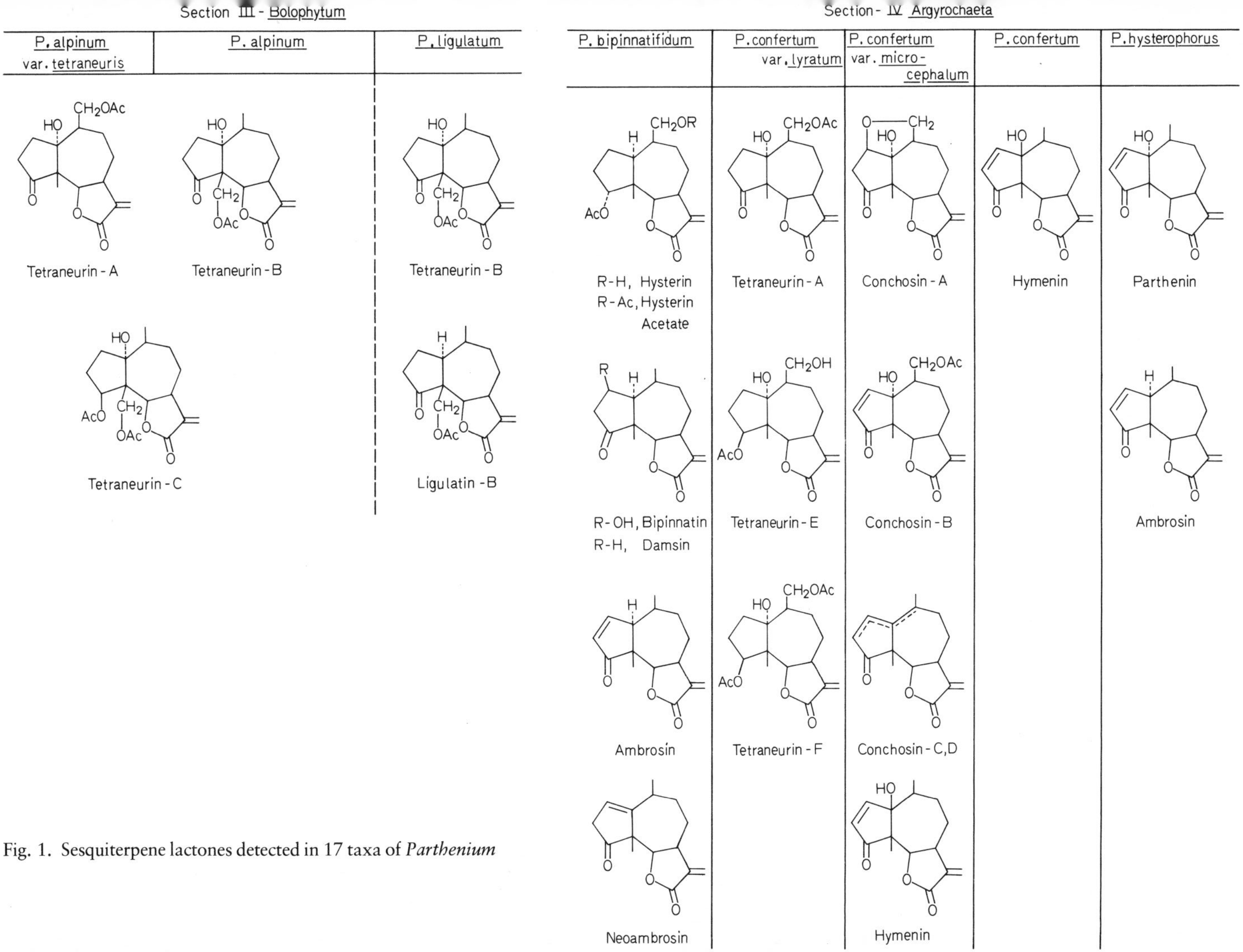

Fig. 1. Sesquiterpene lactones detected in 17 taxa of *Parthenium*

Order Centrospermae

Aizoaceae	Didieraceae
Amaranthaceae	Nyctaginaceae
Basellaceae	Phytolaccaceae
Cactaceae	Portulacaceae
Chenopodiaceae	Stegnospermaceae

Since, so far as it is known, none of the species belonging to the ten betalain families have evolved the pathway to the more common plant pigments, the anthocyanins, it appears that the betalains fill their role in these families. Both anthocyanins and betalains are water-soluble vacuolar pigments, and both types of compounds are considered to be important in the floral colorations associated with attracting pollinators. Also betalains are known to accumulate in leaves and stems and notably at wound and injury sites in plants, thus, it would appear that the unique biogenetic pathway to the betalains (from 3,4-dihydroxyphenylalanine) (MILLER *et al.*, 1968) represents a form of relic chemistry which some species belonging to the phylogenetically-related ten betalain families have now adapted to a variety of problems associated with their survival in different ecosystems.

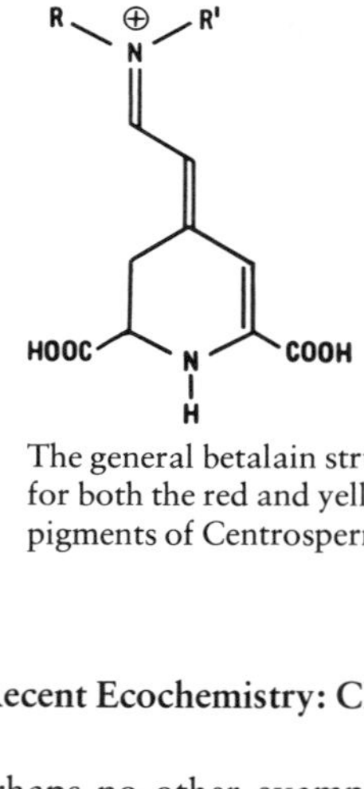

I, The general betalain structure for both the red and yellow pigments of Centrospermae plants.

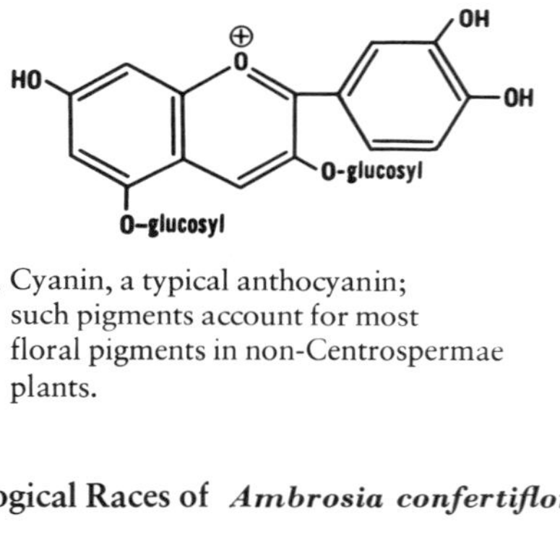

II, Cyanin, a typical anthocyanin; such pigments account for most floral pigments in non-Centrospermae plants.

C. Recent Ecochemistry: Chemical and Cytological Races of *Ambrosia confertiflora*

Perhaps no other example best illustrates the way chemical data representing many populations of a given species permit the recognition of genetically distinct races than do the terpene data for the ragweed *Ambrosia confertiflora* and perhaps no example illustrates the need for correlating the chemical patterns with other elements of the ecosystem.

The sesquiterpene lactone data (MABRY, 1970b; RENOLD, 1970) for more than 250 populations of *Ambrosia confertiflora* indicated the presence of four distinct chemical races in the species (Fig. 2). Populations of *A. confertiflora* from south central Texas are characterized by a chromosome number of n = 66 and by the pseudoguaianolides, confertiflorin and desacetylconfertiflorin (●Fig. 2) while morphologically indistinguishable plants obtained from northeastern Mexico yielded some new germacranolides (★, Fig. 2) along with several eudesmanolides. All of the latter populations which were cytologically examined exhibited a chromosome number of n = 54. A third chemical race based upon sesquiterpene dilactones (✱, Fig. 2) (with mixed chromosome numbers; n=54, n=66) was found to stretch in a narrow band from just east of the Big Bend area of Texas south into Central Mexico along the western slopes of the Sierra Madre mountains. Finally, a fourth major

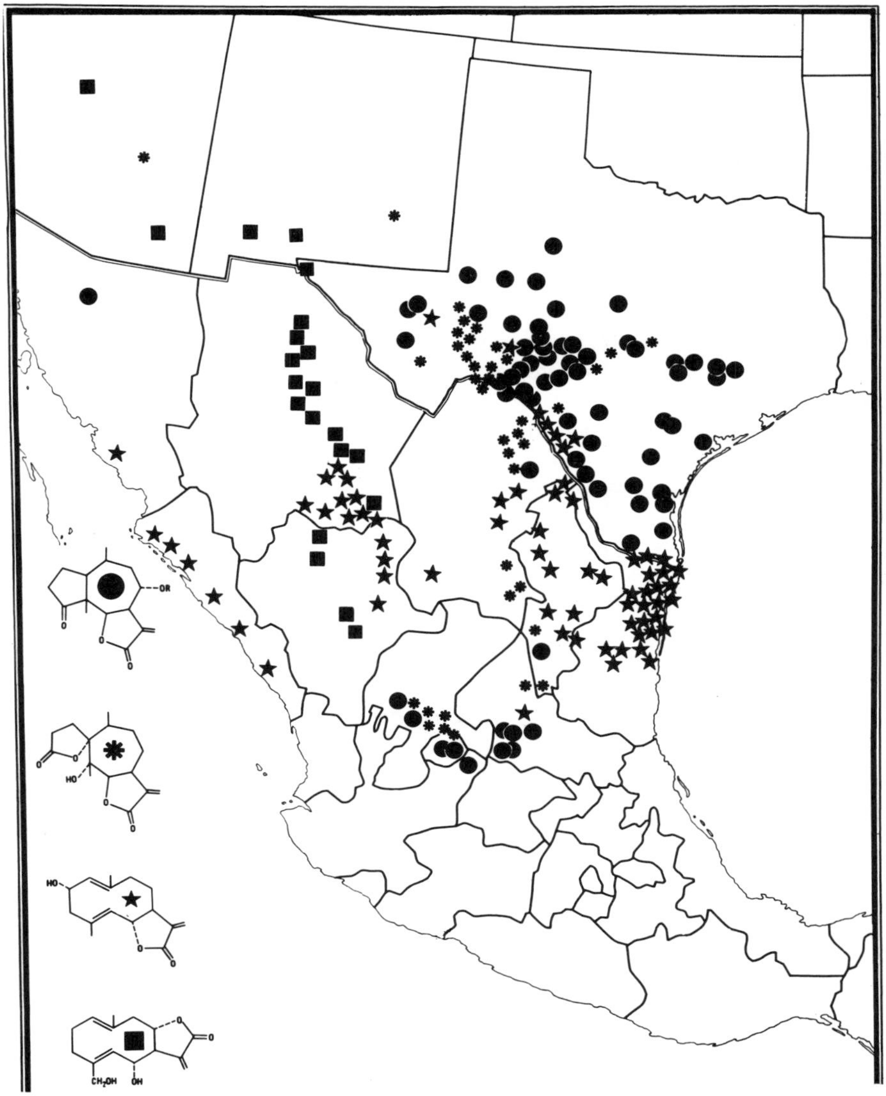

Fig. 2. The distribution of sesquiterpene lactones define four chemical races in the species *Ambrosia confertiflora*. See text for explanation

chemical race was detected in the northwestern part of Mexico; it is characterized by germacranolides (■, Fig. 2) previously encountered in *A. artemisiifolia*.

Despite the wealth of chemical, morphological and cytological information available for *A. confertiflora* nothing is known with regard to the features of the environment which played a role in the selection of the four presumably evolutionarily-recent chemical races. An analysis of the predators on the species along with a study of the climatic and physical features of the four regions associated with the races should be revealing since it is likely that the selection pressures are of recent origin and are still present in the ecosystems.

D. Recent-Ecochemistry: A Quantitative Study of Clinal Variation in *Juniperus virginiana* Using Terpenoid Data

One of the first studies to utilize computerized numerical techniques to analyze terpenoid data was concerned with the question of whether gene flow was occurring between *Juniperus ashei* and *J. virginiana* on a sub-continental scale or whether the morphological variation observed for the species could be explained on the basis of clinal variation and thus reflect clinal selection pressures associated with ecological gradients. FLAKE, VON RUDLOFF and TURNER (1969) analyzed the volatile terpenes in nine populations of *Juniperus virginiana* which were selected at approximately 150-mile intervals along a 1500-mile range from northeastern Texas to Washington, D. C. The resulting data, which were analyzed by computerized numerical classification methods, indicated that clinal variation, rather than hybridization and introgression, could account for the morphological variation observed in *J. virginiana*. Again, the specific features of the ecological gradient which produced by selection processes accounting for the observed morphological and chemical clinal variation remain to be determined.

An Example of the Role of Secondary Compounds in the Co-evolution of Plants and Animals: The Function of Terpenes in the Fragrance of Orchids

In order to determine the chemical bases of plant-plant and plant-insect interactions, more studies such as that of DODSON *et al.* (1969) are required; they dealt with the function of substances which give certain orchid flowers their fragrance:

"Certain orchids in tropical America have become adapted to pollination by male euglossine bees. The bees are attracted by floral fragrances, and the chemical composition of the fragrances determines which species are attracted. The male bees collect the fragrance materials directly from the flower with special tarsal brushes. The bees launch into flight, transfer the fragrance materials, and store them in swollen glandular tibiae of the rear legs. The contents of the tibiae after floral visits were analyzed by gas chromatography; they were the same as the floral fragrance.

Approximately 50 compounds are present in euglossine orchid fragrances, and some species may produce as many as 18 of the compounds. Other species produce fewer compounds. Certain of the compounds, when presented to the bees in tropical America, proved to attract many species of euglossine bees. Other compounds attracted only a few species although in some cases no bees were attracted. Combinations of the compounds attracted markedly fewer species than pure compounds. Appropriate combinations of compounds, in the proportions found in orchid fragrances, attract the same bees which are attracted by the flowers. Speciation and reproductive isolation in euglossine bee-pollinated orchids appears to be based on specific attraction of pollinators to odors produced by the orchid flowers; the substances are believed to play a role in the life cycle of the bees."

Secondary Compounds from the Mediterranean Scrub Vegetation: A Comparison of Micro- and Macro- Chemical and Ecological Gradients

Before it can be determined whether or not the distribution of secondary compounds in the mediterranean scrub vegetation can be correlated with climatic, geological and organismical features of the environment, it is necessary that the chemistry be established on a populational basis for a variety of taxa. In this connection we have tabulated in Table 1 a preliminary survey of the secondary compounds already known from a number of plant species which belong to or are related to this vegetation; a comprehensive literature survey is presently in progress. Although the paucity of the available information is striking, a number of structurally distinct types of chemical constituents do occur.

A. Flavonoids, Other Phenolics and Aromatics

Most phenolic compounds which occur in higher plants are biogenetically related, being derived wholly or in part via the shikimic acid (III) pathway. Thus, it is not surprising that many common phenolics (e.g. the flavonoids IV, VI, VII) are constituents of members of the mediterranean scrub type vegetation; in addition, however, a number of rarer and chemically intriguing types of flavonoids have also been reported; those include the C-glycosides (e.g. V) and the highly methoxylated 6-oxygenated types (e.g. VIII). A variety of other phenolics and aromatics (e.g. IX, X, XIV, XV) have also been detected in species related to the mediterranean scrub vegetation.

III, shikimic acid

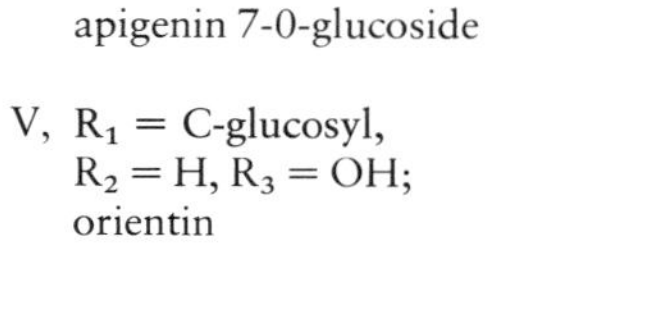

IV, $R_1 = H$, $R_2 =$ glucosyl, $R_3 = H$; apigenin 7-0-glucoside

V, $R_1 =$ C-glucosyl, $R_2 = H$, $R_3 = OH$; orientin

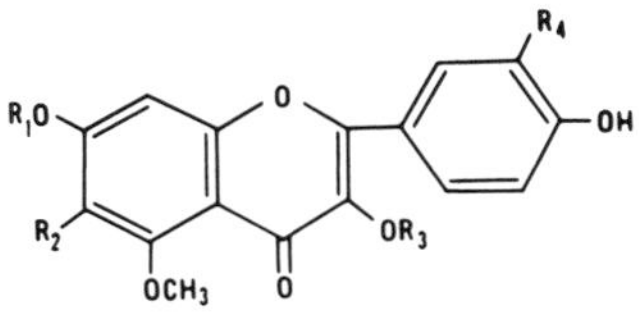

VI, $R_1 = R_2 = R_4 = H$, $R_3 =$ glucosyl; kaempferol 3-0-glucoside

VII, $R_1 = R_2 = H$ $R_3 =$ rutinose, $R_4 = OH$; rutin

VIII, $R_1 =$ glucosyl, $R_2 = R_4 = OCH_3$, $R_3 = CH_3$; jacein

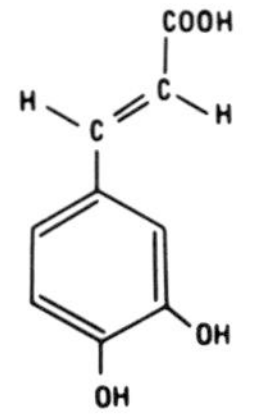

IX, caffeic acid

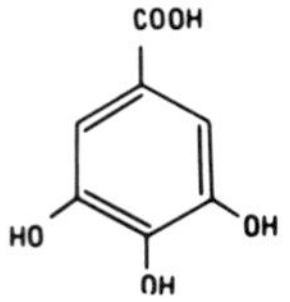

X, gallic acid, a constituent of many tannins

B. Mono and Sesquiterpenes

Perhaps no other group of secondary compounds appear to be so vital to plant evolution and plant development as do the terpenes, all members of which are biogenetically related via the mevalonic acid (XI) pathway. Certain complex mono- and sesqui- terpenes have been used as phylogenetic markers and studies currently underway should reveal more details about the importance of these substances to the role of the species in the ecosystems.

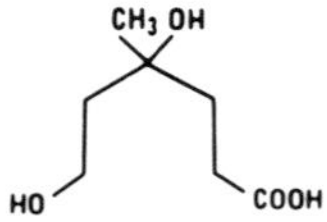

XI, mevalonic acid

Artemisia californica, one of the southern California perennials selected for study in the present program, contains the eudesmanolide-type sesquiterpene lactone artecalin (XII); as noted previously in this chapter, these types of terpenes have been emphasized in our phylogenetic studies of *Ambrosia, Parthenium, Iva,* and other weedy taxa.

It has been reported that *Encelia californica,* another species selected for investigation, produces no sesquiterpene lactones unlike other members of this genus; the major secondary chemical constituents of *E. californica* were two novel aromatic compounds, the benzofuran euparin (XIII) and the chromene encecalin (XIV).

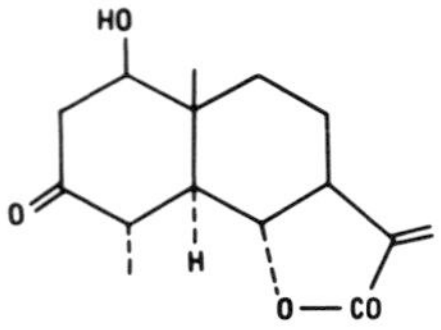

XII, artecalin

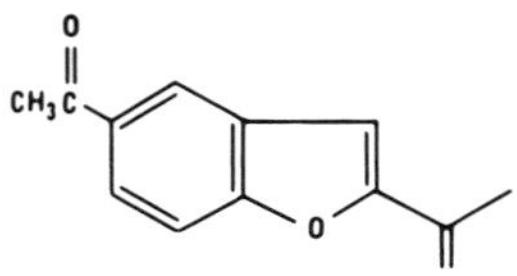

XIII, euparin

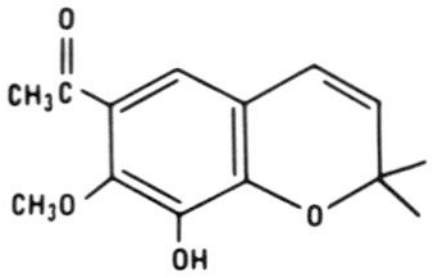

XIV, encecalin

C. Higher Terpenes

Triterpenes and sterols constitute the largest group of higher terpenes and a variety of them have been reported from species related to the mediterranean scrub vegetation (e.g. the sterol XV). Saponins appear to be particularly prevalent in this type of vegetation;

these C_3 -O-glycosides of steroid aglycones such as XVI and triterpenes are water-soluble plant constituents which are distinguished by their ability to form a soapy foam in an aqueous solution. They are commercially important because some of them serve as precursors for synthetic hormones.

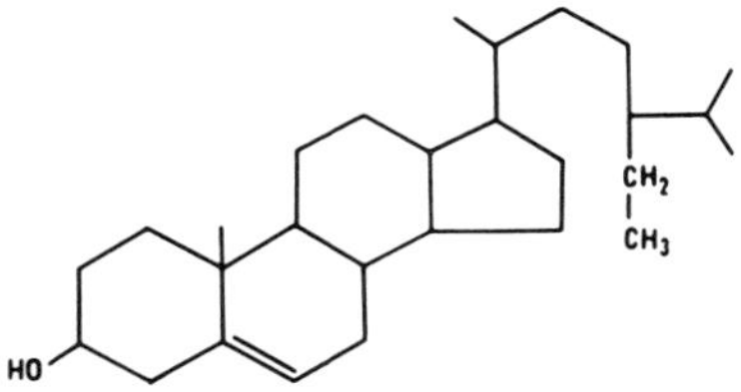

XV, β-sitosterol

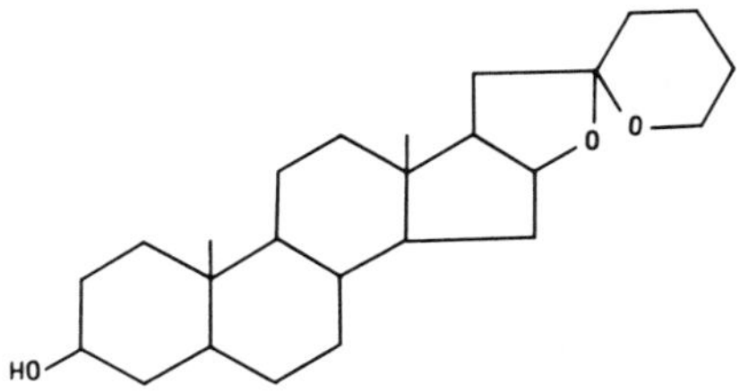

XVI, smilagenin

D. Alkaloids

Most of the members of the mediterranean scrub vegetation either do not contain alkaloids or have not been examined for them. Thus, all the taxa in this vegetation need to be surveyed for alkaloids using a spot test such as the Dragendorff reagent. Among the few nitrogen-containing compounds which have been reported is the modified aporphine alkaloid, aristolochic acid (XVII), which contains a rare (for nature) NO_2 function; this compound and related substances probably play a role in the interaction of certain *Aristolochia* species and some butterflies (such as the relationship between *Aristolochia chilensis* and the butterfly *Papilio psittacos)* (CARBONI *et al.*, 1966; RUVEDA *et al.*, 1968).

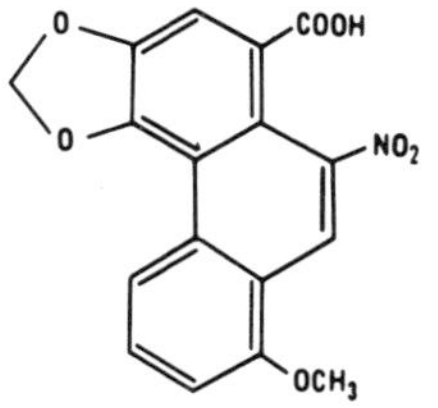

XVII, aristolochic acid

E. Other Secondary Compounds

Finally, a survey for various other classes of secondary compounds such as cyanogenetic glucosides, glucosinolates, quinones, etc. should be conducted on the mediterranean scrub vegetation.

Table 1. Survey of the known chemical constituents from the genera *Artemisia, Bahia, Encelia, Flourensia,* and *Rhus*

(At least one member of these genera occur in the mediterranean-climate regions and has been selected for intensive study; no chemistry has yet been carried out on any member of the genera to which the other species selected for analysis belongs.)

Species	Flavonoids	Other phenolics and aromatics	Volatile terpenes	Sesquiterpene lactones	Higher terpenes including triterpenes	Nitrogen-containing compounds including alkaloids	Miscellaneous
Artemisia abrotanum	rutin artemetin (Oswiecimska *et al.*, 1965)	isofraxidin scopoletin umbelliferone (Nieshulz and Schmersahl, 1968) chlorogenic acid caffeic acid esculetin (Oswiecimska *et al.*, 1965) scopolin calycanthoside scopoletin (Schmersahl, 1966).					
A. absinthium	artemetin rutin (Oswiecimska *et al.*, 1965)	chlorogenic acid caffeic acid (Oswiecimska *et al.*, 1965) absinthin coumarin quebrachitol (Schwaer, 1962)	β-caryophyllene γ-selinene bicabolene (Sorm *et al.*, 1951)	artabsin (Herout *et al.*, 1956)		positive alkaloid test (Allayarov *et al.*, 1965)	

A. annua		pinene cineol artemisia- ketone l-camphor (TAKEMOTO and NAKAJIMA, 1957) camphene cadinene caryophyllene cuminaldehyde (TUCAKOV, 1955)		
A. arborescens	arborescine (MAZUR and MEISELS, 1956) artemetin (MAZUR and MEISELS, 1955)			acetyl-3-hydroxy-5-propynylthiophene, acetyl-5-propynylthiophene-3-methyl ether (BOHLMANN *et al.*, 1962)
A. arbuscula			arbusculin-A, -B, and -E (IRWIN and GEISSMAN, 1969b)	
A. arctica	3′,6-dimethoxy-4′,5,7-trihydroxy-flavone (HERZ *et al.*, 1970)			
A. austriaca			austricin (RYBALKO and BAN'KOVSKII, 1964) leucomysin (RYBALKO, 1963)	

Species	Flavonoids	Other phenolics and aromatics	Volatile terpenes	Sesquiterpene lactones	Higher terpenes including triterpenes	Nitrogen-containing compounds including alkaloids	Miscellaneous
A. balchanorum			citral geraniol l-linaloöl (RAFANOVA *et al.*, 1956)	hydroxycostunolide (SUCHY *et al.*, 1963) costunolide balchanolide isobalchanolide hydroxybalchanolide balchanin (SUCHY, 1962 a)			
A. bigelovi				arbiglovin (HERZ and SANTHANAM, 1965)			
A. brevifolia			camphene cineol thujone (–)-camphor p-cymene (COCKER *et al.*, 1958)				
*A. californica**			(+)-camphor (–)-camphor (BANTHORPE and BAXENDALE, 1970)	artecalin (GEISSMAN *et al.*, 1969)			
A. campestris	rutin (OSWIECIMSKA *et al.*, 1965)	chlorogenic acid caffeic acid (OSWIECIMSKA *et al.*, 1965)	α-pinene β-pinene 1,8-cineol l-thujone thujyl alcohol geraniol				

A. camphorata	isofraxidin (DANIELAK and BORKOWSKI, 1970)		
A. cana			ridentin artecanin canin matricarin deacetylmatricarin (LEE *et al.*, 1969)
A. capillaris	esculetin 6,7-dimethyl ether (IMAI and SAMPEI, 1952)	capillin (HARADA, 1957) capillarin (HARADA *et al.*, 1960) capillon (HARADA, 1956) capillene (HARADA, 1954)	
A. caruthii		chamazulene (PENUNURI and STEELINK, 1962)	matricin (SPITZER and STEELINK, 1967)
A. chamaemelifolia	caffeic acid scopoletin scopolin (BANDYUKOVA and KONOVALOVA, 1970)		
A. chasarica			tauremisin (ABBASOV *et al.*, 1964)
A. cina	cineol l-camphor carvacrol sesquiartemisol (RYBALKO and BAN'KOVSKII, 1959)		artemisin santonin (SABER and KASIM, 1962)

Species	Flavonoids	Other phenolics and aromatics	Volatile terpenes	Sesquiterpene lactones	Higher terpenes including triterpenes	Nitrogen-containing compounds including alkaloids	Miscellaneous
A. compacta				santonin (Rybalko *et al.*, 1963 a)			
A. diffusa		methylumbelliferone (Tarasov *et al.*, 1969)					
A. douglasiana				argalanin (Matsueda and Geissman, 1967 a) douglanin (Matsueda and Geissman, 1967 b)			
A. dracunculus	quercitin hyperoside isorhamnetin-7-0-galactopyranoside (Chumbalov *et al.*, 1969 b) rutin (Ryakhovaskaya *et. al.*, 1970)	scopoletin (Mallabaev *et al.*, 1969) artemidin Mallabaev *et al.*, 1970) chlorogenic acid caffeic acid esculetin (Oswiecimska *et al.*, 1965)	pinene d-sabinene myrcene menthol anethole anisol anisic acid methyl-amyl-ketone (Khisamutdinov and Goryaev, 1959)		β-sitosterol (Mallabaev *et al.*, 1969)		
A. dracunuculoides		7,8-methylenedioxy-6-methoxycoumarin γ, γ-dimethyl ether					

	of scopoletin scopoletin scoparone daphnetin methylene-ether daphnetin 7-methyl ether (HERZ *et al.*, 1970)		
A. feddei		Yomogi alcohol (YAYASHI *et al.*, 1968)	
A. ferganensis		l-camphor α-thujone DL-carvone 1,8-cineol borneol (TIKHONOVA and GORYAEV, 1957)	
A. fragrans			erivanine (RYBALKO and ABBASOV, 1963) tauremisin (ABBASOV *et al.*, 1964)
A. franserioides			artefransin (LEE and GEISSMAN, 1971)
A. halophila			α-santonin artemin (ARKHIPOVA *et al.*, 1970)

Species	Flavonoids	Other phenolics and aromatics	Volatile terpenes	Sesquiterpene lactones	Higher terpenes including triterpenes	Nitrogen-containing compounds including alkaloids	Miscellaneous
A. hanseniana				tauremisin (ABBASOV *et al.*, 1964)			
A. incana				deacetylmatricarin (REVAZOVA *et al.*, 1970) santonin (FU *et al.*, 1963)			
A. jacutica				siversipin ketopelenolide (BENESOVA *et al.*, 1969)			
A. judaica				judaicin (SABER and KASSIM, 1962)			
A. juncea				deacetylmatricarin (KECHATOVA and VLASOV, 1966)			
A. klotzchiana				matricarin deacetylmatricarin chrysartemin-A (ROMO *et al.*, 1970)			
A. kurramensis			camphene cineol	α-santonin β-santonin			

	thujone (-)-camphor (COCKER *et al.*, 1958) α-pinene β-pinene (KURODA, 1963)	(KURODA, 1963) lumisantonin (SATODA *et al.*, 1959)
A. lagocephala	l-camphor l-borneol (BELOVA, 1961)	
A. lercheana		deacetylmatricarin (RYBALKO *et al.*, 1963 a)
A. leucoides		deacetoxymatricarin (HOLUB and HEROUT, 1962) austricin leucomysin (RYBALKO, 1963)
A. ludoviciana		ludovicin-A ludovicin-B ludovicin-C douglanin (LEE and GEISSMAN, 1970)
A. macrocephala	chamazulene (NAZARENKO, 1961)	
A. maritima		l,β-santonin (RYBALKO *et al.*, 1963 a)

Species	Flavonoids	Other phenolics and aromatics	Volatile terpenes	Sesquiterpene lactones	Higher terpenes including triterpenes	Nitrogen-containing compounds including alkaloids	Miscellaneous
A. mendozana			thujone 1-camphor (FESTER *et al.*, 1958)				
A. messershmidiana		esculetin 6-methyl-ether esculetin 7-methyl-ether (HAHN, 1966)					
A. mexicana				arglanin douglanin chrysartemin-A (ROMO *et al.*, 1970) estafiatin (SANCHEZ-VIESCA and ROMO, 1963)			
A. monogyna				mibulactone (KARIYONE *et al.*, 1949) monoginin artemisic acid (KARIYONE *et al.*, 1948)			
A. nova				cumambrin-B cumambrin-A 8-deoxycumambrin-B (IRWIN and GEISSMAN, 1969 a)			

A. pallens			davonone (SIMPA and VAN DER WAL, 1968)	
A. persica		scopoletin (RYBALKO *et al.*, 1964)		
A. princeps				yomogin (GEISSMAN, 1966)
A. rutaefolia			butyric acid thujone cineol α-terpenyl butyrate d-α-pinene l-camphor (GORYAEV and GIMADDINOV, 1959)	
A. sacrorum	genkwanin (CHANDRASHEKAR *et al.*, 1965)	umbelliferone (CHANDRASHEKAR *et al.*, 1965)		
A. santolinaefolia		scopoletin (RYBALKO *et al.*, 1964; BABAKHODZHAEV *et al.*, 1970) umbelliferone (BABAKHODZHAEV *et al.*, 1970)	thujone fenchone camphor thujyl alcohol borneol cineol fenchyl alcohol (GORYAEV *et al.*, 1956)	
A. scoparia		scoparone (TOMOVA, 1964; BAN'KOVSKAYA and	α-pinene β-pinene agropinene	

Species	Flavonoids	Other phenolics and aromatics	Volatile terpenes	Sesquiterpene lactones	Higher terpenes including triterpenes	Nitrogen-containing compounds including alkaloids	Miscellaneous
		BAN'KOVSKII, 1959) chlorogenic acid (HU *et al.*, 1965)	1-phenyl-2,4-hexadiyne (GOL'MOV and AFANAS'EV, 1957) butyraldehyde furfuraldehyde methyl heptenone 1,8-cineol carvone l-thujone α-thujadiacarboxylic-acid l-thujyl alcohol geranyl acetate eugenol cadinene (DAKSHINAMURTI, 1953)				
A. selengensis							trans-1,9-heptadecadien-11,13,15-triyn-8-ol trans, trans-Me(C : C)$_3$-(CH:CH)$_2$-CHOH(CH$_2$)$_3$-CH:CH$_2$ (BOHLMANN and RODE, 1967)

A. sieversiana	rutin isoquercitrin (BEREZOVSKAYA and VELIKHANOVA, 1968)	4-hydroxy-8-acetoxy-guaia-1(2),9(10)-dien-6,12-olide (NAZARENKO, 1965)	sieversinin (NAZARENKO and LEONT'EVA, 1966) artabsin absinthin (NOVOTNY and HEROUT, 1962)
A. spicigera			α-santonin β-santonin (ABBASOV *et al.*, 1965)
A. taurica		sabinol 1,8-cineol camphene myrcene p-cymol α-pinene β-pinene α-thujone farnesene α-betulenol α-kastol (KOZHINA *et al.*, 1968)	tauremisin (vulgarin) (RYBALKO and BANKOVSKII, 1964) artemin (TOLSTYKH *et al.*, 1968) mibulactone (RYBALKO, 1965) taurin (KECHATOVA *et al.*, 1968)
A. tenuisecta			mibulactone (KASYMOV and SIDYAKIN, 1969)
A. tilesii			matricarin artilesin deacetylmatricarin (HERZ and UEDA, 1961)

Species	Flavonoids	Other phenolics and aromatics	Volatile terpenes	Sesquiterpene lactones	Higher terpenes including triterpenes	Nitrogen-containing compounds including alkaloids	Miscellaneous
A. transiliensis	isovitexin violanthin (CHUMBALOV and FADEEVA, 1970) genkwanin acacetin 3-0-methyl-quercetin transilitin (CHUMBALOV and FADEEVA, 1969) quercetin rutin (CHUMBALOV and FADEEVA, 1967)		m-cresol α-pinene cineol thujyl alcohol thujone cumaldehyde (GORYAEV and PUGACHEV, 1955)	transilin (CHUMBALOV *et al.*, 1969 a)			
A. tridentata				ridentin (IRWIN *et al.*, 1969) matricarin (GEISSMAN *et al.*, 1967)			
A. tripartita				cumambrin-A 8-deoxycumambrin-B cumambrin-B (IRWIN and GEISSMAN, 1969 a) colartin (IRWIN and GEISSMAN, 1969 b)			

		ridentin (IRWIN *et al.*, 1969) artecalin (GEISSMAN *et al.*, 1969)
A. vachanica	vachanoic acid (RYBALKO *et al.*, 1963b)	
A. verlotorum		verlotorin artemorin anhydroverlotorin (GEISSMAN, 1970) anhydroartemorin (GEISSMAN and LEE, 1971)
A. vestita	n-caprylaldehyde α-d-phellandrene 1,8-cineol α-terpinene thujone thujyl alcohol citronellal citral citronellol geraniol aromadendrene cadinene chamazulene (SINHA and CHAUHAN, 1970)	

Species	Flavonoids	Other phenolics and aromatics	Volatile terpenes	Sesquiterpene lactones	Higher terpenes including triterpenes	Nitrogen-containing compounds including alkaloids	Miscellaneous
A. vulgaris		chlorogenic acid caffeic acid (OSWIECIMSKA *et al.*, 1965)	(-)-thujone α-amyrin α-pinene β-pinene (KUNDU *et al.*, 1969)	vulgarin (GEISSMAN and ELLESTAD, 1962)	fernenol (KUNDU *et al.*, 1968) stigmasterol β-sitosterol (KUNDU *et al.*, 1969)		
Bahia pringlei				bahia-I bahia-II (ROMO DE VIVAR and ORTEGA, 1969)			
*Encelia californica**				no sesquiterpene lactones (BJELDANES and GEISSMAN, 1969)			euparin encalin euparone methyl-ether (BJELDANES and GEISSMAN, 1969)
E. farinosa		3-acetyl-6-methoxy-benzaldehyde (GRAY and BONNER, 1948)		farinosin encelin (GEISSMAN and MUKHERJEE, 1968)			
Flourensia cernua	cirsimaritin hispidulin (RAO *et al.*, 1970)						
Rhus ambigua	rhoifolin myricitrin (MATSUDA, 1966)	corilagin shikimic acid (MATSUDA, 1966)					

R. chinensis	quercitrin myricetin myricitrin (MATSUDA, 1966)		
R. coriaria	kaempferol myricetin quercetin (EL SISSI *et al.*, 1966)	gallic acid m-digallic acid methyl gallate ellagic acid (EL SISSI *et al.*, 1966)	
R. cotinus		gallic acid (MURKO, 1965)	sulfuretin (KING and WHITE, 1961)
R. javanicus	quercitrin (ARITOMI *et al.*, 1964)	methyl gallate (ARITOMI *et al.*, 1964)	
R. glabra	fustin (KEPPLER, 1957)		
R. striata		1,2-dihydroxy-3-(pentadec-8-enyl)-benzene 1,2-dihydroxy-3-(pentadec-8,11-dienyl)-benzene 1,2-dihydroxy-3-(pentadec-8,11,14-trienyl)-benzene (NAKANO *et al.*, 1970)	
R. toxicodendrons		1,2-dihydroxy-3-pentadecyl benzene (FESCO, 1951)	

Species	Flavonoids	Other phenolics and aromatics	Volatile terpenes	Sesquiterpene lactones	Higher terpenes including triterpenes	Nitrogen-containing compounds including alkaloids	Miscellaneous
R. trichocarpa	fustin (HAYASHI *et al.*, 1950) apigenin-7-0-rutinoside (MATSUDA, 1966)						
R. typhina	myricitroside (PLOUVIER, 1970)	gallic acid ellagic acid (TISCHER, 1960)					
R. vernicifera	fisetin fustin (HASEGAWA and SHIRATO, 1951) luteoloside (PLOUVIER, 1970)						

* -Cited by MOONEY *et al.* (1970) as occurring in the mediterranean regions of California or Chile.

F. Chemical Studies Over Micro- and Macro-ecological Gradients for Evolutionarily Distinct but Either Functionally, Morphologically and/or Niche Alike Species from California and Chile

Current studies in our laboratory are concerned with bringing chemical evidence to bear upon the question: "How did the mediterranean scrub vegetation originate and evolve" by comparing the chemical gradients for the dominant plants (which occur in the mediterranean climatic regions of both Chile and California) with ecological gradients, both on micro (i.e. patterns over selected intensive study sites) and macro levels (i.e. patterns over widely separated and distinct ecosystems). In this connection, a number of systems have been selected for chemical and morphological studies as well as associated predator and pollinator analysis; for example, in the four systems which are noted below, each contains a taxon from California and one from Chile, and, in each case, the two taxa are evolutionarily distinct (but within the same family) and show either morphological, functional or habitat similarities. The variation in the secondary plant chemistry for each species will be correlated with micro- and macro-ecological gradients.

Family	California	Chile
1. Rosaceae (evergreen shrubs):	*Heteromeles arbutifolia*	*Kageneckia oblonga*
2. Anacardiaceae (evergreen shrubs):	*Rhus ovata*	*Lithraea caustica*
3. Compositae (drought deciduous shrubs):	*Artemisia californica*	*Bahia ambrosioides*
4. Compositae (drought deciduous shrubs):	*Encelia californica*	*Flourensia thurifera*

The selected species represent dominant and characteristic members of the different types of California and Chile vegetation; namely, evergreen scrubs, semi-arid coastal scrubs, and semi-arid scrubs with succulents.

Acknowledgements

Much of the research described herein is currently being supported in part by the National Science Foundation Grants (GB-27152 and GB 29576X) and the National Institutes of Health Grant (HD-04488) and the Robert A. Welch Foundation Grant (F-130). T. J. M. wishes to thank his many co-workers for their invaluable research contributions.

References

ABBASOV, R. M., ISMAILOV, N. M., RYBALKO, K. S.: Occurrence of tauremisin in Azerbaidzhan species of worm wood. Izv. Akad. Nauk Azerb. SSR, Ser. Biol. Nauk **4**, 31–35 (1964).

ABBASOV, R. M., ISMAILOV, N. M., RYBALKO, K. S.: The lactone constituents of *Artemisia spicigera*. The dynamics of accumulation of lactones, essential oils, and "phenols" and their interrelations. Vopr. Eksperim. Botan., Sb. 5–11 (1965).

ALLAYAROV, Kh., KHAMIDKHODZHAEV, S. A., KORTKOVA, E. E.: Alkaloids-containing plants of Turkmenian S.S.R. Izv. Akad. Nauk Turkm. SSR, Ser. Biol. Nauk **4**, 62–65 (1965).

ARITOMI, M., MIYAZAKI, K., MAZAKI, T.: Constituents in the leaves of *Rhus javanicus* and *Melia azedarach* var. *subtripinnata*. Yakugaku Zasshi **84**, 894–895 (1964).

ARKHIPOVA, L. I., KASYMOV, Sh. Z., SIDYAKIN, G. P.: Sesquiterpenoid lactones from *Artemisia halophila*. Khim. Prir. Soedin. **6**, 480–481 (1970).

BABAKHODZHAEV, A., KASYMOV, Sh. Z., SIDYAKIN, G. P.: Courmarins, from *Artemisia santolinaefolia*. Khim. Prir. Soedin. **6**, 363–364 (1970).

BANDYUKOVA, V. A., KONOVALOVA, O. A.: Coumarins from *Artemisia chamaemelifolia*. Khim. Prir. Soedin. **6**, 266 (1970).

BAN'KOVSKAYA, A. N., BAN'KOVSKII, A. I.: Chemical study of *Artemisia scoparia*. Trudy Vsesoyus. Nauch. – Issledovatel. Inst. Lekarstv. i Aromat. Rast. **11**, 177–179 (1959).

BANTHORPE, D., BAXENDALE, D.: Terpene biosynthesis. III. Biosynthesis of (+)- and (−)-camphor in *Artemisia, Salvia,* and *Chrysanthemum* species. J. Chem. Soc. c. **19**, 2694–2696 (1970).

BARZ, W., GRISEBACH, H.: Enzymology of flavonoid biosynthesis and metabolism. Phytochemical Section, Botanical Society of America, Newsletter **3**, 4–26 (1970).

BELOVA, N. V.: Preliminary analysis of the composition of the essential oils of *Artemisia lagocephala*. Zhur. Priklad. Khim. **34**, 707–709 (1961).

BENESOVA, V., NAZARENKO, M. V., SLEPTSOVA, L. V.: Sesquiterpenic-lactones from *Artemisia jacutica*. Khim. Prir. Soedin. **5** (3), 186 (1969).

BEREZOVSKAYA, T. P., VELIKHANOVA, V. I.: Phytochemical analysis of *Artemisia sieversiana* grass gathered in the vicinity of Tomsk. Tr. Nauch. Konf. Tomsk. Otd. Vses. Ihim. Obshchest. **1**, 271–273 (1968).

BJELDANES, L. F., GEISSMAN, T. A.: Euparinoid constituents of *Encelia californica* Nutt. Phytochemistry **8**, 1293–1296 (1969).

BOHLMANN, F., RODE, K. M.: Polyacetylene compounds. CXXXII. Components of *Artemisia selengensis*. Chem. Ber. **100**, 1940–1943 (1967).

BOHLMANN, F., KLEINE, K. M., BORNOWSKI, H.: Polyacetylene derivatives. XLI. Two thiophene ketones from *Artemisia arborescens*. Chem. Ber. **95**, 2934–2938 (1962).

CARBONI, S., LIVI, O., SEGNINI, D., MAZZANTI, L.: Constituents of *Aristolochia rotunda*. Gazz. Chim. Ital. **96**, 641–661 (1966).

CHANDRASHEKAR, V., KRISHMAMURTI, M., SESHARDI, T. R.: Chemical investigation of *Artemisia sacrorum*. Current Sci. (India), **34**, 609 (1965).

CHUMBALOV, T. K., FADEEVA, O. V.: Flavonoids of *Artemisia transiliensis*. Khim. Prir. Soedin **3**, 281 (1967).

CHUMBALOV, T. K., FADEEVA, O. V.: Flavonoids of *Artemisia transiliensis*. Khim. Prir. Soedin **5**, 439 (1969).

CHUMBALOV, T. K., FADEEVA, O. V.: Glycoflavonoids from *Artemisia transiliensis*. IV. Khim. Prir. Soedin **6**, 364–365 (1970).

CHUMBALOV, T. K., FADEEVA, O. V., CHANYSHEVA, I. S.: Flavonoids of *Artemisia transiliensis*. II. New flavonoid glycoside, transilin. Khim. Prir. Soedin **5**, 236–239 (1969 a).

CHUMBALOV, T. K., MUKHAMED'YAROVA, M. M., FADEEVA, O. V.: Flavonoids from *Artemisia dracunculus*. Khim. Prir. Soedin. **5**, 323 (1969 b).

COCKER, W., LIPMAN, C., MCMURRY, T. B. H., WHEELER, B. M.: Volatile oils of *Artemisia brevifolia* and *A. kurramensis*. J. Sci. Food Agr. **9**, 826–836 (1958).

DAKSHINAMURTI, K.: Chemical constituents of the oil of *Artemisia scoparia*. Indian Pharmacist **8**, 257–260 (1953).

DANIELAK, R., BORKOWSKI, B.: Isolation of isofraxidin from herbs of *Artemisia abrotanum* and a search for it in herbs of some other *Artemisia* species. Diss. Pharm. Pharmacol. **22**, 231–235 (1970).

DODSON, C. H., DRESSLER, R. L., HILLS, H. G., ADAMS, R. M., WILLIAMS, N. H.: Biologically active compounds in orchid fragrances. Science **164**, 1243–1249 (1969).

EL SISSI, H. I., SALEH, N. A. M., ABD EL WAHID, M. S.: The tannins of *Rhus coriaria* and *Mangifera indica*. Planta Med. **14**, 222–231 (1966).

FESCO, E. J.: Poison ivy, poison oak, poison sumac. Mendel Bull. **16**, 14–17 (1951).

FESTER, G. A., MARTINUZZI, E. A., RETAMAR, J. A., RICCARDI, A. I. A.: Essential oils from Argentine plants. Bol. acad. nacl. cienc. **40**, 189–208 (1958).

FLAKE, R. H., VON RUDLOFF, E., TURNER. B. L.: Quantitative study of clinal variation in *Juniperus virginiana* using terpenoid data. Proc. Nat. Acad. Sci., U.S. **64**, 487–494 (1969).

FU, F., CHEN, Y., SHANG, T.: Isolation of santonin from *Artemisia incana* Keller. Yao Hsueh Hsueh Pao **10**, 140–146 (1963).

GEISSMAN, T. A.: Sesquiterpenoid lactones of *Artemisia* species. I. *Artemisia princeps*. J. Org. Chem. **31**, 2523–2526 (1966).

GEISSMAN, T. A.: Sesquiterpene lactones of *Artemisia: A. verlotorum* and *A. vulgaris*. Phytochemistry **9**, 2377–2381 (1970).

GEISSMAN, T. A., ELLESTAD, G. A.: Vulgarin, a sesquiterpene lactone from *Artemisia vulgaris* L. J. Org. Chem. **27**, 1855–1859 (1962).

GEISSMAN, T. A., LEE, K. H.: Sesquiterpene lactones of *Artemisia*: Artemorin and dehydroartemorin (anhydroverlotorin). Phytochemistry **10**, 419–420 (1971).

GEISSMAN, T. A., MUKHERJEE, R.: Sesquiterpene lactones of *Encelia farinosa*. J. Org. Chem. **33**, 656–660 (1968).

GEISSMAN, T. A., STEWART, T., IRWIN, M. A.: Sesquiterpene lactones of *Artemisia* species. II. *Artemisia tridentata* subspecies *tridentata*. Phytochemistry **6**, 901–902 (1967).

GEISSMAN, T. A., GRIFFIN, T. S., IRWIN, M. A.: Sesquiterpene lactones of *Artemisia*. Artecalin from *A. californica, A. tripartita* ssp. *rupicola*. Phytochemistry **8**, 1297–1300 (1969).

GOL'MOV,V. P., AFANAS'EV, N. M.: Hydrocarbon $C_{12}H_{10}$ from the essential oil of *Artemisia scoparia*. Zhur. Obshchei Khim. **27**, 1698–1703 (1957).

GORYAEV, M. I., GIMADDINOV, Zh. K.: Essential oil of *Artemisia rutaefolia*. Zhur. Priklad. Khim. **32**, 1878–1880 (1959).

GORYAEV, M. I., PUGACHEV, M. G.: Essential oil of *Artemisia transiliensis*. Zhur. Obshchei Khim. **25**, 172–177 (1955).

GORYAEV, M. I., KRUGLYKHINA, G. K., PUGACHEV, M. G., SHABANOV, I. M.: Ethereal oil of *Artemisia santolinaefolia*. Izvest. Akad. Nauk Kazakh. S. S. R., Ser. Khim. (**9**), 33–42 (1956).

GRAY, R., BONNER, J.: Structure determination and synthesis of a plant growth inhibitor, 3-acetyl-6-methoxybenzaldehyde, found in the leaves of *Encelia farinosa*. J. Am. Chem. Soc. **70**, 1249–1253 (1948).

GUVEN, K. C.: Turkish *Artemisia* species. II. *Artemisia campestris*. Folia Pharm. (Istanbul) **5**, 586–591 (1963).

HAHN, D. R.: Biochemical studies on the constituents of *Artemisia messerschmidiana*. I. Identification of esculetin methyl ethers and their cholagogic action. Yakhak Hoeji **10**, 20–24 (1966).

HARADA, R.: Essential oil of *Artemisia capillaris*. I. Chemical structure of capillene. J. Chem. Soc. Japan **75**, 727–732 (1954).

HARADA, R.: Essential oil of *Artemisia capillaris*. II. The structure of capillon. Nippon Kagaku Zasshi **77**, 990–993 (1956).

HARADA, R.: Essential oil of *Artemisia capillaris*. IV. Further studies on the structure of capillin. Nippon Kagaku Zasshi **78**, 415–417 (1957).

HARADA, R., NOGUCHI, S., SOGIYAMA, N.: Essential oil of *Artemisia capillaris*. VI. The structure of capillarin. Nippon Kagaku Zasshi **81**, 654–658 (1960).

HASEGAWA, M., SHIRATO, T.: Phenolic substances of wood. I. The pigments of heartwood of *Rhus vernicifera*. J. Chem. Soc. Japan, Pure Chem. Sect., **72**, 223–224 (1951).

HAYASHI, K., ISAKA, T., SUZUSHINO, G.: Chemical identification of vegetable dyes used on ancient Japanese silk (a preliminary report). Misc. Repts. Research Inst. Nat. Resources **17–18**, 33–42 (1950).

HEROUT, V., DOLEJS, L., SORM, F.: The structure of artabsin, the prochamazulenogen from *Artemisia absinthium*. Chem. and Ind., 1236 (1956).

HERZ, W., SANTHANAM, P. S.: Arbiglovin. A new guaianolide from *Artemisia bigelovi*. J. Org. Chem. **30**, 4340–4342 (1965).

HERZ, W., UEDA, K.: The sesquiterpene lactones of *Artemisia tilesii* Ledeb. J. Am. Chem. Soc. **83**, 1139–1143 (1961).

HERZ, W., BHAT, S. V., SANTHANAM, P. S.: Coumarins of *Artemisia dracunculoides* and 3′,6-dimethoxy-4′,5,7-trihydroxyflavone in *A. arctica*. Phytochemistry **9**, 891–894 (1970).

HOLUB, M., HEROUT, V.: Terpenes. CXLVII. Isolation of deacetoxymatricarin from *Artemisia leucoides*. Collection Czech. Chem. Commun. **27**, 2980–2981 (1962).

HU, J., LI, P., CHEN, M.: Chloretic principles of *Artemisia scoparia*. I. Water-soluble fraction of its extract. Yao Hsueh Hsueh Pao **12**, 289–294 (1965).

IMAI, K., SAMPEI, N.: Seasonal variation of esculetin 6,7-dimethyl ether content of *Artemisia capillaris*. Ann. Rept. Takamine Lab. **4**, 54–59 (1952).

IRWIN, M. A., GEISSMAN, T. A.: Sesquiterpene lactones. Constituents of *Artemisia nova* and *A. tripartita* subspecies *rupicola*. Phytochemistry **8**, 305–311 (1969 a).

IRWIN, M. A., GEISSMAN, T. A.: Sesquiterpene lactones of *Artemisia* species, new lactones from *A. arbuscula* subspecies *arbuscula* and *A. tripartita* subspecies *rupicola*. Phytochemistry **8**, 2411–2416 (1969 b).

IRWIN, M. A., LEE, K. H., SIMPSON, R. F., GEISSMAN, T. A.: Sesquiterpene lactones of *Artemisia*. Ridentin. Phytochemistry **8**, 2009–2012 (1969).

KARIYONE, T., FUKUI, T., KUGUCHI, T., ISHIMASA, M., MIKI, K.: Constituents of "Mibuyomogi", *Artemisia monogyna*. I. J. Pharm. Soc. Japan **68**, 269–270 (1948).

KARIYONE, T., FUKUI, T., ISHIMASA, M., IMAWAKI, T.: Constituents of *Artemisia monogyna*. J. Pharm. Soc. Japan **69**, 310–312 (1949).

KASYMOV, Sh. Z., SIDYAKIN, G. P.: Lactones from *Artemisia tenuisecta*. Khim. Prir. Soedin **5**, 455 (1969).

KECHATOVA, N. A., VLASOV, M. I.: Sesquiterpene lactone from *Artemisia juncea*. Khim. Prirodn. Soedin., Akad. Navk Uz. SSR **2**, 216 (1966).

KECHATOVA, N. A., RYBALKO, K. S., SCHEICHENKO, V. I., TOLSTYKH, L. P.: Sesquiterpene lactones from *Artemisia taurica*. Khim. Prir. Soedin. **4**, 205–207 (1968).

KEPPLER, H. H.: The isolation and constitution of mollisacacidin, a new leucoanthocyanidin from the heartwood of *Acacia mollisima*. J. Chem. Soc. 2721–2724 (1957).

KHISAMUTDINOV, F. S., GORYAEV, M. I.: Chemical composition of the essential oil from *Artemisia dracunculus* subspecies *turkestanica*. Izvest. Akad. Nauk Kazakh. S. S. R., Ser. Khim. (2), 89–97 (1959).

KING, H. G. C., WHITE, T.: Coloring matter of *Rhus cotoinus* wood. J. Chem. Soc., 3538–3539 (1961).

KOZHINA, I. S., KOVALEEVA, V. I., BUKREEVA, T. V.: Essential oil from *Artemisia taurica*. Khim. Prir. Soedin. **4**, 51–52 (1968).

KUNDU, S. K., CHATTERJEE, A., RAO, A. S.: Isolation of fernenol from *Artemisia vulgaris*. Aust. J. Chem. **21**, 1931–1933 (1968).

KUNDU, S. K., CHATTERJEE, A., RAO, A. S.: Chemical investigations of *Artemisia vulgaris*. J. Indian Chem. Soc. **46**, 584–594 (1969).

KURODA, T.: Chemical and botanical properties of *Artemisia kurramensis* cultivated in Japan. V. Seasonal change of contents of santonin and essential oil. Syoyakugaku Zasshi **17**, 19–24 (1963).

LEE, K., GEISSMAN, T. A.: Sesquiterpene lactones of *Artemisia* constituents of *A. ludoviciana* subspecies *mexicana*. Phytochemistry **9**, 403–408 (1970).

LEE, K., GEISSMAN, T. A.: Sesquiterpene lactones of *Artemisia* species: Artefransin from *A. franserioides*. Phytochemistry **10**, 205–208 (1971).

LEE, K., SIMPSON, R. F., GEISSMAN, T. A.: Sesquiterpenoid lactones of Artemisia. Constituents of *Artemisia cana* subspecies *cana*. The structure of canin. Phytochemistry **8**, 1515–1521 (1969).

LOOMIS, W. D.: Biosynthesis and metabolism of monoterpenes. 59–82. In: Terpenoids in plants (J. B. PRIDHAM, Ed.). London: Academic Press 1967.

MABRY, T. J.: The betacyanins, a new class of red-violet pigments, and their phylogenetic significance. 239–254. In: Taxonomic biochemistry and serology (C. A. LEONE, Ed.). New York: Ronald Press 1964.

MABRY, T. J.: The betacyanins and betaxanthins. 231–244. In: Comparative phytochemistry (T. SWAIN, Ed.). London: Academic Press 1966.

MABRY, T. J.: Betalains, red-violet and yellow alkaloids of the Centrospermae. 267–384. In: Chemistry of the alkaloids (S. W. PELLETIER, Ed.). New York: Van Nostrand Reinhold Co. 1970 a.

MABRY, T. J.: Infraspecific variation of sesquiterpene lactones in *Ambrosia* (Compositae): Applications to evolutionary problems at the populational level. 264–300. In: Phytochemical phylogeny (J. B. HARBORNE, Ed.). London: Academic Press 1970 b.

MABRY, T. J., MEARS, J. A.: Alkaloids and plant systematics. 719–746. In: Chemistry of the alkaloids (S. W. PELLETIER, Ed.). New York: Van Nostrand Reinhold Co. 1970.

MABRY, T. J., WYLER, H., SOSSU, G., MERCIER, M., PARIKH, I., DREIDING, A. S.: Die Struktur des Neobetanidins. Helv. Chim. Acta **45**, 640–647 (1962).

MABRY, T. J., TAYLOR, A., TURNER, B. L.: The betacyanins and their distribution. Phytochemistry **2**, 61–64 (1963).

MALLABAEV, A., SAITBAEVA, I. M., SIDYAKIN, G. P.: Scopoletin and β-sitosterol from *Artemisia dracunculus*. Khim. Prir. Soedin. **5**, 320 (1969).

MALLABAEV, A., YAGUDAEV, M. R., SAITBAEVA, I. M., SIDYAKIN, G. P.: Isocoumarin artemidin from *Artemisia dracunculus*. Khim. Prir. Soedin. **6**, 467–468 (1970).

MARKHAM, K. R., MABRY, T. J., SWIFT, W. T. Jr.: Distribution of flavonoids in the genus *Baptisia* (Leguminosae). Phytochemistry **9**, 2359–2364 (1970).

MATSUEDA, S., GEISSMAN, T. A.: Sesquiterpene lactones of *Artemisia* species. III. Arglanine from *Artemisia douglasiana*. Tetrahedron Lett. (21), 2013–2015 (1967 a).

MATSUEDA, S., GEISSMAN, T. A.: Sesquiterpene lactones of *Artemisia* species. IV. Douglanin from *Artemisia douglasiana*. Tetrahedron Lett. (23), 2159–2162 (1967 b).

MATSUDA, H.: Constituents of the leaves of *Rhus* and of some species of related genera in Japan. Chem. Pharm. Bull. **14**, 877–882 (1966).

MAZUR, Y., MEISELS, A.: The isolation of 5-hydroxy-3,6,7,3′,4′-pentamethoxyflavone from *Artemisia arborescens*. Bull. Research Council Israel **5 A**, 67–69 (1955).

MAZUR, Y., MEISELS, A.: The structure of arborescine, a new sesquiterpene from *Artemisia arborescens*. L. Chem. and Ind., 492 (1956).

MEARS, J. A.: The systematics of *Parthenium* section *Bolophytum:* Biochemical, morphological and ecological. Ph. D. dissertation. The Univ. of Texas at Austin (1970).

MILLER, H. S., RÖSLER, H., WOHLPART, A., WYLER, H., WILCOX, M. E., FROHOFER, H., MABRY, T. J., DREIDING, A. S.: Biogenese der Betalaine: Biotransformation von Dopa und Tyrosin in den Betalaminsäureteil des Betanins. Helv. Chim. Acta **51**, 1470–1474 (1968).

MOONEY, H. A., DUNN, E. L., SHROPSHIRE, F., SONG, L.: Vegetation comparisons between the mediterranean climatic areas of California and Chile. Flora **159**, 480–496 (1970).

MURKO, D.: Gallic acid production from tannin of domestic *Rhus cotinus*. Kem. Ind. **14**, 147–150 (1965).

NAKANO, T., MEDINA, J. D., HUTADO, I.: Chemistry of *Rhus striata* ("manzanillo"). Planta Med. **18**, 260–265 (1970).

NAZARENKO, M. V.: Proazulenes from pollen of *Artemisia macrocephala*. Zhur. Priklad. Khim. **34**, 1633–1636 (1961).

NAZARENKO, M. V.: Guaianolides of Sievers wormwood, *Artemisia sieversiana*. Zh. Prikl. Khim. **38**, 2372–2374 (1965).

NAZARENKO, M. V., LEONTEVA, L. I.: Sieversinin, a new sesquiterpenic γ -lactone. Khim. Prir. Soedin. **2**, 399–405 (1966).

NIELSCHULZ, O., SCHMERSAHL, P.: Chloretic agents from *Artemisia abrotanum*. Arzneim.-forsch. **18**, 1330–1336 (1968).

NOVOTNY, L., HEROUT, V.: Plant substances. XV. The composition of *Artemisia sieversiana*. Collection Czech. Commun. **27**, 1508–1510 (1962).

OSWIECIMSKA, M., POLAK, A., SEIDL, O., SENDRA, J.: Comparative study of chromatograms of the flavonoid fractions from herbs of some species of the genus *Artemisia*. Dissertationes Pharm. **17**, 503–511 (1965).

PAYNE, W. W.: A re-evaluation of the genus *Ambrosia* (Compositae). J. Ann. Arb. **45**, 401–430 (1964).

PENUNURI, E., STEELINK, C.: Chamazulene from *Artemisia carruthii*. J. Pharm. Sci. **51**, 598 (1962).

PLOUVIER, V.: Structure of flavone glucosides by nuclear magnetic resonance. Compounds of the genera *Centaurea, Kerria, Rhus,* and *Scabisoa*. C. R. Acad. Sci., Ser. D. **270**, 2710–2713 (1970).

RAFANOVA, R. Ya., STREBEIKO, M. P., KROKHIN, N. G.: Citral, geraniol, and linaloöl. U. S. S. R. **103**, 725, Aug. 25 (1956).

RAO, M. M., KINGSTON, D. G. I., SPITTLER, T. D.: Flavonoids from *Flourensia cernua*. Phytochemistry **9**, 227–228 (1970).

RENOLD, W.: The chemistry and infraspecific variation of sesquiterpene lactones in *Ambrosia confertiflora* DC. (Compositae): A chemosystematic study at the populational level. Ph. D. dissertation, The Univ. of Texas at Austin (1970).

REVAZOVA, L. V., CHUGUNOV, P. V., PAKALNS, D.: Deacetylmatricarin from *Artemisia incana*. Khim. Prir. Soedin. **6**, 372 (1970).

RODRIGUEZ, E., YOSHIOKA, H., MABRY, T. J.: The sesquiterpene lactone chemistry of the genus *Parthenium* (Compositae). Phytochemistry **10**, 1145–1154 (1971).

ROMO, J., ROMO DE VIVAR, A., TREVINO, R., JOSEPH-NATHAN, P., DIAZ, E.: Constituents of *Artemisia* and *Chrysanthemum* species. Structures of chrysartemins A and B. Phytochemistry **9**, 1615–1621 (1970).

ROMO DE VIVAR, A., ORTEGA, A.: Structures of bahia-I and bahia-II: two new guaianolides. Can. J. Chem **47**, 2849–2852 (1969).

RUVEDA, E. A., ALBONICO, S. M., PRIESTAP, H. A., DEULOFEU, V., PAILER, M., GOESSINGER, E., BERGTHALLER, P.: Plant constituents with a nitro group. Montash. Chem. **99**, 2349–2358 (1968).

RYAKHOVSKAYA, T. V., KLYSHEV, L. K., ALYUKINA, L. S.: Flavonoids of *Artemisia* subgenus *Dracunculus*. Tr. Inst. Bot., Akad. Nauk. Kaz. SSR **28**, 194–198 (1970).

RYBALKO, K. S.: Sesquiterpene lactones of *Artemisia leucoides* and *A. austriaca*. Zh. Obsch. Khim. **33**, 2734–2744 (1963).

RYBALKO, K. S.: Isolation of mibulactone from *Artemisia taurica*. Khim. Prirodn. Soedin., Akad. Nauk. Vz. SSr (2), 142–143 (1965).

RYBALKO, K. S., ABBSAOV, R. M.: Erivanine, a new sesquiterpene lactone from *Artemisia fragrans* Wiild. var. *Erivanica* Bess. Zh. Obsch. Khim. **33**, 1700–1701 (1963).

RYBALKO, K. S., BAN'KOVSKII, A. I.: The chemical composition of the essential oil of *Artemisia cina*. Trudy-Vsesoyuz. Nauch.-Issledovatel-Inst. Lekarstv. i Aromat. Rast. (11), 106–151 (1959).

RYBALKO, K. S., BAN'KOVSKII, A. I.: Chemical study of sesquiterpene lactones of certain varieties of *Artemisia*. Izuch. i Ispol'z. Lekarstv. Rastit. Resorsov SSSR Sb. 274–276 (1964).

RYBALKO, K. S., MASSAGETOV, P. S., EVSTRATOVA, R. I.: Sesquiterpene lactones from some forms of wormwood. Med. Prom. SSSR **17**, 41–43 (1963a).

RYBALKO, K. S., GUBANOV, I. A., VLASOV, M. I.: Isolation of a crystalline substance from *Artemisia vachanica*. Zh. Obshch. Khim. **33**, 3781–3782 (1963b).

RYBALKO, K. S., GUBANOV, I. A., VLASOV, M. I.: Isolation of scopoletin from *Artemisia persica* and *A. santolinaefolia*. Med. Prom. SSSR **18**, 19 (1964).

SABER, A. H., KASIM, A. A.: Separation of the crystalline principles of *Artemisia cina* and *A. judaica* by paper chromatography. J. Pharm. Sci. U. Arab. Rep. **3**, 159–166 (1962).

SANCHEZ-VIESCA, F., ROMO, J.: Estafiatin, a new sesquiterpene lactone isolated from *Artemisia mexicana*. Tetrahedron **19**, 1285–1291 (1963).

SATODA, I., YOSHIDA, N., YOSHII, E.: Isolation of lumisantonin from *Artemisia kurramensis*. Yakugaku Zasshi **79**, 267–268 (1959).

SCHMERSAHL, P.: Occurrence of coumarin derivatives in *Artemisia abrotanum*. Planta Med. **14**, 179–183 (1966).

SCHWAER, C.: The effect of isolated components of wormwood on *Foeniculum vulgare*, *Lepidium sativum*, and *Lactuca sativa* var *longfolia*. Flora (Jena) **152**, 509–515 (1962).

SIMPA, G., VAN DER WAL, B.: Structure of davonone. Sesquiterpene from davana *(Artemisia pallens)*. Recl. Trav. Chim. Pays-Bas **87**, 715–720 (1968).

SINHA, C. K., CHAUHAN, R. N. S.: Essential oils from odoriferous plants of Kumaon. Indian Oil Soap J. **35**, 229–233 (1970).

SORM, F., SUCHY, M., VONASEK, F., PLIVA, J., HEROUT, V.: Terpenes. XXVII. Sesquiterpenic and diterpenic components of wormwood oil. Chem. Listy **45**, 135–139 (1951).

SPITZER, J. C., STEELINK, C.: Isolation of matricin from *Artemisia caruthii*. J. Pharm. Sci. **56**, 650–651 (1967).

SUCHY, M.: Terpenes. CXLVI. Structure of balchanin, a sesquiterpenic lactone of santonin type from *Artemisia balchanorum*. Collection Czech. Chem. Commun. **27**, 2925–2928 (1962a)

SUCHY, M., HEROUT, V., SORM, F.: Terpenes. CLIII. The proof of existence and structure of hydroxy costunolide, a sesquiterpenic lactone of the germacrane type in *Artemisia balchanorum*. Collection Czech. Chem. Commun. **28**, 1620–1622 (1963).

TAKEMOTO, T., NAKAJIMA, T.: Essential oil of *Artemisia annua*. I. Isolation of a new ester compound. Yakugaku Zasshi **77**, 1307–1309 (1957).

TARASOV, V. A., KASYMOV, Sh. Z., SIDYAKIN, G. P.: Herniarin from *Artemisia diffusa*. Khim. Prir. Soedin. **5**, 436–437 (1969).

TIKHONOVA, L. K., GORYAEV, M. J.: The chemical composition of the essential oil from *Artemisia ferganensis*. Izvest. Akad. Nauk. Kazakh. SSR., Ser. Khim. (**2**), 65–74 (1957).

TISCHER, J.: Some constituents of the fruits of staghorn sumach (*Rhus typhina*). Pharmazie **15**, 83–89 (1960).

TOLSTYKH, L. P., SHEICHENKO, V. I., BAN'KOVSKII, A. I., RYBALKO, K. S.: Artemin, a new sesquiterpene lactone from *Artemisia taurica*. Khim. Prir. Soedin. **4**, 384–385 (1968).

TOMOVA, M.: Isolation of scoparone from *Artemisia scoparia*. Farmatsiya **14**, 18–21 (1964).

TUCAKOV, Y.: Determination and knowledge of the essential oil of *Artemisia annua*. Perfumery Essent. Oil Record **46**, 75–78 (1955).

WHITTAKER, R. H., FEENEY, P. O.: Allelochemics: Chemical interactions between species. Science **171**, 757–770 (1971).

YAYASHI, S., YANO, K., MATSUURA, T.: Yomogi alcohol, a new monoterpene alcohol from *Artemisia feddei*. Tetrahedron Lett. (**59**), 6241–6243 (1968).

4

Comparative Anatomy of Sclerophylls of Mediterranean Climatic Areas

JOCHEN KUMMEROW

Introduction

Vast territories in the Mediterranean Basin, in southern California, South Africa, southwestern Australia and central Chile are characterized by a physiognomically uniform vegetation. In general terms the predominant vegetation of these regions is termed "xerophytic" according to the concept of SCHOUV (1822), who so designated plants of dry habitats. Later this concept has been narrowed and precisely defined. THODAY (1933) proposed that the term "xerophyte" should not carry any particular functional or structural implication, but should be used for plants of dry areas, irrespective of their modes of adaptation to such habitats. WALTER and KREEB (1970) state that xerophytes are those plants growing in arid zones and on dry habitats without access to ground water. The aboveground organs of these plants remain alive through the entire year with the exception of the foliage of deciduous species. Excluded from this group are the water accumulating species, the succulents. Nevertheless, the remaining group of xerophytes is still extraordinarily heterogeneous and has been further subdivided. WALTER and KREEB (1970) distinguish the poikilohydrous, malakophyllous, sclerophyllous, aphyllous, and stenohydrous xerophytes.

Our interest here will be focused on the sclerophyllous xerophytes or, briefly, "sclerophylls" which are represented as the characteristic evergreen hard-leaved trees and shrubs all over the mediterranean climatic areas of the world.

The Problem

The anatomical structure of sclerophylls has long been investigated, with the objective of finding those structures which control transpiration and which ultimately would explain the existence of the respective species in their corresponding dry habitats. STOCKER (1960) stated that this problem has been approached from an observational as well as an experimental point of view. Numerous observations have shown that many plants growing in arid conditions, but by no means all of them, have structural peculiarities in their leaf anatomy. To describe these "adaptations" as mechanisms for decreasing transpiration, a concept characteristic of the age of Darwinism, MONTFORT (1918) coined the term "xeromorphien" (xeromorphism in English). Much work, generally speculative, was dedicated to strengthening this view. NEGER, as early as 1913, stated that "Extraordinary is the number of investigations which attempt to correlate the structure of the vegetative organs with the process of transpiration".[1]

Physiological experiments form the basis for the second analytical approach. The influence of soil and atmospheric humidity on transpiration was studied and as a general result it became clear that stress on the water metabolism of plants produces structural changes in leaf anatomy ("xeromorphosis" [STOCKER, 1960], or "xeroplastics" [THODAY, 1933]). Xeromorphisms and xeromorphosis are very similar, and in specific cases it may be difficult or impossible to distinguish between them. The necessity for careful interpretation is reflected by OPPENHEIMER (1960), who writes, "if the true xerophytes are not all xeromorphous, conversely xeromorphous plants are not all xerophytes. Thus sclerophyllous (hard-leaved) plants, widespread in countries with a mediterranean climate are also found in wet tropical associations and even beyond the arctic circle, e.g., *Rhododendron lapponicum* in Lapland".

The ecophysiological properties of sclerophylls have been frequently analyzed (KILLIAN and LEMÉE, 1956; STOCKER, 1956). There is, however, a lack of modern comparative anatomical studies of these species. The physiological properties which enable the sclerophylls to thrive in a region with substantial water stress during the summer months are somehow related to the anatomical structure of their leaves and/or other vegetative organs.

Without any claim to completeness, this chapter will attempt to describe the characteristic leaf structures of evergreen sclerophylls and will discuss the significance of these structures.

Anatomy

As mentioned above the anatomical structure of leaf xeromorphisms has long been investigated, and the amount of literature dealing with this topic is enormous. SHIELDS's (1950) review on "leaf xeromorphy as related to physiological and structural influences", cites more than 250 pertinent papers. The main conclusions from these early studies can be summarized as follows. Water deficit, high light intensity, and nitrogen deficiency are the principal factors causing a lower ratio of surface to volume. This reduction of external area results in structural modifications of the leaf anatomy. Smaller cell size, thicker cell walls, more veins per surface unit area, a thicker cuticle, higher stomata frequency, stronger mechanical tissue, and a more differentiated palisade parenchyma, are some of the most prominent xeromorphic leaf structures. All of these features were recently reaffirmed by PYYKKÖ (1966), who studied the leaf anatomy of 284 East Patagonian plants.

But, as abundant as the information is on xeromorphisms in general, there are few specific anatomical data available on leaves of evergreen sclerophylls. The information that is available is discussed here on a regional basis.

---

1 Translated from German

Table 1. Summary of the structural characteristics of 7 broad-sclerophyll forest and 18 climax chaparral species. (After COOPER, 1922)

	Broad-sclerophyll forest, 7 species	Climax chaparral, 19 species	Total, 25 species
Leaf more than 300 μ thick	1	12	13
Mesophyll:			
Bifacial	4	12	16
Imperf. bifacial	3	4	7
Isolateral	–	3	3
Entirely palisade	–	2	2
More than 2 palisade layers	3	15	18
Epiderm:			
Lower papillate	2	8	10
Upper cuticle more than 4 μ thick	1	13	14
Stomata:			
Lower side only	7	15	22
On both sides	–	4	4
With exterior chambers	1	6	7

Species analyzed

A. Broad sclerophyll forest
Castanopsis chrysophylla
Quercus agrifolia
Lithocarpus densiflora
Arbutus menziesii
Myrica californica
Quercus chrysolepis
Umbellularia californica

B. Climax chaparral
Quercus durata
Dendromecon rigidum
Adenostoma fasciculatum
Pickeringia montana
Rhamnus crocea
Garrya elliptica
Arctostaphylos hookeri
A. pumila
A. tomentosa
A. vestita
Berberis pinnata
Heteromeles arbutifolia
Prunus ilicifolia
Rhamnus californica
Ceanothus cuneatus
C. papillosus
C. sorediatus
Eriodictyon californicum

California

The broad sclerophyll vegetation of California has been the subject of an intensive study by COOPER (1922). The average Californian broad sclerophyll leaf is characterized by COOPER as follows: "it is moderately small (averaging 2–3 cm in length), simple, unlobed, elliptic, and in a majority of cases entire and glabrous. Important groups are toothed, spiny toothed, revolute or pubescent on the lower or on both surfaces.The leaf is thick, averaging 314 microns, while the deciduous species average only 127 microns. The mesophyll is most commonly bifacial, though often imperfectly so, but a few are isolateral. The palisade tissue is several layers deep. The epiderm is nearly always single, and often papillate. The cuticle is very thick, the upper averaging 5.69 microns as against 1.58 microns in the deciduous species. In the case of the lower the difference is even greater: 4.43 microns and 0.91 microns. The stomata in a large majority are on the lower surface only". These data are summarized in Table 1. Even considering that mean values from anatomical structure measurements frequently lead to erroneous conclusions, there are some interesting details resulting from the comparison of the leaves of broad-sclerophyll forest species with those of the climax chaparral. The leaves from the last group are 50% thicker than those from the first one. The chaparral group includes all the isolateral and the majority of the imperfectly bifacial leaves. All species with stomata on both sides (isolateral leaves) are found in the chaparral. COOPER (1922) concluded, therefore, that special features protective against water loss are confined to the broad chaparral sclerophylls.

Detailed information on leaf anatomy in *Ceanothus* is given by NOBS (1963). These data are especially interesting as they are confined to the comparison of several species from the same genus. The typical *Ceanothus* leaf has a thick cuticle, abundant sclerenchymatous tissue, and stomata on the lower (abaxial) surface of the leaves. An interesting feature is that the stomata of the section *Cerastes* appear to be localized in crypts—deep invaginations of the lower leaf surface (Fig. 1). These crypts are typical for adult leaves. In the seedling plant the first pair of leaves shows the basic organization of the cryptless *Euceanothus*-section. In the subsequent leaf pairs, a sequence of developing crypts appears which terminates in the mature crypt of the *Cerastes*-section. These crypts appear very stable in their anatomy. NOBS (1963) has utilized these anatomical features as taxonomic characters.

Chile

Unfortunately, central Chile, related to California by very similar climatic conditions and many convergent plant species, has not been the subject of a survey study such as has been described for California by COOPER (1922). Only recently has the leaf anatomy of two characteristic evergreen sclerophylls from central Chile, *Cryptocarya alba* and *Peumus boldus*, been analyzed (HURTADO, 1969; HOMANN, 1968).

Cryptocarya alba, which is ecologically analogous to the Californian *Quercus agrifolia*, has a smooth leaf surface on both sides. Stomata are found on the lower epidermis only, in an abundance of 690/mm^2 on sun and 610/mm^2 on shade leaves. The general anatomy is strictly dorsiventral (Fig. 2). The palisade parenchyma is well formed, 1- to 4-layered, and interrupted by groups of sclerenchymatous cells. These latter cells form characteristic ribs of mechanical tissue which are in contact with the upper and lower epidermis. It may be assumed that these ribs are important in preventing the leaves from collapsing upon wilting. In Table 2 the more conspicuous anatomical details are summarized.

The second representative of this group of central Chilean sclerophylls is *Peumus*

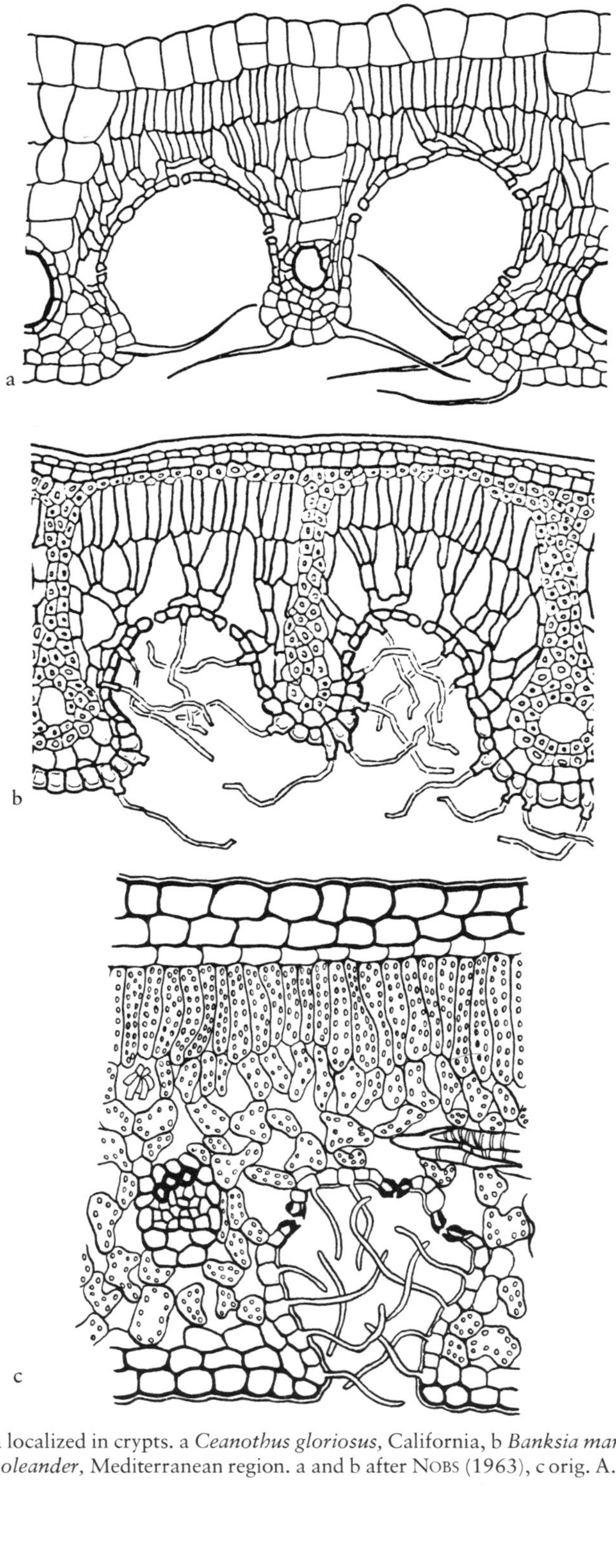

Fig. 1. Stomata localized in crypts. a *Ceanothus gloriosus,* California, b *Banksia marginata,* Australia, c *Nerium oleander,* Mediterranean region. a and b after NOBS (1963), c orig. A. HOFFMANN

boldus. This evergreen tree or shrub has large branched trichomes on both the upper and lower epidermis, although they are somewhat more frequent on the lower side (ratio 5 ÷ 7). The upper epidermis is one- to multi-layered and covered by a thick cuticle. Stomata are found exclusively on the lower epiderm with an average frequency of 570/mm^2 and are somewhat elevated over the epidermis level. The palisade tissue is strongly developed, compact and 2- to 3-layered in sun leaves. The spongy parenchyma has large intercellular spaces with ample substomatal chambers. These leaves thus combine typically mesophytic and xerophytic characters.

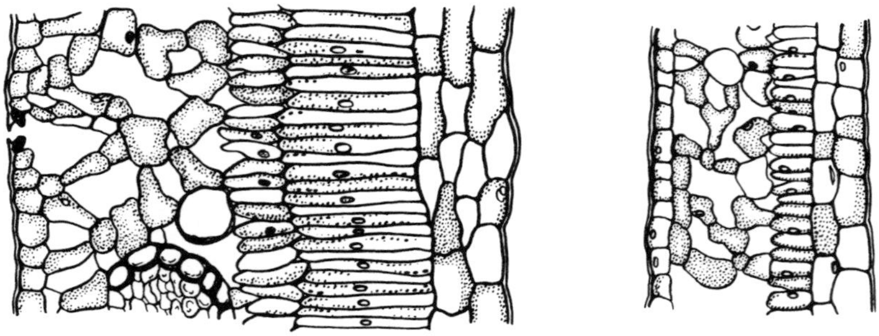

Fig. 2. Cross section of shade leaf (right) and sun leaf (left) of *Peumus boldus*, Chile. Orig. A. HOFFMANN

Table 2. Anatomical structure of sun and shade leaves of *Cryptocarya alba*. Mean values of 50 leaves. (After HURTADO, 1969)

	Sun leaf	Shade leaf
Leaf surface area	9.7 cm^2	18.7 cm^2
Leaf thickness	321.7 μ	216.8 μ
Thickness of upper epiderm	21.0 μ	15.2 μ
Thickness of lower epiderm	13.5 μ	8.0 μ
Cuticle thickness, upper	4.8 μ	2.7 μ
Cuticle thickness, lower	1.5 μ	1.1 μ
Stomata frequency	690/mm^2	610/mm^2
Contribution of palisade tissue to mesophyll	47%	39%

Although not belonging directly to the mediterranean climatic region of South America, the eastern Patagonian vegetation should be mentioned here. The above-cited work of PYYKKÖ (1966) includes some species which can be counted under the true evergreen sclerophylls, e.g., *Maytenus boaria, Lomatia obliqua*, and *Schinus patagonicus*. The leaves of these species are broad, thick and coriaceous. The epiderm is very small-celled and covered with a thick cuticle. The upper epiderm is reported to be largely mucilaginous. Stomata occur only on the lower side of the leaves. The mesophyll is compact, has a dorsiventral structure, and the multi-layered palisade parenchyma is usually well developed. It is obvious that the anatomical details described here fit perfectly into the scheme of leaves from true mediterranean areas.

Australia

Recent information on western Australian evergreen sclerophylls has been communicated by GRIEVE and HELLMUTH (1970). Their anatomical data are summarized in Table 3.

Among 17 species, including Acacias and Eucalypts, but also *Hibbertia* and *Casuarina*, a great diversity in anatomical structure and morphology was observed. Leaf sizes range from leptophyll through nanophyll to microphyll. The cuticle was thick in all observed leaves. The stomata of 4 species were sunken, in 8 species they were found at epidermal level and 5 species showed an intermediate condition. None of these 17 species had particularly hairy coverings on each surface of their leaves, 13 showed no hairs at all, and 5 had hairs on the lower surface only. For all plants examined, the intercellular spaces were small and mechanical tissue was well developed. Reduction in leaf size is characteristic of most sclerophylls in western Australia (GRIEVE, 1955). It may be mentioned that species of the genus *Banksia* (Proteaceae) (EWART, 1930) have their stomata arranged in crypts as has been noted for the *Ceanothus* section *Cerastes* in California (Fig. 1 b).

Table 3. Leaf size and anatomical features of western Australian hard-leaved evergreen sclerophylls. (After GRIEVE and HELLMUTH, 1970)

	Thick cuticle	Stomata	Hair cover	Name of species
Leptophyll	+	±	±	*Hibbertia hypericoides*
0–25 mm^2	+	−	−	*Bossiaea eriocarpa*
	+	in grooves	−	*Casuarina campestris*
	+	+	−	*Malaleuca uncinata*
	+	−	−	*Olearia muelleri*
Nanophyll	+	±	±	*Banksia attenuata*
25 mm^2 –	+	−	−	*Stirlingia latifolia*
9×25 mm^2	+	+	−	*Acacia acuminata*
	+	+	−	*Acacia craspedocarpa*
	+	−	−	*Dodonaea viscosa*
	+	+	−	*Eremophila* sp.
Microphyll	+	−	−	*Hardenbergia comptoniana*
9×25 mm^2 –	+	±	±	*Banksia grandis*
$9^2 \times 25$ mm^2	+	±	±	*Banksia menziesii*
	+	±	±	*Banksia prionotes*
	+	−	−	*Eucalyptus marginata*
	+	−	−	*Eucalyptus calophylla*
	+	−	−	*Eucalyptus redunca*

Cuticle: Presence or absence of a thick cuticle is represented by + or − respectively; ± represents some intermediate development of cuticle.

Stomata: + and − represent sunken or superficial stomata, respectively; ± represents an intermediate condition.

Hair cover: − represents a hairless leaf, + is a hairy covering on both sides of the leaf; ± indicates presence of hairs on lower surface only.

Mediterranean Region

Kamp (1930) studied the problem of the relation between epidermal and cuticular structure and transpiration. He included in his study the typical mediterranean evergreen scerophylls, *Laurus nobilis, Nerium oleander, Olea europaea, O. lancea, Quercus coccifera, Q. ilex, Rhamnus alaternus,* and *R. glandulosus.*

In taxonomically related species, e.g., *Quercus coccifera* and *Q. ilex,* the thickness of the cuticle is greater in smaller leaves. Further, for species within the same genus, cuticular thickness is positively correlated with the length of the cuticular ribs in the anticlinals of the epidermal cells. This is significant, considering that in all of the observed species the epidermal cells had more or less the same height. When the author arranged these species according to the pattern of the cuticular thickness, it became clear that species with a thick cuticle generally had much smaller cuticular ribs than those with

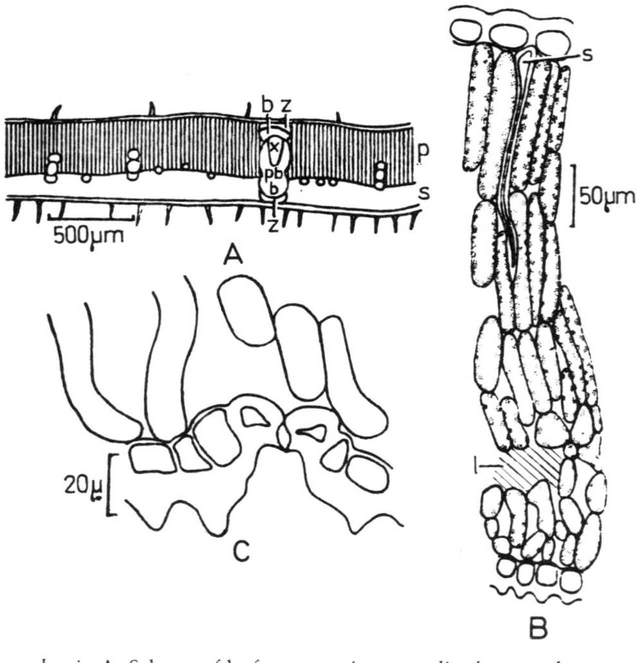

Fig. 3. *Boscia senegalensis.* A: Scheme of leaf cross section. p: palisade parenchyma, s: spongy parenchyma, x: xylem, ph: phloem, b: fibers, z: cells with cellulose-strengthened walls. B: Leaf cross section with sclereid. 1: bundle. C: Stoma on the lower epidermis. (After Stocker, 1970)

relatively thin cuticles. When the structure of the sclerophylls was compared, two groups could be distinguished: 1) species with a relatively thin cuticle but thick outer cellulose layers, which were much thicker than the cuticle, and 2) species with a much thicker cuticle and a very thin cellulose layer (1/10 the thickness of the cuticle). Finally, Kamp concluded that the leaves of different species from the same habitat, e.g., *Quercus ilex, Olea europaea, Rhamnus alaternus,* may show very pronounced differences in the structure of their outer cell walls.

Special attention has always been paid to the stomatal frequency of evergreen leaves. The published figures, which include data on evergreen sclerophylls from all over the world, cover the range from 150–700 stomata/mm^2 (GINDEL, 1969; BRECKLE, 1966).

It may be mentioned that the species of the more arid habitats generally have their stomata on both sides of their leaves while the species from the more moderate mediterranean maqui have them on the lower epidermis only. This is the case for *Pistacia lentiscus* and *P. palaestina,* while *P. atlantica,* with stomata on both sides of the blade, grows naturally also in the semi-desert and desert, often on rocky slopes (GINDEL, 1969).

Two evergreen sclerophylls are described by STOCKER (1970) from North Africa: *Salvadora persica* and *Boscia senegalensis.* Both species grow in very arid conditions and are taxonomically related to the hard-leaved evergreens of the equatorial rainforests. They should be considered as intermediate between these evergreens and the evergreen sclerophylls of the mediterranean climate. In both cases the basic anatomical structure is similar to that described above (Fig. 3). But in *Salvadora* there is a very conspicuous lack of sclerenchymatous tissue. Only the thick-walled epidermis and some collenchyma under the main bundles maintain the mechanical resistance of the leaf. The case of *Boscia* is different in that not only do strong fiber strands surround the bundles, but also the multilayered palisade tissue is strengthened by frequent sclereids.

Discussion

The data given above are selective and by no means summarize all that is known of the anatomical structures of evergreen sclerophylls. Rather, anatomical information is given for different species which grow in distant localities under very similar macro-climatic conditions. The limitation of this comparison to leaf anatomy is somewhat arbitrary, considering that the root and stem anatomy could yield additional and perhaps, in specific cases, significant information. But the basic question is to elucidate whether there exists, and if so to what degree, a relation between specific xeromorphous structures and the arid habitat of the species. For this purpose leaf anatomy seems to be the most adequate parameter because of the relative abundance of available information.

A thick cuticular layer has always been considered a feature of plants of arid climates. All of the data summarized above confirm this point. The general conclusion has been that a thick cuticle is important in preventing excessive water loss. The situation is, however, more complex than this. Old leaves apparently exhibit higher cuticular transpiration than younger ones. KAMP (1930) speculated that this increasing permeability to water with age was brought about by a process of dissolution and disintegration of the cuticle, which had itself been brought about by atmospheric factors (see also OPPENHEIMER, 1960). His final conclusion was that plants of the same ecological type, growing in the same place, may be structurally completely different. EVENARI (1938), who analyzed the leaf structure of some desert and semi-desert plants, attempted to quantitatively assess the various anatomical characteristics and to classify the species accordingly. Again, the results were negative. The observed and measured characters were overlapping and the grouping of different species according to their anatomical structures did not form a pattern.

A more recent and modern approach to this problems was published by NOBS (1963). This author compared the leaf anatomy of *Ceanothus* species which bear their stomata in crypts (section *Cerastes*) or on the plain lower surface (section *Euceanothus*). Species of both sections are found growing in similar habitats, and no selective advantage of one or the other leaf type has been demonstrated. The same may be said regarding the abovementioned *Banksia* species from Australia.

Considering the whole range of evergreen sclerophylls, it appears that we have a group of moderate xeromorphic plants characterized by similar leaf structures. But this cannot mean that the whole plant group looks alike. On the contrary, it seems that these species form a very "phytoplastic" group (GINDEL, 1957) designed to adapt to a wide spectrum of environmental conditions. This is reflected by the structures enumerated above: the stomatal frequency varies in a broad range, and stomata are found on both sides of the epidermis or on the lower sides only or in crypts. The cuticle may be very thick or as thin as in many mesophytic species; the palisade tissue can be strongly developed and multi-layered or even single-layered. And, finally, these differences can be discerned, at least partially, in comparing sun and shade leaves of the same individual plant.

In the introduction it was noted that most of the early anatomical studies sought transpiration-inhibiting structures which would help to explain the presence of determinate plant types in arid habitats. A careful comparison of these structures makes it clear that this approach was much too simplistic for an understanding of the significance of evergreen sclerophylls. Hence, a high stomatal frequency with these stomata being sometimes even elevated over the epidermal level, or a well-developed spongy parenchyma, cannot be considered as transpiration-limiting structures. These structures can be judged only by whether their combination as a whole provides any advantage for the species. Thus the presence of an individual feature such as high stomatal frequency in a situation where water should be saved remains an unresolved problem.

As a group the evergreen sclerophylls seem to be composed of a number of physiologically similar plant species. They may have in common the fact that their genetic potential allows a high degree of plasticity, producing structures which ensure a high photosynthetic efficiency as long as water is available. The compact palisade parenchyma and the generally well-developed sclerenchymatic tissue provide the leaves with a certain degree of protection against irreparable damage during severe drought. It is clear that protoplasmic drought resistance must be genetically based.

Further anatomical structure research, particularly on the long-neglected cortex as well as the roots, in combination with careful microclimatic measurements, could provide more insight into the complex of structural and functional relationships in these plants.

Summary

The mediterranean climatic areas of the world, i.e., the true Mediterranean Basin, southern California, central Chile, South Africa, and southwestern Australia, are characterized by an evergreen sclerophyllous vegetation. A literature summary shows that the basic anatomical features of the leaves from these species are very similar. The available information suggests that it is not possible to conclude from anatomical structures alone the adaptive advantages of distinct features. Nevertheless, relations between structure and habitat of the evergreen sclerophylls exist. These must be studied by means of careful and detailed anatomical analysis of each species in relation to the microclimate data of the respective habitat.

References

BRECKLE, S.: Oekologische Untersuchungen im Korkeichenwald Kataloniens. Diss., Naturw. Fak., Landw. Hochsch. Hohenheim 1966.

COOPER, W. S.: The broad-sclerophyll vegetation of California. Carneg. Inst. Wash. Publ. **319**, 1–124 (1922).

EVENARI, M.: The physiological anatomy of the transpiratory organs and the conducting systems of certain plants typical of the wilderness of Judaea. Journ. Linn. Soc. Lond. Bot. **51**, 389–407 (1938).

EWART, A. J.: Flora of Victoria, Univ. Press, Melbourne, Australia 1930 (not seen in original, cited after NOBS, 1963).

GINDEL, J.: Acclimatization of exotic woody plants in Israel. The theory of Phyto-Plasticity. Materiae Vegetabiles **2**, 81–101 (1957).

GINDEL, J.: Stomatal number and size as related to soil moisture in tree xerophytes in Israel. Ecology **50**, 263–267 (1969).

GRIEVE, B. J.: The physiology of sclerophyll plants. J. Roy. Soc. W. Austr. **39**, 31–45 (1955).

GRIEVE, B. J., HELLMUTH, E. O.: Eco-physiology of Western Australian plants. Oecol. Plant. **5**, 33–68 (1970).

HOMANN, C.: Estudio sobre reproducción y anatomía de hojas y frutos en Boldo *(Peumus boldus* Mol.). Thesis, Univ. de Chile, Fac. de Agronomia, Esc. Ing. Forestal 1968.

HURTADO, P.: Observaciones sobre la anatomıa foliar y la transpiración en Peumo (*Cryptocarya alba* (Mol.) Looser). Thesis, Univ. de Chile, Fac. de Agronomía, Esc. Ing. Forestal 1969.

KAMP, H.: Untersuchungen über Kutikularbau und kutikuläre Transpiration von Blättern. Jb. wiss. Bot. **72**, 403–465 (1930).

KILLIAN, CH., LEMÉE, G.: Les xérophytes: leur économie d'eau. Hdb. der Pflanzenphysiol. **III**, 787–824, herausgeg. von W. RUHLAND. Berlin-Göttingen-Heidelberg: Springer 1956.

MONTFORT, C.: Die Xeromorphie der Hochmoorpflanzen als Voraussetzung der physiologischen Trockenheit der Hochmoore. Z. Bot. **10**, 257–352 (1918).

NEGER, F. W.: Biologie der Pflanzen auf experimenteller Grundlage (Binomie). Stuttgart: F. Enke 1913.

NOBS, M. A.: Experimental studies on species relationships in *Ceanothus*. Carnegie Inst. Wash. Publ. **623**, 1–94 (1963).

OPPENHEIMER, H. R.: Adaptation to drought: Xerophytism. 105–138. Arid Zone Res. XV. Plant-water relationships in arid and semi-arid conditions. Paris: UNESCO 1960.

PYYKKO, M.: The leaf anatomy of East Patagonian xeromorphic plants. Ann. Bot. Fennici **3**, 453–622 (1966).

SCHOUV, J. F.: Grundtrak tilen almindelig Plantegeografi. Kjobenhavn 1822 (not seen in original, cited after WALTER and KREEB, 1970).

SHIELDS, L. M.: Leaf xeromorphy as related to physiological and structural influences. Bot. Rev. **16**, 399–447 (1950).

STOCKER, O.: Die Dürreresistenz. Hdb. der Pflanzenphysiologie **III**, 696–741, herausg. von W. RUHLAND. Berlin-Göttingen-Heidelberg: Springer 1956.

STOCKER, O.: Physiological and morphological changes in plants due to water deficiency. 63–104. Arid Zone Res. XV. Plant-water relationships in arid and semi-arid conditions. Paris: UNESCO 1960.

STOCKER, O.: Der Wasser- und Photosynthese-Haushalt von Wüstenpflanzen der mauretanischen Sahara. Flora **159**, 539–572 (1970).

THODAY, D.: The terminology of "xeromorphism". J. Ecol. **21**, 1–6 (1933).

WALTER, H., KREEB, K.: Die Hydratation und Hydratur des Protoplasmas der Pflanzen und ihre oeko-physiologische Bedeutung. Protoplasmatologia II C 6, Wien-New York 1970.

Section IV: Soil Systems in Mediterranean Climate Regions

A number of chapters of this volume, in addition to that of ZINKE which deals specifically with soil and to the three chapters included in this section, make some reference to the genesis and the characteristics of soils of the several mediterranean regions. Various aspects of this subject are treated by THROWER and BRADBURY, PASKOFF, MOONEY and PARSONS, SPECHT, NAVEH and DAN. However, no attempt is made in this volume to classify the "mediterranean" soils. It could even be argued that in these regions, considered in their broadest definition, almost any type of soil might be found; in some parts as zonal or climax soil in balance with the present environment, in other places as young azonal soils or as intrazonal soils determined by local conditions, and very often as paleosols reflecting relict conditions or truly fossil buried soils.

This remarkable heterogeneity of soils can be easily understood, when one considers the many paleoclimatic events which took place in these transitional areas and the several phases of pedogenesis which occurred accordingly, the great differences in the geological substrata, and the rough topography of most of these regions which modelled a mosaic of almost punctual situations. Also the long history of occupation by man of these territories has played an important pedogenetic role, in "rejuvenating" certain soils through regression because of moderate processes of erosion, or in degrading other soils to skeletal conditions. For background information on these very complicated topics, reference should be made to three recent books by DUCHAUFOUR (1970), RUELLAN (1970) and YAALON (1971). Concerning the semantic difficulties in the nomenclature of these soils, hopefully the series of volumes which are being published by FAO-UNESCO presenting soil maps of the world at a scale 1:5,000,000 can help in providing a basis for more comparability.

In this section soil is not treated in a classical pedological way, but is discussed mainly as a habitat for bacteria, fungi, algae and a large range of different animal groups which are extensively considered here. Taking a more functional view, soil is also approached in this section as a subsystem coupled with the above-ground vegetation subsystem, between which a reciprocal transfer of energy and information takes places.

While the characteristics of soil systems in mediterranean climates are obviously determined by both historical and contemporary factors, the weight of the "historical heritage" is probably more marked than in the case of vegetation and surface animals. The influences of the present seasonal climate on soil organisms are dampened to a certain extent by the "filtering" action of vegetation and superficial humus layers. Also the impact by man on the deeper inhabited horizons of soil is less marked. In other words, the rate of change through time seems to be slower in the soil than in the above-ground subsystem.

Soil is by essence a very conservative environment in which ancient phylogenetic lines of invertebrates having had a long common evolutionary history can be found in a scarcely modified form. This is particularly true for the soils of the regions which at present have a mediterranean type of climate, where the highest occurrence of "living

---

The introduction to this section was prepared by FRANCESCO DI CASTRI.

fossil" species of soil arthropods have been discovered. It is not by chance that faunistic surveys aimed at the *chasse à l'endogé* (the "hunting" of rare soil-living species), carried out by professional as well as by amateur naturalists, have mainly been initiated and developed in the Mediterranean regions of Europe. SAIZ and VITALI-DI CASTRI discuss elsewhere in this volume the biogeographical and phylogenetic links in selected animal groups which are typical of the soils of mediterranean climate areas.

In this section three chapters are presented, each taking a different aspect or approach. DI CASTRI's chapter is devoted to animal communities, mainly arthropods, inhabiting the litter-soil subsystem. It focuses on structural aspects and directs special attention to the evolutionary strategies adopted by different animal taxa in response to past and present selective forces. To the extent that the inadequate state of knowledge allows for generalization, it seems that the soil-animal communities of several mediterranean regions share a number of similar structural features, reflected in their deep stratification in the soil, in the composition of acarofauna which changes in a similar way along climatically or topographically controlled gradients, and in comparable values of species diversity.

SCHAEFER's chapter deals with another component of the soil biota, the microbial populations, from a more biochemical and functional standpoint. It is not centered on different taxa of microflora, but on physiological or metabolic groups of these soil organisms. In spite of these differences in subject and approach, both DI CASTRI and SCHAEFER emphasize the central role of humus in the regulation processes of mediterranean soil systems, as well as the radical changes which take place in the humus after agricultural manipulation.

Finally, LOSSAINT presents an overview of the results obtained from an integrated research programme on evergreen forest climax and degraded *garrigues* in southern France. Data are given on the inputs of matter to these soils through litter fall, canopy washings and root decomposition, on the turnover and cycling of nutrients (both macronutrients and oligoelements) through the whole ecosystem, on the water cycle and soil water deficits, and on the total biological activity in soil. His chapter provides an appropriate link therefore between this section on soil and the previous section on mediterranean vegetation, and illustrates some of the functional couplings between the two subsystems.

References

DUCHAUFOUR, P.: Précis de Pédologie. Third edition. Paris: Masson 1970.

RUELLAN, A.: Contribution à la connaissance des sols des régions méditerranéennes: les sols à profil calcaire différencié des plaines de la Basse Moulouya (Maroc Oriental). Thèse Sciences Strasbourg: 1970.

YAALON, D. H. (Ed.): Paleopedology. Origin, Nature and Dating of Paleosols. Jerusalem: International Society of Soil Science and Israel Universities Press 1971.

1

Soil Animals in Latitudinal and Topographical Gradients of Mediterranean Ecosystems

FRANCESCO DI CASTRI

Introduction

Particularly from a structural viewpoint, the litter-soil subsystem in regions having a mediterranean climate is far less well known than the vegetation or the animal populations living in the ground surface and on plants. Almost all the data on community convergences in mediterranean climates pertain to vegetation.

This gap in present knowledge is due to numerous causes. Firstly, quantitative studies on soil fauna are relatively recent, and most of these have been carried out in the humid temperate or cold ecosystems of Europe and North America, especially in deciduous forests, coniferous forests, tundras and grasslands. Secondly, there are great intrinsic difficulties in the study of soil communities, due to the enormous abundance of these populations, the very high species diversity and the acute lack of suitable information on the taxonomy of these animals. Thirdly, research designs, and more particularly the methods of extraction used for estimating the density of soil animals, have been so varied that there is not as yet a reliable basis for intercontinental comparisons. This is especially true for arid, mediterranean and tropical ecosystems. Other difficulties in the comparison of soil faunas have been discussed by DI CASTRI (1970).

With this as background, data from Chile were taken as a basis for the considerations in this chapter on soil animals communities. Chilean findings were then compared with the results obtained in other mediterranean regions using a similar approach and a relatively similar methodology. Two indispensable criteria for comparison were adopted: that the soil fauna was extracted using BERLESE-TULLGREN funnels (the most widely used extraction method in mediterranean zones) and that estimates were available of the entire mesofauna, or at least of all populations of mites and springtails, and preferably also of other taxa such as Protura, Symphyla, Pauropoda, Psocoptera and small Coleoptera.

Of the work on soil fauna undertaken in Chile special reference should be made to DI CASTRI (1963 a, 1963 b), COVARRUBIAS, RUBIO and DI CASTRI (1964) and HERMOSILLA and MURUA (1966). Recent, as yet, unpublished observations on the Chilean soil fauna will also be taken into account here, as well unpublished data by GERTRUDE FRANZ on soil bacteria and fungi.

There is almost total absence of published information on soil animal communities of California. Use was therefore made of the preliminary results obtained in 1970–1971 by the author and coworkers in California. This work formed part of the preparatory phase of the integrated project on the origin and structure of mediterranean ecosystems, undertaken within the framework of the International Biological Programme. For the Mediterranean basin, particular attention was given to the findings of DI CASTRI (1960), MAR-

CUZZI (1968), MARCUZZI and DI CASTRI (1967), MARCUZZI and DALLA VENEZIA (1972) and MARCUZZI, LORENZONI and DI CASTRI (1970). A number of these works deal with "true" mediterranean ecosystems. For South Africa, relevant publications include those by DEN HEYER and RYKE (1966), LOOTS and RYKE (1967), THERON and RYKE (1969) and VAN DEN BERG and RYKE (1967–1968), though some of these publications do not deal specifically with mediterranean climate areas. Finally, in relation to Australia, particular attention should be drawn to the work of WOOD (1970).

There are, of course, a number of other publications devoted to soil animals in mediterranean regions. Note might be taken, for example, of the work of ATHIAS-HENRIOT (1967) in Mediterranean Sea areas, of LEE and WOOD (1968) in Australia, and of PARIS and PITELKA (1962) in California. Most of these are in-depth studies which focus on the ecology of a single animal group and which are not at this stage comparable with similar investigations in other mediterranean areas. Recently there has been an upsurge of intensive work on the soil fauna communities of all of these regions (see for example DELAMARE DEBOUTTEVILLE and VANNIER 1966, and LOSSAINT 1967, for southern France); in a few years' time, there will be a more solid data base for intercontinental comparison.

Objectives

The present study attempts to provide some partial answers to three basic questions:
a) To what extent do morphological convergences really exist in the soil fauna of disjunct areas having a mediterranean climate?
b) Are there peculiar patterns of structure which are characteristic of the soil systems of mediterranean regions?
c) If morphological and structural similarities are found, are they due to evolutionary convergence, or to phylogenetic links maintained by very old lines which are little differentiated in spite of their present geographical disjunction?

From *a priori* analysis of these problems, it might be expected that true convergence in soil communities would be less marked than in mediterranean vegetation. In fact, soil communities, and especially those of the deeper soil layers, are affected to a lesser degree by the particular characteristics of the mediterranean climate (which represent the main selective force on the vegetation), because of the "filtering" action of the above-ground vegetation and litter. The soil microclimate is consequently more stable. The ecological conditions of soils at depth are fairly similar even in climatically very different zones, except when the aridity is extreme or when the drainage is very poor. Historically the soil has been a constant and permanent environment, and, like the deep sea, has favoured the survival of very ancient lines. For example, fossil springtails and psocids show a very great similarity to some contemporary species.

On the other hand, there are other selective forces in mediterranean regions which act particularly on soil communities and which would tend to favour convergence within soil faunas. Most of these forces are a consequence of the peculiar adaptive strategies of mediterranean plants. Mention might be made, for instance, of the coriaceous nature of sclerophyll leaves which limits attack by litter arthropods, the presence of secondary plant substances (for allelopathic effects of Californian plants, see MCPHERSON and MULLER, 1969), the patterns of distribution and deep penetration of roots in soil (HELLMERS *et al.*, 1955), and the water trapping devices of some mediterranean plants.

Methodological Approach of Gradient Analysis

Adoption of a gradient approach facilitates the analysis of the structural organization of these mediterranean systems. This approach entails measurement of changes in system variables in relation to well defined and readily quantifiable changes in the major environmental factor of control.

The most typical gradient is the *latitudinal* one. In Chile especially, due to the great thermic uniformity of all its mediterranean-climate territory, this gradient is mainly controlled by hydric conditions. A schematic outline of this gradient (Fig. 1) illustrates a North-South transect of the climatic and vegetational features in 8 ecological zones of

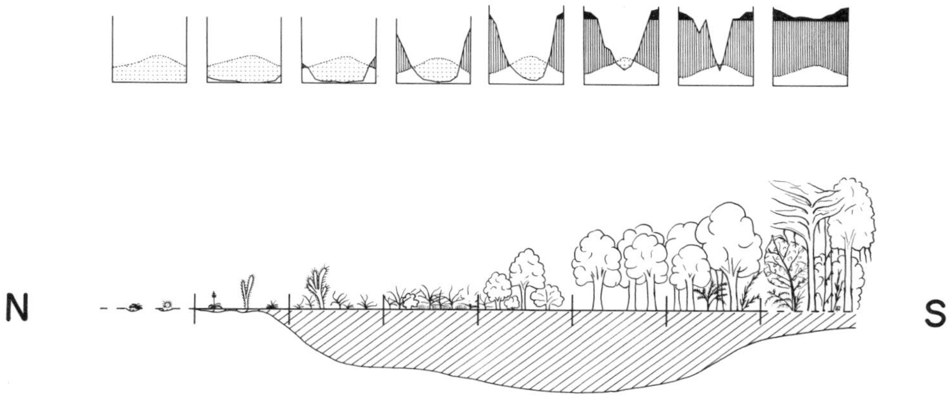

Fig. 1. Schematic outline of climate and vegetation across a North-South latitudinal gradient in Chile. For explanation of the climatic diagrams, see DI CASTRI's chapter on climatographical comparisons. From left to right, non-mediterranean desertic climate (true desert), perarid mediterranean type (desert with succulents), arid mediterranean type (succulent scrub), semi-arid mediterranean type (chaparral), subhumid mediterranean type (mixed high chaparral), humid mediterranean type (sclerophyllous forest), perhumid mediterranean type (mesophilous forest) and non-mediterranean oceanic climate (temperate rain forest). The shaded part below, which is not at the same scale of vegetation, symbolises the depth of colonization of soil layers by edaphic arthropods. See text for further explanation

Chile (see also DI CASTRI, 1968 a). This broad approximation shows a progression from a desertic non-mediterranean climate and an extreme desert vegetation through perarid (succulent desert vegetation), arid (succulent scrub), semiarid (chaparral), subhumid (mixed high chaparral), humid (evergreen sclerophyllous forest), and perhumid (mesophilous, prevalently evergreen forest, closely related to the Valdivian formations) mediterranean types. The gradient terminates with an oceanic non-mediterranean climate and with the typical Valdivian temperate rainforest. A fairly similar sequence, though with some differences, could be shown for the transect between Baja California and British Columbia (see DI CASTRI's chapter on climatographical comparisons).

A microcosm of this long latitudinal gradient is found in the valleys of the Coastal Range in Central Chile. Here the hydric conditions are not determined by latitudinal progressions, as in the previous gradient, but by edaphic and topographical factors.

Examples of these factors are the inclination and exposure of slopes, the concentration of rainwaters towards the central part of valleys, or the existence of superficial underground water near valley bottoms. This *topographical* gradient is broadly outlined in Fig. 2. At the left (north facing slope), there are near desertic conditions; at the right side,

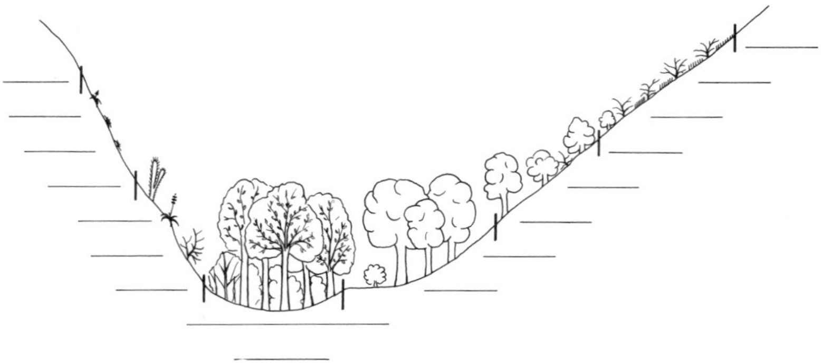

Fig. 2. Outline of a topographical gradient in a typical valley of Central Chile. From left to right, north-facing slope with semi-desertic *Puya-Trichocereus* formations, valley bottom with relict hygrophilous forest, central part with sclerophyllous forest, south-facing slope with different types of chaparral

different forms of chaparral and sclerophyllous woodlands can be found, according to soil type, inclination and elevation; in the central part, relicts of hygrophilous forest frequently occur with plants (e.g. *Drimys winteri*) having a disjunct distribution up to the Valdivian and even the Magellanic regions. This gradient cannot, of course, mirror exactly the latitudinal one. Links of the latitudinal *catena* are sometimes missing; an edaphic site factor may have a very specific action. Nevertheless, the small scale of this ecological "recapitulation" enables adoption of interesting research approaches.

A third type is the *altitudinal* gradient, from the lowland up to the mountains of the mediterranean climate regions. ZINKE's chapter gives many examples of altitudinal soil *catena*, although with a different approach. In this gradient, changes in thermic conditions become the major factor of ecological control. However, little attention will be given here to montane soil communities, most of which in the Chilean Central Andes are clearly not of a mediterranean character. Their affinities, both from a species and a community structure viewpoint, lie mainly with the fauna of the Chilean-Bolivian plateaux, where precipitation is in summer due to a high mountain tropical regime. In fact, summer is also the favourable biological period for soil organisms in the high mountains of Central Chile with winter precipitation, mainly because of the melting of snow. This contrasts with the situation of corresponding lower altitudes, where maximal biological activity in soil is recorded in winter or in the equinoctial seasons.

Finally, there is the so-called *anthropogenic* gradient. Taking as baseline as "natural" an ecosystem as possible, this entails measurement of the same soil variables in systems which have been subjected to agricultural activities over a different number of years, or are subjected to different degrees of manipulation or impact by man. The degree of manipulation may range from relatively small modifications, such as in well controlled grazing lands, to stronger modifications due to overgrazing or intensive use of fire, up to complete replacement of the natural ecosystems by irrigated cultures or plantations of fast-growing exotic trees. Fig. 3 shows in a very schematic way the situations studied in Central Chile; namely, two modified ecosystems (through overgrazing by goats on slopes and through sheep grazing in *Acacia caven* savanna), a relict of the "natural" ecosystem

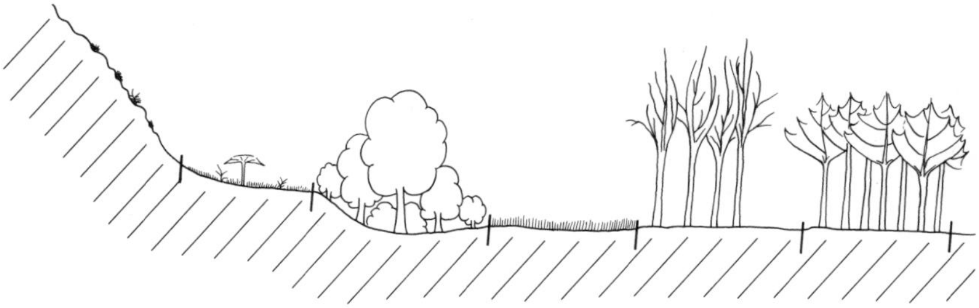

Fig. 3. Outline of the man-manipulated ecosystems studied in an anthropogenic gradient in Central Chile. From left to right, overgrazed eroded slope, grazed savanna of *Acacia caven*, remains of the natural sclerophyllous forest, irrigated fields with *Medicago sativa*, plantation of *Eucalyptus globulus*, plantation of *Pinus radiata*

(sclerophyllous forest) and three replacement ecosystems (*Medicago sativa* culture in irrigated lands, *Eucalyptus globulus* and *Pinus radiata* plantations). Convergence of the agricultural aptitude of soils in the different mediterranean regions seems to be at least as great as the convergence of the native mediterranean plants. In spite of its importance, there have been very few comparative and basic studies on this subject. Consequently, it is only possible here to present some indications on the structure of soil communities in relation to cultivated fields in mediterranean areas of Chile. This problem of anthropogenic changes in soil systems has been discussed in a general way by DI CASTRI (1966, 1968 b).

Structure of Soil-Animal Communities

In illustrating the structural patterns of communities of soil animals in relation to mediterranean climate conditions, particular attention is given to physiognomy, stratification, density, aggregation, affinity, species diversity and redundance, phenology.

Physiognomy of Soil Communities

Initially, attempts were made to adopt a similar morphological approach to that used by plant ecologists. In arid mediterranean climates the dominant surface animals are chitinized species of Coleoptera (mainly Tenebrionidae), Prostigmata, Oribatei, Opilionida, Formicidae, and very hairy species of Diplopoda Pselaphognata, larvae of Dermestidae, Thysanura, Solifugae, Araneida. On the other hand, animals living in deeper layers of compacted soils with small pore spaces are fragile, elongated species of Prostigmata, Acaridiae, Protura, Parajapygidae, while among bacteria the sporulated germs are dominant and among fungi the *Aspergillus* like forms. However, it would be difficult to argue that these are typically mediterranean characteristics; they are found in other arid regions, irrespective of the precipitation regime.

More comparable physiognomic information is given by consideration on the relative abundance and dominance of some taxa in different mediterranean conditions, as follows:

a) in arid mediterranean climates: Prostigmata, Coleoptera Tenebrionidae, Solifugae, Pseudoscorpionida Olpiidae, Parajapygidae, Caeculidae (Acarina, Trombidiformes), Orthoptera, Formicidae, fungi of *Aspergillus* type and Actinomycetes;

b) in true mediterranean climates: Oribatei, Psocoptera, Diplopoda Pselaphognata, Thysanoptera, Japygidae, Pseudoscorpionida Chernetidae, Entomobryomorpha in litter, Poduromorpha in deeper soil, Protura and Symphyla at depth, Coleoptera Lathridiidae and Cryptophagidae in litter, Coleoptera Ptiliidae and Pselaphidae in humus layers, Staphylinidae *Leptotyphlinae* at depth;

c) in humid mediterranean climates: Oribatei, Gamasides, Uropodina, Collembola Poduromorpha, Pseudoscorpionida *Gymnobisiinae* and Chthoniidae, Isopoda, Blattaria, larvae of Diptera, edaphic Copepoda, Enchytraeidae, Campodeidae, Coleoptera Carabidae, Mucoraceae fungi such as *Circinella* and *Zygorhynchus*.

The abundance of earthworms is very variable. Populations may be high in coastal mediterranean arid steppes and in humid and perhumid mediterranean climates, but they are generally low in the true mediterranean ecosystems. There are, of course, differences in species composition in these different ecosystem types. For recent information on lumbricid populations in southern mediterranean France, see BOUCHÉ (1972).

The above faunistic characterization is mainly based on research on Chilean ecosystems. Nevertheless, observations in comparable European and Californian environments tend to confirm that the faunistical composition is fairly similar, though in California there is a greater abundance of termites, favoured perhaps by a tropical bioclimatic influence (see DI CASTRI's chapter on climatography). However, this approach for faunal characterization of different mediterranean types is not entirely satisfactory, because of the large interpenetration of different elements and changes in their abundance, in response to even small local variations in environmental conditions.

A roughly quantitative method comprises measurement of the variations in density ratio between representative groups in relation to change in environmental conditions. Two such ratios are Oribatei / Acaridiae (DI CASTRI, 1963a) and Oribatei / Prostigmata (COVARRUBIAS, RUBIO and DI CASTRI, 1964). Taking Oribatei/Prostigmata, the ratio is about 1.0 in the central part of sclerophyllous forests, increases towards the more hygrophilous formations (dominance of Oribatei) and decreases strongly towards such arid formations as chaparral and steppes where Prostigmata are dominant (ratio as low as 0.1). Similar changes in the ratio are found along topographical as well as latitudinal gradients. The value of this index is confirmed by the author's observations in California and Italy, and particularly by the extended research of LOOTS and RYKE (1967) in South Africa.

For small transects especially, use has been made of a similar index on the density ratio of two Pseudoscorpion families, Chthoniidae/Olpiidae; hygrophilous forms belong to the Chthoniidae, xerophilous forms to the Olpiidae.

Considering the anthropogenic gradient (Fig. 3), the main physiognomic changes which are found in man-manipulated soil systems compared with "natural" systems are as follows:

a) in slopes eroded and degraded by overgrazing, most of the typical soil animal groups disappear and are replaced by very xerophilous and heliophilous populations, such as Caeculidae and Tenebrionidae;

b) in the grazed chaparral and savanna, as the vegetation becomes more xerophilous and more open, the density of humicolous groups decreases, the ratio Oribatei/Prostigmata is lower, and in general the composition of soil communities progressively changes towards a physiognomy corresponding to a more arid climatic type;

c) in irrigated and cultivated fields, the ratio Oribatei/Acaridiae is very markedly reduced because of the dominance of Acaridiae; the populations of Collembola are in general higher (sometimes reaching outbreak proportions), though it is impossible to establish any reliable rule on the ratio Acarina/Collembola as has been proposed by some authors (see BRAUNS, 1968). Uropodina, Pseudoscorpionida and other humicolous groups disappear almost completely, and the proportion of predatory species is also depressed. Concerning the microflora, the number of *Azotobacter* is much greater;

d) in the plantations which replace the natural woodlands, physiognomic changes in the soil fauna are less evident. On the other hand, microfloral populations are completely different in plantations with *Pinus radiata,* due to the new dominance by fungi. The ratio Oribatei/Prostigmata is often very high in *Eucalyptus globulus* plantations, due to the abundance of a few very specialized species of Oribatei.

Stratification of Animal Communities in Soil and Related Environments

One striking characteristic of ecosystems in mediterranean climates is the well-defined vertical stratification of the soil fauna. In describing the main aspects of this stratification in a typical ecosystem with a true mediterranean climate, as a basis for evaluation of the principal differences along the latitudinal gradient, a somewhat arbitrary distinction is made between deep soil, humus layers, litter layers and the under-bark environment.

The main ecological characteristic of the *deep soil layers* is the constancy of the thermic and hygric conditions. These layers are inhabited by strictly hygrophilous populations of arthropods, including endogeous beetles (Staphylinidae *Leptotyphlinae* and some Pselaphidae), Pauropoda, Symphyla, Protura, small species of Gamasides and Collembola, and sometimes Palpigradi.

In the *humus layers* there is the greatest abundance of different animal groups. Oribatei, Uropodina, Collembola, and many families of Coleoptera of intermediate ecological tolerance are dominant.

The *litter layers* are populated mainly by saprophagous and herbivorous forms (Psocoptera, Thysanura, some Isopoda, larvae of many groups of insects) specialized in attacking the hard mediterranean plant leaves and other wastes, as well as by species of other trophic levels (mycetophagous, predatory, etc.), most of them having a clear xerophilous adaptation.

Most species in the *under-bark environment,* as well as in moss and lichens on bark, have an undoubted edaphic affinity. The most common groups are Collembola (Entomobryomorpha), Prostigmata, Diplopoda (Pselaphognata), Psocoptera, Homoptera, Heteroptera and Pseudoscorpionida (Chernetidae, Cheiridiidae and Olpiidae). Their ecological tolerance is variable according to the type of bark, its exposition and the direction in which rainwater runs along the trunk; in general, however, xerophilous forms are dominant.

Regarding changes in stratification in other mediterranean climates, there tends to be an upward shifting of the superposed strata as one moves towards the more humid zones. Deep soil is less populated, partly because the main ecological factor impelling life at depth is no longer acting (in fact, even the more superficial soil layers present here a constant humidity), partly because of insufficient drainage which results in anaerobic conditions in deeper horizons. In addition, the excess of water in some soils (or in some seasons) forces many ground or humicolous groups to seek refuge in arboreal microhabitats (under bark, moss, hepatica, lichens on trees), which support a very rich fauna having great affinities with the humus and litter inhabitants of less humid mediterranean regions. It is not unusual to find on trees the so-called "true soil animals" (Protura, Pauropoda, edaphic Collembola), together with Campodeidae and "edaphic" Copepoda. Endogeous beetles such as the Staphylinidae *Leptotyphlinae* are often present here in the superficial humus layers. In the distant semi-flooded forests of the Valdivian and Magellanic regions, one can find the "suspended soils" habitat, very similar to that described by DELAMARE DEBOUTTEVILLE (1951) in tropical ecosystems.

Towards more arid mediterranean formations, on the other hand, the under-bark environment disappears, the humus stratum is strongly reduced, but in deep soil (and at even greater depth than in more typical mediterranean areas) it is possible to find populations of Protura, Pauropoda, Parajapygidae, Symphyla, small Gamasides and Collembola (Poduromorpha and Isotomidae) and larvae of Tenebrionidae. The true endogeous Coleoptera do not reach this region, except in more humid relict formations (as the Fray Jorge forest in Chile, 30° 38′ S). Finally, in the perarid mediterranean zones, the very compact soils do not favour animal life at depth. The soil related populations inhabit mainly the thin rough humus layers beneath chamaephytes and the under-stone environment.

These changes in soil stratification of animal communities are symbolised in the lowest part of Fig. 1. Soil depth and tree height are, of course, shown on different scales in this figure, since the soil was sampled at most to a depth of 75 cm. The changes described are similar, to a certain extent, to the phenomena observed in other regions of the world, by GHILAROV (1964), who formulated the "rule of the zonal change of strata".

In man-modified ecosystems, this stratification is strongly disturbed, mainly as a result of ploughing. The humus-inhabiting forms suffer the greatest damage, while the edaphic Collembola and Protura remain very numerous. Irrigation in perarid areas increases the possibility of colonisation by soil animals, even in the deep layers which sometimes support the most abundant populations (DI CASTRI and VITALI-DI CASTRI, 1971). The under-bark environment in *Eucalyptus* is fairly rich in species, that in *Pinus radiata* is very poor.

Density of Soil-Animal Populations

Density is a most unstable variable, on which it is particularly difficult to derive reliable conclusions, except perhaps the general trend of increase in total density from the arid towards the humid parts of the latitudinal gradient, and of decrease in density from the humus towards the deep soil layers. In irrigated fields, the total density, of the fauna as well as the microflora, is often much higher (but also more variable) than in the corresponding natural soils.

Comparisons of density with other mediterranean or non-mediterranean regions are almost impossible, due to the different methods of extraction used, to the fact that density is expressed either in surface or in volume units, to the often inadequate sampling regime adopted. As a rough indication, the average density in true mediterranean ecosystems is somewhat intermediate between the values obtained in dry savannas on the one hand and in temperate forests on the other. DI CASTRI (1963 a) suggested that density in similar environments might be higher in Italy than in Chile, but there are so many uncontrolled variables (human intervention, climatic changes, different geological history of the site) that valid generalization in this respect is not tenable.

It seems more relevant to establish whether changes in total density, and particularly in the abundance of the dominant animal groups, along these gradients with different hydric conditions are really determined by the amount of water in soil or by other ecological factors. Correlations between densities of different taxa in various seasons and different climatic and edaphic factors in Italy and Chile (DI CASTRI and ASTUDILLO, 1966 b) show that the main cause of density variations in soil arthropodes in mediterranean climates is not soil moisture, but the content of organic matter, N and C. These results are supported by the work of LOOTS and RYKE (1967) in South Africa on the composition of acarofauna. The importance of organic matter can be readily appreciated, because it constitutes the main input of energy and source of food for soil communities and determines the nature of the soil pores and thus of the living space for soil animals. Organic matter levels are generally low in mediterranean soils, and its amount depends largely on the density and characteristics of the vegetation, which is itself greatly influenced by hydric status; this illustrates the at least indirect significance of this factor on the soil fauna.

Aggregation of Soil-Animal Populations

Very few data are available on the aggregation of soil arthropod populations in mediterranean regions. Most of these data refer to species of Collembola in natural and cultivated soils in Italy and Chile. Values on aggregation in mediterranean woodlands are intermediate between those for tropical soils in Paraguay and those for Antarctic tundras in the South Shetland Islands (DI CASTRI, unpublished). In irrigated soils of Central Chile, a sort of "pulsatile" tendency was observed in the same given species, there being a range from extremely high values of aggregation to the absence of aggregation (expansion). For further information on aggregation in soil animals, reference should be made to ASTUDILLO, MORALES and LOYOLA (1966), DEBAUCHE (1962), and VANNIER and CANCELA DA FONSECA (1966).

Affinity among Soil-Animal Communities

The biocenotic affinity among soil-animal communities, and sometimes also among the corresponding plant associations, was derived using the index of SØRENSEN (1948), the results being ordered by means of flux diagrams (MOUNTFORD, 1962). Work near Recoaro, Vicenza, in Italy illustrates the different factors which determine the affinities among soil communities, as compared to plant associations in the same localities (DI CASTRI, unpublished). While the grouping of plant associations depended mainly on the present climate and of the kind of recent manipulations and impacts by man, the closest affinities in the soil communities were a consequence of factors related to the geological substratum and to the characteristics that the "natural ecosystems" would have had if left undisturbed. In other words, the response of the vegetation subsystem to the present ecological conditions precedes that of the soil subsystem, the composition and structure of which is mainly determined by historical factors. Frequently, and especially in modern agricultural systems in mediterranean regions, a young above-ground subsystem is coupled to an old, often relict, soil subsystem.

Species Diversity of Soil-Animal Communities

It is not the purpose here to discuss the relations of diversity to ecosystem stability and productivity. Various aspects of this problem have been extensively reviewed in a Brookhaven Symposium in Biology (see Brookhaven National Laboratory, 1969). However, measures of the diversity of soil communities in ecosystems ranging from the Tropics of South America to the Antarctica (DI CASTRI, 1969; DI CASTRI and ASTUDILLO, 1966 a and 1967; DI CASTRI, COVARRUBIAS and HAJEK, 1970; HERMOSILLA, COVARRUBIAS and DI CASTRI, 1967), demonstrate that species diversity is a most stable parameter, which can define in a synthetic way the community structure. Most of this work involved use of SHANNON's formula (SHANNON and WEAVER, 1949), adopted after the results given by this method had been compared (DI CASTRI, ASTUDILLO and SAIZ, 1964) with the results obtained by using BRILLOUIN's measure (BRILLOUIN, 1962) and with a method proposed by MARGALEF (1957).

The average species diversity value per sample is about 3,80–4,30 bits per individual in ecosystems of the true mediterranean climate region of Chile (sclerophyllous woodland). This value is nearly 5 bits per individual in the humid mediterranean ecosystems, and decreases to 2,60–3,20 bits in the chaparral, 1,80–2,30 bits in the open scrubs of the arid region, and 1,50 or less in the deserts of the perarid zone. It should be noted that in the Valdivian temperate rainforests the average values are often over 5 bits (SAIZ and DI CASTRI, 1971), and that in tropical forest values of 6 bits or more are frequently reached. On the other hand, the sub-antarctic tundras show values from near 0 to near 2 bits (SCHLATTER, HERMOSILLA and DI CASTRI, 1968). All these values refer to the species diversity of the soil arthropod component of the mesofauna extracted by BERLESE-TULLGREN funnels.

Species diversity decreases in manipulated ecosystems. Taking as a basis the species diversity in soil of a sclerophyllous forest (about 4,30 bits per individual), values of 2,20–2,80 bits are recorded in non-irrigated grazed lands, 1,30–2,50 bits in irrigated fields (though there are great variations in the values in different stages of irrigation and harvesting) and only 0,30–1,20 bits in eroded slopes.

These observations compare closely with those of MARCUZZI (1964) in Puglia, southern Italy. Though dealing with different taxa (ground beetles), MARCUZZI reported a similar sequence of decreasing species diversity from the climax formation of sclerophyllous forest towards lands with a progressive degree of human intervention.

Following the approaches described by PIELOU (1966 a, 1966 b, 1966 c), these measurements of species diversity have been complemented by calculation of the accumulated diversity, the maximum diversity and the redundance. For further information on these indices, see also CANCELA DA FONSECA (1969) and LLOYD, ZAR and KARR (1968).

Table 1 presents values for average diversity per sample (H), accumulated diversity

(H_k), maximum diversity (EH_{max}) and redundance (Re); the first four columns represent BRILLOUIN's measures, the latter four (H', H'_k, H'_{max}, Re') correspond to the relevant SHANNON's measures. Values are given for the soil fauna of a sclerophyllous forest, of a grazed savanna with *Acacia caven*, of a plantation of *Eucalyptus globulus* and of a plantation of *Pinus radiata*. Samples from the four ecosystem types were taken within a single area, of maximum diameter 100 m, near Valparaíso, Central Chile. The four ecosystem types occurred at the same elevation and on a similar soil type. Comparable data are also included for a sclerophyllous forest in Southern California (near San Diego).

Table 1. Values of average species diversity ($\bar{H}$) per sample, accumulated diversity (H_k), maximum diversity (EH_{max}) and redundance (Re), according to BRILLOUIN's and SHANNON's measures, for soil arthropod communities in a gradient from natural to man-modified ecosystems in Central Chile (near Valparaíso), and in a sclerophyllous forest in Southern California (near San Diego). The measures of diversity are in bits per individual

Ecosystems	BRILLOUIN's measures				SHANNON's measures			
	$\bar{H}$	H_k	EH_{max}	Re	$\bar{H}'$	H'_k	H'_{max}	R'e
Central Chile								
Sclerophyllous forest	3,76	4,83	6,32	0,24	4,18	4,99	6,54	0,24
Savanna with *Acacia caven* (grazed)	2,13	4,16	5,24	0,21	2,57	4,41	5,55	0,21
Plantation of *Eucalyptus globulus*	1,93	2,77	5,95	0,53	2,15	2,81	6,02	0,53
Plantation of *Pinus radiata*	1,78	3,43	5,33	0,36	2,33	3,61	5,58	0,35
Sclerophyllous forest in California	3,47	4,68	6,43	0,27	3,66	4,78	6,57	0,27

A principal conclusion from these findings is the great similarity of these structural parameters in the two "natural" ecosystems of Chile and California. This is particularly evident for the most stable parameters, such as the accumulated and maximum diversity and the redundance.

There is also a clear decrease in diversity from the natural to the modified (grazed land) to the replaced (plantations with exotic trees) ecosystems. The values of redundance in plantations are high, due to the strong dominance of a few species of soil animals. In this sense, the simplification and specialization represented by tree monocultures is reflected in the structure of the soil communities. This phenomenon is especially marked in the *Eucalyptus* plantation, where the values of redundance are the highest recorded to date in a number of mediterranean systems. It might reasonably be assumed that some secondary plant substance from *Eucalyptus* is inhibiting the colonization of these soils by a number of soil species. Related information on this topic is given by DEL MORAL and MULLER (1969, 1970) and POCHON, DE BARJAC and FAIVRE-AMIOT (1959).

In concluding these remarks on species diversity, the main factors which exert or have exerted a positive or negative influence on this parameter should be noted.

Factors which have tended to promote high diversity in the soil communities of areas with a true mediterranean climate include:

a) the transitional position, during the paleoclimatic changes and the cliseral shifts of ecosystems, of the zones which at present have a mediterranean climate. This permitted the continued presence in these regions of elements of different biogeographical origin and of diverse ecological tolerance. Advance and withdrawal of plant formations, and alternating changes in dominant environmental conditions, led to a strict segregation of small populations of arthropods in isolated habitats (deep soil, caves, disjunct valleys, mountain tops where there is condensation of marine fog). The very high rate of endemics in these mediterranean communities, as well as the "pulverization" in these regions of phylogenetic lines into many differentiated species, are partially explained by these evolutionary phenomena. This problem has been discussed in depth by JEANNEL (1965). The great abundance of species in mediterranean soils does not seem to be confined to the arthropod fauna, since a similar richness has been observed for soil fungi in Central and Southern Chile (GERTRUDE FRANZ, unpublished data). Furthermore, the progressive retraction of hygrophylous formations, produced a gradual concentration of already differentiated species in a few relict areas, where humid conditions persevered because of edaphic or atmospheric factors. The high intrageneric sympatry of some groups in a few Chilean coastal localities (Paposo, Fray Jorge, Zapallar) has probably originated in this way.

b) the great spatial heterogeneity of these habitats, referable to the special topographical features and the large range of microclimatic conditions which can occur even in very restricted areas. This fact, as well as the variety of soil of different type and different pedogenesis, enables settlement by species of very dissimilar ecological tolerance.

c) the favourable thermic conditions throughout the year and, in regard to the coastal fringe, the frequency and persistence of marine fog.

A number of factors have tended to act against high species diversity, as follows:

a) the strong seasonal fluctuations of precipitation, and especially the long period of drought in summer. Also, the fact that the length of the drought period is very variable from one year to another, particularly in regions with an arid mediterranean climate.

b) the special nature of the energy input from the vegetation subsystem (plant tissues of a very coriaceous nature, sometimes releasing secondary chemical products having an inhibitory action). Soil humus is consequently scarce.

c) the poor "filtering" effect of the open mediterranean vegetation.

d) in an opposite sense to that discussed above, the frequent paleoclimatic and geological changes which occurred in these regions probably eliminated a large number of species.

e) finally, the so intensive impact by man on mediterranean ecosystems ("anthropogenic retrogression").

Phenological Aspects

Though present data are insufficient for conclusive remarks, it seems that the maxima in the total density of soil arthropods fit fairly well with the favourable climatic periods, as deduced from climatographical analysis (DI CASTRI, 1964). Maxima are recorded in winter in the perarid and arid mediterranean regions, and in spring and autumn in the semi-arid and subhumid regions. Lesser variations of density occur through the year in humid and perhumid regions, though here there is also a tendency to equinoctial maxima.

For soil bacteria, FRANZ (unpublished) has found that the maximal density in the arid mediterranean region of Chile occurs a month after the first rainfall following the long summer drought.

In relation to seasonal migrations, preliminary observations in Chile seem to corroborate the hypothesis that during the drought period there are vertical migrations of soil animals towards the deeper layers (interstrata turnover), while the ground animals migrate towards the central humid parts of the valleys (interhabitat turnover). Return movements occur during the humid season.

Vertical migrations in Protura populations in Central Chile have been induced experimentally through controlled irrigation (HAJEK *et al.*, 1967). There was change in the strata position of Protura in the soil according to the water content of the different layers.

Evolution of Soil Animals in Mediterranean Climate Regions

In summarizing information on the structural characteristics of mediterranean soil communities, an evolutionary approach may usefully be adopted. Three main elements are:

a) the selective forces acting on organisms that inhabit soil in mediterranean climate areas.

b) the origin of the phylogenetic lines upon which these selective forces act.

c) the evolutionary strategies followed by soil organisms as adaptive response to these selective forces.

Selective Forces on Soil Organisms

Reference has already been made to certain selective forces in regions with mediterranean climate. Only the principal features will be enumerated here, as below:

1) extended drought in summer, with moisture available in deep soil horizons during drought but not in litter and superficial humus layers;

2) favourable temperature throughout the year, and thermic constancy at depth;

3) rainfall in the cold winter season, thus ultimately diminishing the occurrence of asphyxial phenomena in soil;

4) high frequency of fire;

5) very accentuated seasonality in climate;

6) high atmospheric humidity in coastal strips, which is condensed on tall vegetation, epiphytes, mosses and lichens;

7) bedrock generally in deeply fractured form;

8) high topographical heterogeneity in most mediterranean regions, that favouring heterogeneity of vegetation;

9) water trapping devices in bark and branches;

10) water conserving devices in soil;

11) deep roots of shrubs and trees, thus extending rhizosphere effects and providing input of organic matter to very deep layers;

12) evergreen sclerophyll leaves of a very coriaceous nature, and secondary toxic plant substances that make more difficult and more specialized the degradation of litter and the formation of humus.

It is noteworthy that MOONEY and DUNN (1970), in their evolutionary model for the mediterranean climate shrub form, consider the first four selective forces given above for soil animals (1–4) are also the main selective forces for mediterranean vegetation; the final four forces for soil animals (9–12) are considered by MOONEY and DUNN to be adaptive evolutionary strategies adopted by mediterranean plants. This interplay of selective forces and evolutionary strategies in relation to different components of the mediterranean ecosystem is a striking example of parallel evolution at a community level.

Biogeographical Origin of Mediterranean Soil Animals

This topic can only be discussed here in very broad terms. The problem, apart from terminological difficulties, is too complex and controversial to be dealt with in summary fashion. Also two other chapters in this volume (those by SÁIZ and VITALI-DI CASTRI) discuss in detail the biogeography of two animal groups inhabiting mediterranean soils and related environments. Relevant works have been reviewed in RAPOPORT (1968), which also provides a detailed bibliography on the subject.

Two features should however be accorded special emphasis. Firstly, because of their present and past transitional positions, soils in mediterranean climate regions are populated by forms of extremely different origin, both in space and in time, to a degree probably much greater than any other ecological-geographical zone. Secondly, very ancient phylogenetic lines persist in these soils. This faunistic "conservation" in soil can be used as a basis for diagnosis in soil pedogenesis (GHILAROV, 1956). In addition, the evolutionary concept of "living fossil" has a particular significance in soil (DELAMARE DEBOUTTEVILLE and BOTOSANEANU, 1970).

As brief illustration of the biogeographical heterogeneity of the lines of soil animals, the main types of distribution of organisms found in Chilean mediterranean soils may be cited, as below:

1) an austral (or antarctic, paleoantarctic, holoantarctic) type of distribution, this mainly including austral South America, Australia, New Zealand, South Africa and subantarctic islands. The most important biogeographical component of Chilean soils has undoubtedly this type of distribution. It comprises most lines of Gamasides, Collembola, humicolous and edaphic Coleoptera, Turbellaria, Copepoda Harpacticida, Oligochaeta *Acanthodrilinae,* Isopoda, Pseudoscorpionida, Psocoptera, and many others. Among the numerous publications dealing with this topic, reference should be made to the works of BRUNDIN (1965), DELAMARE DEBOUTTEVILLE and ROUCH (1962), JEANNEL (1962), KUSCHEL (1963), OMODEO (1963) and RAPOPORT (1971);

2) a pantropical (holotropical) type of distribution, a rare type of distribution for Chilean soil animals. Groups having this distribution are the Pseudoscorpionida *Cheiridiinae,* some Diplopoda and some parasite Nematoda;

3) a neotropical type of distribution. There are a number of elements with this biogeographical affinity, including Solifugae, many spiders, Trombidiformes Caeculidae, probably many Coleoptera Tenebrionidae, some Pseudoscorpionida, and some Formicidae. Most of these are xerophilous species living on the soil surface, under stones or under bark in arid habitats. In general, they are strongly related with Andean elements, and it is probable that neotropical lines (as well as a few nearctic lines) penetrated into Chile from the North along the Cordillera. In an opposite sense, it is also probable that the Andes represented the "bridge" followed during the migration of paleoantarctic lines towards the North. Differentiation of these two different ways of migration and types of distribution is sometimes difficult;

4) a disjunct worldwide type of distribution, frequently of a very fragmentary nature.

This distribution type includes most of the primitive "conservative" lines which in ancient times probably had a worldwide continuous distribution. In spite of their poor capacity for either active or passive dispersal, there are genera and even species which still exist in widely disjunct and ecologically diverse regions. This category includes most of the Protura, Pauropoda, Symphyla and Palpigradi. Nearly all of these groups are represented by strictly hygrophilous species which, in arid or semi-arid mediterranean climates, are restricted to deeper layers of soil. The concept of strict edaphic preference ("euedaphism") particularly when applied to these groups, is though a very relative one. Thus, the author has personally collected Protura from the humus on trees in Colombian tropical rainforests and even from the fronds of ferns in magellanic-like tundras, both ecosystems with a constantly moist climate. Similar examples could be reported also for the other groups;

5) a cosmopolitan type of distribution. Two groups must be distinguished under this heading. Firstly, there are those forms whose worldwide distribution is referable to their capacity for passive dispersal, facilitated by the existence of resistance forms and very small size. This group includes the Rotifera, Tardigrada, many Protozoa and many elements of the microflora. Secondly, there are forms introduced accidentally by man during the transfer and spread of particular agricultural practices. This group, characterized by plastic versatile species having a great aptitude for colonization, includes the Acaridiae, many Collembola and Oribatei, Nematoda, European lumbricids and some Isopoda. They constitute the main component of the soil communities of cultivated soils, particularly of the irrigated ones, and their relative abundance is constantly increasing. Nevertheless, it appears that, in mediterranean regions of Chile, the proportion of cosmopolitan elements in soil communities is lower than in weed associations, where native species are now almost completely absent.

Evolutionary Strategies of Soil Animals

Most of the adaptive mechanisms of soil animals in mediterranean regions can be grouped into three principal strategies:

a) a strategy of almost permanent habitation in deep soil strata;

b) a strategy of permanent habitation at the soil surface or in superficial soil layers, in spite of the drought in summer;

c) a mixed strategy, with alternation of life at depth and in the surface layers.

Life at Depth. The main factors leading to life at depth were doubtless of a historical nature. Paleoclimatic changes such as glaciation and desiccation, having acted on thermophilous or hygrophilous lines, progressively impelled some elements with poor capacity for active dispersal to search for more stable micro-climatic conditions in such environments as deep soil, caves and interstitial waters. This provoked a number of morphological and physiological adaptations in hypogeous animals, particularly marked in cavernicolous species.

On the other hand, there are many contemporary selective forces in mediterranean regions which favour or necessitate life at depth. Reference should be made, for instance, to selective forces 1, 2, 3, 4, 7, 10, and 11 enumerated above. As a consequence of these factors, the deep soil habitat has a high and constant moisture content and constant conditions of temperature, and is in general well oxygenated with good drainage. Furthermore, the relative shortage of direct food input from above-ground is not particularly crucial, because of root decomposition, the abundant microbial populations of the rhizosphere, and also the introduction of organic matter to deep strata through the migrations of earthworms.

Within forms having an edaphic preference, two groups can be distinguished:

– A very old hygrophilous component, described above as having a disjunct biogeographical distribution (category 4 above, which includes the Protura, Pauropoda, etc.). In a certain sense, and avoiding the finalistic connotation of the term, these lines were "preadapted" to life at depth. In spite of their hygrophily, they are fairly versatile elements; they can occupy different horizons according to the hydric regime, and they are also relatively resistant to agricultural practices.

– A component with a secondary specialization to life at depth, whether in caves (Pseudoscorpionida, Isopoda, Coleoptera Carabidae) or in deep soil. Among endogeous forms, typical examples are found in many families of Coleoptera (see COIFFAIT, 1958, and SAIZ's chapter). Cavernicolous and endogeous species, and more especially the former, have a more restricted ecological tolerance. In spite of this, "euedaphism" is even in this case not an absolute concept, since even the most euedaphic group (Staphylinidae *Leptotyphlinae*) is frequently found in superficial humus layers in regions such as the Valdivian rain forests, where soil moisture is not critical.

Life in Surface and Ground Layers. This grouping comprises animals living in litter, in the most superficial layers of humus and on the ground, and, to a certain extent, the underbark inhabitants. Two types of strategies can be distinguished: those adopted by animals which show some activity throughout the year, in spite of the seasonal drought, and those of organisms which are dormant during the unfavourable period.

Among the adaptations of the first group, mention should be made of the strongly chitinized cuticle and the many physiological mechanisms in regard to water metabolism for surviving periodical hydric stress. These features have been well studied in species of Coleoptera Tenebrionidae, Isopoda, Orthoptera and Arachnida. EDNEY (1957) has described a general approach to this subject. Other special adaptations in this group are those for attacking coriaceous and even woody plant tissues (mouth structures, enzymatic adaptations, symbiosis with Protozoa in the digestive tract), and, in many species, migrations from one habitat to another along the topographical transect. These evolutionary strategies are related to selective forces 1, 2, 5, 8 and 12 detailed above.

The arboreal life of soil-related animal groups is favoured by selective forces No. 6 and 9. In animals living permanently under bark (such as some mites and some pseudoscorpions), clear morphological adaptations, such as dorso-ventral body flattening and mimetism to the particular colours of the bark of mediterranean trees, frequently occur.

Organisms which pass the unfavourable dry periods in dormant or diapause phases possess different kinds of resistance forms (cysts, spores). They are characterized by very rapid biological cycles. Normal activity patterns and reproductive processes are readopted shortly after the start of the humid season. These strategies are mainly typical of the so-called "aquatic soil animals", those that live in the thin water films surrounding small soil particles (such as Rotifera, Protozoa and Tardigrada), and most Bacteria, Fungi and Actinomycetes.

Mixed Strategies. Animal groups adopting these strategies take advantage of surface living (greater trophic availability) as well as life at depth (greater microclimatic uniformity) by means of periodical vertical migrations, of a daily, seasonal, occasional or ontogenetic nature. Examples are the large vertical migrations of earthworms, the small occasional migration of hygrophilous groups in response to higher soil moisture near the soil surface, and the most classical type of mixed strategy based on ontogenetic life cycles. Characteristic of this last named strategy are groups of holometabolous insects (many Coleoptera, Diptera and Lepidoptera) whose larvae live at depth or in litter (having generally a hygrophilous preference), while adults emerge at the soil surface and are frequently very xerophilous in nature.

Final Remarks

At the beginning of this chapter, three main questions were posed. It seems reasonable to purport that there are particular structural patterns in the soil communities of mediterranean climate regions, referable mainly to the type of soil stratification, to parameters of species diversity and redundance, and to phenological behaviour. On the other hand, the existence of true morphological convergences in soil animals is more controversial given the present state of knowledge. The undoubted similarities between the soil communities of several mediterranean areas are referable to the similar paleoclimatic events to which these communities were submitted and to the affinities between phylogenetic stocks, as well as to the comparable evolutionary strategies adopted in response to similar stringent selective forces. Historical heritage appears to be more pervasive in soil communities than in the corresponding vegetation.

Humus seems to be the crucial factor in the regulation and functioning of the litter-soil subsystem. Because of the nature of the mediterranean vegetation, production of humus is both poor and very specialized. However, the humus of mediterranean soils largely determines the patterns of distribution of soil animals, their aggregation, the relative abundance of dominant groups and the values of species diversity, as well as the physical characteristics of the soil habitat (pore spaces, water adsorption, filtering and dampening of atmospheric stresses). On the other hand, humus characteristics depend ultimately on the activities of soil organisms. As the centre of biochemical information and of biological complexity, humus is strongly altered, up to almost complete depletion, by agricultural practices in mediterranean regions. Priority attention should be given to changes in humus formation, and their consequences on soil organisms, under different management regimes. Special attention should be devoted to the extensive plantations that are progressively changing the present mediterranean landscape, particularly on slopes. This is particularly true for *Eucalyptus* in several mediterranean regions and to *Pinus radiata* plantations mainly in Australia and Chile.

References

Astudillo, V., Morales, M. A., Loyola, R.: Problemas en el análisis estadístico de poblaciones con distribución contagiosa. 359–369. In: Progresos en biología del suelo, Monografías I. Montevideo: UNESCO 1966.

Athias-Henriot, C.: Observations sur les *Lasioseius spathuliger* Méditerranéens (Parasitiformes, Laelapoidea). Rev. Ecol. Biol. Sol **4**, 143–154 (1967).

Bouché, M. B.: Lombriciens de France. Ecologie et systématique. Paris: Institut National Recherche Agronomique 1972.

BRAUNS, A.: Praktische Bodenbiologie. Stuttgart: G. Fischer 1968.

BRILLOUIN, L.: Science and information theory. New York: Academic Press 1962.

Brookhaven National Laboratory: Diversity and stability in ecological systems. Brookhaven Symposia in Biology Number 22. Springfield: U.S. Department of Commerce 1969.

BRUNDIN, L.: On the real nature of transantarctic relationships. Evolution **19**, 469–505 (1965).

CANCELA DA FONSECA, J.–P.: L'outil statistique en biologie du sol. VI. – Théorie de l'information et diversité spécifique. Rev. Ecol. Biol. Sol **6**, 533–555 (1969).

CASTRI, F. DI: Prime osservazioni sulla fauna del suolo di una regione delle Prealpi Venete (Monte Spitz, Recoaro). Atti Ist. Ven. Sc. Lett. Arti **118**, 475–493 (1960).

CASTRI, F. DI: Estado biológico de los suelos naturales y cultivados de Chile Central. Bol. Prod. anim. (Chile) **1**, 101–112 (1963a).

CASTRI, F. DI: Etat de nos connaissances sur les biocoenoses édaphiques du Chili. 375–385. In: DOEKSEN, J., DRIFT, J. VAN DER (Eds.), Soil Organisms. Amsterdam: North-Holland Publ. Co. 1963b.

CASTRI, F. DI: Interpretación bioclimática de las biocoras de Chile de acuerdo a su período de actividad biológica. Bol. Prod. anim. (Chile) **2**, 173–186 (1964).

CASTRI, F. DI: Consideraciones sobre el estado de disclímax en las zoocenosis edáficas. 333–341. In: Progresos en biología del suelo, Monografías I. Montevideo: UNESCO 1966.

CASTRI, F. DI: Esquisse écologique du Chili. 7–52. In: DELAMARE DEBOUTTEVILLE, C., RAPOPORT, E. (Eds.), Biologie de l'Amérique Australe, Vol. IV. Paris: C.N.R.S. 1968a.

CASTRI, F. DI: Interferencias del hombre en los sistemas edáficos. 133–143. In: Progressos em biodinâmica e productividade do solo. Santa Maria: Universidade Federal de Santa Maria 1968b.

CASTRI, F. DI: Effects of climate and weather on the structure of soil arthropod communities. 47–48. In: TROMP, S. W., WEIHE, W. H. (Eds.), Biometeorology, Vol. 4. Amsterdam: Swets and Zeitlinger 1969.

CASTRI, F. DI: Les grands problèmes qui se posent aux écologistes pour l'étude des écosystèmes du sol. 15–31. In: PHILLIPSON, J. (Ed.), Methods of study in soil ecology, Ecology and Conservation 2. Paris: UNESCO 1970.

CASTRI, F. DI, ASTUDILLO, V.: Revisión crítica de las aplicaciones de la teoría de la información en zoología del suelo. 313–331. In: Progresos en biología del suelo, Monografías I. Montevideo: UNESCO 1966a.

CASTRI, F. DI, ASTUDILLO, V.: Análisis de algunas causas abióticas de variación en la densidad de la fauna del suelo. 371–377. In: Progresos en biología del suelo, Monografías I. Montevideo: UNESCO 1966b.

CASTRI, F. DI, ASTUDILLO, V.: Relationship between bioclimatic conditions and the amount of information in soil biocenosis. 255. In: TROMP, S. W. and WEIHE, W. H. (Eds.), Biometeorology, Vol.3. Amsterdam: Swets and Zeitlinger 1967.

CASTRI, F. DI, ASTUDILLO, V., SAIZ, F.: Aplicación de la teoría de la información al estudio de las biocenosis muscícolas. Bol. Prod. anim. (Chile) **2**, 153–171 (1964).

CASTRI, F. DI, COVARRUBIAS, R., HAJEK, E.: Soil ecosystems in subantarctic regions. 207–222. In: Ecology of the subarctic regions, Ecology and Conservation 1. Paris: UNESCO 1970.

CASTRI, F. DI, VITALI-DI CASTRI, V.: Colonización por organismos edáficos de territorios desérticos sometidos a riego (Estancia Castilla, Provincia de Atacama, Chile). Bol. Mus. Nac. Hist. Nat. Chile **32**, 17–40 (1971).

COIFFAIT, H.: Les Coléoptères du sol. Paris: Hermann 1958.

COVARRUBIAS, R., RUBIO, I., DI CASTRI, F.: Observaciones ecológico-cuantitativas sobre la fauna edáfica de zonas semiáridas del Norte de Chile (Provincias de Coquimbo y Aconcagua). Monografías sobre Ecología y Biogeografía de Chile. Santiago: Universidad de Chile 1964.

DEBAUCHE, H. R.: The structural analysis of animal communities of the soil. 10–15. In: MURPHY, P. W. (Ed.), Progress in soil zoology. London: Butterworths 1962.

DELAMARE DEBOUTTEVILLE, C.: Microfaune du sol des pays tempérés et tropicaux. Paris: Hermann 1951.

DELAMARE DEBOUTTEVILLE, C., BOTOSANEANU, L.: Formes primitives vivantes. Paris: Hermann 1970.

DELAMARE DEBOUTTEVILLE, C., ROUCH, R.: Sur la présence de quelques lignées paléantarctiques en Patagonie andine. C. R. Acad. Sc. Paris **254**, 1336–1338 (1962).

DELAMARE DEBOUTTEVILLE, C., VANNIER, G.: La recherche coopérative sur programme en écologie du sol, ou R.C.P. 40. Rev. Ecol. Biol. Sol **3**, 523–531 (1966).

DEL MORAL, R., MULLER, C. H.: Fog drip: a mechanism of toxin transport from *Eucalyptus globulus*. Bull. Torrey Bot. Club **96**, 467–475 (1969).

DEL MORAL, R., MULLER, C. H.: The allelopathic effects of *Eucalyptus camaldulensis*. Amer. Midl. Nat. **83**, 254–282 (1970).

DEN HEYER, J., RYKE, P. A. J.: A mesofaunal investigation of the soil in a thorn-tree *(Acacia karroo)* biotope. Revista de Biologia **5**, 309–364 (1966).

EDNEY, E. B.: The water relations of terrestrial arthropods. Cambridge: Cambridge University Press 1957.

GHILAROV, M. S.: Soil fauna investigation as a method in soil diagnostics (the South Crimean Terra Rossa taken as an example). Boll. Lab. Zool. Gen. Agr. F. Silvestri Portici **33**, 574–585 (1956).

GHILAROV, M. S.: Connection of insects with the soil in different climatic zones. Pedobiologia **4**, 310–315 (1964).

HAJEK, E., ASTUDILLO, V., DI CASTRI, F., COVARRUBIAS, R., RUBIO, I., and SAIZ, F.: Efectos microclimáticos sobre la diversidad de la mesofauna del suelo. Arch. Biol. Med. Exper. **4**, 210 (1967).

HELLMERS, H., HORTON, J. S., JUHREN, G., O'KEEFE, J.: Root systems of some chaparral plants in southern California. Ecology **36**, 667–678 (1955).

HERMOSILLA, W., COVARRUBIAS, R., DI CASTRI, F.: Estudio comparativo sobre la estructura de zoocenosis edáficas en el trópico y en la Antártida. Arch. Biol. Med. Exper. **4**, 210 (1967).

HERMOSILLA, W., MURUA, R.: Estudio ecológico-cuantitativo de la fauna hipogea en las dunas de Concon-Quintero. Bol. Prod. anim. (Chile) **4**, 69–102 (1966).

JEANNEL, R.: Les Trechides de la Paléantarctide occidentale. 527–655. In: DELAMARE DEBOUTTEVILLE, C., RAPOPORT, E. (Eds.), Biologie de l'Amérique Australe, Vol. I. Paris: C.N.R.S. 1962.

JEANNEL, R.: La genèse du peuplement des milieux souterrains. Rev. Ecol. Biol. Sol. **2**, 1–22 (1965).

KUSCHEL, G.: Problems concerning an austral region. 443–449. In: Proceedings Tenth Pacific Science Congress. Honolulu: Bishop Museum Press 1963.

LEE, K. E., WOOD, T. G.: Preliminary studies of the role of *Nasutitermes exitiosus* (HILL) in the cycling of organic matter in a yellow podzolic soil under dry sclerophyll forest in South Australia. Proc. 9th. Int. Cong. Soil Sci. Adelaide **2**, 11–18 (1968).

LLOYD, M., ZAR, J. H., KARR, J. R.: On the calculation of information-theoretical measures of diversity. Amer. Midl. Nat. **79**, 257–272 (1968).

LOOTS, G. C., RYKE, P. A. J.: The ratio Oribatei: Trombidiformes with reference to organic matter content in soils. Pedobiologia **7**, 121–124 (1967).

LOSSAINT, P.: Etude intégrée des facteurs écologiques de la productivité au niveau de la pédosphère en région méditerranéenne dans le cadre du P. B. I. Programme et description des stations. Oecol. Plant. **2**, 341–366 (1967).

MARCUZZI, G.: Osservazioni biocenologiche sulla coleotterofauna pugliese. Atti Acc. Naz. It. Ent. **11**, 1–11 (1964).

MARCUZZI, G.: Osservazioni ecologiche sulla fauna del suolo di alcune regioni forestali italiane. Ann. Centro Econ. Mont. Venezie **7**, 209–331 (1968).

MARCUZZI, G., DI CASTRI, F.: Osservazioni ecologico-quantitative sulla fauna del suolo di Recoaro (Prealpi Venete). Mem. Mus. Civ. St. Nat. Verona **15**, 159–172 (1967).

MARCUZZI, G., DALLA VENEZIA, L.: First results of the study of the soil fauna of two Italian artificial ecosystems. Rev. Ecol. Biol. Sol **9**, 229–233 (1972).

MARCUZZI, G., LORENZONI, A. M., DI CASTRI, F.: La fauna del suolo di una regione delle Prealpi Venete (M. Spitz, Recoaro). Aspetti autoecologici. Atti Ist. Ven. Sc. Lett. Arti **128**, 411–567 (1970).

MARGALEF, R.: La teoría de la información en ecología. Mem. Real Acad. Ciencias Artes Barcelona **32**, 373–449 (1957).

MCPHERSON, J. K., MULLER, C. H.: Allelopathic effects of *Adenostoma fasciculatum*, "chamise", in the California chaparral. Ecol. Monogr. **39**, 177–198 (1969).

MOONEY, H. A., DUNN, E. L.: Convergent evolution of mediterranean-climate evergreen sclerophyll shrubs. Evolution **24**, 292–303 (1970).

MOUNTFORD, M. D.: An index of similarity and its application to classificatory problems. 43–50. In: MURPHY, P. W. (Ed.), Progress in soil zoology. London: Butterworths 1962.

OMODEO, P.: Distribution of the terricolous Oligochaetes on the two shores of the Atlantic. 127–151. In: North Atlantic Biota and their History. Oxford: Pergamon Press 1963.

PARIS, O. H., PITELKA, F. A.: Population characteristics of the terrestrial isopod *Armadillidium vulgare* in California grassland. Ecology **43**, 229–248 (1962).

PIELOU, E. C.: SHANNON's formula as a measurement of specific diversity: its use and misuse. Amer. Natur. **100**, 463–465 (1966 a).

PIELOU, E. C.: The measurement of diversity in different types of biological collections. J. Theor. Biol. **13**, 131–144 (1966 b).

PIELOU, E. C.: The use of information theory in the study of the diversity of biological populations. Proc. 5th Berkeley Symp. Math. Statist. Probability **4**, 163–177 (1966 c).

POCHON, J., DE BARJAC, H., FAIVRE-AMIOT: L'influence de plantations d'*Eucalyptus* au Maroc sur la microflore et l'humus du sol. Ann. Inst. Pasteur **97**, 403–406 (1959).

RAPOPORT, E. H.: Algunos problemas biogeográficos del Nuevo Mundo con especial referencia a la región neotropical. 53–110. In: DELAMARE DEBOUTTEVILLE, C., RAPOPORT, E. (Eds.), Biologie de l'Amérique Australe, Vol. IV. Paris: C.N.R.S. 1968.

RAPOPORT, E. H.: The geographical distribution of neotropical and antarctic Collembola. Pacif. Ins. Monogr. **25**, 99–118 (1971).

SAIZ, F., DI CASTRI F.: La fauna de terrenos naturales e intervenidos en la región valdiviana de Chile. Bol. Mus. Nac. Hist. Nat. Chile **32**, 5–16 (1971).

SCHLATTER, R., HERMOSILLA, W., DI CASTRI, F.: Estudios ecológicos en Isla Robert (Shetland del Sur). 2-Distribución altitudinal de los Artrópodos terrestres. Publ. Inst. Antártico Chileno N. 15, 1–32 (1968).

SHANNON, C. E., WEAVER, W.: The mathematical theory of communication. Urbana: Univ. Illinois Press 1949.

SØRENSEN, T.: A method of establishing groups of equal amplitude in plant sociology based on similarity of species content and its application to analysis of the vegetation on Danish commons. Vidensk. Selsk. Biol. Skr. **5**, 1–34 (1948).

THERON, P. D., RYKE, P. A. J.: The family Nanorchestidae GRANDJEAN (Acari: Prostigmata) with descriptions of new species from South African soils. J. ent. Soc. sth. Afr. **32**, 31–60 (1969).

VAN DEN BERG, R. A., RYKE, P. A. J.: A systematic-ecological investigation of the acarofauna of the forest floor in Magoebaskloof (South Africa) with special reference to the Mesostigmata. Revista de Biologia **6**, 157–234 (1967–1968).

VANNIER, G., CANCELA DA FONSECA, J.-P.: L'échantillonnage de la microfaune du sol. Terre et Vie **1**, 77–103 (1966).

WOOD, T. G.: Micro-arthropods from soils in the arid zone in Southern Australia. Search **1**, 75 (1970).

2

Microbial Activity under Seasonal Conditions of Drought in Mediterranean Climates

ROGER SCHAEFER

Introduction

The nature of microbiological and biochemical activity in the soils of mediterranean regions is still largely unknown. Some literature exists on the physiology and taxonomy of soil microorganisms in mediterranean areas; but this information is generally sketchy, both from temporal and spatial aspects.

Further, we have almost no information on the activity and metabolism of microbial communities and the biochemical aspects of humus formation in mediterranean climates. Little is known of the microbial communities across the range of mediterranean sub-climates. Thus, we are not yet able to predict the most probable effect of microbiological activity on the evolution of biotopes in the mediterranean zone, nor the trends which present-day stages of man-created successions are undergoing.

The litter/soil subsystem represents a central part of the total ecosystem. It is an ecotone between the autotrophic, energy-storing vegetation and the microbial community. The latter not only mineralizes part of the synthesized organic material but reorganizes these substrates and stores them in the form of microbial biomass and humified material. We now know that soil humus acts as a source of stored information and as a biochemical code, thus exerting a powerful role in system regulation.

The epigeous subsystem is directly influenced by seasonal variations of the climate. Higher plants react to climatic variations by regulating their vegetative and reproductive cycles. To the hypogeous subsystem, roots and particularly the microorganisms, the impact of the external environment is filtered, dampened and displaced in time by the successive screens of vegetation cover, litter layer, and soil itself. The response of the microorganism and of microbial communities to the soil microclimate takes place only after transmission and reception of signals from the aerial environment which pass through these screens or stages.

Little is known about the relationships between the external macro-, meso- and micro-climate on the one hand, and the soil micro-climate on the other. The precise effects of plant cover, litter type, and soil, in their functions as screens or filters, has been largely ignored. Moreover, because of its dynamic equilibrium, the characterization of the telluric microclimate is generally incomplete and uncertain. A large number of frequent measurements, often of a technically difficult nature, are still required.

The understanding of the mechanisms of ecosystem regulation is based in part on knowledge about the modalities of interaction and the periodic change in different environmental variables. Areas with a mediterranean climate provide a good example of convergence of ecosystems. A detailed study of the similarities of these areas, and particularly of their distinctive features, will show the decisive effect of this type of climate on the organization, the functioning and the evolution of a constellation of ecosystems which are shaped on a similar pattern. The dynamic status of humus integrates the activity of the microbial communities with the changes in their environment. Thus, the ecological and biochemical study of humification and humus degradation is the key to the understanding of the participation of microorganisms in ecosystem regulation.

Microbial Activities in Mediterranean Environments

The alternation of a generally hot and dry season with a humid cool season results in a peculiar pattern of microbial activity in mediterranean regions. Two growth flushes or peaks take place, one at the onset of the drying phase and the other when moistening starts. One of these transitory peaks is due to thermal activation, the other to the removal of drought as a limiting factor.

Few studies are available which consider the characteristics of mediterranean climate and vegetation in relation to microbial ecology. The more arid the environment, the less is known about the dynamic state of activity of microbial types. It is precisely these arid regions where future research ought to demonstrate the inter-relationships between microbial activities, seasonal evolution of climate and physico-chemical characters of soils.

Water Requirements

A seasonal phase of aridity characterizes the mediterranean climate. The microflora of mediterranean soils is thus adapted to a temporary water deficit. Although ammonifying bacteria and *Azotobacter* may be found at considerable depths in arid soils, the most representative and active microorganisms are located within the superficial or subsuperficial horizons. Algae, mainly Chlorophyceae and Cyanophyceae, play the role of pioneers and are often localized in particular micro-habitats: under translucent stones or salt crusts. Fungi are relatively abundant *(Penicillium* and *Aspergillus)* and sporulating bacilli (*Bacillus subtilis, B. licheniformis, B. megatherium*) and Actinomycetes are frequently present. Microorganisms are generally concentrated at the level of the rhizosphere or in the litter layer formed under cushion-forming plants (VARGUES, 1952).

The productivity of higher vegetation is rather low and follows a marked seasonality. This, in addition to the distinctive chemical composition of the litter, exerts a strong influence on humus formation. The low level of accumulation is attributable to a climate which provides for two periods of intense mineralization each year.

The existence of microbial groups with very different water requirements has been shown by DOMMERGUES (1962, 1964a, 1964b).

– Hyperxerophilic groups, active at a pF higher than 4,9, which ammonify tyrosine and caseine, mineralize glucose, glycerophosphate, starch, and complex organic compounds, and promote sulfhydrization.

– Xerophylic groups, active between pF 4,9 and 4,2, which comprise cellulolytic, sulfoxidating and nitrifying bacteria.

– Hygrophilic groups, active below pF 4,2, which are composed of nitrogen fixing bacteria.

These pF thresholds determine the growth and physiological activity of different microbe groups. If there is sufficient humidity, microorganisms are active in the process of mineral cycling. However, beyond pF 5,5, physico-chemical processes predominate. Intense solar irradiation and prolonged drought cause active physico-chemical mineralization which follows the period of biological activity, as has been shown by DROUINEAU, LEFEVRE and BLANC-AICARD (1952, 1953 a, 1953 b), BIRCH (1958, 1959, 1960), and others.

Variation in threshold levels, according to the water requirements of the different microbial types, results in the accumulation of certain substrates during the drying of the soil. The shift between the threshold of the ammonifying and the nitrifying bacteria thus brings about an accumulation of ammonia. Also, the succession of phases of mineralization and immobilization is regulated by the proliferation of physiological groups with different water requirements as the soil moisture changes. It is evident that soil type, composition of the microflora, soil moisture content, and the nature of the organic matter present, exert their influences on the level of the thresholds of activity.

The microorganisms of arid and semi-arid soils are remarkable for their xerophily and xerotolerance, itself increased in the presence of kaolin which exerts a protective effect against drying-out (DOMMERGUES, 1964a, 1964b). The mechanisms of microbial adaptation to soil desiccation are diverse. They include the production of spores and cysts. These microorganisms are remarkable also for their thermophily and thermotolerance. The thermophiles are essentially proteolytic bacteria which are also halophilous and halotolerant. Indeed, the soluble salt content of the soil varies greatly according to the season, sometimes reaching extremely high values, temporarily or permanently. KILLIAN (1936) and KILLIAN and FEHER (1939) have indicated a stimulation of activity of the microflora when the mineral salts accumulated during the dry season are put into solution and become progressively diluted.

Several authors, among them SASSON and DASTE (1963), have reported an intrinsic seasonal periodicity in activity, especially of *Azotobacter,* which is not modified by irrigation or other cultivation practices. During the course of the year, changes in physiological state are linked to differences in the capacity of *Azobacter* to utilize different substrates.

Effect of Rewetting and Freeze-Thaw Cycles

The consequences on the composition and activities of the microflora of alternating cycles of wetting and drying have been reported by several investigators. It has been shown, mainly by BIRCH (1958, 1959, 1960) that such alternation causes periodic change in microbe populations, not only in numbers but also in metabolic characteristics. There is a succession of different physiological groups, depending on group requirements and tolerance to different temperatures and humidities, and to the nature and concentration of substrates and metabolites.

An analogous effect is produced by freeze-thaw cycles. The effect of freezing is similar to that of desiccation, in that an essential element of the environment reaches a minimum, reducing metabolic activity to extremely low levels. The concomitant release of mineral nitrogen, due to the death of part of the population, results in an accumulation of ammonia and of nitrate, which, in spite of subsequent dilution, stimulates vegetation growth on thawing. The mechanical pressure of ice within soil aggregates increases interstices and the extent of capillary pores, and crushes particles to smaller sizes. Thus the active surface for nitrification and general biological activity increases, new sources of nutrients become available, and oxygen supply and humidity distribution are improved.

In addition to the above effects, part of the microbe population dies during freezing, as it does during desiccation. Easily degradable material becomes available as energy and nitrogen to the surviving population, whose activity is greatly increased for a short period while conditions are again optimal.

Mineralization of Humus and Evolution of Organic Matter

The flush of humus decomposition that follows the moistening of dry soil is due to the high activity of the logarithmic growth phase of an increasing microbial population. The high initial rate of organic C and N mineralization following the moistening of a dry soil, and the subsequent rapid decline in this rate, are due to changes in the total metabolic capacity and size of the microbial community. There are also physical reasons for the flush. The longer a soil remains in an air-dry condition, the greater is the accumulation of water-soluble organic matter which goes into solution upon remoistening. Drying leads to changes in organic gels, fragmentation and increased porosity enlarging their surface area. If gel structure changes are still reversible, the surface area will be reduced on moistening. This explains in part why the flush of activity, especially mineralization of N, soon returns to normal background values.

Humic compounds, adsorbed on clay minerals, are released upon drying; photochemical depolymerization processes take place in the upper layer of the soil exposed to solar radiation and produce new sources of available energy.

Successive treatments of dry heat and moist cold lead to successive flushes of decomposition. As a similar effect is obtained by treatment with organic gases (ether, chloroform), the changes in population size and activity are of primary importance.

The frequency and effectiveness of drying-wetting cycles are determined by the type of climate and by agricultural practices. Any activity that intensifies soil drying, such as burning or certain types of cultivation (bare fallowing), leads to an increased loss in organic carbon. Because such practices also lead to more nitrate production when rains or irrigation begin, they are often shortsightedly used in land management. These practices rapidly deplete soil humus, which is already very low in arid or semi-arid climates.

DUCHAUFOUR (1968) has summarized the knowledge of fersiallitic and chestnut-brown soils which are frequent in mediterranean climates. The former correspond to the most humid stages, with a forest climax; the latter are found within the driest stages, presently supporting steppe vegetation or cleared forest.

Fersiallitic soils are typical of a mediterranean climate with a strong contrast between a relatively cold humid season and a warm, very dry summer season. As in subtropical climates of low humidity, silica-rich soils of the illite or montmorillonite type predominate. The wet season brings about decarbonation, if parent rocks are calcareous, which induces fersiallitic alteration and rubefaction. The free iron is immobilized. Lixiviation of bases takes place during the humid phase. During the long dry season this process is slowed down and the soil is maintained in a state of base saturation by capillary rise.

Chestnut brown soils currently support steppe vegetation which is probably a secondary formation. These soils develop in a climatic range of transition steppe/cleared forest, or under shrubby vegetation with grasses. They formerly had a very humiferous superficial horizon which was removed by erosion after forest clearing. The present humiferous horizon results from the contemporary grass vegetation. The organic matter is much less polymerized than that of chernozems and vertisols. This is related to the fact that the cold season humid phases do not result in the marked temporary anaerobiosis as is found in the latter two soils. But the humified material is equally stabilized by a narrow binding with clay minerals, and is much more polymerized than isohumic soils with a desaturated exchange complex.

The type of experiments carried out by NGUYEN KHA and DUCHAUFOUR (1969) and NGUYEN KHA and DOMMERGUES (1970), involving, during the incubation of soils, a reproduction of various climatic patterns under laboratory conditions, would yield important information on mediterranean soils. These investigators found that tropical conditions with strong microclimatic contrasts markedly activate the decomposition of fresh organic matter on one hand, and the process of humification on the other.

The effect of a humid and moderately hot summer is an acceleration of degradation of labile humic acids formed during winter. This contrasts to a dry and hot summer which promotes the formation of stable humic acids which resist microbial attack. This is a very significant aspect of regulation of the humus balance under mediterranean climate, tending to adjust humus catabolism to the relatively low production of litter. Cyclic wetting and drying increases the susceptibility of organic matter to microbial oxidation and simultaneously induces polymerization of humic acids.

Since convergence in soil organic matter properties is experimentally obtained under cyclic effects from different pedological types, very strong mechanisms are involved. The transfer, under field conditions, of a certain volume of humic horizon from soils taken from areas differing in climatic conditions should yield interesting results on the direct influence of seasonal changes on soil humus properties. THALMANN (1967) has described such an experiment, in which plots of $2 \times 6 \times 0.6$ m were exchanged between four localities in Germany along a N–S axis. KILLIAN and VARGUES (1953), STEUBING (1960) and STEBAEVA (1967) have reported on experiments with a similar design. No attempt at such an exchange has yet been made in mediterranean climates. Such an exchange along catenas of sub-types and reaching into areas adjacent to, and different from, the mediterranean type of climate would certainly furnish a better understanding of the peculiar effects of the seasonally contrasted changes on the functioning of mediterranean ecosystems.

Status of Available Nitrogen

Seasonal variations in nitrate fluctuations depend on several factors, predominantly: 1) the extent of the dry periods and the intensity of drying after each wet period, as controlled by temperature and humidity; 2) the pattern of rainfall at the onset and termination of the rains, which may determine the extent of nitrate accumulation before leaching or redrying can act against further accumulation; 3) the soil type, with particular reference to total N status, the texture and structure of the topsoil, and the status of other nutrients essential for bacterial nitrification.

Nitrate accumulation in the field depends upon a progressive drying period after rain, in which the topsoil gradually dries out to an increasing depth. Under these conditions, aeration and soil temperature are favorable for rapid bacterial nitrification and promote the retention of nitrate ions in the surface layers. As long as the drying cycle continues, nitrate will accumulate and remain "fixed" in the dry topsoil. When drying starts, the concentration of soluble nutrients in the soil solution increases, thus stimulating the general activity and respiration of the microflora and inducing immobilization of mineral N. This explains the existence of a growth flush upon moistening as long as soluble substrates and organic nitrogen are nonlimiting.

The sudden entry of heavy rain into the warm dry topsoil, rich in nitrate and available organic matter, sometimes produces reducing conditions, leading to nonassimilative nitrate reduction. Generally, nitrates are removed biologically through assimilative reduction as well as physically by leaching. A flush of nitrate accumulation is observed on progressive remoistening.

In medium and heavy soils, nitrate is absorbed within structural aggregates; percolating rainfall merely passes around the soil crumbs, allowing nitrates to persist in these soils throughout periods of heavy rain. Only when the aggregates collapse is leaching more complete. The initial drainage water consequently has a lower nitrate concentration than does subsequent water (WEBSTER and GASSER, 1959). Increased mineralization in an air-dried, remoistened soil appears akin to partial sterilization processes which inhibit bacterial activity temporarily; also, drying followed by rewetting shatters soil particles, exposing new and larger interior surfaces to bacterial action.

Anthropogenic Influence

"Natural" ecosystems are seldom encountered in mediterranean climates because of the age-old cultivation of soils. Removal of native plant cover has caused erosion and changes in micro-climate. Soil management practices, mainly irrigation and plowing, alter to a great extent the mediterranean character of the microbial environment. Repeated exposure to solar radiation, wetting-drying cycles or constant moisture applied during the normally dry and hot season, promote an increase in the mineralization of humus. This tends to exhaust the already low amounts of accumulated organic matter unless the loss is compensated by an input of organic fertilizer. Burning of cultivation residues acts deleteriously, less through direct action on the microbe population than through removal of energy sources. For these reasons, cultivated soils in mediterranean climates are generally not very fertile. Soil humus has been depleted to a point critical to the natural equilibrium. Spontaneous control of plant pathogens is weak because the size and diversity of the microbial population has narrowed, limiting the probability of antagonistic activities.

Deterioration in the balance of soil organic matter, due to excessive stimulation of microbial catabolism, causes a decrease in the direct influence of humic substances on higher vegetation, especially crops. With less humic colloids, soil is less resistant to drying-out, and there is reduction in the length of the period favorable for microbial activity. This phenomenon may be regarded as a mechanism of regulation. The effect of monocultures (wheat, *Pinus radiata, Eucalyptus*) on humus quality and balance in mediterranean climates should be investigated much more thoroughly.

Conclusions

In general terms, the pattern of microbial and biochemical activity in mediterranean climates is unique in that, instead of one strong peak of activity such as occurs under warm, humid summer conditions, two short flushes take place. Relatively high activity during winter favors the conservation of a stock of stable humus, whereas during summer the microbial population remains in a cryptobiotic state.

There are many aspects of the ecology of microorganisms in mediterranean regions which need further investigation. For example, detailed research is needed on the chemical nature, distribution, and behaviour of canopy-drip and stem-flow immediately after the onset of rain. A differentiation between salts formerly deposited as atmospheric dust on leaves and lixiviated organic and mineral compounds seems necessary. Rhizosphere effects could be triggered by this short cycling, which may have a particular influence upon plant productivity.

Great variation in temperature over a 24 hour period of air near the soil surface gives rise to a movement of water vapor through the upper part of the profile. Little is known of the effects of these changes on the activity of the microbe populations. The influence of dew in causing a minor but daily fluctuation in activity has received little attention; it may well affect soil metabolism under mediterranean climatic conditions.

More information is also required about the relations in such climates between modulation of microbial metabolism, rhythmicity of plant production, and allelopathic manifestations. Strong regulatory mechanisms are at work: if plant growth is favored by sufficient humidity, mineralization is increased by the enhanced microbial activity. This means that total biomass does not vary as much as prevalent climatic irregularity would suggest, particularly because higher production during favorable years is passed on to less favorable periods. The high content of lipids, waxes, essential oils, terpenes (MULLER and DEL MORAL, 1966), soluble tannin and phenols in mediterranean vegetation certainly controls to some extent the nature of microbial activity in the litter subsystem.

References

BIRCH, H. F.: The effect of soil drying on humus decomposition and N-availability. Plant Soil **10**, 9–31 (1958).

BIRCH, H. F.: Further observations on humus decomposition and nitrification. Plant Soil **11**, 262–286 (1959).

BIRCH, H. F.: Nitrification in soils after different periods of dryness. Plant Soil **12**, 81–96 (1960).

DOMMERGUES, Y.: Contribution à l'étude de la dynamique microbienne des sols en zone semi-aride et en zone tropicale sèche. Ann. Agro. **13**, 265–324, 379–469 (1962).

DOMMERGUES, H.: Etude de quelques facteurs influant sur le comportement de la microflore du sol au cours de la dessication. Sci. Sol. 141–155 (1964a).

DOMMERGUES, H.: Influence létale de la dessication sur la microflore. 8. Congr. Intern. Sci. Sol, Bucarest III (10), 627–635 (1964b).

DROUINEAU, G., LEFEVRE, G., BLANC-AICARD, D.: Estimation de la richesse des sols en N et aspects particuliers de ce problème dans la région méditerranéenne. Sci. Sol. **1**, 13–21 (1952).

DROUINEAU, G., LEFEVRE, G., BLANC-AICARD, D.: Observations on summer localization of mineral N in the Mediterranean climate. Ann. Agro. **4**, 245 (1953a).

DROUINEAU, G., LEFEVRE, G., BLANC-AICARD, D.: Minéralisation de l'N organique du sol durant la saison sèche sous climat méditerranéen. C.R. Séanc. Acad. Sci., Paris, **236**, 524–526 (1953b).

DUCHAUFOUR, P.: L'évolution des sols, essai sur la dynamique des profils. Paris: Masson 1968.

KILLIAN, C.: Etude sur la biologie des sols des Hauts-Plateaux algériens. Ann. Agro. 6, 595–614, 702–722 (1936).

KILLIAN, C., FEHER, D.: Recherches sur la microbiologie des sols désertiques. Encyclop. Biol. Paris: Lechevalier 1939.

KILLIAN, C., VARGUES, H.: Etude microbiologique de quelques banquettes de restauration des sols en Algérie. Bull. Soc. Hist. Nat. Afr. **44**, 149–163 (1953).

MULLER, C. H., DEL MORAL, R.: Soil toxicity induced by terpenes from *Salvia leucophylla*. Bull. Torrey bot. Club **93**, 130–137 (1966).

NGUYEN KHA, DOMMERGUES, Y.: Influence de l'hygrothermopériodisme sur la stabilité de la matière organique du sol mesurée par respirométric. Sci. Sol. **1**, 53–62 (1970).

NGUYEN KHA, DUCHAUFOUR, P.: Etude comparative de l'évolution de la matiére organique du sol en conditions tempérées et tropicales. Pédologie **19**, 49–64 (1969).

SASSON, A., DASTE, P.: Observations nouvelles concernant l'écologie de l'*Azotobacter* dans certains sols arides du Maroc. C.R. Acad. Sci. **257**, 3516 (1963).

STEBAEVA, S. K.: Pedobiologische Experimente mit ausgetauschten Bodenblöcken. Pedobiologia **7**, 172–191 (1967).

STEUBING, L.: Bodenökologische Untersuchungen in Trockenrasen des westkroatischen Karstgebietes. Oecol. Plant. **2**, 175–186 (1960).

THALMANN, A.: Über die mikrobielle Aktivität und ihre Beziehungen zu Fruchtbarkeitsmerkmalen einiger Ackerböden. Giessen: Dissert. Univ. Giessen 1967.

VARGUES, H.: Etude microbiologique de quelques sols Sahariens en relation avec la présence d'*Anabasis aretioides*. Proc. Int. Symp. on Desert Research, Israel, 318–324 (1952).

WEBSTER, R., GASSER, J. K. R.: Soil Nitrogen, V. Leaching of N from soils in laboratory experiments. J. Sci. Fd Agr. **11**, 584–588 (1959).

3

Soil-Vegetation Relationships in Mediterranean Ecosystems of Southern France

PAUL LOSSAINT

Study of the primary productivity of terrestrial ecosystems can be undertaken at two complementary levels. Attention may be focussed either at the interactions between the lower atmosphere and the photosynthetic organs or at the relationships between the plant cover and the mineral environment, including its microfloral and faunal populations. In the second approach efforts are usually directed to the study of what can be called the plant-litter-soil subsystem. This second approach has been followed in considering the evergreen sclerophyllous forest and its principal degradation stages on the french Mediterranean coast.

Study of these environments present a double interest. From the bioclimatic viewpoint, these systems have a certain unity and constitute a transitional zone between typically temperate and typically tropical conditions. This zone is characterized by a xerophyll type of vegetation in which the principal species is *Quercus ilex* (evergreen oak) often replaced in disturbed areas by *Quercus coccifera* (evergreen shrub). These two oaks, together with *Pinus halepensis*, are distributed over the entire Mediterranean Sea basin region. Other regions of the world which have a similar type of climate and vegetation with similar morphological and physiological characteristics as the Mediterranean basin are California (chaparral), central Chile, South West Australia (mallee) and South Africa (fynbosch).

The investigations described here have been undertaken across the classical degradation sequence in these regions. As described by phytosociologists, this sequence comprises the climax vegetation of *Quercus ilex (Quercetum ilicis)*, the lower "garrigue" of *Quercus coccifera (Quercetum cocciferae)*, the Aleppo pine *(Pinus halepensis)* forest and the *Brachypodium ramosum* pasture.

The study programme includes work on the physical characteristics of the environment (GHAFOURI, 1968; RAPP and ROMANE, 1968; ETTEHAD, LOSSAINT and RAPP, 1973), the distribution and dynamics of organic material (RAPP and LOSSAINT, 1967; RAPP, 1969b; BOTTNER, 1970; BILLES, CORTEZ and LOSSAINT, 1971; LOSSAINT and RAPP, 1971a, 1971b; RAPP, 1971), turnover of mineral elements (RAPP, 1967; RAPP, 1969a, 1969b; LOSSAINT and RAPP, 1971b; RAPP, 1971) and soil microbial ecology (FOGUELMAN, 1966; BILLES, 1971; BILLES, CORTEZ and LOSSAINT, 1971; BILLES, LOSSAINT and CORTEZ, 1971; CORTEZ, LOSSAINT and BILLES, 1972).

This on-going work has been developed by an integrated research group that began in 1964. The study forms part of a research programme, undertaken within the auspices of the Centre National de Recherche Scientifique (CNRS), in which similar techniques and procedures are being used in the study of various ecosystem types in France. The work at Montpellier is part of the French IBP-PT programme, under the title "Ecology of Mediterranean Soils". Details of this programm of research are given in LOSSAINT (1967) and LOSSAINT and RAPP (1970).

The present condensed chapter is intended to provide an overview of the approach employed, the principal problems encountered and some of the findings to date. Descriptions of the methods used and the results obtained are given in the various publications listed in the bibliography.

The Study Area

Climate

The climate of the Montpellier region, where this study has been carried out, corresponds to the subhumid mediterranean type of EMBERGER (1955), according to whom the pluviothermic quotient is 80. The annual mean temperature is 14.40° C, the mean minimum temperature of the coldest month (January) is 1.08° C, the mean maximum temperature of the warmest month (July) is 31.4° C, the number of days with frost per year (over 28 years) is 40, and the annual mean precipitation (over 60 years) is 770 mm. The hydric deficit of the soil was estimated, after TURC, to extend from April to September. The mean annual total radiation for the period 1947–69 was 129.210 cal/cm²/year.

Soils

The soils of the region are set either on soft or hard calcareous rock. In the second case, the actual soils have been formed on the very old (probably upper Tertiary) terra rossa. The evolution of these slightly carbonated and strongly calcic soils continues today under forest cover, the soils tending to become more brown in character.

According to the French system of soil classification, these are "brown fersiallitic soils having a reserve of calcium".

Vegetation

The climax formation of the region is *Quercus ilex* forest. This formation is now found only very locally. Due to human intervention it has been replaced over large areas by the "garrigue" of *Quercus coccifera,* itself transformed for grazing into *Brachypodium ramosum* pastures. Plantations of *Pinus halepensis* are also present in the region. The greater part of the cultivated land is occupied by vineyards and by fruit orchards of apples, peaches and apricots.

Some Characteristics of Quercus Study Sites

Le Rouquet (altitude 180 m): *Q. ilex*. The trees in this study area have a density of 1440 stems/ha, a basal area of 38.80 m²/ha and a mean height of 11 m. The stand age is about 150 years, most trees are formed from stump shoots. The total crown projection area is 75 % of the total surface, and herbaceous undergrowth is not well developed. The soil is fersiallitic in character, with a saturated absorbant complex lacking active calcium carbonate. It is a loamy clay in texture, has a pH of 7.5–7.8, and contains 8–9% of organic matter.

La Madeleine (altitude 10 m): *Q. ilex*. The stand is less dense that at Le Rouquet with 527 stems/ha. The basal area is 34.11 m^2/ha, and the mean height of trees is 15 m. The age of the stand is not known. The total crown projection area is 85 % of the total plot. Some 90% of the herbaceous layer is formed by *Hedera helix*. The soil is of a forest rendzina or brown calcareous type, with a loamy-clay texture comprising 15–20 % of CaCO3. The pH is 7.6–8.0, and the organic matter content 9.1%.

Le Puech du Juge (altitude 130 m): *Q. coccifera*. The oaks at this study area are 70 to 100 cm in height, and have a crown projection of 80–90%. The soil is of a loamy-clay, brown calcareous type with saturated complex. It is rich in stones and has an active calcium carbonate content of 2%. The pH is 7.9 and the organic matter content 5.6%.

Grabels (altitude 100 m): *Q. coccifera*. This stand is 140 cm in height with a crown projection of 100 %. The loamy-clay soil is very stony with 2 % active calcium carbonate and 22% total CaCO3. The pH is 8.0.

Distribution of Organic Matter

Quercus ilex Evergreen Oak (Le Rouquet)

A population of 150 year-old evergreen oaks at Le Rouquet had an estimated aerial biomass of 269 t/ha (Fig. 1). This total biomass is composed of: 235 t of wood (diameter> 7 cm), 27 t of branches and twigs, and 7 t of leaves, of which 4.5 t corresponds to new growth and 2.5 t to growth in the previous year.

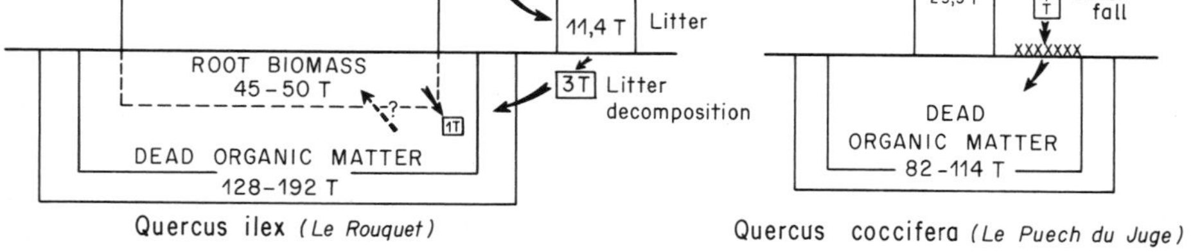

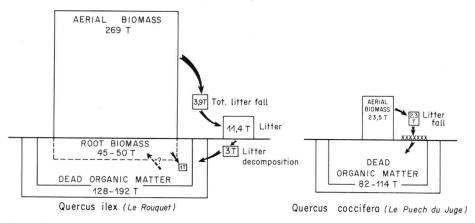

Fig. 1. Distribution of organic matter in two different oak ecosystems (*Quercus ilex* and *Q. coccifera*)

The use of two different methods gave estimates for the net primary production of the 150 year stand of 6.5–7 t/ha/year. The annual wood increment is 1.7–2.2 t/ha. The leaf area index is between 4 and 5.

The rocky nature of the subsoil and the depth of the root system precluded the accurate determination of root biomass, though measurements to a depth of 40 cm gave an estimate of 45 t/ha. The total biomass is therefore in the order of 315 t/ha. Observations over a 4-year period showed that litter fall occurred throughout the year. However, there is a peak in litter fall (50% of total annual fall) during the period of spring growth (April to June) which precedes the summer drought (Fig. 2). The average return of organic material to the soil is 3.8 t/ha/year, including leaves, wood and inflorescences. In addition to the seasonal pattern of litter fall, there is a strong difference between years, a year with a high litter fall alternating with a year of low leaf fall. This phenomenon is now well known in *Q. ilex* but has not been satisfactorily explained.

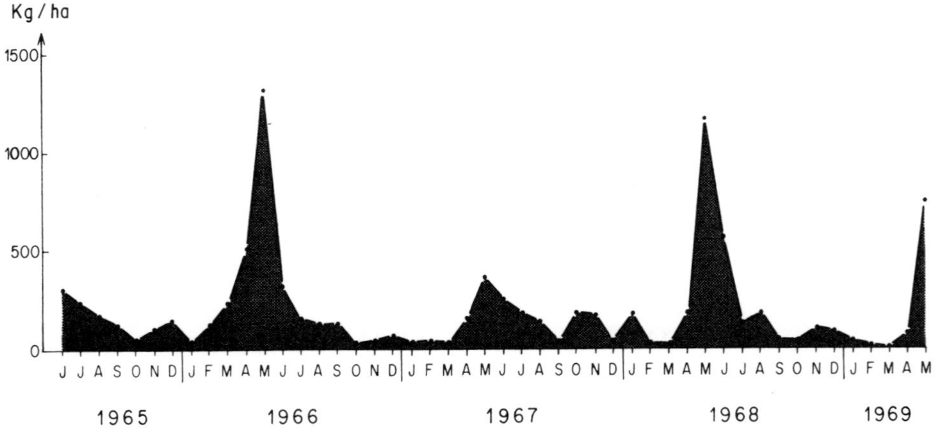

Fig. 2. Variations in monthly leaf fall in the *Quercus ilex* ecosystem at Le Rouquet

The organic matter temporarily accumulated in the upper soil layer is 11.4 t/ha. Its decomposition coefficient, estimated after the formula of JENNY, is equal to 26%.

Estimates of the organic reserve in the soil varies according to the proportion of fine particles to stones in the soil samples. The estimated organic reserve to a soil depth of 40 cm is 128 t/ha for samples with 50% of fine particles and 192 t/ha for samples with 90% of fine particles. Taking the latter estimate, the total organic matter in this ecosystem is estimated to be 518 t/ha (51.8 kg/m^2).

Quercus coccifera "garrigue" (Le Puech du Juge)

A total above-ground biomass of 23.5 t/ha was estimated for a 17 year-old low "garrigue" 1 m in height with 100% crown projection. This "garrigue" had been burned 17 years ago. The total aerial biomass is comprised of 19.5 t of wood and 4 t of leaves. The net primary productivity is in the order of 3.4 t/ha/year, with a mean annual production of litter of 2.3 t/ha. The organic matter reserve of the soil to a depth of 30 cm varies between 82 and 114 t (Fig. 1).

The Mineral Cycle

Quercus ilex at Le Rouquet

The aerial mineral mass composed of Na, K, Ca, Mg, P, N, Fe, Mn, Zn, Cu is equal to 5,698 kg/ha. Calcium is responsible for 69% of this mass. Other important elements are N (14%), K (11%), P (4%) and Mg (2%). The other elements represent less than 1.5% of the total mineral mass. Some 82% of the elements are accumulated in the wood. Assuming a root biomass of 45 t/ha, it is possible to estimate the mineral mass in roots to 1.547 kg/ha. This provides an estimate of the total mineral mass for this evergreen oak forest of 7.245 kg/ha.

The mineral reserve of the soil present in an exchangeable or assimilable form and thus readily available to plants is composed of 21–38 t/ha of Ca; 0.7–1.2 t/ha of K; 0.3–0.6 t/ha of Mg; 0.2–0.4 t/ha of P; 0.02–0.05 t/ha of Na; 0.18–0.38 t/ha of mineral N, assuming an annual mean net mineralization of 3 %. These figures are valid for soil depth of 40 cm and with a fine particle proportion of 50–90 %.

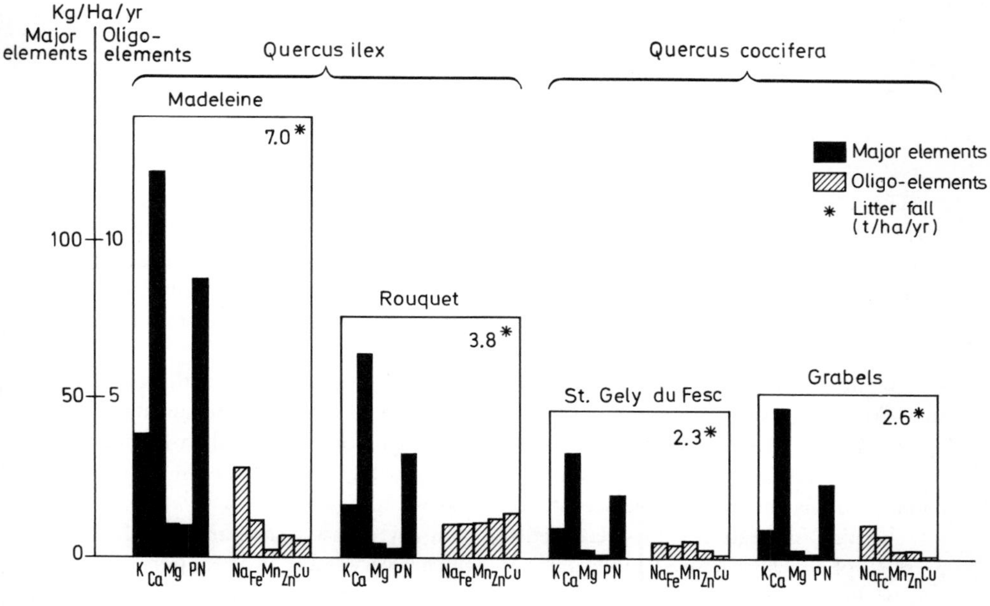

Fig. 3. Input to soil of mineral elements through litter in four oak ecosystems

Considering the turnover and cycling of elements, the annual input to the soil through the litter is 124 kg/ha. The order of importance of the different elements is similar to these recorded for the mineral mass (Ca, N, K, Mg, P, Na). Fig. 3 shows the differences in the litter input to the soil in the four study areas.

Some 511 kg/ha are stored temporarily in the surface layers of the soil, and 93,7 kg/ha are leached away each year. According to the results of laboratory experiments and lysimetric observations *in situ*, elements are released in the following order: Na, K, Ca, Mg, P, N.

Another important mineral input to the soil is from precipitation and from washings from the canopy (Fig. 4). The significant input of Na through precipitation is attributable to the proximity of the sea. The N input value is of 14 kg/ha. The high K input is referable to leaching from the canopy, and not to the composition of the rainfall. This element is in fact excreted by the leaves.

Taken together, the input of these two sources to the soil system (105.8 kg/ha) is almost as great as that of the litter (124 kg/ha). This might seem surprising, but similar findings have been recorded for certain temperate deciduous forest ecosystems.

Quercus coccifera at Le Puech du Juge

The total aerial mineral mass is estimated to be 773.7 kg/ha, of which 629.5 kg/ha is contained in wood and 144.2 kg/ha in leaves. As for the *Q. ilex* forest, Ca is the most important element with 485 kg/ha, followed by N (159 kg/ha), K (85 kg/ha), Mg (21 kg/ha) and P.

With a stand age of 17 years, there is an estimated mean annual incorporation of biogenic elements into perennial organs of 37 kg/ha. Calcium represents two thirds of this incorporation. The annual return of elements to the soil through litter is 72.5 kg/ha (Fig. 5).

The Water Cycle

The following parameters of the water cycle have been studied during 5 consecutive years in two *Q. ilex* stands by ETTEHAD, LOSSAINT and RAPP (1973): incident precipitation, throughfall, stem flow, available soil water, infiltration. Though the 60-year average annual rainfall for the Montpellier region is 770 mm, during the 5 years of observation the annual rainfall ranged between 315 and 1628 mm. The study period therefore included years of very different character.

The interception rate, calculated from data on incident precipitation, throughfall and stem flow, in the two oaks communities averaged 30 to 33 %, with extreme annual values of 26.2% and 41.4%. These values are relatively high compared to data for temperate deciduous forests. The most important reason for this high interception in the evergreen communities is undoubtedly the persistence of leaves.

The high infiltration values recorded for both dry and wet soils are referable to the presence of a well structured, mollic horizon, which is very stable and very porous. Intense and prolonged precipitation can readily infiltrate such soils; infiltration is only reduced where there is a large surface area of rocky outcrops and where the reduced surface area of fine earths becomes supersaturated by water.

Study of the water retention capacity of larger structural elements, composed principally of hard limestone, shows that participation in the constitution of the soil reserve is not important (0.2%).

The establishment of water profile curves, using the plaster blocks of Bouyoucos, allowed calculation of the periods during which the soil was under different capillary potentials. Summation of data provided weekly values for the hydric conditions of these soils. These values are of interest in consideration of the capacity of vegetation for uptake

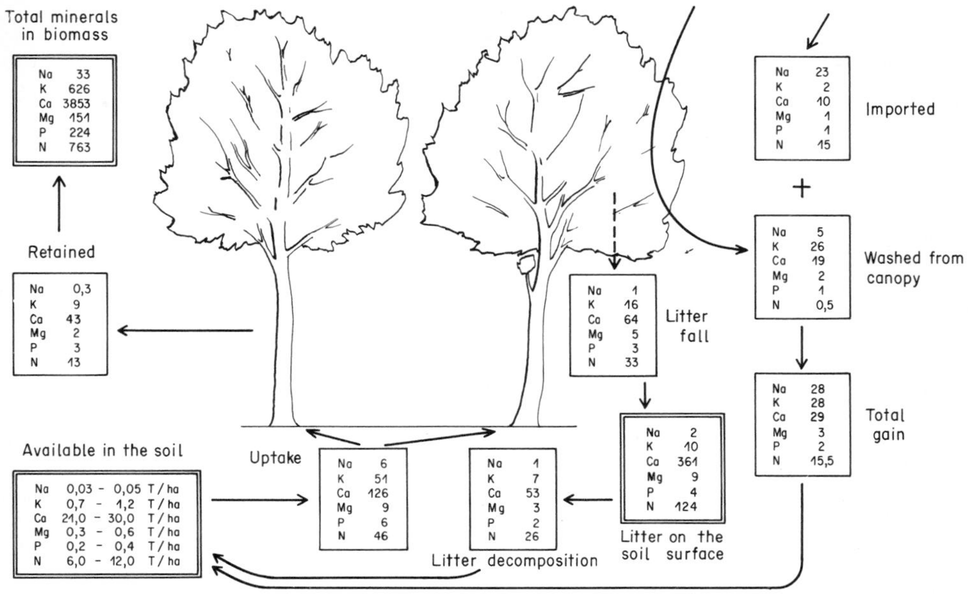

Fig. 4. Annual turnover (in kg/ha) of macronutrients in the *Q. ilex* forest at Le Rouquet

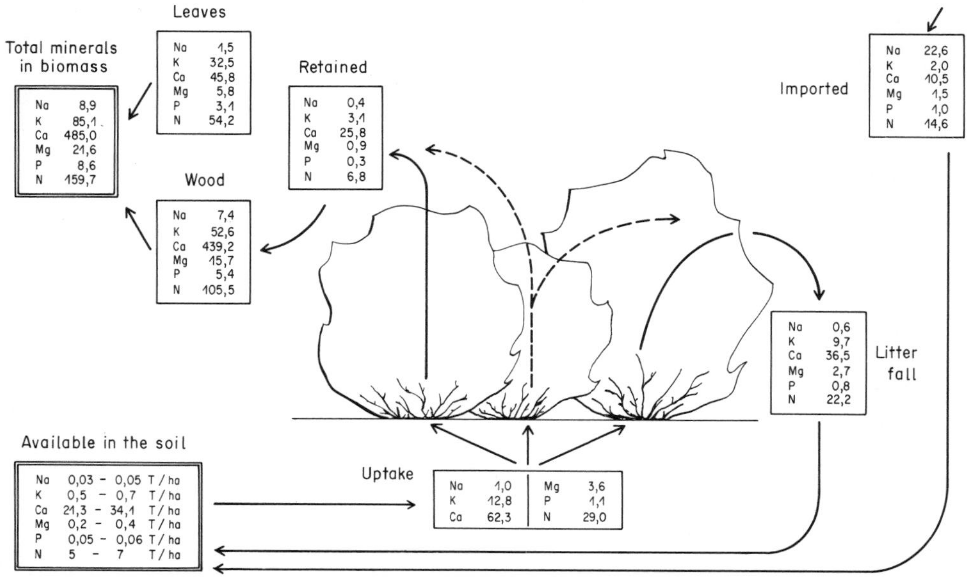

Fig. 5. Annual turnover (in kg/ha) of macronutrients in the *Q. coccifera* shrub ("garrigue") at Le Puech du Juge

of water. As illustration of the data obtained, the humidity of the whole soil profile was below the wilting point during 78 days in the summer and autumn of 1967 at Le Rouquet.

Another approach to the evaluation of the water balance has been used: evapotranspiration has been calculated using TURC's formula which is particularly adapted for french climatic conditions.

$$ETP = 0.40 \frac{t}{t + 15} (Ig + 50)$$

t = mean air temperature outside of the forest
0.40 = valid for every month, except February when the coefficient is 0.37
Ig = total solar radiation, direct and diffuse (cal/cm^2/day)
Ig = IgA (0.18 + 0.62 h/H)
IgA = radiation energy which would reach the soil if there was no atmosphere
h = duration of sunshine in hours
H = duration of the day in hours.

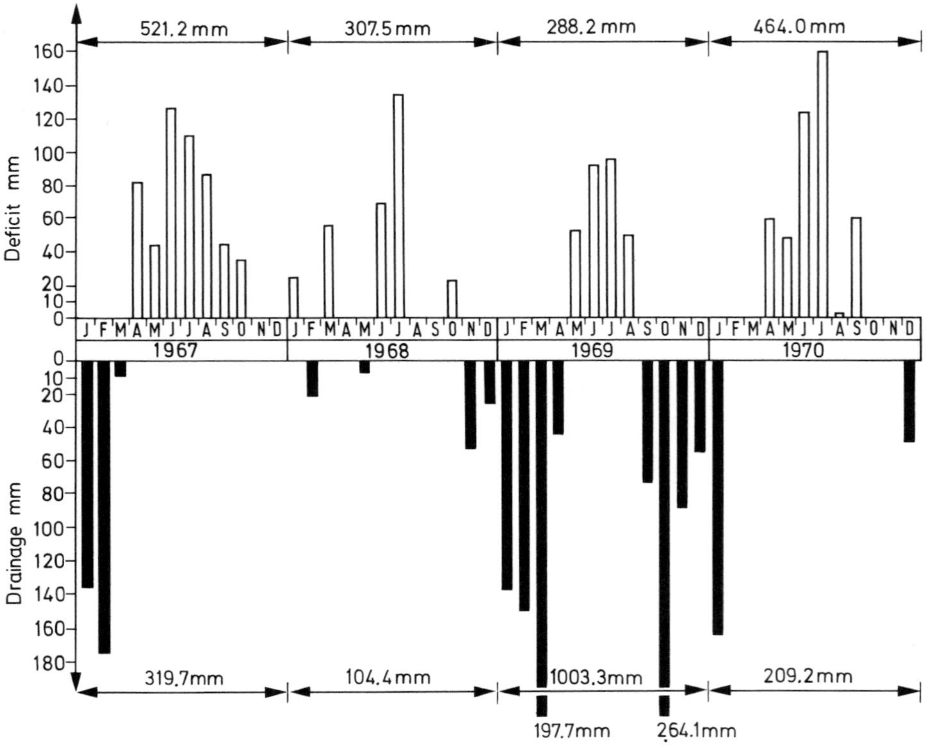

Fig. 6. Monthly values for soil water deficit, equilibrium and drainage at Le Rouquet, calculated from TURC's formula

This method has been used to evaluate the months during which there is water deficit, an equilibrium or a loss of water through drainage. The values obtained are summarised in Fig. 6 and Table 1. It is clear that in certain years, even when the rainfall is well distributed, drainage is deficient. Under these forest communities, water shortage is very frequent.

It should be noted that though these balances are based on theoretical considerations, comparison with lysimeter findings is envisaged.

Knowledge of the water cycle presents a double interest. Firstly, it provides information on a factor often limiting to plant growth in the Mediterranean region. Secondly it helps clarify the present trends in evolution of Mediterranean forest soils. Results to date show that these soils can scarcely be leached, thus explaining the saturated absorbant complex, a certain blockage of calcium and, in some cases, the presence of calcium carbonate in the surface of the soil profile.

Table 1. Annual water deficit, equilibrium and drainage in the soils of two evergreen oak ecosystems

Site	Year	Deficit months	mm	Drainage and equilibrium	Drainage months	mm
Rouquet	1967	7	521,2	5 months	3	319,7
	1968	5	307,5	7 months	4	104,4
	1969	4	288,2	8 months	8	1003,3
	1970	6	464,0	6 months	2	209,2
Madeleine	1966	7	455,6	5 months	3	66,9
	1967	8	694,5	4 months	2	89,2
	1968	7	426,8	5 months	3	22,9
	1969	4	378,0	8 months	7	595,7
	1970	7	499,6	5 months	2	71,2

Soil Biological Activity

Among the different methods employed for evaluation of the biological activity in soils, we have concentrated on those for measuring CO_2 exchange, for studying the importance of ammonification and nitrification, and for examining the activity of telluric enzymes. Details of the techniques used and results obtained are given in LOSSAINT (1967), LOSSAINT and RAPP (1970), BILLES, CORTEZ and LOSSAINT (1971), BILLES, LOSSAINT and CORTEZ (1971), and CORTEZ, LOSSAINT and BILLES (1972).

Total Activity

Measurement of CO_2 release were made *in situ* using a modified KOEPF method (BILLES, 1971). This method entails the retention of 4 cylinders in the soil throughout the period of observation. An example of seasonal release of CO_2 is shown in Fig. 7 concerning the O. *ilex* site La Madeleine. Winter values are very low, about 50 mg $CO_2/m^2/hr$. Highest values are recorded during the warm summer months, with values averaging about 350 $mg/m^2/hr$. At the La Madeleine site, extreme values of 20 $mg/m^2/hr$ in winter and 600 $mg/m^2/hr$ in summer were obtained. There is a highly significant correlation between CO_2 release and soil temperature.

Water has, evidently, an importance equal to that of temperature. However, this factor does not become limiting until the summer drought. This explains the reduction in CO_2 release every year during July and August.

The most important part of CO_2 production is attributable to microfloral activity, though root respiration is also important. The proportional contribution of microflora and roots to CO_2 production is not yet known.

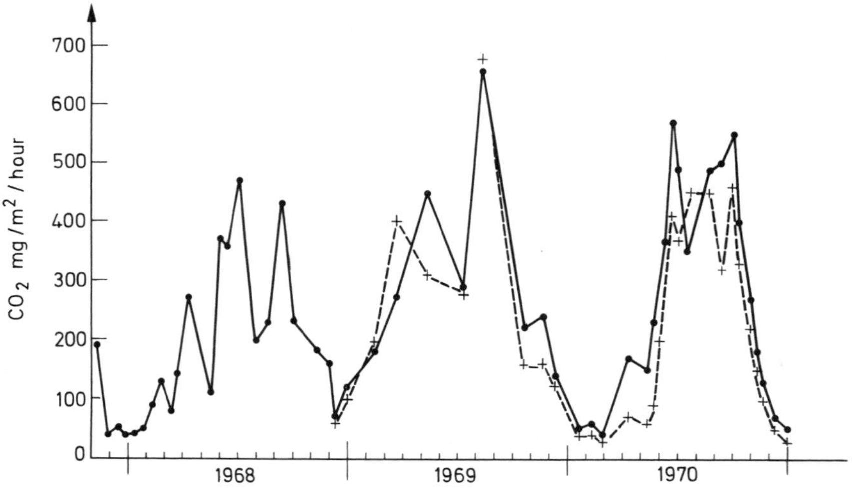

Fig. 7. *In situ* CO_2 release in the *Q. ilex* ecosystem at La Madeleine. Solid line: with litter. Dashed line: without litter

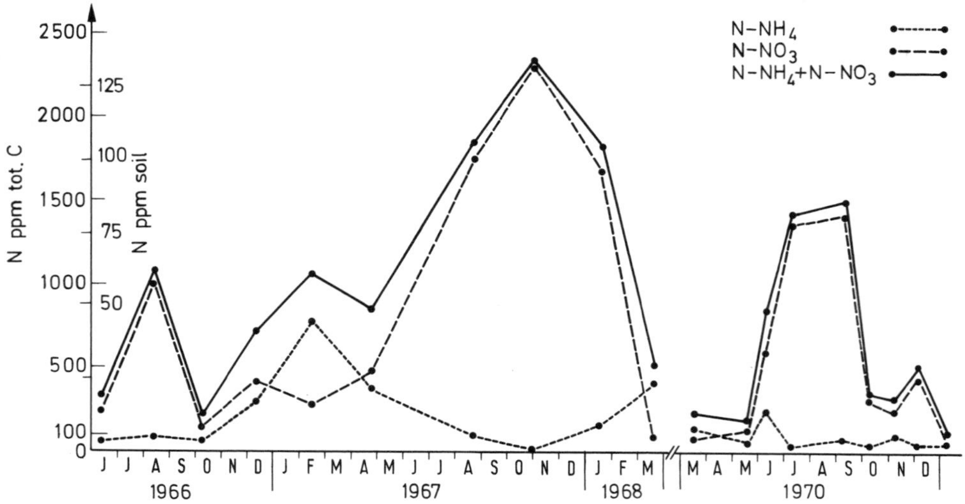

Fig. 8. Variations in levels of NH_4, NO_3 and total mineral N in the soil at La Madeleine

These *in situ* observations have been complemented by *in vitro* incubations that provide useful indications on the capacity of organic matter for mineralization. These data can, moreover, prove useful in consideration of the carbon cycle, discussed earlier in this chapter.

Nitrogen Mineralization

In the general balance of individual elements described by RAPP (1971), one aspect not taken in account is the transformation of organic nitrogen into ammonia and nitrates. Here it is necessary to continue observations on mineralization throughout several seasons. Data on the available NH_4 and NO_3 during a 4-year period at the La Madeleine site are given in Fig. 8.

The amounts of total mineral N show important fluctuations, ranging between some ppm to 2400 ppm according to total C content. Taking the average C content of soil to be 5.7%, this range is between almost 0 to 140 ppm of dry soil.

The curve for nitrates is very similar to that for total mineral N, except during certain cold periods when NH_4 levels rise above those of nitrates. In general terms, however, the nitrate rates are very high and are responsible for about 90% of the total mineral N, indicating an important activity of the nitrifying autotrophs. Attention should be drawn to the remarkably high summer levels, attributable to high temperatures and to the absence of leaching due to rainfall.

Enzyme Activity

Study of biological activity in soils can also be approached through evaluation of the activity of certain enzymes. Thus, urease is important in nitrogen mineralization, saccharase in the hydrolysis of saccharose in fructuose and glucose, amylase in the breakdown of starch. Bimonthly data on the status of these enzymes at the various study sites are given in CORTEZ, LOSSAINT and BILLES (1972).

Acknowledgements

We are very grateful to Dr. M. HADLEY for assistance in translating this chapter.

References

BILLES, G.: Améliorations techniques de la méthode de KOEPF pour la mesure du dégagement de CO_2 du sol. Rev. Ecol. Biol. Sol. **8**, 235–241 (1971).

BILLES, G., CORTEZ, J., LOSSAINT, P.: L'activité biologique des sols dans les écosystèmes méditerranéens. I. Minéralisation du carbone. Rev. Ecol. Biol. Sol. **8**, 375–395 (1971).

BILLES, G., LOSSAINT, P., CORTEZ, J.: L'activité biologique des sols dans les écosystèmes méditerranéens. II. Minéralisation de l'azote. Rev. Ecol. Biol. Sol. **8**, 532–552 (1971).

BOTTNER, P.: La matière organique des principaux types de sols sous l'étage bioclimatique du Chêne vert dans le midi de la France. Sci. du Sol **2**, 3–18 (1970).

CORTEZ, J., LOSSAINT, P., BILLES, G.: L'activité biologique des sols dans les écosystèmes méditerranéens. III. Activité enzymatique. Rev. Ecol. Biol. Sol. **9**, 1–19 (1972).

EMBERGER, L.: Une classification biogéographique des climats. Rev. Trav. Fac. Sci. Montpellier, Bot. **7**, 3–43 (1955).

ETTEHAD, R., LOSSAINT, P., RAPP, M.: Recherches sur la dynamique et le bilan de l'eau des sols de deux écosystèmes méditerranéens à chêne vert. 195–288. In: Nouveaux documents pour une étude intégrée en écologie du sol. Série R. C. P. 40, Vol. III. Paris: C. N. R. S. 1973.

FOGUELMAN, D.: Etude de l'activité biologique, en particulier de la minéralisation de l'azote, de quelques sols du Languedoc et du Massif de l'Aigoual. Doctorat d'écologie: Montpellier 1966.

GHAFOURI, H.: Etude des caractéristiques hydrodynamiques et thermiques des sols de quelques écosystèmes du Midi de la France. Doctorat de pédologie: Montpellier 1968.

LOSSAINT, P.: Etude intégrée des facteurs écologiques de la productivité au niveau de la pédosphère en région méditerranéenne dans le cadre du P.B.I. Programme et description des stations. Oecol. Plant. **2**, 341–366 (1967).

LOSSAINT, P., RAPP, M.: Un exemple d'installation d'une station expérimentale en milieu naturel pour une étude intégrée d'Ecologie du sol. 81–89. In: Methods of study in soil ecology (J. PHILLIPSON, Ed.). Paris: UNESCO 1970.

LOSSAINT, P., RAPP, M.: Le cycle du carbone dans les forêts de *Pinus halepensis*. 213–216. In: Productivity of forest ecosystems (P. DUVIGNEAUD, Ed.). Paris: UNESCO 1971 a.

LOSSAINT, P., RAPP, M.: Répartition de la matière organique, productivité et cycle des éléments minéraux dans des écosystèmes de climat méditerranéen. 597–617. In: Productivity of forest ecosystems (P. DUVIGNEAUD, Ed.). Paris: UNESCO 1971 b.

RAPP, M.: Etude expérimentale de la libération d'éléments minéraux lors de la décomposition de litières d'essences méditerranéennes. C. R. Acad. Sc., Paris **264**, 797–800 (1967).

RAPP, M.: Apport d'éléments minéraux au sol par les eaux de pluviolessivage sous des peuplements de *Quercus ilex* L., *Quercus lanuginosa* Lamk et *Pinus halepensis* Mill. Oecol. Plant **4**, 71–92 (1969 a).

RAPP, M.: Production de litière et apport au sol d'éléments dans deux écosystèmes méditerranéens: la forêt de *Quercus ilex L.* et la garrigue de *Quercus coccifera L.* Oecol. Plant. **4**, 377–410 (1969 b).

RAPP, M.: Cycle de la matière organique et des éléments minéraux dans quelques écosystèmes méditerranéens. 19–184. In: Série R. C. P. 40, Vol. II. Paris: C. N. R. S. 1971.

RAPP, M., LOSSAINT, P.: Apports au sol de substances organiques et d'éléments minéraux par la litière dans une futaie de *Quercus ilex* et une garrigue de *Quercus coccifera* dans le sud de la France. C.R. Conf. Sols Médit. Madrid, 269–275 (1967).

RAPP, M., ROMANE, F.: Contribution à l'étude du bilan de l'eau dans les écosystèmes méditerranéens. 1) Egouttement des précipitations sous des peuplements de *Quercus ilex* L. et de *Pinus halepensis* Mill. Oecol. Plant. **4**, 271–284 (1969).

Section V: Plant Biogeography

In the five regions of the world characterized by a mediterranean climate there exist structurally and functionally similar ecosystems. The major question concerning these ecosystems has been whether they originated independently from one another – presumably then derived from other, different ecosystems on each of the five continents where they occur – or whether they represent relictual occurrences of ancient floras that were once in direct contact. There are two principal ways in which this question can be analyzed: 1) by evaluating the floristic and faunistic composition of the present areas with a mediterranean climate and making deductions about the ease of migration between them; 2) by studying the geological record in each area, and in neighboring areas, in an effort to assess past migrations of plants and changes in climate.

The two chapters presented in this section take these different approaches to an understanding of the relationship between areas with a mediterranean climate, and thus make possible a more accurate and complete answer to the major question than has hitherto been available. The mediterranean climate itself is one in which the summers are hot and dry and the winters, the principal growing season, are cool and moist. Such climates occur between the deserts and more temperate vegetation types, and owe their existence in the modern world to the presence of offshore cold ocean currents along the western sides of the continents in the zones of prevailing westerlies. These currents are associated with the melting of the polar icecaps, and the climates they produce cannot be much more than a million years in age, and perhaps less. Mediterranean climates, on general meteorological grounds, can never have existed in the tropics; if there is only one annual peak in precipitation there it occurs during the hot part of the year. The same is true today everywhere in the world except in the regions of mediterranean climate.

Annual and other herbaceous plants that belong to groups largely or entirely confined to regions with mediterranean climates have presumably had a very recent origin in the regions that have developed these climates during the latter part of the Pleistocene. Although annual and other herbaceous plants are very well represented in regions of mediterranean climate, most of those that are presently common to more than one such region are not native to both. They have been introduced as weeds by man during the past three centuries. A very few groups of closely related annuals seem to be common to the Mediterranean region of Eurasia and North Africa on the one hand and to the California mediterranean region on the other; their presence in both areas must be explained by chance, long-distance dispersal. More than a hundred such species or species groups are common to western North and South America in regions of mediterranean climate; their disjunct ranges too must be explained by long-distance dispersal, probably during the recent past, and certainly not before the latter half of the Pleistocene.

The woody plants of regions with mediterranean climates (like the herbaceous ones found in the same areas) are entirely distinct in Chile, South Africa, and Australia. What similarities they possess may be attributed to differentiation from common tropical or tropical-margin ancestors. Neither are the woody plants of any of these three areas more closely related to the woody plants of either area in the Northern Hemisphere than would

---

The introduction to this section was prepared by PETER H. RAVEN.

be the case if they had differentiated from common tropical ancestors. The lack of similarity between the woody vegetations of the three Southern Hemisphere areas with a mediterranean climate strongly suggests that semiarid or subhumid climates were either unrepresented or very poorly represented in Antarctica at the close of the Cretaceous, when South America, Antarctica, and Australia were in direct contact.

There is some similarity between the woody vegetation of the areas with a mediterranean climate in and around California and around the Mediterranean basin itself. This similarity was much greater in the early Tertiary, when Eurasia and North America were in direct contact, as revealed by fossil floras. The ancestors of the plants occupying the present-day regions of mediterranean climate began to differentiate in the broad zone of subarid to semihumid climate flanking the tropics in the early Tertiary. Still under a regime of abundant summer rainfall, communities similar in their floristic composition to the modern ones had been constituted in both western North America and in Eurasia/North Africa by the late Miocene, more than 12 million years ago. These communities still survive in a relatively unaltered form in regions with summer rainfall, as at the margins of the Chihuahuan Desert in the southwestern United States and northwestern Mexico from Arizona to Nuevo León. Where mediterranean climates developed nearer the oceans, in both Old and New Worlds, many genera of woody plants were eliminated. Those which have survived in mediterranean areas at present are the relatively few that by virtue of unusually deep root systems, sclerophyllous leaves, or other ecological adaptations were able to survive even under these extreme conditions.

In general, therefore, the woody floras found in the regions of the world with a mediterranean climate did not evolve in this climate. Rather, they are the survivors of a richer, tropical-margin vegetation that developed under conditions of summer rainfall through the Tertiary, and which still survives in areas not far removed from those characterized by a mediterranean climate at present. The rich and largely endemic annual and herbaceous flora of these areas, on the other hand, probably evolved largely during and since the Pleistocene in response to the mediterranean climate itself. The development of these climates likewise promoted evolution and speciation in a few genera of woody plants, such as *Arctostaphylos, Ceanothus,* and possibly *Cistus.*

The climate of the world is changing, and either a warming or a cooling trend will ultimately result in the elimination of the cold ocean currents that result in the summer-dry climates of mediterranean regions. When these currents have been eliminated, the mediterranean climate will disappear from the world as summer rainfall is restored. Many genera of woody plants will be added to the floras of the regions where a mediterranean climate now exists. Relicts that are unable to stand the competition afforded by these new immigrants (*Lyonothamnus, Dendromecon, Heteromeles, Pickeringia* are possible examples) may be eliminated. Finally, thousands of species comprising dozens of genera of annual and other herbaceous plants will become extinct. These trends will clearly reveal the mediterranean climates of the present world as phenomena that are transient on a geological scale, just as are most of the species and genera of plants that have evolved under these conditions.

1

The Evolution of Mediterranean Floras[1]

PETER H. RAVEN

Introduction

In five widely separated regions of the world, lying for the most part between 30° and 40° N and S of the Equator, are areas of similar climate. These regions closely resemble the lands bordering the Mediterranean Sea not only in climate, but also in the gross aspect of their vegetation; hence, both climate and vegetation are said to be "mediterranean". When European explorers first reached the Cape region of South Africa, central Chile, California, and southwestern and southern Australia, they were surprised to find landscapes that reminded them of those they knew in Spain, southern France and Italy. They were also delighted to find that the grapes, olives, and citrus that they brought with them flourished in these distant lands. Later, they began to wonder just how closely the floras and faunas of these regions really resembled one another, and whether they had ever been connected by areas of climate so similar that plants and animals could have moved directly from one to another.

A century ago GRISEBACH (1872) considered the physiognomic similarity between the sclerophyll vegetation of different regions with mediterranean climate an outstanding example of the dependence of life form on climate. But does this similarity stem from a common origin: have the five areas of mediterranean climate ever been connected by others with a similar climate?

In the following pages, we shall first consider what is the climate we call "mediterranean"; how and where mediterranean vegetation evolved in the North Temperate region; and what we can say about specific floristic resemblances between California and the Mediterranean region. We shall then review briefly similarities between the North Temperate regions with a mediterranean climate and those in the Southern Hemisphere. Finally, in an effort to explain why there are so many species and so many endemics in these regions, we shall offer some hypotheses on the modes of evolution prevalent in regions with a mediterranean climate.

---

1 This chapter is a partial revision of RAVEN (1971), which was prepared for a symposium—"Plant Life of SW Asia"—held at Edinburgh in 1970. The current paper is based upon additional information that has come to my attention since the completion of that manuscript in April 1970. I am grateful to D. I. AXELROD and to HAROLD MOONEY for their critical reviews of the entire paper, and to D. M. BATES, J. F. M. CANNON, L. CONSTANCE, W. F. GRANT, M. C. JOHNSTON, H.A. MOONEY, G. WAGENITZ, and S. M. WALTERS for special information provided during the course of its preparation. The work was supported by grants from the U. S. National Science Foundation.

The Mediterranean Vegetation and Climate

Broad sclerophyll evergreen trees and shrubs dominate the vegetation in regions with a mediterranean climate. There seems to be little environmental demand for the deciduous habit, since neither the low temperatures of temperate regions nor the severe drought and extremely high temperatures of the savannas are encountered. In addition, annual plants reach the zenith of their diversity in areas with a mediterranean climate. They often constitute half of the total species (e.g., ZOHARY, 1962), yet rarely amount to as much as a tenth in other climatic regions. The combination of mild, moist winters and summer drought is especially suitable for the growth of annual plants, and it is not surprising that they have achieved such preeminence here.

In the simplest terms, a mediterranean climate is one in which the summers are dry and hot and the winters are cool and moist. Unlike most other temperate regions, plant growth is active during the winter, inactive and often completely suspended during the summer. This climatic regime developed for the first time in the Pleistocene, with the development of relatively stable air masses over the region during the summer, owing to a strengthening of the major cold-water currents (AXELROD, this volume). In the Northern Hemisphere, most of the rainfall occurs from October to March in regions with a mediterranean climate; in the Southern Hemisphere, it occurs mainly from April to September. It is of Pleistocene origin, and will disappear when the polar ice-caps melt, returning summer rainfall to the regions where this climatic type occurs at the present time.

Mediterranean climates occupy limited regions between the deserts – which we may with WALTER (1968) define as the edge of the subtropics – and areas with a truly temperate climate. They occur only along the western sides of continents onshore from cool ocean currents. There are vast local differences from place to place both within the Mediterranean basin itself and in the other areas with a mediterranean climate. These depend upon the patterns of air circulation, the proximity of the bordering oceans and mountains, and other more local features such as degree of human disturbance. Portions of the neighboring deserts are affected by the cool winter rains of the mediterranean regions, in contrast to the remainder of the deserts, where, as in the tropics, precipitation falls mainly during the warm part of the year.

Mediterranean climates have never existed in the tropics, even as local islands, because it is not possible to develop a pattern of dry summers and moist winters in these regions. Islands of drought do occur in all parts of the tropics, and probably always have; they also occur on the margins of the tropics. But their climates are not, and can never have been, mediterranean. Mediterranean climates developed in their present areas during the Pleistocene and have attained their greatest extent at present. It is in this context that we must interpret the relationships between their plants.

Northern Hemisphere Mediterranean Areas

The appearance of mediterranean floras must be viewed as part of the continuing evolution of subhumid to semiarid plant communities at the margins of the tropics. In western North America, these communities made up the Madro-Tertiary Geoflora (AXELROD, 1958), which began to differentiate at least as early as the Cretaceous. It drew species and genera both from the warm, humid Neotropical Tertiary Geoflora to the south and the cool, temperate Arcto-Tertiary Geoflora (CHANEY, 1936, 1947) to the north. Similar events were occurring and leading to the evolution of analogous plant communities at the same time in the Mediterranean region. In considering the relationships between the plants of the now widely separated Northern Hemisphere areas with Mediterranean and other semiarid climates, it is important to remember, as reviewed by AXELROD (this volume) that until the middle Cretaceous Eurasia and North America were in direct contact.

Many genera of woody plants are common to the Mediterranean region and California, playing conspicuous roles in the vegetation of both regions. Among these are *Acer, Aesculus, Alnus, Arbutus, Cercis, Clematis, Crataegus, Cupressus, Fraxinus, Juniperus, Lonicera, Platanus, Populus, Prunus* subg. *Emplectocladus, Rhamnus, Rosa, Rubus, Smilax, Staphylea, Styrax, Viburnum, Vitis,* and especially *Pinus* (with "closed-cone", fire-type species in each area) and *Quercus* (with a number of very important and conspicuous evergreen oaks in each region). Many of these genera are distributed continuously through the temperate regions of Eurasia and North America and the species found in the two disjunct regions with a mediterranean climate are unrelated. These genera have clearly been associated with the Arcto-Tertiary Geoflora and its derivatives since at least the Cretaceous, and during this time endemic species have evolved around the Mediterranean and in western North America. Many of the genera on the above list are in fact still associated with Arcto-Tertiary vegetation and are deciduous trees or shrubs found where this association interfingers with drier, more sclerophyllous plant communities. In some of the instances cited, they are now absent from the eastern sides of their respective continents, but are nevertheless thought to have had a more nearly continuous distribution in the Tertiary: their present distributions appear to be relict.

The same may well be true for some non-woody groups such as *Datisca* and the Cistaceae, with apparently disjunct ranges at present. Lamiaceae are well represented in both areas, but only a few genera (such as *Salvia*) are common, and the species found in California and the Mediterranean basin have different subtropical relatives within the genus from which they were presumably independently derived.

Lavatera (Malvaceae) has often been interpreted as having an analogous disjunct distribution. In the Mediterranean region it seems to be made up of two poorly defined elements; one, represented by *L. cretica* L. and *L. arborea* L., appears very closely related to and perhaps even congeneric with *Malva;* the second, represented by *L. thuringiaca* L. and *L. cachemiriana* Cambess., is perhaps related to the South African genus *Anisodontea.* Suffrutescent species of *Lavatera* are found on the Canary Islands and on the islands off the coast of southern California and northern Baja California (where they are the only species of the Western Hemisphere) but the relationship of the New World species with either of the groups found in the Old World is a matter of conjecture at present (D.M.

BATES, pers. comm.). They may have eventually proved to have entirely different affinities within the family, and to have been evolved independently in the New World.

Some Mediterranean woody plants appear to have been derived from the tropics and to have been associated with semiarid vegetation in western Eurasia for a very long time. Here are included such genera as *Ceratonia, Chamaerops, Cotinus, Laurus, Myrtus, Olea, Paliurus,* and *Phillyrea,* which have not become members of the temperate flora. They do not in any case extend to the New World, which suggests that the semiarid floras of the Old and New Worlds have not been in direct connection, for a number of these genera are isolated taxonomically and must have had a long evolutionary history on the semiarid fringes of the tropics in the Old World.

Another group of distinctive woody genera is found in and near areas of mediterranean climate in and near California, such as *Adenostoma, Calocedrus, Carpenteria, Cercocarpus, Chrysolepis* (*Castanopsis* p.p.), *Cneoridium, Comarostaphylis, Dendromecon, Fremontodendron, Garrya, Heteromeles, Lyonothamnus, Ornithostaphylos, Pickeringia, Sequoia, Sequoiadendron, Simmondsia, Umbellularia, Xylococcus,* and, above all, *Arctostaphylos* and *Ceanothus.* Most of these genera are highly isolated taxonomically and have presumably evolved in semiarid regions in western North America and been associated with these plant communities since at least early Tertiary times. Some, like *Cercocarpus, Garrya, Arctostaphylos,* and *Ceanothus,* range widely in Mexico but have their major concentrations of species in California. One species of *Arctostaphylos, A. uva-ursi* (L.) Spreng., even extends across Eurasia; but in general these plants are confined to western North America. In such a widespread Northern Hemisphere genus as *Berberis,* there are likewise groups that are very well developed and rich in species in western North America. They differ only in degree from the situation in *Arctostaphylos.* Certain genera on this list, like *Heteromeles,* have obvious Arcto-Tertiary affinities; others, like *Umbellularia,* are peripheral representatives of tropical groups. A number had a wide range in the Arcto-Tertiary Geoflora but are now restricted to its remnants along the borders of the present areas of Madro-Tertiary Geoflora.

About 10% of the genera, and probably 40% of the species found in those parts of California and neighboring states (southwestern Oregon, northwestern Baja California) with a mediterranean climate are endemic (HOWELL, 1957a, 1957b; NOLDECKE and HOWELL, 1960; SMITH and NOLDECKE, 1960). Certain primarily herbaceous families have important centers of evolution in this area: e.g., Hydrophyllaceae, Polygonaceae, Onagraceae, Polemoniaceae, Boraginaceae, Brassicaceae, and Liliaceae. In the Mediterranean basin, on the other hand, Caryophyllaceae, Rosaceae, Fabaceae, and Apiaceae are proportionally more important, and the first three families mentioned above are much more poorly represented. Even though genera such as *Delphinium, Lotus,* and *Trifolium* have many species in each of the two areas, these species are only distantly related.

A special word must be said about genera of more arid areas, such as *Prosopis, Frankenia, Gossypium, Pilostyles, Fagonia, Zizyphus, Thamnosma, Menodora,* and *Hoffmanseggia,* which occur both in the Old World and the New. There is no evidence to support the contention that plants of arid areas or mediterranean climates could ever have migrated directly between Eurasia and North America via the Bering Straits, as envisioned by JOHNSTON (1940) and STEBBINS and DAY (1967) among others. To postulate such an occurrence is to overlook the preponderant dissimilarities between the biota of subhumid to semiarid regions of the Old and New Worlds and to overemphasize their rather few similarities. The reasons why direct migration or even sweepstakes dispersal of the plants of arid lands across the Bering Straits must be ruled out have been cogently outlined by AXELROD (1970, p. 309). He proposes as an alternative that some of the plant groups with disjunct ranges may be remnants from a dry flora that inhabited Gondwanaland prior to its breakup by ocean-floor spreading near the end of the Cretaceous. Such an ancient dispersal would be consistent with the great differences between the floras of the arid lands of Eurasia-Africa and North America and thus may provide a reasonable alternative to direct long-distance dispersal, which appears to be the only other possibility for explaining the disjunctions.

The extent to which direct migration may have been possible between the plants of semiarid North America and Eurasia during the Cretaceous or early Tertiary is also discussed by AXELROD (this volume). I believe that, in the absence of a pertinent fossil record from eastern North America and in view of the preponderant dissimilarities between these two areas both now and in the Tertiary, that the disjunctions require further investigation. Nevertheless, Eurasia and North America were in direct contact into the late Cretaceous (TARLING, 1971) and islands were scattered along the mid-Atlantic ridge into Paleogene times. Thus, even if migration between the two areas was interrupted by regions of unfavorable climate, it was much more direct than at present. These facts are consistent with the observation that the woody plants of these two areas are much more similar to one another than are those of any other two regions of arid or semiarid climate, as we shall discuss below.

In summary: 1) about 10% of the genera and more than 40% of the species in each of the northern Hemisphere areas with a mediterranean climate are endemic; 2) most of the common genera have had an independent origin from ancestral groups that have (or had) a continuous distribution through the area occupied by the Arcto-Tertiary Geoflora in early Tertiary time or earlier; 3) each of the areas with a mediterranean climate represents one facet of an ecotone between tropical and temperate floras that has been "accumulating" derivatives from both sides since at least Cretaceous time; and 4) most close resemblances at the specific level, including the "desert disjuncts" mentioned in the last paragraph, must therefore be explained by long-distance chance dispersal, or, in some cases, by dispersal from a southern continent where they evolved during Cretaceous time.

Related Species in California and the Mediterranean

In view of the many weeds which have spread from the Mediterranean basin, where agriculture is of such considerable antiquity (NAVEH, 1967), it should not be surprising that some of the instances of supposed disjunction between these two areas have recently been shown to have been accidental human introductions. Among these are *Plantago insularis* Eastw. (*P. fastigiata* Morris), discussed at length by STEBBINS and DAY (1967), and shown recently by BASSETT and BAUM (1969) to be conspecific with the Mediterranean *P. ovata* Forssk. The same is likely to be true for *Oligomeris linifolia* (Vahl) Macbr. (*O. subulata* Del.), probably also an early introduction into southern California (ABDALLAH, 1967). Neither therefore can be taken as evidence for a slow rate of evolution.

Notwithstanding this, there are certain genera and species groups for which a critical reevaluation of the relationship between Mediterranean species and those in California is needed. These include *Aphanes,* which has also supposedly native species in South America and Australia; *Cicendia (Microcala); Antirrhinum,* with a cluster of species in the

Mediterranean and another in California; *Caucalis, Valerianella* and *Plectritis; Specularia* s. l.; *Erodium* sect. *Barbata* subsect. *Malacoides; Equisetum telmateia* Ehrh.; and perhaps *Lupinus* and *Erysimum* (MEUSEL, 1969). A particularly noteworthy group consists of Asteraceae, tribe Inuleae, with several interesting problems in the reduced annual genera of the two regions (WAGENITZ, 1969). Among these, *Filago* sect. *Oglifa* comprises seven species of western Eurasia and north Africa, three of California; they are closely related but distinct and native in their respective areas. One species of the western Eurasian and north African genus *Micropus* occurs in and near California. The relationships of the Californian species currently assigned to *Evax* (= *Filago* subgen. *Evax,* a group of about 10 species in the Mediterranean), the single Old World species of *Stylocline* (there are four in and near California), and the supposed New World genera *Psilocarphus* and *Filaginopsis* remain to be worked out in detail. It will be necessary eventually to determine not only the taxonomic relationships but also the interfertility and breeding systems for these supposed disjuncts, as has been done for many of the plants common to California and Chile (RAVEN, 1963; MOORE and RAVEN, 1970). To the extent to which these plants are members of annual groups entirely restricted to areas of summer dry climate, their evolution and the formation of their disjunct ranges must be more recent than the early Pleistocene, when such climates first appeared.

Southern Hemisphere Mediterranean Areas

As reviewed by WALTER (1968) and discussed by numerous earlier authors, the greatest difference biologically between the Northern and Southern Hemispheres consists in the extremely small land area in the south that can be classified as temperate. Only the southern tip of South America, some of the mountains of southern Africa, and a very small portion of southern Australia together with Tasmania, and New Zealand, are comparable in any sense with the vast stretch of temperate forest, taiga, and tundra that occupies the northern regions of the globe. Thus each Southern Hemisphere area with a mediterranean climate is totally cut off, not only from every other region with a similar climate, but also from any well developed temperate flora. Even though Antarctica was populated by plants in Cretaceous time, and provided a direct path for migration for some (e.g., *Nothofagus,* podocarps) between at least South America and Tasmania-New Zealand, this could not have had any effect upon the composition of the mediterranean floras of the three southern continents, as will be shown in the following paragraphs.

In the regions of central Chile with a mediterranean climate, the woody vegetation has apparently had a tropical origin, but now exists in areas of reasonably equable climate outside of the tropics. Families such as Monimiaceae, Lauraceae, Myrtaceae, Loganiaceae, Anacardiaceae, Sapotaceae, Elaeocarpaceae, Euphorbiaceae, Flacourtiaceae, Caricaceae, Icacinaceae, Bignoniaceae, Solanaceae, Cunoniaceae, and Rhamnaceae are prominent. In the drier parts of the region are found other groups associated with semiarid regions in and flanking the tropics, such as Cactaceae, Agavaceae, Zygophyllaceae, Asteraceae (for example, *Baccharis, Haplopappus,* and *Gutierrezia*), and Fabaceae (for example, *Acacia, Dalea,* and *Prosopis*). Isolated groups such as *Myzodendron, Ercilla, Anisomeria, Quillaja,* Malesherbiaceae, and Nolanaceae clearly are of tropical origin also.

These plants are in general the dominant elements in the vegetation of the mediterranean portions of Chile. In addition to them, there are certain basically North Temperate genera which are not especially associated in Chile with regions of mediterranean climate in North America and show no direct relationships at the specific level. These have evidently moved into South America along the Andes, jumping from mountain to mountain at the appropriate elevation and thus remaining in a temperate climate the whole while. The gaps in the distribution of some of these originally northern genera are not wide even at the present day. Among them may be mentioned *Alnus, Astragalus, Berberis, Castilleja, Epilobium, Geranium, Lathyrus, Myrica, Ribes, Salix, Sambucus, Sanicula, Saxifraga, Trifolium* and *Veronica* (but not *Hebe*) as well as the families Brassicaceae and Caryophyllaceae. Whether or not the present Arctic-Antarctic (bipolar) disjuncts reviewed by RAVEN (1963, p. 168) represent the result of stepwise migration along the Andes, or achieved their widely disjunct distributions by long-distance dispersal, will have to be considered separately for each species.

In the mediterranean region of Chile there is also a considerable element consisting of more than 130 species of identical or closely related species with widely disjunct ranges in temperate North and South America (RAVEN, 1963). Most of these appear to be plants of North American origin; all are herbs; almost all are autogamous; all occur in open habitats; and it appears highly likely that they achieved their remarkably disjunct ranges by long-distance dispersal across the tropics, perhaps carried inadvertently by one of the several species of migratory birds, especially the semipalmated plover, *Charadrius vulgaris,* known to move regularly between these two areas (CRUDEN, 1966). Their disjunctions must, at least in those instances in which the groups involved in the disjunctions are entirely restricted to regions of summer-dry climate, have arisen during approximately the past million years, when such climates first appeared. In most cases the evolution of the species themselves must also have taken place within this period of time.

Most important for us here, however, is the fact that, of the woody plants predominant in the regions of Chile and California with a mediterranean climate, there is not a single genus in common. The characteristic genera of woody plants of temperate western North America are almost totally lacking from Chile, and many characteristically Chilean groups – *Escallonia, Tropaeolum, Adesmia, Calceolaria,* and *Fuchsia* – are totally lacking in the western United States. This is exactly what would have been predicted from the climatic evidence that areas of mediterranean climate could never have extended through or even exist in the tropics. They were in fact formed during the Pleistocene when summer rainfall was eliminated or greatly restricted in various portions of the ecotone between tropical and temperate floras. Although some temperate genera have migrated along the Andes, these represent a very limited sample of the floras of the two regions concerned. In contrast to the direct migration possible throughout the Tertiary between Eurasia and North America for the plants of temperate regions, the plants of temperate North and South America, and especially those of the mediterranean regions, have never been in direct contact with one another. They have evolved in isolation, with little subsequent opportunity for interchange between the two areas. Their similarities are to be explained almost entirely by differentiation from common tropical (including semiarid, tropical-margin) ancestors. But as mentioned above, there are still more direct similarities at the *specific* level than between any two other areas with a mediterranean climate, suggesting that the possibilities for long-distance dispersal between California and Chile have been relatively great during the Pleistocene and recent times.

Those groups which show circum-Antarctic distributions – the plants common to

Tasmania or New Zealand on the one hand and southern South America, occasionally also southern Africa, on the other – have not contributed significantly to the vegetation of the mediterranean regions of the southern continents. These include such conspicuous and interesting plants as the southern conifers, Winteraceae, Proteaceae, Epacridaceae (a single species in the New World), *Eucryphia, Nothofagus, Laurelia, Acaena, Coprosma,* and the tribe Hydrocotyloideae of Apiaceae. In general, their distributions are more temperate, and most of them are restricted to moist temperate forests south of the mediterranean areas in Chile and in Australia. This strongly implies that, even when fairly direct migration across Antarctica was possible, semiarid to subhumid plant communities were lacking in this region, while warm temperate rain forest was well represented.

At the southwestern corner of Africa exists another mediterranean flora, one sharply distinct from all those we have considered earlier. As summarized by LEVYNS (1964), the Cape flora is characterized by the rich development of a number of peculiar groups, among them Proteaceae, Restionaceae, Penaeaceae, Bruniaceae, Geissolomataceae, Stilbaceae, Retziaceae, Grubbiaceae, and Rutaceae (tribe Diosmeae). Other outstanding features are the huge genus *Erica,* with almost 600 species, and *Cliffortia* (Rosaceae) and *Muraltia* (Polygalaceae), each with over 100 species. Only a few genera of woody plants are found both in the Mediterranean basin and in the Cape Region–among them the extremely polymorphic groups *Olea* and *Rhus*–but the species are not closely related to those of the Mediterranean. Many elements of the flora are clearly tropical in origin: for example, species of Sterculiaceae, Flacourtiaceae, Meliaceae, Rutaceae, Anacardiaceae, and Araliaceae. Even though some herbaceous plants, even species, extend down through the mountains of East Africa to the Cape, they are few, and opportunities for migration do not seem to have been as direct as between North and South America. Clearly the plants of semiarid and subhumid communities in the Cape region have evolved from tropical sources isolation over a period of at least 100 million years.

The flora of mediterranean regions in southwestern and southern Australia is equally peculiar. Proteaceae and Restionaceae are very richly represented, with over 400 species of the former in temperate West Australia alone. Characteristic Australian groups such as *Eucalyptus* and *Acacia,* with more than 100 and 200 species, respectively, in temperate West Australia, contribute greatly to the flora of the mediterranean areas. Other very well represented families are Goodeniaceae, Epacridaceae, Compositae, Sterculiaceae, Stylidiaceae, and Rhamnaceae. There are many unique, isolated genera, among them *Casuarina, Cephalotus, Xanthorrhoea, Anigozanthus, Actinostrobus,* and *Callitris,* but only a few that are represented in the Northern Hemisphere. Families such as Fabaceae, Liliaceae, and Menthaceae are represented here by entirely different genera and groups of genera as compared with the northern mediterranean regions; and it is clear that there have never been any direct connections between them.

In general, the floristic relationships amongst the Southern Hemisphere mediterranean areas discussed in this section, and between them and the Northern Hemisphere areas mentioned earlier, point to four major conclusions: 1) the flora of each area evolved in isolation; 2) it evolved almost entirely from border tropical, not temperate, predecessors; 3) the climate in Antarctica in the Cretaceous was never sufficiently arid to allow the passage of the plants and animals of semiarid or subhumid regions between South America and the Old World; and 4) the similarities between the floras of southern Africa and

Australia might best be explained on the basis of differentiation of such families as Proteaceae (JOHNSON and BRIGGS, 1963) from common tropical ancestors, perhaps originally on a Mesozoic southern continent.

Mode of Evolution in Mediterranean Areas

Each of the five areas of the world with a mediterranean climate is relatively rich in species, and especially in local species. The reasons for this are various, and we shall now enumerate some of them.

First, rainfall is limiting to plant growth. This not only means that rain-shadow effects will be greatly enhanced, and the differences between nearby habitats accentuated; it also means that weathering of the parent rock will be relatively slow, and its composition will be reflected with unusual fidelity in the composition of the soils. Very different soil types can occur in close proximity, each with its own particular complement of local plant species. The rapid Quaternary mountain building that took place in some of the areas of mediterranean vegetation provided a diversity of new habitats in close proximity to one another. Speciation of annuals, and in some instances *(Arctostaphylos, Ceanothus, Cistus?)* shrubs, was rapid as these new habitats were occupied.

Second, regions of the world with a mediterranean climate are by definition regions of climatic stress. They lie on the balance between the semiarid margins of the tropics or the deserts and the temperate zone, with its prevailing westerlies and better distributed precipitation. Major climatic changes and year-to-year fluctuations will exert major effects on the populations of plants in this area and may lead repeatedly to the severe reductions in population size that give rise to the sort of catastrophic selection envisioned by LEWIS (1962).

Third, bees are best developed in terms of numbers of species in semiarid to subhumid regions. Since they are of all groups of pollinators the ones most specific in their flower-visiting habits, they may have accelerated the rate of speciation in the plants of these areas.

Fourth, the environment favors the development and growth of annual plants, and rates of evolution in such plants are apparently faster than in long-lived perennials. If annual plants do not acquire reproductive isolation, they will not be seen as distinct species because their annual turnover will cause the swamping of the parental populations by hybrids. If barriers to intercrossing separate populations with different adaptive modes, they will be highly favored; and collectively the process should lead to a more rapid multiplication of annual than of perennial or woody species. In trees and shrubs, turnover of the population is slow; species, even if essentially interfertile, can coexist by responding to the continual selection of the environment which determines which individuals will become established at a particular site. Hybrids and unsuitable recombinants can be eliminated, and only a very small fraction of the seeds produced may become established: in other words, the variability of the zygote population may be much greater in perennials and woody plants than the variability of the adult population.

Given a similar environmental mosaic, woody plants and perennials, which are modally outcrossing, may exploit it by their variability, while still maintaining genetic contact within the population as a whole. Annual plants, which are modally inbreeding, may exploit it by their ability to form relatively homozygous races which efficiently occupy particular niches. There is no room for intermediates between these races, since there are no intermediate niches, and annual plants must adjust to changes in climate by the production of new homozygous species suited to the new conditions; perennials and woody plants may adjust merely by calling into play some of the variability that is continually present in the population of zygotes.

As a concrete example of this phenomenon we may mention the genus *Eschscholzia*, in which a single perennial species, *E. californica* Cham., which is self-incompatible and highly variable, is grouped with about a dozen annual, modally autogamous species. Individual annual species extend far beyond the range of the perennial, but collectively they do not exhibit more morphological variability; there may be as much variability in the 16 species of *Quercus* in California as in the approximately 100 annual and short-lived perennial species of *Eriogonum* (Polygonaceae). The annual habit in itself seems to lead to a great multiplication of species.

An interesting sidelight has been provided by WELLS (1969), who has pointed out that the only two large evergreen sclerophyll genera in the chaparral of California are *Arctostaphylos* and *Ceanothus*. Some members of these genera, and all members of the other chaparral genera in California, crown-sprout following fires, so that the individuals are very long-lived. But in these two genera, 59 out of 75 and 46 out of 58 taxa, respectively, have lost the ability to crown-sprout, reproducing from seed following the periodic fires that ravage these areas. Presumably the rate of speciation is accentuated among the species that reproduce by seed, this being correlated with their short generation time. Closer and more rapid adjustment to the diverse environments of the area is presumably possible in the populations that have lost the ability to crown-sprout, a clear analog to the relationship between annual and perennial plants in general.

These are some of the reasons, then, why the mediterranean regions of the world each have such distinctive and interesting floras, containing so many species and such a high proportion of endemic ones. Although they resemble one another to a great extent in their vegetation, this vegetation has been built up by similar evolutionary processes from different sources, temperate and tropical in the Northern Hemisphere and mainly tropical in the Southern Hemisphere. Although they are generally highly disturbed by human activities, and share an ever-increasing number of widespread weeds, areas of mediterranean climate afford not only repositories for relict plant families but great natural laboratories for students of evolution, particularly among their annual plants.

Summary

Mediterranean climates, characterized by hot, dry summers and cool, moist winters, occur along the western sides of the continents at the poleward margins of the subtropical deserts: in western North America, the Mediterranean basin, central Chile, the Cape region of South Africa and southwestern and southern Australia. All five areas support a physiognomically similar broad sclerophyll vegetation. In the two Northern Hemisphere areas, it is estimated that about 50% of the species are annuals; there are probably fewer in the Southern Hemisphere. Perhaps 10% of the plant genera and at least 40% of the species are endemic in each of the two Northern Hemisphere areas, and these figures are probably even higher for the Southern Hemisphere areas. Precipitation always is heaviest in the warm season in the tropics, and mediterranean climates could never have existed there or on the humid, eastern sides of the continents where tropical and warm temperate forests intergrade smoothly into one another. Consequently, each mediterranean type area has evolved its own distinctive flora and fauna in isolation, and there has never been direct exchange between them. In the Northern Hemisphere areas, the mediterranean floras, evolving since the Cretaceous in the ecotone between temperate and tropical floras, have been selected from a mixture of temperate and tropical predecessors; whereas in the Southern Hemisphere they have come almost entirely from the tropics. In each area, the summer-dry climates that we now call "mediterranean" appeared only in Pleistocene times. Their plants were recruited from the richer communities that occurred in subarid areas during the Tertiary. Many of the genera common to the Mediterranean basin and California seem to have evolved from temperate ancestors associated with the Arcto-Tertiary Geoflora; many of the similarities between the mediterranean floras of southern Africa and southwestern Australia are held to be because of differentiation from common tropical ancestors. In every case of disjunction, however, the possibilities of differentiation from Cretaceous plants of arid lands and dispersal through the south need to be evaluated. Resemblances at a specific level, or close resemblances between the plants of deserts in the Old and New Worlds may have resulted from change long-distance dispersal (including accidental introduction by man) or parallel differentiation from common tropical ancestors. In some cases, the resemblances might be the result of more or less direct migration in the Cretaceous, when Eurasia and North America were in direct contact; but the overwhelming dissimilarities between the plants of arid regions in the Old World and the New (Cactaceae, succulent *Euphorbia* species) seem to militate against any direct contact between the arid areas where they now occur since these groups evolved.

References

ABDALLAH, M. S.: The Resedaceae. A taxonomical revision of the family. Meded. Landbouwhogesch. Wageningen **67**, (1967).

AXELROD, D. I.: Evolution of the Madro-Tertiary Geoflora. Bot. Rev. **24**, 433–509 (1958).

AXELROD, D. I.: Mesozoic paleogeography and early angiosperm history. Bot. Rev. **36**, 277–319 (1970).

BASSETT, I. J., BAUM, B. R.: Conspecificity of *Plantago fastigiata* of North America with *P. ovata* of the Old World. Canad. J. Bot. **47**, 1865–1868 (1969).

CHANEY, R. W.: The succession and distribution of Cenozoic floras around the northern Pacific basin. 55–85. In: Essays in Geobotany in Honor of William Albert Setchell (T. H. GOODSPEED, Ed.). Berkeley: Univ. of California Press 1936.

CHANEY, R. W.: Tertiary centers and migration routes. Ecol. Monogr. **17**, 139–148 (1947).

CRUDEN, R. W.: Birds as agents of long-distance dispersal for disjunct plant groups of the temperate Western Hemisphere. Evolution **20**, 517–532 (1966).

GRISEBACH, A.: Die Vegetation der Erde nach ihrer klimatischen Anordnung. Leipzig: Engelmann 1872.

HOWELL, J. T.: The California flora and its province. Leafl. West. Bot. **8**, 133–138 (1957a).

HOWELL, J. T.: The California flora province and its endemic genera. Leafl. West. Bot. **8**, 138–141 (1957b).

JOHNSON, L. A. S., BRIGGS, B.: Evolution in the Proteaceae. Aust. J. Bot. **11**, 21–61 (1963).

JOHNSTON, L. M.: The floristic significance of shrubs common to North and South American deserts. Jour. Arnold Arb. **21**, 356–363 (1940).

LEVYNS, M. R.: Migration and origin of the Cape flora. Trans. Roy. Soc. S. Afr. **37**, 85–107 (1964).

LEWIS, H.: Catastrophic selection as a factor in evolution. Evolution **16**, 257–271 (1962).

MEUSEL, H.: Beziehungen in der Florendifferenzierung von Eurasien und Nordamerika. Flora, Abt. B. **158**, 537–564 (1969).

MOORE, D. M., RAVEN, P. H.: Cytogenetics, distribution and amphitropical affinities of South American *Camissonia* (Onagraceae). Evolution **24**, 816–823 (1970).

NAVEH, Z.: Mediterranean ecosystems and vegetation types in California and Israel. Ecology **48**, 445–459 (1967).

NOLDECKE, A. M., HOWELL, J. T.: Endemism and A California Flora. Leafl. West. Bot. **9**, 124–127 (1960).

RAVEN, P. H.: Amphitropical relationships in the floras of North and South America. Quart. Rev. Biol. **38**, 151–177 (1963).

RAVEN, P. H.: The relationship between 'mediterranean' floras. In: Plant Life of South West Asia. (DAVIS, P. H., HARPER, P., HEDGES, I. C., Eds.) Edinburgh: Botanical Society of Edinburgh 1971.

SMITH, G. L., NOLDECKE, A. M.: A statistical report on A California Flora. Leafl. West. Bot, **9**, 117–123 (1960).

STEBBINS, G. L., DAY, A.: Cytogenetic evidence for long continued stability in the genus *Plantago*. Evolution **21**, 409–428 (1967).

TARLING, D. H.: Gondwanaland, paleomagnetism and continental drift. Nature, Lond., **229**, 17–21 (1971).

WAGENITZ, G.: Abgrenzung und Gliederung der Gattung *Filago* L. s. l. (Compositae-Inuleae). Willdenowia **5**, 395–444 (1969).

WALTER, H.: Die Vegetation der Erde in öko-physiologischer Betrachtung. Band II. Die gemäßigten und arktischen Zonen. Jena: Gustav Fischer Verlag 1968.

WELLS, P. V.: The relation between mode of reproduction and extent of speciation in woody genera of the California chaparral. Evolution **23**, 264–267 (1969).

ZOHARY, M.: Plant Life of Palestine. New York: Ronald Press 1962.

2

History of the Mediterranean Ecosystem in California

DANIEL I. AXELROD

This Study is Dedicated to the Memory of RALPH WORKS CHANEY, *Paleobotanist and Conservationist*

Introduction[1]

Similar physical environments, acting on organisms of dissimilar origins in different parts of the world, have produced structurally and functionally similar ecosystems. The fossil record provides a reliable basis for understanding how this occurred because all modern ecosystems are the result of the interaction between evolving lineages and changing environments during long spans of geologic time. Since many woody plants similar to those still living have left a fossil record, it is possible to reconstruct the ecosystems they represent, and to discern the development of the modern descendant vegetation which has survived in modified form.

In evaluating the record, we must keep in mind that mixed evergreen forest, oak woodland-savanna, and chaparral, which constitute the typical broadleaved sclerophyllous vegetation zones of the mediterranean climate of California, are not restricted to it. They also inhabit regions with ample summer rainfall, extending from Arizona to Texas and far south into Mexico. Not only is the vegetation similar structurally, but the floras are also similar. Identical and vicarious species link these areas with similar kinds of vegetation in California. For example, species common to the chaparral of southern California and central Arizona include *Arctostaphylos pungens, Ceanothus greggii, Cercis occidentalis, Cercocarpus betuloides, Fremontia californica, Garrya flavescens, Quercus palmeri, Q. turbinella, Rhamnus californica, R. ilicifolia, Ribes quercetorum,* and *Schmaltzia (Rhus) ovata.* There are, of course, species in the Arizona chaparral that are not now in California. The Arizona and Coahuila chaparral also have many taxa in common and some of these are in California, though there are others that are distinctive of each area. Similar relations are displayed by mixed evergreen forest and oak woodland-savanna vegetation. Clearly, the vegetation of the typical mediterranean climate of California is not unique to it, nor are all of its taxa.

The fossil record indicates that the summer rainfall regime from Arizona to Texas and southward into Mexico approximates the conditions under which these kinds of vegetation lived during most of their recorded history, which can be traced back into Oligocene time at least. The evidence suggests that the homologous vegetation zones in

---

1 This chapter incorporates the results of current and earlier work on Tertiary floras from Nevada and border areas, a research project made possible by grants from National Science Foundation which are gratefully acknowledged.
Thanks are also extended to Profs. MICHAEL BARBOUR, RALPH W. CHANEY, JACK MAJOR and PETER RAVEN whose helpful suggestions improved an early draft of this chapter.

each region were segregated from ancestral Tertiary communities that were richer in taxa. The gradual development of regional differences in the distribution of seasonal rainfall and in temperature relations over southwestern North America during Miocene and later times, and changes in taxa to produce both ecotypes and vicarious species, account for the differences now seen in the mixed evergreen forest, oak woodland-savanna, and chaparral vegetation of California (winter rain, summer drought), Arizona to west Texas (winter and summer rain), and Mexico (summer rain), and also in other vegetation zones marginal to them.

The thesis to be developed here is that the mediterranean climate of California is not old, but very young. The sclerophyllous plants that now typify the area are survivors of a richer flora that persisted here as summer rainfall gradually disappeared in the late Cenozoic. To show that mixed evergreen forest, oak woodland-savanna and chaparral have lived here under regional summer-dry climate only since the first glacial, the history of sclerophyllous vegetation in California and adjacent areas from Miocene down to the present will be reviewed. Following this, the history of sclerophyllous vegetation in the Mediterranean area will be considered because the fossil floras there indicate a similar history for its sclerophyllous vegetation. By contrast, the origins of sclerophyllous vegetation in the mediterranean climates of Chile, South Africa and southern Australia can not be assessed accurately because the fossil record there is too incomplete. However, the general sequence of events that applies to the evolution of sclerophyllous vegetation in the California and Mediterranean areas appears applicable to the austral regions of mediterranean climate, though the taxa are very different in each of those areas inasmuch as they have largely originated in isolation.

Mixed Evergreen Forest

Today, mixed evergreen forest[2] is distributed from southwestern Oregon through the Coast Ranges to southern California (San Marcos Pass), and it also occurs discontinuously on the west slope of the northern to central Sierra Nevada at moderate altitudes. With increased moisture it grades into Douglas fir-redwood *(Pseudotsuga-Sequoia)* forest in the milder parts of the north Coast Ranges, and into mixed coniferous *(Pinus-Abies-Calocedrus)* forest in the cooler, higher parts of the Coast Ranges and in the Sierra Nevada. In drier areas it is replaced by oak woodland-savanna and chaparral vegetation, as in the inner Coast Ranges, the foothills of the Sierra Nevada and in southern California.

A number of fossil floras from Oregon, Idaho, Nevada and California provide evidence of the gradual development of the modern vegetation from a richer, ancestral community. In the following paragraphs, some of the representative floras that reveal its history are referred to briefly, commencing with sites in the north and progressing southward and thence to the west (Fig. 1, Table 1).

---

2 Also termed border-redwood forest, oak-madrone forest, tanoak-madrone forest, mixed evergreen forest.

Oregon-Idaho

Mixed evergreen forest is well developed in the Blue Mountains flora of northeastern Oregon, where it occupied warm, southerly-facing slopes adjacent to a rich conifer-hardwood forest (CHANEY and AXELROD, 1959). Among the representative members of the mixed forest were *Arbutus* (aff. *xalapensis*), *Quercus* (aff. *chrysolepis*), *Lithocarpus (densiflora,* Sierran form), and probably *Ilex* (extinct)[3]. Associated with these evergreens were

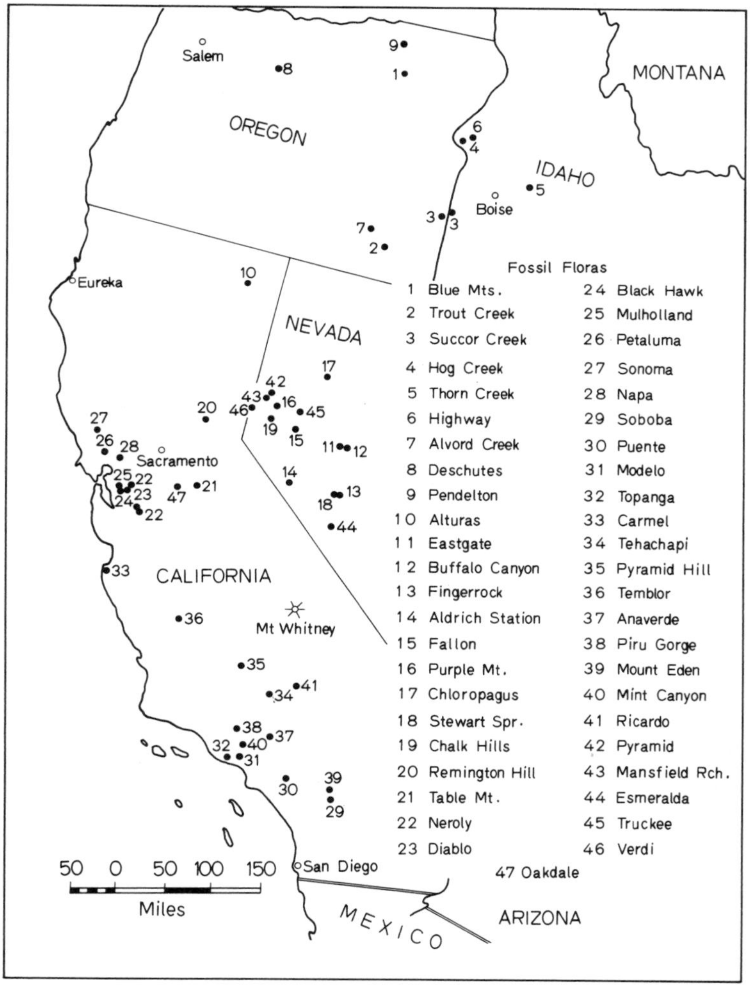

Fig. 1 Location of fossil floras in western United States that contain sclerophyllous vegetation, and which are discussed in the text

3 Throughout this chapter, modern species most similar to the fossil plants are indicated in parenthesis; names of fossil species are given in the respective papers to which reference is made.

deciduous hardwood trees and shrubs that have closely similar species in the modern tanoak-madrone forest, notably *Quercus (kelloggii), Acer (macrophyllum), Rhamnus (purshiana), Vaccinium (parviflorum),* and *Alnus (rhombifolia).* The Blue Mountains flora also has a well-developed conifer-hardwood forest, composed of *Abies (concolor), Picea (breweriana), Pseudotsuga (menziesii), Sequoia (sempervirens), Tsuga (heterophylla),* and *Thuja (plicata)* as well as the deciduous trees and shrubs noted above. Significantly, all of the nearest descendants of these fossils live on cooler, moister slopes bordering broad-leaved sclerophyllous forest in northwest California and adjacent Oregon today. However, the Miocene forest also included numerous deciduous hardwoods whose closest relatives now live only in areas of summer rainfall, notably *Ailanthus (altissima), Acer (saccharinum, pictum, rubrum), Crataegus (hupehensis), Fagus (americana), Quercus (prinus, imbricaria), Juglans (nigra), Liquidambar (stryaciflua),* and *Sassafras (officinale).* They contributed to both the conifer-hardwood forest and mixed evergreen forest. In the latter case, the oaks, walnut and red gum and others probably occurred chiefly in moister sites on the warmer exposed slopes where the oak-madrone forest attained optimum development.

Table 1. Showing approximate ages of fossil floras depicted in Fig. 1. The heavy bars represent the age range of the grouped floras

Age	Epoch	WESTERN NEVADA	CENTRAL SIERRAN SLOPE	CALIFORNIA WESTERN	SO. CALIFORNIA	OREGON IDAHO
0	QUAT.					
2	QUAT.			Santa Clara	Soboba	
4	PLIOCENE			Sonoma Napa		Deschutes
6	PLIOCENE	Verdi Truckee	Oakdale	Petaluma Mulholland	Mt. Eden	Highway
8	PLIOCENE	Mansfield Rch. Esmeralda		Black Hawk Diablo	Piru Gorge Anaverde	
10	MIOCENE	Chalk Hills	Table Mt. Remington Hill	Carmel Neroly	▌Puente ▌Ricardo Mint Canyon	Hog Creek
12	MIOCENE	Stewart Spr. ▌Chloropagus ▌Fallon Aldrich Sta.				Thorn Creek
14	MIOCENE	▌Pyramid ▌Middlegate Fingerrock		Temblor	▌Modelo ▌U. Topanga	Trout Creek ▌Succor Creek Blue Mts.
16–18	MIOCENE	Buffalo Canyon			Tehachapi	
20						

Similar relations are shown by the Trout Creek flora in southeastern Oregon, with the differences primarily reflecting its higher altitude as judged from the more prominent role of conifers in it (CHANEY and AXELROD, 1959). Mixed evergreen forest on warmer exposed slopes included *Arbutus (xalapensis), Quercus (chrysolepis), Lithocarpus (densiflora),* and *Machilis kusanoi.* Among the conifers that dominated the conifer-hardwood forest on moister slopes were *Abies (magnifica), Keteleeria (davidiana), Picea (breweriana), Pseudotsuga (menziesii), Tsuga (heterophylla), Chamaecyparis (lawsoniana), Thuja (plicata).* A number of trees and shrubs entered the matrix of both forests, and their descendants occur in them today, notably *Alnus (rhombifolia), Amelanchier (alnifolia), Acer (macro-*

phyllum, glabrum, negundo), Fraxinus (oregona-americana), Holodiscus (discolor), Mahonia (nervosa), and *Rosa* (spp.). Deciduous hardwoods in the flora that now live only in areas with ample summer rainfall include *Acer (saccharinum, rubrum, pictum), Ailanthus (altissima), Carpinus (caroliniana), Hydrangea (bretschnederi), Juglans (nigra), Tilia (americana)* and *Ulmus (fulva).* They contributed to both forests, with the walnut, elm and basswood occurring in the oak-madrone forest chiefly in the moister sites on the exposed, warmer slopes where it attained best development.

At lower altitudes to the east, the Succor Creek flora shows that broadleaved sclerophyllous forest dominated the drier, exposed slopes over the rolling volcanic lowlands, and that deciduous hardwoods were confined chiefly to the moister valleys and slopes (CHANEY and AXELROD, 1959). At most localities the dominant sclerophyllous trees are *Quercus (chrysolepis)* and *Lithocarpus (densiflora,* Sierran form). Their sclerophyllous associates included *Arbutus* (aff. *xalapensis), Persea (borbonia), Oreopanax (floribundum* group) and *Cedrela (odorata).* Taxa similar to those that contribute to the oak-madrone forest today in California include *Acer (macrophyllum, negundo), Alnus (rhombifolia), Amelanchier (alnifolia), Populus (trichocarpa), Fraxinus (oregona-americana), Mahonia (nervosa),* and *Symphoricarpos (albus).* Conifers are present but rare, with *Glyptostrobus (pensilis)* confined to riverbanks and lake margins, and *Ginkgo (biloba)* and *Pinus (ponderosa)* on the better-drained cooler slopes. There are no megafossils of *Abies, Picea, Sequoia, Tsuga, Thuja, Chamaecyparis* and other conifers that lived in the cooler mountains to the west. Deciduous hardwoods are abundant, notably *Acer (saccharinum, rubrum, pictum), Ailanthus (altissima), Betula (lenta), Castanea (dentata), Diospyros (virginiana-kaki), Fagus (americana), Gymnocladus (dioicus), Nyssa (aquatica), Ostrya (virginiana-japonica), Platanus* (aff. *occidentalis), Ptelea (trifoliata), Quercus (montana), Sassafras (officinale), Tilia (americana), Ulmus (alata-fulva),* and *Zelkova (serrata).* The more drought resistant members of this group probably also contributed to the mixed evergreen forest in moister sites on the warmer, exposed slopes where it presumably attained optimum development.

Available evidence thus suggests that during the middle Miocene in this region, mixed evergreen forest covered warmer exposed slopes adjacent to conifer-deciduous hardwood forest in the mountains, and to deciduous hardwood forest in the lowlands. The presence of numerous genera and species in these (and other) floras that are represented by equivalents that now live only in areas with precipitation well distributed through the year indicates that there was ample summer rain in this region which is now largely summer-dry over the lowlands, and receives only minimum summer showers in the mountains. The Miocene broadleaved sclerophyllous forest evidently included some of the deciduous hardwoods whose descendants now occur in areas of summer rainfall. They probably were confined chiefly to the moister sites, much as the western hardwoods (maple, alder, dogwood) and deciduous shrubs (serviceberry, cherry, snowberry) whose close fossil relatives lived with them in the Miocene, occur in the tanoak-madrone forest today.

In response to the general trend toward drier climate, forest had become impoverished over the region by late Miocene time. For example, the Hog Creek flora from southwestern Idaho is dominated by species representing tanoak-madrone forest (DORF, 1936; SMITH, 1938). It included *Lithocarpus (densiflora,* Sierran form), *Quercus (chrysolepis, lobata), Arbutus (xalapensis)* and *Cedrela (odorata).* Regular associates of the dominants whose nearest relatives contribute to the modern forest include *Alnus (rubra), Amelanchier (alnifolia), Amorpha (californica), Fraxinus (oregona), Prunus (demissa), Mahonia (pinnata),* and *Salix (lasiandra).* Higher hills directly to the north supported a conifer-deciduous hardwood forest. Among the conifers are *Abies (amabilis), Keteleeria (davidiana), Picea (breweriana), Pinus (ponderosa), Chamaecyparis (lawsoniana),* and *Thuja (plicata).* Deciduous hardwoods included *Acer (pictum, saccharinum), Cercis (canadensis), Cercidiphyllum (japonicum), Gymnocladus (dioicus), Ostrya (virginiana-japonica), Pterocarya (stenophylla), Quercus (montana), Hydrangea (aspera), Ptelea (trifoliata)* and *Ulmus (fulva).* It is emphasized that they are comparatively rare in this flora which was dominated by broadleaved sclerophylls. Some of them no doubt entered into the composition of the oak-madrone forest, presumably in the moister sites much as the hardwoods and deciduous shrubs do at the present time in the related modern vegetation.

The rich Thorn Creek flora, situated about 80 miles northeast of the Hog Creek flora, provides additional evidence with respect to the composition of the late Miocene broadleaved sclerophyllous forest (SMITH, 1941). The Thorn Creek basin was surrounded by granitic hills, and bounded by an active fault scarp on its east margin, as shown by the coarse fanglomerate which intergrades basinward into conglomerate and thick arkose that contains the thin lacustrine beds. Broadleaved sclerophyllous plants are dominant, and the oak-madrone forest occupied the warmer, more southerly- and westerly-facing slopes where *Lithocarpus (densiflora,* Sierran form), *Quercus (kelloggii, chrysolepis, lobata), Arbutus (xalapensis),* and *Cedrela (odorata)* lived under optimal conditions. Their associates included *Acer (negundo), Alnus (rhombifolia), Populus (trichocarpa), Salix (lasiandra, hookeriana),* and *Symphoricarpus (albus),* all of which closely resemble plants in the forest today along stream margins and in moister sites. They also contributed to the conifer-hardwood forest that inhabited moister slopes and valleys, and included *Abies (grandis), Picea (breweriana), Pinus (monticola), Pseudotsuga (menziesii), Tsuga (heterophylla), Sequoia (sempervirens),* and *Thuja (plicata).* Deciduous hardwoods that occur today in regions where rainfall is rather evenly distributed through the year are also present, notably *Acer (saccharinum, pictum, rubrum), Betula (lutea), Carpinus (caroliniana), Fagus (americana), Quercus (prinus),* and *Ulmus (alata).* Most of these are relatively rare as compared with their representation in the middle Miocene floras of the region, suggesting reduced summer rainfall. This is also implied by the much smaller leaves of the Thorn Creek species of *Acer, Fagus, Populus, Quercus* and *Ulmus* as compared with related (ancestral) species in the middle Miocene. Leaves of the deciduous hardwoods in the Hog Creek flora show the same relations, and also suggest reduced summer rainfall as well as somewhat higher summer temperatures.

By early Pliocene time, to judge from the few floras now known from the region,vegetation had become impoverished in response to lowered precipitation and to the decrease in summer rainfall. In the Weiser area, the small Highway flora (Loc. 636) from a diatomite section 12 miles northeast of the Hog Creek flora, is dominated by a sclerophyllous oak, *Quercus (chrysolepis). Lithocarpus (densiflora)* is present, but *Arbutus* is not recorded (DORF, 1936). Bordering slopes were covered with a *Pinus-Abies* forest that included *Quercus (prinus), Ulmus (alata), Acer (saccharinum),* and *Mahonia (japonica),* as well as the broadleaved evergreens. The leaves of the oak and elm are reduced in size as compared with those of the middle Miocene, apparently in response to lowered precipitation. Riparian- and lake-border plants included *Glyptostrobus, Salix, Typha, Trapa* and *Fraxinus (americana).*

More recently, two new sites were discovered in this younger sequence which rests unconformably on the beds that yield the Hog Creek flora (Prof. CHARLES J. SMILEY, written communication, Jan. 1971). Live oak and tanoak dominate both florules, and their regular associates included *Arbutus (xalapensis)* and *Quercus (lobata)*. In moister areas these mingled with riparian-border species, notably *Alnus (rhombifolia), Platanus* (cf. *occidentalis), Betula (papyrifera)*, and *Salix (argophylla)*. There are rare records of *Abies (concolor)* and *Tsuga (heterophylla)* in the flora, and also of taxa that probably lived with them on moist nearby slopes, notably *Betula (papyrifera), Populus (tremuloides)*, and *Salix (scouleriana)*. Rare records of *Ailanthus (altissima), Cercidiphyllum (japonicum)*, and *Robinia (hispida)* indicate that summer rainfall was reduced in amount as compared with that indicated by the underlying section which yields the Hog Creek, Succor Creek and related floras.

The small, early Pliocene Alvord Creek flora from southeastern Oregon, situated in an upland region where conifer-deciduous hardwood forest dominated, has only a small representation of broadleaved sclerophyllous forest (AXELROD, 1944f). *Arbutus* (cf. *arizonica)* is present, but *Lithocarpus* and evergreen *Quercus* are not recorded. Associates of the madrone included *Acer (macrophyllum), Amorpha (californica), Heteromeles (arbutifolia), Prunus (demissa)* and *Rosa* (spp.). They lived on warmer, more exposed sites together with a few chaparral plants, as suggested by *Ceanothus (cuneatus), Cercocarpus (betuloides, paucidentatus), Rhus (glabra)*, and *Heteromeles (arbutifolia)*. Moist slopes and valleys supported conifer-hardwood forest, composed of *Abies (shastensis, recurvata), Picea (breweriana), Pinus (murrayana, ponderosa), Pseudotsuga (menziesii), Chamaecyparis (lawsoniana)*, with their associates including *Amelanchier (alnifolia), Prunus (demissa), Populus (tremuloides), Rosa* (spp.), *Mahonia (repens)*, and *Sorbus (scopulina)*. Taxa whose nearest equivalents occur now in areas of summer rainfall are rare, and include *Juglans (nigra), Prunus (serotina)* and *Abies (recurvata)*. To judge from the evidence available, broadleaved sclerophyllous forest probably occupied lower warmer altitudes. Inasmuch as the deciduous hardwoods that require summer rainfall are greatly reduced, and the leaves of the angiosperms *(Acer, Amelanchier, Rosa)* are smaller than related Miocene species, a general lowering of summer rainfall seems indicated.

Forest had virtually been eliminated from the lowlands east of the Cascades by middle Pliocene time. The record provides no evidence of forest, and the only trees recorded at sites of plant deposition over the lowlands are those that required a high water table, and hence are confined to the margins of streams and lakes. The small Deschutes flora of eastern Oregon represents riparian-border vegetation like that found today below forest in eastern Oregon (CHANEY, 1938). It includes *Populus (tremula-davidiana, trichocarpa* – both small-leaved ecotypes), *Salix (caudata)* and *Acer (negundo)*. It is evident that rainfall was greatly reduced, that summer rainfall (implied by *Populus tremula*) was near a minimum, and that forest was no longer present over the lowlands. Similar relations are suggested by a small, undescribed flora from McKay Reservoir south of Pendelton, Oregon (U.C. Mus. Pal.). Forest is absent and the woody vegetation is limited to riparian-border trees and shrubs. Included in the small flora are *Salix (goodingii), Ulmus (alata,* small leaves), and *Populus (trichocarpa,* small leaves). The Alturas flora from the northeast corner of California reveals similar relations (AXELROD, 1944g). The collection is dominated by *Populus (tremuloides, trichocarpa* – both small-leaved ecotypes), *Salix (caudata)*, and *Ulmus (americana* – small-leaved ecotype). Rainfall was greatly reduced as compared with the Miocene, summer rainfall was very low, and forest must have been restricted to higher, moister levels in this region.

West-central Nevada

Tanoak-madrone forest is well represented in the middle Miocene floras of western Nevada (Table 1). Among the taxa in the Eastgate, Buffalo Canyon and Fingerrock floras are fossil plants similar to *Arbutus (xalapensis), Lithocarpus (densiflora), Quercus (chrysolepis, kelloggii),* and *Acer (macrophyllum).* Not only are most of these among the regular dominants of the living forest (COOPER, 1922), they are abundantly represented in the fossil floras. Associated shrubs that presumably formed a part of the fossil community, as judged from their present occurrence in the tanoak-madrone forest, are *Amelanchier (alnifolia), Prunus (demissa), Mahonia (pinnata, aquifolium), Garrya* (cf. *elliptica),* and *Heteromeles* (cf. *arbutifolia).* An extinct, small-leafed species of *Lyonothamnus* represented in these floras probably was also a part of the broadleaved sclerophyll forest. The *Lithocarpus-Arbutus-Quercus* forest occupied relatively drier, more exposed slopes adjacent to a rich conifer-deciduous hardwood forest composed of *Abies-Picea-Pinus-Pseudotsuga, Chamaecyparis-Sequoiadendron* and their regular associates – an ecologic relation that still exists today.

Broadleaved sclerophyll forest and conifer-hardwood forest both included plants whose nearest descendants live in areas with ample summer rainfall. Species similar to those in the mixed evergreen forest of Mexico and bordering regions included *Arbutus (xalapensis), Juglans (major),* and *Eugenia* (extinct). Contributing chiefly to the conifer-hardwood forest that reached down to the margins of the Miocene lake basins, and also inhabiting cooler slopes, were *Betula (lenta, papyrifera), Acer (pictum, saccharinum), Aesculus (octandra), Diospyros (virginiana), Carya (ovata), Ulmus (alata, fulva)* and *Zelkova (serrata),* all with their nearest descendants in either the eastern United States or in eastern Asia. These species are quantitatively less abundant in the Nevada floras than in those to the north Oregon and Idaho, and the diversity of taxa that represent these eastern elements is also lower in western Nevada.

By late Miocene time conifer-hardwood and mixed evergreen forest had become impoverished over western Nevada (AXELROD, 1956; WOLFE, 1964). The Aldrich Station, Fallon, Purple Mountain, Chloropagus and Stewart Spring floras show that broadleaved sclerophyll forest included *Arbutus (menziesii, xalapensis), Lithocarpus (densiflora,* Sierran form), *Quercus (chrysolepis),* and *Lyonothamnus* (extinct). Oak is predominant at all localities, and madrone, tanoak and ironwood are rare to locally common. The mixed evergreen forest, which included species of *Acer (macrophyllum, negundo), Fraxinus (oregona), Populus (trichocarpa), Salix (lasiandra, lasiolepis),* and *Amorpha (californica)* in moister sites, probably was confined to moister slopes over the lowlands. This is consistent with the general rarity of conifers in these floras. Such species as *Abies (concolor), Chamaecarpis (lawsoniana), Pinus (ponderosa), Picea (breweriana),* and *Sequoiadendron (giganteum)* are not as well represented as in the middle Miocene floras, and their regular forest associates are also less common. Only a few taxa in these younger floras indicate summer rainfall, as judged from the present occurrence of their nearest relatives in summer-wet regions today. Not only are they rare, but some have smaller leaves than the similar species in the middle Miocene. Among those recorded are *Acer (saccharinum), Betula (papyrifera, lenta), Populus (grandidentata), Sophora (japonica), Ulmus (alata),* and *Zelkova (serrata),* the others having been eliminated from near the sites of plant deposition.

The change from the relatively luxuriant middle Miocene vegetation to more impoverished forests of the late Miocene resulted from two factors. First, the general trend toward increased aridity and decreased summer rainfall was responsible in part for the elimination of most mesic species from the lowlands near sites of deposition, and hence for their comparatively poorer representation in the lowland deposits. Some of them no doubt inhabited moister upland areas, but in general forest vegetation was being restricted to more favorable environments farther west – to California. The second factor that accounts for reduction of forest over the lowlands and for the lower species diversity is related to middle and late Miocene volcanism in the northern to central Sierra Nevada and adjacent Nevada that built up a broad volcanic field. The rainshadow to the east not only resulted in somewhat less rainfall (AXELROD, 1956), and hence in the further restriction of forest at the expense of oak woodland-savanna, but also in a sunnier climate. With the resultant higher evaporation rate, summer rainfall was less effective than on the windward flanks of the range. This accounts for the poor representation of taxa in the Nevada floras that require summer rainfall. By contrast, they make up a much higher percentage of the floras of the same age to the north in Oregon, Idaho or Washington. Taxa that indicate summer rainfall are also more abundant in interior southern California (Tehachapi, Mint Canyon) floras to the lee of the Sierra Nevada-Peninsular Range barrier during the middle and late Miocene than to the east of the low Sierran divide in western Nevada. The lower effective summer rainfall over western Nevada gave to the middle and late Miocene floras of that area the aspect of younger floras, as noted earlier (AXELROD, 1956). Clearly, environmental change was proceeding here at a more rapid rate than elsewhere in the Pacific border region: vegetation was modernized here at an earlier date than in other areas. This evidence provides strong support to theoretical considerations that point to the importance of aridity in evolution (STEBBINS, 1952). Dry climates appear first in local areas, and in a very brief time may provide the impetus for change at a greatly accelerated tempo – either in terms of evolution of taxa, or of vegetation (AXELROD, 1967b).

By Pliocene time, broadleaved sclerophyll forest had been eliminated from near sites of deposition in lowland basins over most of western Nevada. Oak woodland-savanna and juniper-oak woodland now assumed dominance over the lowlands. However, mixed evergreen forest still inhabited moister sites near the west margin of the province. The Chalk Hills flora has *Chrysolepis [= Castanopsis] (chrysophylla)*, *Arbutus (menziesii)*, and *Lithocarpus (densiflora)* (AXELROD, 1962). A number of their regular associates that contribute also to conifer forest, notably species of *Amelanchier*, *Ceanothus*, *Prunus*, *Mahonia*, and *Rosa*, are also present. Tanoak-madrone forest presumably inhabited warmer, drier slopes adjacent to the mesophytic Sierra redwood forest composed of *Abies (concolor)*, *Pinus (ponderosa)*, *Chamaecyparis (lawsoniana)*, and *Sequoiadendron (giganteum)* at Chalk Hills. Members of the tanoak-madrone community also contributed to conifer forest much as their descendants do today. Evidence of summer rainfall at Chalk Hills is provided by only a few taxa, and all of them are rare. Some of the conifers and angiosperms, notably species of *Chamaecyparis*, *Picea* and *Betula* are represented today by similar species in Oregon or Washington, in areas where there is 4 or more inches summer rainfall, and where temperatures are lower in summer. In addition, the flora has a single leaflet of *Carya (ovata)* and leaves of sawtoothed-aspen *Populus (grandidentatum)* that indicate some summer rainfall was still present in western Nevada. However, it was insufficient for most typical members of the eastern elements (*Betula*, *Diospyros*, *Gymno-*

cladus, Taxodium, Ulmus, Zelkova, etc.) which occurred here in larger numbers in the middle Miocene. Members of the alliance were still abundant in the coastal strip, indicating more effective summer rainfall there than in the interior.

Central California

Tanoak-madrone forest is represented in the late Miocene Remington Hill and Table Mountain floras from the middle and lower slopes of the Sierra, respectively (CONDIT, 1944a, 1944b). The Remington Hill flora has *Arbutus (menziesii), Quercus (chrysolepis), Lithocarpus (densiflora), Quercus (kelloggii), Umbellularia (californica)* and *Persea (borbonia).* They formed a broadleaved sclerophyll forest which graded into a *Sequoia-Chamaecyparis* forest in moister sites and on cooler slopes. Drier slopes supported oak-savanna woodland as shown by the presence of *Quercus (lobata, wislizenii),* and chaparral was present to judge from species of *Arctostaphylos (manzanita)* and *Ceanothus (cuneatus).* Evidence of summer rain is provided by species of *Acer (grandidentatum), Aesculus (glabra), Karwinskia (humboldtiana), Liquidambar (styraciflua), Persea (borbonia), Ulmus (americana).* The Table Mountain flora, from a site 100 miles south and at least 500 feet (150 m) lower in altitude during the Miocene, also had a broadleaved-sclerophyll forest, as shown by *Arbutus (menziesii), Umbellularia (californica), Persea (podadenia), Magnolia (grandiflora),* and *Mahonia (lanceolata).* This flora was situated below the mixed conifer forest that occupied the middle and higher parts of the range, as shown by the absence of conifer-forest species, and by the presence of chaparral. Evidence for ample summer rainfall is provided by *Carya (cordiformis), Berchemia (scandens), Acer (saccharum), Gleditschia (aquatica), Ilex (opaca), Magnolia (grandiflora), Nyssa (sylvatica), Persea (borbonia)* and *Ulmus (americana).*

Some of these probably contributed to the mixed evergreen forest, chiefly in moister sites together with the species of *Acer, Cornus, Platanus,* and *Salix* that lived chiefly along stream banks and at seepages. These late Miocene floras from the windward slopes of the Sierra Nevada differ from those in the east in Nevada in having larger numbers of taxa that live in areas with summer rainfall. They indicate that the rainshadow cast by the low Sierran divide was sufficient to reduce effectiveness of summer moisture there to a level so low that only a few deciduous hardwoods remained. This was no doubt due in part to the higher evaporation rate in the interior which resulted in decreasing the effectiveness of summer moisture there as compared with the milder, more equable coastward slope.

Evidence of later change in the Sierra is not now available because fossil floras representing the lower forest zone are not now known. Nonetheless, it is apparent that by the later Miocene broadleaved evergreen sclerophyllous forest inhabited the west flank of the range and included many species scarcely distinguishable from those that are in the same area today. The late Miocene forest was more diverse in taxa, chiefly because it lived under a mild equable climate characterized by ample summer as well as winter precipitation.

In west-central California, mixed evergreen forest first appears in the middle Pliocene Mulholland and Petaluma floras – the older Black Hawk (AXELROD, 1944a), Diablo (UC Mus. Pal.) and Neroly (CONDIT, 1938) floras representing lowland floodplain deciduous hardwood forest and swamp forest. The Mulholland flora (AXELROD, 1944b) includes *Lithocarpus (densiflora), Arbutus (menziesii),* and *Umbellularia (californica)* together with their regular riparian associates, notably *Alnus (rhombifolia), Amelanchier (alnifolia), Cornus (californica), Fraxinus (oregona), Platanus (racemosa), Populus (trichocarpa), Rhus (diversiloba),* and *Salix (lasiandra).* The small Petaluma flora north of San Francisco Bay also has mixed evergreen forest, as shown by the presence of *Arbutus (menziesii)* and *Umbellularia (californica).* Live oaks are abundant in each flora (see below), but conifer forest species *(Abies, Picea, Pinus, Sequoia, Pseudotsuga)* are absent (AXELROD, 1944d, p.187; 1944e, p.215). The relations suggest that rainfall was so low that conifer forest was confined to higher hills, to sites where precipitation was greater than over the lowlands. In this connection, it is recalled that the middle Pliocene was the driest part of the Tertiary, and forest was confined throughout the region. Summer rainfall was reduced, but the floras do contain a few taxa related to those that no longer live in California, *notably Nyssa (sylvatica), Sapindus (drummondii), Karwinskia (humboldtiana), Acer (grandidentatum),* and *Populus (grandidentata,* small-leaved ecotype) in the Mulholland, *Ulmus (americana,* small-leaved ecotype) in the Petaluma, and *Robinia (neomexicana)* and *Sapindus (drummondii)* in the Oakdale flora at the western base of the central Sierra. As discussed below, liveoak woodland-savanna vegetation was dominant over the lowlands during the middle Pliocene. It is apparent that the middle Pliocene vegetation of this area differed significantly from that in northeastern California (Alturas) and Oregon (Deschutes, McKay Reservoir), where trees were now largely limited to streambanks over the lowlands.

Late Pliocene floras of central California provide evidence for a substantial increase in precipitation over that of the middle Pliocene: this rise heralds the growth of the first ice sheet (AXELROD, 1944e, p.215). In response to increased rainfall, mixed evergreen forest and mixed conifer forest invaded the lowlands where earlier there was chiefly oak woodland-savanna and chaparral vegetation. The Sonoma flora near Santa Rosa has *Lithocarpus (densiflora), Quercus (chrysolepis)* and *Umbellularia (californica)* as representative members of mixed evergreen forest, as well as other evergreens (AXELROD, 1944d). Of these, *Garrya (elliptica), Chrysolepis [Castanopsis] (chrysophylla), Myrica (californica)* and *Quercus (agrifolia)* are regular members of the mixed evergreen forest, and *Ilex (brandegeana)* and *Persea (borbonia)* are relicts that probably lived with the forest on warmer slopes in the region. Regular stream-bank associates of the mixed evergreen forest were *Alnus (rhombifolia), Fraxinus (caudata), Platanus (racemosa), Populus (trichocarpa), Rhamnus (purshiana), Salix (lasiolepis), Symphoricarpos (albus)* and *Vaccinium (parvifolium).* The latter were also associates of the conifer forest that inhabited the moister sites in the area, and was composed of *Abies (grandis), Chamaecyparis (lawsoniana), Picea (breweriana), Pseudotsuga (menziesii), Sequoia (sempervirens)* and *Tsuga (heterophylla).* The sclerophyllous forest inhabited warmer slopes adjacent to the conifer forest, a relation found today in the north Coast Ranges and western Siskiyou Mountains. That summer rain was still present is indicated by two of the evergreens, *Ilex (brandegeana)* and *Persea (borbonia),* and by two deciduous hardwoods, *Castanea (americana)* and *Ulmus (alata).*

The Napa flora, situated 26 miles southeast of the Santa Rosa locality and 16 miles east of the Petaluma flora, also has mixed evergreen forest, as shown by the occurrence there of *Castanopsis (chrysophylla), Quercus (agrifolia, wislizenii, chrysolepis), Persea (borbonia)* and *Umbellularia (californica)* (AXELROD, 1950a). Associated with them in moister sites were *Amorpha (californica), Holodiscus (discolor), Platanus (racemosa)* and *Salix (laevigata).* The cooler, north-facing slopes supported a redwood-yellow pine forest, as shown by the occurrence of *Abies (concolor), Pinus (ponderosa),* and *Sequoia (sempervirens).* Associated with them were some of the plants noted above, as well as *Mahonia (nervosa), Castanopsis (sempervirens), Populus (tremuloides)* and *Salix (scouleriana)*. Evidence of summer rainfall is provided by *Persea (borbonia)* and by a leaflet of *Pterocarya (stenoptera)*. Drier slopes in the region supported oak woodland-savanna vegetation, and rocky sites with poorer soil had chaparral vegetation (see below).

Southern California

The moister cooler climate of the early Pleistocene enabled yellow pine forest and tanoak-madrone forest to reach into interior southern California, as shown by the Soboba flora from the Bautista formation near San Jacinto (Axelrod, 1966). *Arbutus* is well represented, and other species that occurred with it and contributed to sclerophyllous evergreen forest included *Quercus chrysolepis, Q. wislizenii,* and *Q. morehus* – a hybrid of *wislizenii* and *kelloggii,* the latter not recorded there as fossil. The broadleaved sclerophyllous forest lived adjacent to a conifer forest of big-cone spruce *(Pseudotsuga macrocarpa)* and Coulter pine *(Pinus coulteri),* and mixed coniferous forest occupied cooler slopes in the adjacent area, as shown by the remains of *Abies concolor, Pinus ponderosa, P. lambertiana* and *Calocedrus decurrens.* Chaparral and oak woodland are well represented. Streamways in the area supported species of ash *(Fraxinus),* willow *(Salix),* cottonwood *(Populus),* and box elder *(Acer).* The presence of *Magnolia grandiflora* and *Acer brachypterum* in the Soboba flora indicates summer rain was still effective into the early Pleistocene. Both of them probably contributed to the broadleaved sclerophyll forest, and well as the more equable parts of the conifer forest.

The Soboba flora is well inland from the southern occurrences of *Arbutus* in southern California today. These are on the seaward slopes of the Agua Tibia Mountains (Nellie, Roderick Mt.) and Santa Ana Mountains (Trabuco Canyon), and in the foothills of the San Gabriel Mountains in Santa Anita Canyon. Today, tanoak-madrone forest attains its southern limit in the Santa Ynez Mountains (San Marcos Pass) in the hills above Santa Barbara. This occurrence probably is relictual from the late Pleistocene, when tanoak-madrone forest, together with *Sequoia* and *Pseudotsuga,* ranged southward to Carpinteria (Axelrod, 1967a, p. 265–6). The present discontinuous range of the forest in the south Coast Ranges and in the central Sierra Nevada probably is the result of range restriction that accompanied the warm dry Xerothermic period, a factor that also appears to account for the discontinuous distribution of other mesic forests in California (Axelrod, 1966; 1967a, p. 298).

Summarizing, mixed evergreen forest *(Lithocarpus-Arbutus-Quercus)* inhabited the region from central California and central Nevada northward into Oregon and Idaho during Miocene time. In moister, cooler areas it merged with mixed conifer-hardwood forest that included taxa whose nearest relatives are now in the eastern United States and eastern Asia. In drier, warmer areas it was replaced by oak-laurel forest, oak woodland-savanna, and chaparral vegetation composed of taxa whose descendants are now in California, as well as the southwestern United States and Mexico. Most of the species are scarcely separable from living plants. The Miocene forest was richer in taxa than the modern descendant vegetation which has survived in California and southwestern Oregon. It lived under mild temperature and a precipitation regime with rainfall well distributed through the year. Greater diversity resulted from the mingling of taxa from both Arcto-

Tertiary and Madro-Tertiary sources, for the Miocene forest included more deciduous hardwoods of the former and broadleaved evergreens of the latter, than does the modern derivative vegetation.

Broadleaved sclerophyllous forest gradually disappeared from the interior during late Miocene and Pliocene times as rainfall decreased and as temperatures became more extreme. Thus it was gradually confined to the region west of the Sierra-Cascade axis, to areas of adequate precipitation and mild temperature. In the latter area, the forest progressively lost broadleaved evergreens and deciduous hardwoods during the Pliocene as summer rainfall was reduced, though a few exotics persisted down to the close of the epoch in central California, and into the early Pleistocene in southern California. The forest probably invaded southern California at the close of the Tertiary and in the early Quaternary, as precipitation increased. Its present discontinuous distribution in the Sierra Nevada, the south Coast Ranges and southern California seems attributable to restriction of range during the Xerothermic period.

The fossil record thus shows that a) the Neogene mixed evergreen forest had a much wider distribution than does the descendant, modified forest, that b) the Neogene forest included more numerous deciduous hardwoods and broadleaved evergreens than does the modern related community, that c) the degree of dominance by evergreens increased as summer rainfall decreased, that d) the modern forest is a relict that came into existence chiefly by a process of climatic selection which eliminated taxa that require summer rain, and – above all – that e) the taxa of the broadleaved-sclerophyll forest did not evolve in response to mediterranean climate.

Oak-Laurel Forest

Oak-laurel forest occupied southern California during the Miocene, as indicated by the Puente, Modelo and Topanga floras, ranging north along the outer coastal strip at least to Carmel. These floras are similar in composition, and formed a natural province as shown by the greater resemblances between them as compared with floras to the east or north (AXELROD, 1956, p. 262). They are dominated by liveoaks *(Quercus.* cf. *virginiana, chrysolepis, potosiana, eduardii)* and laurels *(Persea, Ocotea, Nectrandra, Umbellularia),* and included other broadleaved evergreens, notably palm *(Sabal)* and ironwood *(Lyonothamnus).* Closed-cone pine forest *(Pinus* aff. *radiata, muricata)* lived on nearby sheltered slopes close to the sea. The oak-laurel forest shows relationship to forests now in Mexico, notably on the seaward slopes of the Sierra Madre south of Monterrey, and on the Pacific slope from southern Sonora southward. In physiognomy, it also resembled the present hammock-flora of Florida, where palm, liveoaks, and laurels are dominants of the vegetation. Oak-laurel forest lived under a frostless climate to judge from the plants, and also from the shallow-water marine invertebrate faunas that suggest winter seasurface temperatures were not less than 18° C in southern California, nor below 16° C in central California (Monterey), according to HALL (1960).

Oak-laurel forest extended into the interior, to judge from the Middle Miocene Tehachapi flora with live oaks *(Quercus* cf. *chrysolepis, emoryi, brandegeei, arizonica),* laurels *(Persea, Umbellularia),* and ironwood *(Lyonothamnus)* (AXELROD, 1939). In addition, the Tehachapi flora has other genera that are now in the broadleaved-evergreen forests of

Mexico which are dominated by laurels and live-oaks, notably species of *Arbutus, Clethra, Myrica, Cedrela* and *Sabal*. The forest has a poor quantitative representation at Tehachapi, no doubt reflecting the interior position of the basin and the drier climate there. The forest probably inhabited more distant cooler, moister slopes from which leaves, fruits and other structures were contributed less frequently to the record. In this drier area, oak woodland-savanna vegetation dominated over the lowlands, chaparral covered well drained bordering slopes, and thorn forest inhabited the warmer dry sites where species of *Acacia, Bursera, Colubrina, Ficus, Pithecolobium, Trichilia, Randia* and others are recorded. The flora provides clear evidence for ample summer rainfall, and also for winters without frost.

A small collection from middle Miocene marine deposits on the west front of the Sierra Nevada north of Bakersfield provides evidence for the altitude of the topographic barrier that separated the Tehachapi flora from the sea that then flooded the west front of the range. The collection from Pyramid Hill contains endocarps of *Juglans*, a few liveoak leaves similar to those of *Q. virginiana*, and a cast of a trunk of *Dioon* occurs at a nearby locality of similar age. The cycad and walnut are typical members of the oak-laurel forest in northern Mexico. Also represented in the collections are pine cones that were transported from the Sierra Nevada into the marine basin. The cones resemble those of *P. ponderosa* and *P. lambertiana,* and suggest that at a minimum the forest from which they were derived probably had an altitude near 2,500–3,000 feet (750–900 m). This seems adequate to account for the semiarid climate indicated by the Tehachapi flora 40 miles southeast across the low Sierran ridge.

An ecotone between the oak-laurel forest of the coastal region and mixed deciduous hardwood forest is recorded by the Temblor flora near Coalinga (UC Mus. Pal). Members of the oak-laurel forest are predominant, with species of liveoak *(virginiana, chrysolepsis), Arbutus, Persea, Ilex* and *Lithocarpus*. They are associated with deciduous hardwoods and conifers like those recorded to the north in typical middle Miocene floras. Among the taxa represented are *Pinus* (aff. *strobus), Keteleeria (davidiana), Quercus (borealis, falcata), Nyssa (aquatica),* and *Castanea (americana)*. The latter lived chiefly in the bordering hills to the west, where climate was more temperate and rainfall higher. This is consistent with their representation, for the live oaks are predominant and the deciduous hardwoods and their associates are less abundant and they also show more evidence of transport.

Inasmuch as only a few Pliocene floras have been described from California, the later history of the oak-laurel forest is not well understood today. Available evidence indicates that the taxa which formed a conspicuous part of it were gradually eliminated from the area as summer rainfall diminished, and also as winter temperature was lowered. The known records of oak-laurel forest include the Anaverde flora from the west margin of the Mohave Desert, which includes *Persea (podadenia), Quercus (wislizenii)* and a palm (AXELROD, 1950c). The forest occupied cooler slopes in the bordering uplands, adjacent to oak savanna-woodland, chaparral and thorn forest vegetation that dominated the lowlands there. Summer rainfall indicators include *Bumelia, Colubrina, Dodonaea, Eysenhardtia,* the palm, and also a species of *Populus (brandegeei)*. In the Piru Gorge flora 30 miles west, oak-laurel forest is represented by *Quercus (engelmannii, wislizenii), Persea (podadenia), Laurocerasus (lyonii),* and *Sabal (uresana)* which covered moist slopes of the basin directly west of the fossil locality (AXELROD, 1950d). Adjacent vegetation on drier slopes included chaparral and oak woodland-savanna. As compared with the Anaverde, the Piru Gorge flora lived farther west and hence in a moister region, which accounts for the absence of thorn forest in it. Summer rain indicators include *Persea* and *Sabal*, as well as *Acer (brachypterum)* and *Populus (brandegeei)*. The Mount Eden flora from interior southern California also had oak-laurel forest, for it includes *Quercus (agrifolia, chrysolepis, engelmannii)*, *Arbutus (glandulosa)* and *Persea (podadenia)* (AXELROD, 1937, 1950b). It covered nearby cooler slopes adjacent to a *Pseudotsuga (macrocarpa)-Pinus (coulteri)* forest which inhabited the bordering hills. Chaparral was also present, and there were patches of thorn scrub persisting in warmer, drier sites. Summer-rain indicators in this area include *Arbutus (glandulosa)*, *Persea*, *Cercidium*, *Dodonaea*, *Eysenhardtia*, *Ficus* and *Populus (brandegeei)*.

Avocado and live oaks are abundant in the upper Jacalitos formation south of Coalinga, and avocado was common into the late Pliocene as shown by its abundance in the San Joaquin flora (UC Mus. Pal.). Oak-laurel forest must therefore have covered slopes in the Coast Ranges to the west, and probably persisted into the first pluvial at least. The oak-laurel woodland near Coalinga was associated with riparian-border trees similar to those that still live in the region, notably, *Populus (fremontii)*, *Platanus (racemosa)*, *Salix (lasiolepis)*, and *Alnus (rhombifolia)*. Summer-rain indicators include *Persea* and *Sapindus*, as well as *Celtis* which ranges from Arizona southward into oak-laurel country today.[4]

From the preceding evidence, it appears that a highly modified descendant of the Neogene oak-laurel forest has survived in California. This is the community composed of *Umbellularia californica* which locally in the central coastal strip forms nearly pure stands, but more commonly is associated with live oaks, *Q. agrifolia* in the coastal region and with *Q. wislizenii* and *Q. chrysolepis* in the interior and at higher altitudes. It regularly inhabits moist canyons in equable sites where it is associated with riparian-border trees similar to those in the richer Neogene forest, for instance *Acer macrophyllum*, *Alnus rhombifolia*, *Platanus racemosa*, *Populus trichocarpa* (lowland form), and *P. fremontii*. This oak-laurel forest occupies sites below mixed evergreen (tanoak-madrone) forest where they are in proximity, for its taxa can endure lower rainfall and higher temperature. This is its usual occurrence in the Coast Ranges, and also in the Sierra Nevada where it ranges south of tanoak-madrone forest. It occurs chiefly above oak savanna country, reaching up to the margins of yellow pine forest where *Quercus chrysolepis* replaces *Q. wislizenii*, and where *Q. agrifolia* is absent. In southern California it is more widely spread than tanoak-madrone forest, occupying deep canyons in the Santa Monica and Santa Ana Mountains, and also scattered in canyons along the south and west front of the San Gabriel, San Bernardino and San Jacinto Mountains.

Summarizing, during Miocene time oak-laurel forest inhabited southern California and reached northward along the outer coastal strip into central California. It occupied moister uplands over interior southern California, and probably ranged far to the southeast. The Miocene forest was composed of taxa similar to those now in the oak-laurel forest of Mexico which lives under ample summer rainfall and mild winter temperature, and in the derivative, impoverished forest that persisted in California.

In southern California, the Neogene oak-laurel forest was bordered at lower, warmer

---

4 In California, *Celtis* occurs locally in the Kern River Canyon east of Bakersfield, and also near Campo close to the Mexican border.

and drier levels by oak woodland-savanna, chaparral, and thorn scrub vegetation, but in central California it was in ecotone with mixed evergreen forest and mixed deciduous hardwood forest.

As summer rainfall decreased, and as winter temperatures were lowered following the Miocene, oak-laurel forest gradually lost a number of taxa in California, though exotics were a part of it into the close of the Pliocene. The surviving, modified community that persisted in California occupies warmer, drier areas below the tanoak-madrone forest, or ranges south of it where it is replaced by oak woodland-savanna and chaparral vegetation at lower levels, and by pine forest at higher altitudes.

Oak Woodland-Savanna

Southern California

Vegetation dominated by live oaks, and often with nut pine and juniper and various sclerophyllous and some drought-deciduous shrubs, is first recorded in abundance in the Miocene of interior southern California. Included in the middle Miocene Tehachapi flora (AXELROD, 1939) are *Pinus (monophylla), Cupressus (arizonica),* and *Quercus (arizonica, chrysolepis, emoryi).* Shrubby associates included *Amorpha (californica), Ceanothus (cuneatus, verrucosus), Arctostaphylos (glandulosa, pungens), Cercocarpus (betuloides), Quercus (turbinella)* that also contributed to chaparral (see below). Moist swales and stream banks supported *Acer (brachypterum), Erythea (armata), Lyonothamnus* (extinct), *Populus (fremontii, brandegeei), Platanus (racemosa)* and *Sabal (uresana).* At higher, moister levels oak woodland graded into a rich oak-laurel forest (see above) which, in addition to the oaks, included *Clerhra, Myrica, Persea, Umbellularia* and other evergreens that lived under a more equable climate. At lower, drier levels the dense oak woodlands opened out into a grassland region with scattered oaks and nut pine. Warmer, drier sites in that area supported thorn scrub, composed of *Acacia, Bursera, Cardiospermum, Colubrina, Eysenhardtia, Ficus, Lysiloma, Pithecolobium,* and others. Some of these no doubt entered into the matrix of the oak woodland-savanna as scattered shrubs, much as they do today in Sonora and bordering areas. The Tehachapi flora provides ample evidence of summer rainfall, and for mild, frostless winters.

The upper Miocene Mint Canyon flora is similar in composition, but its vegetation reflects a drier climate than the Tehachapi (AXELROD, 1940b). This resulted in part from the general trend to aridity, and also from its interior position, having been displaced from the Chocolate Mountains 140 miles southeast by movement along the San Andreas fault (EHLIG and EHLERT, 1972). Oaks are predominant and are represented by species that are similar to *Q. agrifolia, emoryi, engelmannii, lobata* and *wislizenii.* Stream- and lake-margin sites supported *Celtis (reticulata), Juglans (major), Platanus (racemosa), Populus (fremontii),* and *Salix (lasiolepis).* Remains of the oak-laurel forest that inhabited the higher hills are not as common, with *Ilex (brandegeana), Persea (podadenia, americana),* and *Laurocerasus (lyonii)* represented. Thorn scrub vegetation is more prominently developed here than in the Tehachapi flora, with species of the following genera represented: *Acacia, Bursera, Castelea, Celtis, Colubrina, Cardiospermum, Eysenhardtia, Euphorbia, Lysiloma, Piscidia, Pachycormus, Pithecolobium* and *Thouinia.* Chaparral is also well developed (see below), and shrubs from both plant formations no doubt were scattered in the oak woodland-savanna. It is evident that the flora lived under semiarid climate. with ample summer rainfall and with temperatures well above freezing in winter – a conclusion consistent with shallow-water molluscan faunas of this region which indicate minimum sea-surface temperatures were near 18 °C (HALL, 1960).

A small, early Pliocene flora represented chiefly by fossil wood is in the lower Ricardo formation in the Mohave Desert (WEBBER, 1933). It includes *Quercus* (liveoak), *Pinus* (cf. *monophylla)*, and *Cupressus (arizonica)* that represent woodland-savanna vegetation, and also a legume *(Robinia?)* that probably was a part of it. There is a *Sabal* palm in the flora that inhabited lake-and stream-margins in this region which was semiarid and open as judged from the large grazing mammals in the Ricardo fauna. A few fossil leaves have been found at the petrified forest locality. They include *Acacia, Lycium*, and *Ceanothus*, and suggest that thorn scrub and chaparral vegetation were present in the area. This is consistent with the composition of the middle Pliocene Anaverde flora situated 60 miles south, on the southwest margin of the Mohave Desert (AXELROD, 1950c). Here the oak woodland-savanna was composed of *Pinus (sabiniana)* and *Quercus (wislizenii)* chiefly. Inhabiting streambanks in the area were palm (indet.), *Bumelia (lanuginosa), Populus (fremontii, brandegeei), Salix (lasiolepis), Platanus (racemosa)* and *Sapindus (drummondii)*. Shrubs that were scattered in the oak woodland-savanna vegetation and that contributed to chaparral on shallow soils included *Quercus (palmeri, dumosa), Peraphylum (ramosissimum), Dodonaea (viscosa), Ceanothus (cuneatus)* and *Rhamnus (crocea, californica)*. Shrubs that appear to have contributed to a relict thorn scrub are represented by *Bumelia (lanuginosa), Colubrina (californica), Dodonaea (angustifolia)*, and *Eysenhardtia (polystachya)*, and some of them probably were scattered in the drier parts of the woodland-savanna with the chaparral species. The abundance of *Persea (podadenia)* leaves in the flora suggests that oak-laurel forest probably was well developed at higher, moister levels in the mountains to the west.

This inference finds support in the nature of the Piru Gorge flora, situated in the mountains 30 miles west (AXELROD, 1950d). Oak woodland-savanna vegetation is represented by *Quercus (douglasii, engelmannii, wislizenii)* and *Aesculus (californica)*. Their regular riparian- and lake-margin associates included *Platanus (racemosa), Populus (fremontii, brandegeii), Salix (exigua, lasiandra, lasiolepis)*, and *Sabal (uresana)*. Remains of *Persea (podadenia)* and *Laurocerasus (lyonii)* indicate a rich oak-laurel forest in the nearby hills to the west where rainfall was somewhat higher. Among the shrubs that were scattered in the woodland-savanna vegetation were *Arctostaphylos (glauca), Ceanothus (spinosus), Fremontia (californica)* and *Schmaltzia (ovata)*, and they also contributed to chaparral on rocky slopes. Mild winters are indicated by *Persea, Sabal, Populus (brandegeei), Laurocerasus* and *Ceanothus*, and the palm, avocado and cottonwood *(brandegeei)* imply summer rainfall.

To the south, oak woodland-savanna vegetation is well represented in the middle Pliocene Mount Eden flora of interior southern California (AXELROD, 1937; 1950b). Among the species typical of this vegetation are *Quercus (agrifolia, douglasii, engelmannii), Juglans (californica, rupestris)*, and *Pinus (sabiniana)*. Stream- and lake-border sites in the oak-savanna region supported *Forestiera (neomexicana), Platanus (racemosa), Populus (brandegeei), Salix (goodingii, exigua, lasiolepis)* and *Sapindus (drummondii)*. At higher, moister levels liveoak woodland merged with a relict oak-laurel forest, composed of *Quercus (chrysolepis), Arbutus (xalapensis), Persea (podadenia), Pinus (radiata)*,

Cupressus (forbessi), and *Laurocerasus (lyonii),* with a *Pseudotsuga (macrocarpa)-Pinus (coulteri)* forest on cooler slopes. On rocky slopes with shallow soil woodland gave way to a rich chaparral (see below), and the drier, warmer sites over the lowlands supported a relict thorn scrub composed of *Cercidium (floridum), Dodonaea (angustifolia), Eysenhardtia (polystachya),* and *Ficus (cotinifolia).* That climate was one with mild temperature is apparent from the occurrence of taxa whose nearest modern derivatives (e.g. *Pinus radiata, Malosma laurina, Schmaltzia integrifolia, Laurocerasus lyonii)* now live only in the mild coastal strip, or on the offshore islands. Summer rain indicators include such genera as *Cercidium, Dodonaea, Ficus, Persea, Eysenhardtia,* and *Sapindus,* as well as certain species of *Populus (brandegeei), Arbutus (xalapensis),* and *Juglans (rupestris).*

Western Nevada

Whereas typical oak savanna-woodland vegetation dominated interior southern California during the middle Miocene, it was scarcely represented at this time in western Nevada. Conifer-deciduous hardwood forest and mixed evergreen forest (tanoak-madrone) were dominant in a region where yearly rainfall totalled 35 inches (890 mm) or more. Only locally in the region is there evidence of typical woodland-savanna vegetation, and it was of restricted occurrence. The Middlegate flora, situated at the south front of the low volcanic hills that afforded dry sites suitable for colonization includes *Quercus (wislizenii)* in some abundance (AXELROD, 1956). Along stream- and lake-margin sites it was associated with *Lyonothamnus* (extinct sp.), *Platanus (racemosa), Populus (angustifolia),* and *Salix (goodingii, lasiolepis).* Scattered shrubs included *Ceanothus (cuneatus), Cercocarpus (betuloides, paucidentatus), Fraxinus (anomala), Sumac (glabra)* and *Styrax (californica)* and they probably formed a chaparral on poorer sites. The other middle Miocene floras now known from this region, the Buffalo Canyon, Eastgate, Pyramid and Fingerrock, contain very few fossils representing oak woodland vegetation, indicating that it was not yet important at those sites. This was chiefly because these areas supported a mesophytic forest composed of conifers *(Abies, Picea, Pinus, Tsuga, Pseudotsuga, Sequoiadendron, Chamaecyparis)* and deciduous hardwoods *(Alnus, Betula, Carya, Diospyros, Gymnocladus, Liquidambar, Taxodium, Ulmus, Zelkova),* many of the latter indicating ample summer rainfall.

Oak woodland-savanna vegetation commenced to spread in western Nevada during later Miocene time, when it occupied sites marginal to broadleaved evergreen (oak-madrone) forest and conifer-deciduous hardwood forest. In the Aldrich Station, Fallon, Chloropagus and Purple Mountain floras (AXELROD, 1956), oak woodland-savanna is represented chiefly by *Quercus (wislizenii)* and (locally) *Juniperus (osteosperma).* Scattered through the woodland were *Aesculus (parryi), Amorpha (californica), Amelanchier (utahensis), Arbutus (xalapensis), Rhamnus (californica), Robinia (neomexicana),* and *Cercis (occidentalis).* Some of these shrubs also contributed to a localized chaparral, as noted below. Moister sites, streambanks and lake borders supported *Platanus (racemosa), Bumelia (lanuginosa), Populus (brandegeei)* and *Salix (exigua, goodingii, lasiandra).* In western Nevada, woodland-savanna vegetation and associated shrubs lived in the more exposed, drier sites, in areas where effective rainfall was near 20 to 25 inches (508–625 mm) yearly. Cooler, wetter slopes supported mixed evergreen forest and mixed conifer-hardwood forest. The latter show that summer rains were reduced in amount, for taxa that now find their nearest relatives in summer-wet regions are less important from both a quantitative and a qualitative standpoint as compared with middle Miocene floras in this area.

To the south and east, the Stewart Spring flora (WOLFE, 1964) of late Miocene age indicates a drier climate. This was due in part to its position farther to the east, and to the increased height of the Sierran barrier farther south. Taxa recorded here are *Quercus (wislizenii), Juniperus (osteosperma)* and a few woodland plants that have not yet been found elsewhere in central Nevada, notably *Juglans (major), Sapindus (drummondii), Colubrina* and *Astronium*. They suggest that the oak woodland in this area was ecotonal with the Miocene woodland vegetation in interior southern California. On drier slopes, associates of the oak woodland-savanna included *Garrya (elliptica), Ribes (cereum), Cercocarpus (betuloides), Holodiscus (dumosus), Peraphyllum (ramosissimum), Schmaltzia (integrifolia)* and others. Streambanks and exposed sites in the lake shore area supported more mesic members of the oak woodland-savanna vegetation, including *Juglans (major), Lyonothamnus (floribundus), Sapindus (drummondii),* and *Arbutus (xalapensis)*. Cooler slopes near at hand supported mixed conifer-hardwood forest *(Abies, Picea, Pinus, Larix, Tsuga, Chamaecyparis, Betula, Populus, Amelanchier, Mahonia, Prunus, Rosa)*, as well as mixed evergreen (oak-madrone) forest, as discussed above. Summer rainfall had decreased appreciably as compared with the underlying middle Miocene Fingerrock flora of this area, for there are fewer taxa in the Stewart Spring flora whose modern equivalents are restricted to such regions.

By Pliocene time, conifer-deciduous hardwood forest and mixed evergreen forest had largely been eliminated from lowland areas over the central and southern Great Basin, and liveoak woodland-savanna vegetation was now predominant. Only along the west margin of the province (Chalk Hills and Mansfield Ranch floras) is there evidence of forest vegetation. The Esmeralda flora from near Coaldale in southwestern Nevada is representative of early Pliocene vegetation (AXELROD, 1940a). *Quercus (arizonica, chrysolepis)* is abundant, and *Juniperus (osteosperma)* is also represented. Associated with them near the lake margin were *Celtis (reticulata), Populus (angustifolia, trichocarpa), Salix* (spp.) and *Umbellularia (californica). Lyonothamnus,* the wood of which is recorded from a different locality some miles farther south (PAGE, 1964), probably had a similar ecologic occurrence in this region. Shrubs that were scattered in the juniper-oak woodland included *Arctostaphylos (glauca), Cercocarpus (betuloides), Mahonia (fremontii),* and *Prunus (andersonii),* and in favorable areas they probably formed chaparral. Total yearly rainfall was near 20 inches (508 mm), and only a small amount appears to have been effective in the summer season – probably because of a high evaporation rate in this relatively interior area.

Several small collections have been recovered from the early to middle Pliocene Truckee formation, situated 130 to 140 miles northwest of the Esmeralda area (UC Mus. Pal.). *Quercus (wislizenii, chrysolepis)* is predominant, and *Juniperus (osteosperma), Lyonothamnus* (extinct sp.), and *Juglans (californica)* (BERRY, 1928) are also recorded. Although an occasional seed of *Picea (breweriana)* is represented at one of the early Pliocene localities, it probably lived in the distant hills and was transported by currents to the lacustrine site of deposition. Shrubs recovered from the Truckee include *Sumac (lanceolata), Purshia (tridentata), Cercocarpus (betuloides),* and *Mahonia (fremontii),* all of which probably were scattered within oak-juniper woodland. It is apparent that precipitation was near 20–25 inches (508–625 mm), and summer rainfall was reduced as compared with that of the late Miocene when forest occupied this same area (Fallon, Chloropagus, Purple Mountain floras). Temperatures were still relatively mild as judged from the records of walnut, ironwood, and also the live oaks.

Finally, in the area 50 miles west of the Truckee floras there is evidence of oak woodland-savanna vegetation near Verdi, Nevada, situated close to the California border (AXELROD, 1958 a). The middle Pliocene Verdi flora shows that *Quercus (engelmannii, lobata, wislizenii)* formed a woodland-savanna at the lower margin of forest. The latter, which reached down from the bordering hills, was composed of *Abies (concolor), Pinus (attenuata, ponderosa, lambertiana), Arctostaphylos (nevadensis), Prunus (emarginata)* and *Ribes (roezlii)*. Riparian-border trees included small-leafed ecotypes of *Populus (trichocarpa, tremuloides, tremula-davidiana),* and *Salix (gooddingii, scouleriana)*. Rainfall was near 25 inches (625 mm) at the forest margin, and about 20 inches (508 mm) in areas where the oak savanna covered drier slopes in the lower parts of the basin. Effective summer rainfall was at a minimum as judged from the small-leafed ecotype of *Populus (tremula-davidiana)* in the flora. But winters were still comparatively mild, as indicated by the 3 oaks, and also by the composition of the Truckee floras to the east. Liveoak woodland-savanna vegetation had largely disappeared from the Great Basin by the late Pliocene as judged from the trend to moister, colder climate which enabled yellow pine forest to enter the lowlands as the first glacial stage commenced (AXELROD and TING, 1960).

Central California

Oak woodland-savanna vegetation was already present on the south-western flank of the southern Sierra Nevada during the early Miocene, and gradually moved northward on dry, south-facing slopes as aridity increased. By the late Miocene it is recorded in the north-central part of the range in the Remington Hill flora which had a moderate altitude at the time (CONDIT, 1944 a). In this area, oak woodland-savanna was subordinate to conifer-deciduous hardwood forest and mixed evergreen forest. The oak woodland-savanna at Remington Hill included 3 species of *Quercus (douglasii, lobata, wislizenii)* which formed a woodland on moister sites and a savanna on the drier ones. Regular shrubby associates in the flora include *Arctostaphylos (manzanita), Ceanothus (cuneatus),* and *Laurocerasus (ilicifolia),* and riparian trees in the community included *Acer (negundo), Umbellularia (californica), Salix (lasiandra),* and *Platanus (racemosa)*. Evidence for summer rainfall at Remington Hill is provided by genera *(Berchemia, Liquidambar, Persea, Ungnadia)* that are confined to such areas today, and several of its species are related to taxa that are found only in regions with summer rain.

The Table Mountain flora shows generally similar relations though woodland oaks evidently are not present. Although *Q. convexa* Lesquereux appears to represent an extinct species of *Chrysolepis (= Castanopsis),* it may have contributed to woodland vegetation. Typical species in the area included *Celtis (reticulata), Forestiera (neomexicana), Mahonia (lanceolata), Persea (podadenia),* and *Robinia (neomexicana).* Drier slopes supported fossil knobcone pine *(Pinus* cf. *attenuata),* and shrubs such as *Cercocarpus (betuloides)* and *Quercus (turbinella)* that locally formed chaparral. Adjacent cooler slopes in valleys were covered with a mixed deciduous-hardwood forest with scattered broadleafed evergreens. Among the species present were *Acer (grandidentatum-saccharum),*

Berchemia (multinervis), Carya (cordiformis), Cercis (canadensis), Ilex (opaca), Magnolia (grandiflora), Nyssa (sylvatica), Persea (podadenia) and *Ulmus (americana)*, all of which are summer rain indicators. Both the Remington Hill and Table Mountain floras include riparian-border trees and shrubs, notably species of *Magnolia, Persea, Platanus, Populus, Salix* and others that evidently ranged to lower levels across the floodplains of the Sierran rivers to the margin of the sea which was then in the Mount Diablo area, for they occur there. The swamp and floodplain forest represented by the Neroly flora of the Mount Diablo region also includes plants that are reliable summer-rain indicators, nobably *Castanea, Ilex, Magnolia, Nyssa* and *Taxodium* (CONDIT, 1938). Clearly, the late Miocene oak woodland-savanna of central California – which includes species that survived down to the present without important change, and makes up the characteristic species of the woodland-savanna vegetation today–lived under a climate with rainfall distributed during both summer and winter months.

The small Oakdale flora from the lower foothill belt of the central Sierra Nevada provides evidence of the nature of woodland-savanna vegetation during the middle Pliocene (AXELROD, 1944c). The flora is dominated by leaves of *Quercus (wislizenii)*, and *Quercus (douglasii)* is also recorded. They suggest that the interfluves were primarily a savanna, with the oaks scattered across widespread grasslands. Shrubs scattered in the woodland-savanna included *Arctostaphylos (mariposa), Ceanothus (cuneatus), Heteromeles (arbutifolia), Quercus (dumosa)* and *Ribes (quercetorum)*, and they probably formed a limited chaparral on sites with poorer soil. The moist floodplain supported *Celtis (reticulata), Populus (acuminata, trichocarpa), Robinia (neomexicana), Salix (lasiolepsis), Sapindus (drummondii)*, and *Umbellularia (californica)*. Rainfall was reduced over that of the late Miocene. Summer precipitation was low but still present, as indicated by the occurrence of *Robinia* and *Sapindus*, genera no longer native to California, but to areas with ample summer rainfall.

Liveoak woodland-savanna vegetation is not recorded in abundance over the lowlands of west-central California until after the early Pliocene. The region was sufficiently low and well-watered prior to the middle Pliocene that it supported mixed deciduous hardwood forest on the floodplains, and a swamp forest in oxbow lakes and swampy sites. This is shown by the late Miocene Neroly which includes floodplain forest (*Alnus, Betula, Populus, Platanus, Persea, Magnolia, Castanea, Ilex*) and also swamp forest (*Taxodium, Nyssa*). The succeeding Diablo flora includes many of the same species, though the swamp-cypress forest is not recorded (UC Mus. Pal. coll.). The Black Hawk flora of early Pliocene age (AXELROD, 1944a) from a slightly higher level also represents floodplain vegetation, as shown by the dominance of *Platanus, Populus, Salix* and *Ulmus*. There are a few leaves that represent plants of oak woodland areas, notably *Quercus (agrifolia, engelmannii)*. Two shrubs – *Cercocarpus (betuloides)* and *Schmaltzia (ovata)* – indicate drier sites in the region, and suggest that nearby hills supported chaparral as well as oak woodland-savanna. Taxa that indicate summer rainfall are in all these floras. They show a progressive decrease in numerical abundance, and also in diversity, indicating summer rainfall was decreasing during the Pliocene.

It is the succeeding middle Pliocene Mulholland flora that provides evidence of a rich oak woodland-savanna composed of *Quercus (lobata, tomentella, wislizenii)* (AXELROD, 1944b). They formed a dense woodland on sheltered slopes in the lowlands, and a savanna on the exposed ones. Associated stream-border trees and shrubs in the savanna-woodland included *Alnus (rhombifolia), Celtis (reticulata), Fraxinus (oregona), Lyonothamnus (floribundus), Platanus (racemosa), Populus (fremontii), Toxicodendron (diversiloba), Salix (lasiandra), Sapindus (drummondii),* and *Umbellularia (californica).* Bordering slopes also supported a rich chaparral (see below). As noted above, moister, higher hills and cooler slopes were covered by a mixed evergreen (tanoak-madrone) forest. Included in it were a few plants, notably the genera *Berchemia* and *Nyssa* and species of *Acer (grandidentatum)* and *Populus (grandidentata)* which indicate that there was still some summer rainfall in the region. That winters were milder than those now in the area is implied by several plants whose nearest descendants occur only in the coastal strip of southern California, or on the offshore islands, for instance *Ceanothus (spinosus), Lyonothamnus (floribundus), Malosma (laurina)* and *Quercus (tomentella).*

About 32 miles northwest of the Mulholland flora is the small Petaluma flora, also of middle Pliocene age (AXELROD, 1944d p. 186–187; AXELROD, 1944e, p. 215). Oaks are abundant and include *Quercus (lobata, vaseyana, wislizenii).* They appear to have formed an open oak-savanna, associated with *Alnus (rhombifolia), Platanus (racemosa), Salix (lasiolepis, lasiandra), Fraxinus (caudata), Populus (trichocarpa),* and *Umbellularia (californica)* along stream margins. Higher hills in the nearby area were covered with mixed evergreen forest, as shown by the presence of *Arbutus (menziesii)* and *Quercus (chrysolepis)* in the flora. Summer rainfall was still present, though in minor amount, as indicated by *Ulmus (americana*—small-leaved ecotype) which probably was a constituent of both the floodplain forest and the mixed evergreen forest.

Evidence provided by the nearby floras from the Sonoma volcanics to the north and east of the Petaluma flora indicates that oak woodland vegetation was well developed during the late Pliocene. In view of the increased rainfall, which heralds the building up of the first ice sheet (AXELROD, 1944e, p. 215), we find that forest assumed dominance over the lowlands at the expense of woodland-savanna. The Sonoma floras near Santa Rosa include a number of taxa that represent oak savanna-woodland notably *Quercus (engelmannii, agrifolia, douglasii, tomentella),* with associates of *Amorpha (californica), Garrya (elliptica), Ilex (brandegeana), Myrica (californica),* and probably *Umbellularia (californica)* and *Persea (borbonia)* which also contributed to oak-laurel forest (AXELROD, 1944d). Stream- and lake-border sites in the woodland area supported *Alnus (rhombifolia), Fraxinus (oregona), Platanus (racemosa), Salix (lasiolepis, scouleriana),* and *Umbellularia (californica).* The woodland was rather restricted in occurrence, because the mesophytic forest includes *Abies (grandis), Tsuga (heterophylla), Chamaecyparis (lawsoniana), Sequoia (sempervirens), Picea (breweriana), Pseudotsuga (menziesii),* as well as *Alnus (rubra), Lithocarpus (densiflora), Quercus (chrysolepis), Mahonia (nervosa), Umbellularia (californica), Rhamnus (purshiana), Rhododendron (californicum)* and *Vaccinium (parvifolium).* Emphasis must be placed on the fact that this flora also has plants that are indubitable indicators of summer rainfall at this late date, for instance, *Castanea (americana), Ilex (brandegeana), Ulmus (alata), Persea (borbonia)* and *Mahonia (lomariifolia).*

Some 20 miles southeast, the small Napa flora from the Sonoma volcanics indicates oak-woodland savanna vegetation in a region marginal to conifer forest and mixed evergreen forest (AXELROD, 1950a). Among the taxa that represent the woodland-savanna are *Quercus (agrifolia, lobata, wislizenii),* and probably with *Umbellularia (californica)* scattered on the rolling volcanic plain. Moister slopes supported *Sequoia (sempervirens), Abies (concolor), Pinus (ponderosa)* and their usual associates, notably *Chrysolepis [= Castanopsis] (chrysophylla), Holodiscus (discolor), Mahonia (nervosa),* and *Populus (tremuloides)*. Slopes of intermediate moisture-temperature relations were covered with oak-laurel forest and mixed evergreen forest. Driest sites in the area with poor soil were covered with a limited chaparral composed of *Arctostaphylos (viscida), Ceanothus (leucodermis), Cercocarpus (betuloides)* and others. Summer rainfall is indicated by *Persea (borbonia)* and *Pterocarya (stenoptera)*.

Summarizing, oak woodland-savanna vegetation similar to that now in California first appeared in the southeastern interior during the Oligocene, and spread widely over southern California during the Miocene. It was more diverse in composition than the modern descendant vegetation, for it included taxa that have survived only in areas with summer rainfall to the southeast. Oak woodland-savanna vegetation invaded the lowlands of west-central Nevada during the late Miocene, where it bordered mixed evergreen forest and conifer-deciduous hardwood forest. It was dominant there during the early Pliocene, was restricted in the middle Pliocene as aridity increased, and was largely eliminated from the province in the late Pliocene as moister, colder climate developed.

Oak woodland-savanna vegetation ranged north along the lower western foothills of the Sierra Nevada during the Miocene, and was well developed there late in the epoch. It is first recorded in west-central California near the close of the early Pliocene, it was dominant in the middle Pliocene, but was restricted in the late Pliocene as precipitation increased and forest spread more widely. In each area, oak woodland-savanna vegetation lived under mild temperature and a regime of winter and summer rainfall, with the latter decreasing during time. Thus, the present oak woodland-savanna vegetation not only occupies a smaller area, it is less diverse in composition than the middle and late Tertiary vegetation, and its species are much older than the mediterranean climate in which they have survived in scarcely modified form.

Chaparral

Southern California

Evergreen, sclerophyllous shrubland is an ancient type of vegetation. It is well developed in the middle Miocene Tehachapi flora of interior southern California (AXELROD, 1939). Here it is represented by *Arctostaphylos (glandulosa), Ceanothus (crassifolius, cuneatus), Cercocarpus (betuloides), Condalia (lycioides), Dodonaea (angustifolia), Fremontodendron (californicum), Mahonia (fremontii), Prunus (fremontii), Quercus (turbinella, palmeri), Rhamnus (californica), Schmaltzia (chondroloma, virens)*. It lived on slopes adjacent to oak woodland-savanna vegetation, and was bordered at lower, warmer levels by thorn scrub vegetation. The late Miocene Mint Canyon flora of the nearby region to the south, which had an interior position during the Miocene, includes more numerous chaparral species (AXELROD, 1940b). Among these are *Arctostaphylos (glauca), Ceanothus (arboreus, fendleri, perplexans, verrucosus, tomentosus), Cercocarpus (betuloides), Condalia (parryi), Fendlera (rupicola), Fraxinus (dipetala), Fremontodendron (californicum), Laurocerasus (ilicifolia), Quercus (palmeri, turbinella), Rhamnus (crocea)*, and *Schmaltzia (microphylla, virens)*. As in the Tehachapi area, it was bordered by oak woodland-savanna and thorn scrub vegetation. It seems probable that some of the thorn scrub taxa were scattered in the chaparral, for they have such an occurrence today in Arizona, New Mexico, Coahuila and elsewhere. On this basis, the Miocene chaparral probably included occasional shrubs representing *Acacia, Euphorbia, Eysenhardtia, Lysiloma, Pithecolobium* and others. When we recall that these floras have genera that are found today only in areas with summer rainfall (e.g. *Bursera, Cardiospermum, Clethra, Eysenhardtia, Ilex, Pachycormus, Persea, Passiflora, Pistacia, Pithecolobium, Randia, Sabal, Thouinia),* then it is evident that chaparral lived under a climate with adequate summer rainfall during Miocene time in southern California—much as it does today in the southwestern United States and Mexico.

Chaparral is also represented in the Pliocene of southern California. It is well developed in the middle Pliocene Mount Eden flora, where *Arctostaphylos (glauca, pungens), Ceanothus (cuneatus, spinosus, leucodermis), Cercocarpus (betuloides), Fraxinus (dipetala), Laurocerasus (lyonii), Quercus (palmeri), Rhamnus (ilicifolia), Condalia (parryi), Prunus (fremontii),* and *Schmaltzia (ovata)* are recorded (AXELROD, 1937, 1950b). Here also there were a few members of thorn scrub vegetation, and some probably contributed to the chaparral in drier areas, notably *Dodonaea, Cercidium, Eysenhardtia,* and possibly *Ficus.* Chaparral is recorded in other Pliocene floras in the region, but it is not as diverse in composition. This probably reflects the fact that the floras are smaller in terms of the sample which could be recovered, though the topographic setting would also be important in controlling the representation of shrubs in the accumulating record. In the Anaverde flora, where chaparral was adjacent to oak woodland-savanna vegetation, we find *Ceanothus (cuneatus), Quercus (dumosa, palmeri),* and *Rhamnus (californica, ilicifolia).* Thorn scrub was also present, as shown by *Colubrina, Dodonaea* and *Eysenhardtia,* and they may have entered the community—at least marginally on its drier borders (AXELROD, 1950c). The middle Pliocene Piru Gorge flora also includes a few chaparral species that lived adjacent to an oak-laurel forest and oak woodland-savanna (AXELROD, 1950d). Among the shrubs represented are *Arctostaphylos (glauca), Ceanothus (spinosus), Fremontodendron (californicum),* and *Schmaltzia (ovata).* All of these Pliocene floras include species that are now in the coastal strip, indicating milder climate over the interior. Summer rains were still present, as shown by the composition of the oak-laurel forest, oak woodland-savanna, and thorn forest vegetation in the region at this time.

Finally, mention must be made of chaparral in the early Pleistocene Soboba flora of interior southern California. Here the community included *Ceanothus (cuneatus, leucodermis, tomentosus), Cercocarpus (betuloides), Garrya (flavescens), Laurocerasus (ilicifolia), Quercus (dumosa, palmeri), Rhamnus (crocea), Schmaltzia (ovata),* all of which live in the bordering mountains (AXELROD, 1966). This chaparral occupied rocky slopes adjacent to big cone spruce-Coulter pine forest and mixed coniferous forest. Of special significance is the presence of two summer-rain indicators in this flora—*Magnolia grandiflora* and *Acer brachypterum.*

Central Nevada

Chaparral is represented in the Miocene and Pliocene floras of Nevada, but is not as diverse in composition as in southern California. This is chiefly because in this higher, cooler area the community was adjacent to conifer-hardwood forest and mixed evergreen forest, and hence well removed from the warm semiarid climate in which it attained optimum development. It is also to be noted that in some of the Miocene floras certain taxa (e.g. *Cercocarpus*) are sufficiently abundant to suggest that, in contrast to the diverse community which lived in southern California, the shrubland often was dominated by a single species, and that the other shrubs were quite subordinate. This is not surprising, for a generally similar relation may be seen today at the moister borders of chaparral where it lives on slopes adjacent to forest.

Among the species in the Middlegate flora are *Cercocarpus (betuloides, paucidentatus)*, *Ceanothus (cuneatus)*, *Sumac (glabra)* and *Styrax (californica)*, and *Cercocarpus* is relatively common (AXELROD, 1956). Across the basin 5 miles to the southeast, the rich Eastgate flora (undescribed) has *Cercocarpus (betuloides)* as a subdominant species. Here it probably formed pure colonies on exposed slopes adjacent to the rich tanoak-madrone and *Sequoiadendron* forests whose species dominated at this site. *Cercocarpus* also is a subdominant of a new middle Miocene flora in the Fingerrock area, where its associates included *Garrya, Heteromeles, Amelanchier* and *Ribes* on dry rocky slopes where it was adjacent to a mixed hardwood forest and tanoak-madrone forest. The later Miocene Chloropagus, Aldrich Station, Fallon and Purple Mountain floras (AXELROD, 1956) include species of *Amelanchier (utahensis)*, *Cercocarpus (betuloides, paucidentatus)*, *Rhamnus (californica)*, and *Symphoricarpos (oreophilus)*. They probably formed small brush patches on rocky slopes bordering the oak woodland-savanna and conifer-hardwood forest in these areas. In the Stewart Spring flora to the south, chaparral species of *Amelanchier, Cercocarpus, Holodiscus, Garrya, Peraphyllum* and *Schmaltzia* are recorded (WOLFE, 1964). Its better representation here is consistent with the warmer climate which resulted from a somewhat lower altitude and a more southerly position. The record suggests that in western Nevada Miocene chaparral included deciduous shrubs like those found today in the Sierran and Petran forest-chaparral. This is consistent with the occurrence in these floras of many plants that have their nearest relatives in both the Sierra Nevada and Rocky Mountain forests, where they contribute to the understory and also to forest-chaparral which is generally seral in relation as compared with the climax nature of the sclerophyllous community. Chaparral lived under summer rainfall during the Miocene, as indicated by the presence in these floras of genera (e.g. *Carya, Diospyros, Gymnocladus, Ulmus, Zelkova*) that survive now only in regions with summer rain.

There are only a few indications of chaparral vegetation in the Pliocene floras of Nevada. This is partly because very few Pliocene floras are known, most of them are not represented by adequate samples, and some represent forest-border vegetation where chaparral was at or near its altitudinal and climatic limits. The early Pliocene Esmeralda flora from southwestern Nevada has species of *Arctostaphylos (glauca)*, *Cercocarpus (betuloides)*, *Mahonia (fremontii)* and *Peraphyllum (ramosissimum)*. It lived on rocky slopes with poor, shallow soils where it was adjacent to oak woodland-savanna vegetation (AXELROD, 1940a). Apart from an occasional leaf of *Cercocarpus* and *Ceanothus*, chaparral is not represented in the small Truckee floras from the Carson Sink region. This is probably a reflection of the sample, for the diatomites that yield the plants were laid down well out in the lake, some distance from the shore area. The middle Pliocene Verdi flora has forest-chaparral, represented by *Arctostaphylos (nevadensis)*, *Ceanothus (cuneatus)*, *Ribes (roezlii)* and *Prunus (emarginata)*, which lived on slopes near the ecotone between yellow pine forest and live-oak woodland vegetation (AXELROD, 1958a). Forest chaparral is also in the Chalk Hills flora, as shown by *Amelanchier (alnifolia)*,

Ceanothus (integerrimus, velutinus), Holodiscus (dumosus), Mahonia (pinnata), Prunus (emarginata), and *Rhamnus (californica).* In this area, the shrubby community occupied drier rocky slopes adjacent to a rich *Sequoiadendron* forest (AXELROD, 1962).

Central California

Chaparral attained greatest diversity in central California during the middle Pliocene, when climate was drier than at any other time during the Tertiary. It finds optimum development in the Mulholland flora, where it occupied well drained slopes adjacent to the sclerophyllous tanoak-madrone forest and oak woodland-savanna vegetation (AXELROD, 1944b). Among the species recorded are *Arctostaphylos (glandulosa), Ceanothus (spinosus), Cercocarpus (betuloides), Dendromecon (rigida), Fremontia (californica), Mahonia (fremontii), Heteromeles (arbutifolia), Quercus (vaseyana, palmeri), Rhamnus (californica, crocea), Schmaltzia (ovata)* and *Malosma (laurina).* This assemblage includes species whose nearest modern equivalents are now in southern California, which is consistent with the presence of *Lyonothamnus (floribundus)* and *Quercus (tomentella)* in the flora, as well as other plants that are now found in regions far to the southeast, notably *Sapindus (drummondii)* and *Quercus (vaseyana).* It is apparent that the Pliocene chaparral in the Mulholland area was more diverse in taxa than the chaparral in central California today.

As for the other floras, chaparral has only a poor representation in the Oakdale flora, probably because of the low terrain out on the lower foothills of the Sierra Nevada where it was in oak woodland-savanna country somewhat below areas of its optimum development (AXELROD, 1944c). Among the species represented that probably formed chaparral on rocky slopes with shallow soil were *Arctostaphylos (mariposa), Ceanothus (cuneatus), Mahonia (fremontii), Heteromeles (arbutifolia), Quercus (dumosa)* and *Ribes (quercetorum).* All of their modern relatives contribute to chaparral in the foothills of the range today. The late Pliocene Sonoma floras also have chaparral species, but they are poorly represented because rainfall was sufficient to support conifer forest and mixed evergreen forest, as well as oak woodland-savanna vegetation which was more restricted in distribution. In the Sonoma flora near Santa Rosa (AXELROD, 1944d), chaparral species include *Amorpha (californica), Ceanothus (cuneatus), Garrya (elliptica), Mahonia (lomariifolia), Heteromeles (arbutifolia, Rhamnus (purshiana)* and *Symphoricarpos (albus),* several of which are deciduous. These shrubs probably formed a forest-chaparral. The small Napa flora to the southeast has *Amorpha (californica), Ceanothus (leucodermis), Chrysolepis [= Castanopsis] (sempervirens), Cercocarpus (betuloides), Heteromeles (arbutifolia), Holodiscus (discolor)* and *Mahonia (pinnata)* (AXELROD, 1950a). In view of the drier climate in this area, the shrubs may well have contributed locally to a chaparral on nearby slopes, but they probably attained best development as forest-chaparral, with the shrubs entering into the understory of the forest—much as their descendants to today. As noted in discussing tanoak-madrone forest and oak woodland-savanna vegetation, there was sufficient summer rainfall in this area into the late Pliocene to support genera that are no longer native to this part of the continent.

Summarizing, chaparral was well established in southern California in the middle Miocene, and spread northward into Nevada and central California as drier climates developed there. It lived under mild temperature and summer as well as winter rainfall during later Neogene time. During the Pliocene chaparral had a wider distribution than it does today, occupying areas now desert or covered with broadleaved evergreen forest. As with the other broadleaved sclerophyllous vegetation types, Neogene chaparral included groupings of taxa whose modern descendants contribute to different segregate communities which have emerged more recently (see below).

As wetter, colder climates developed during the later Pliocene and early Pleistocene, chaparral disappeared from west-central Nevada and became subordinate to forest over central California. It was living under some summer rainfall into the early Pleistocene of southern California. The available evidence suggests that most of the woody taxa that contribute to chaparral are much older than the climate in which they now live: they evolved under a non-mediterranean climate.

One important difference between the Tertiary chaparral of southern California—the area of its optimum development—and that of the present day, is that the former included deciduous shrubs that now occur in regions far to the south, and contribute to thorn forest and desert-border vegetation. The modern chaparral of Arizona, New Mexico and Coahuila shows similar relations. As the deciduous shrubs were eliminated in response to decreased summer rainfall and lowered winter temperature, the modern community dominated more thoroughly by sclerophylls came into existence. As with the other types of sclerophyllous vegetation, the taxa that make up chaparral did not evolve in response to mediterranean climate: rather, they were preadapted to it.

Summary

The preceding review of the Neogene record of fossil plants whose closest living descendants contribute to three different types of sclerophyllous vegetation in California and border areas leads to the following general conclusions.

1) Mixed evergreen (tanoak-madrone) forest covered exposed slopes bordering mixed conifer-deciduous hardwood forest, and sheltered slopes bordering oak woodland-savanna vegetation, during Miocene and Pliocene times. It differed from the modern forest in having more numerous broadleaved evergreens and deciduous hardwoods scattered through the canopy and in moister sites. The Miocene forest lived under a climate of mild temperature and precipitation well distributed through the year. During later Miocene and Pliocene times the forest was confined gradually to areas west of the Sierra-Cascade axis as precipitation and equability decreased over the interior. The progressive reduction in summer rainfall gradually eliminated deciduous hardwoods and broadleaved evergreens from the forest. The modern forest appeared in its present form only during the Pleistocene. The modern broadleaved sclerophyllous forest is new in terms of dominance by evergreens, but the taxa are ancient, hardy survivors that adapted to the new climate.

2) Oak-laurel forest dominated coastal southern California and central California during the Miocene and ranged into the bordering hills where it was in ecotone with mixed evergreen forest. It was composed of taxa similar to those now in the oak-laurel forests of Mexico. These were gradually eliminated from the region during the Pliocene as summer rainfall decreased. Although some (e.g. *Persea)* were abundant into the latest

Pliocene, they disappeared in the early Pleistocene as summer rainfall was eliminated and as winter temperatures were lowered. The surviving oak-laurel forest which inhabits moist, equable sites below broadleaved sclerophyll forest and moister sites in the upper parts of oak woodland-savanna, is thus composed of ancient taxa, but the community composition and structure is new, and the modern forest is less diverse than the Neogene.

3) Oak woodland-savanna vegetation spread out from interior southern California as climate became drier during the Miocene, it assumed dominance over Nevada (with juniper) in the Pliocene, and it appeared in west-central California in the middle Pliocene. The Neogene vegetation was composed of many sclerophyllous plants that have survived with scarcely any change in California, as well as others whose nearest relatives occur only in areas with summer rainfall in the region from Arizona to western Texas and south into Mexico. Oak woodland-savanna vegetation disappeared from Nevada as temperatures were lowered, and numerous taxa were eliminated from Nevada and California as summer rainfall diminished during the Pliocene. Most of the woody plants that make up the relatively impoverished woodland-savanna vegetation are ancient taxa. These have been regrouped into new associations whose composition and distribution is controlled by the more localized climates that developed during the Quaternary (see below).

4) Chaparral was well established over interior southern California by Miocene time. It was rich in taxa, and included species that are similar to those now in California as well as others that are now in chaparral vegetation in areas of summer rain to the south-east. Chaparral spread northward into Nevada with expanding dry climate during the later Miocene and was locally prominent in the Pliocene. In areas where it bordered mixed evergreen forest and conifer-hardwood forest, deciduous shrubs also entered into its make-up. Chaparral ranged northward into the northern Sierra Nevada foothills during the late Miocene, and attained greatest diversity in west-central California in the middle Pliocene—the driest part of the Tertiary. Chaparral disappeared from Nevada in response to increasing cold, and taxa that earlier were present in central California also were restricted southward in response to lowered temperature, but others disappeared as summer rainfall was eliminated. The present chaparral is composed chiefly of ancient taxa, and it is more impoverished as compared with that of the Tertiary. The diverse modern associations are of Quaternary age chiefly, and some are dominated by species *(Arctostaphylos, Ceanothus)* that are new.

Evolution under Mediterranean Climate

As outlined above, diverse types of sclerophyllous vegetation have occupied California and bordering areas since early Miocene time. They were composed of many sclerophyllous plants that have persisted here with scarcely any change, and also included plants similar to those that now live only in areas with summer rainfall. As the latter gradually decreased during late Miocene and Pliocene times, the modern sclerophyllous vegetation came into existence. It is less diverse in woody plants and is more conspicuously dominated by sclerophylls, for many of the exotics were deciduous. The modifications in composition, structure, and distribution of sclerophyllous vegetation that occurred during the Neogene were followed by changes during the Quaternary which gave rise to segregate communities of the modern sclerophyllous vegetation. Although *Arctostaphylos* and *Ceanothus* provide notable exceptions, most of the evolution under mediterranean climate, which only appeared after the first glacial, was among herbaceous plants of primarily open areas. Since this was a response to fundamental changes in physical factors that also affected the distribution and composition of vegetation, their role in evolution is now reviewed (see AXELROD, 1957; 1966, p. 55–60; 1967b).

Flora

The appearance of mediterranean climate alone could not possibly have resulted in much evolution had not certain other environmental factors been involved. Fundamental changes in topographic, edaphic, climatic and biologic factors – all acting in concert – explain the rapid evolution of new taxa under mediterranean climate.

1) Intense mountain building and regional uplift greatly increased topographic diversity over the area during the Quaternary. As mountain ranges were rapidly elevated, hundreds of new canyons and gorges were incised into them, steep slopes came into existence, and divergent localized microenvironments (north-south or east-west slopes) developed in proximity—where earlier there were comparatively few. As mountain ranges were uplifted, extensive new areas suitable for grassland, woodland, chaparral, sclerophyll forest and other (yellow pine, fir, subalpine forest, alpine meadows) vegetation zones appeared. Since climatic conditions in these zones are (and were) slightly different in the Coast Ranges, the Sierra Nevada, the Transverse Ranges and the Peninsular Ranges, they must have been favorable breeding grounds for new populations. Elevation of the Coast Ranges produced rainshadows over the inter-range valleys, the Central Valley, and the interior valleys of southern California, providing wholly new environments (or subzones) to exploit. And major uplift of the low Sierra Nevada-Peninsular Range barrier brought more extreme temperatures to the lee, intensified the dry zone there, and provided more localized new environments (subzones) favorable for invasion by populations. And within the desert area itself, uplift of the desert ranges increased topographic-climatic diversity and provided new areas for living there.

2) As mountain ranges were deformed, uplifted, and eroded, diverse rocks were rapidly exposed. They present a formidable array, ranging from newly-exposed serpentine areas to varied sedimentary, metamorphic and igneous terranes that often are juxtaposed as a result of complex folding and faulting. This made possible the development of very different soils in local areas, and provided populations with new opportunities for living under edaphic conditions far more diverse than those which had existed previously. In addition, as mountain ranges were rapidly elevated, drainage was locally altered and impeded over the lowlands, producing unique edaphic sites—vernal pools, gypsiferous soils, saline zones around playas, as well as bogs, swamps and marshes in both estuarine and freshwater environments. Along the fronts of the rising ranges in southern and interior areas, rivers built large alluvial fans that gradually coalesced into continuous range-front aprons – all available for exploitation by plants.

3) Mediterranean climate was present only intermittently during the Quaternary. At times of continental glaciation, areas of mediterranean climate were experiencing pluvial conditions. Severe summer drought was greatly alleviated at least, and was certainly absent early in the period to judge from the fossil record in the Mediterranean basin. It was chiefly during the early interglacial ages that typical mediterranean conditions were

present. The scheme of climatic change was essentially as follows (ADAM, 1969): temperatures are rather stable during the long interglacials; as the oceans gradually warm, rate of evaporation increases to a maximum, precipitation rises, a glacial age commences, and this is the wettest and mildest part of the glacial stage; temperatures gradually decrease (with some oscillation) as glaciation progresses; the coldest climates develop at the end of a glacial age, and are followed by a relatively rapid rise in temperature which starts the next interglacial; the early interglacial is marked by warm dry climate, which gradually returns to a more stable interglacial age. Since topographic diversity and relief increased throughout the world during the Quaternary, and since the coldwater west-coasts in areas of mediterranean climate have become progressively colder, each successive mediterranean (interglacial) climate has been marked by increasingly greater drought. As precipitation increases at the start of a glacial, areas of mediterranean climate decrease and summer drought is less pronounced because cyclonic storm tracks shifted below the present southern borders of these transitional climates.

In this connection, it is recalled that west-coast mediterranean climates have alternate wet and dry seasons because they lie between wet marine west-coast climate on the pole-ward side and dry tropical desert climate equatorward. In winter, the regime of middle-latitude cyclones and moist maritime polar air masses brings ample precipitation. As the continents gradually warm up in spring the thermal low over the desert region increases, and the subtropical high pressure system strengthens offshore and shifts northward. This prevents summer cyclonic storms, such as occur in the marine west-coast climates, from moving much south of Lat. 42 ° N. Since all areas of mediterranean climate are bordered by cold-water currents (California, Humboldt, Benguela), as moist marine air from the offshore subtropical high pressure area moves toward the thermal low in the interior it can not provide precipitation even though the air is so saturated it forms a persistent fog deck in summer along the coast. There can be no summer precipitation because in its movement into land the air is being heated and it is therefore inherently stable. Hence there is a long summer drought, though proximity to the coast and the cold-water current results in mild temperatures during summer. However, during Tertiary time sea-surface temperatures were much higher inasmuch as ice-caps were not present. During the Miocene and Pliocene sea-surface temperatures were sufficiently warm to support tropical and subtropical shallow-water marine molluscan faunas in areas now under mediterranean climate, and bathed by cold water currents. It was the gradual decline in seasurface temperature that led finally to the appearance of mediterranean climates, and their summer-dry conditions (Fig. 2). Clearly, as the ice disappears from the polar seas and the oceans gradually warm up, mediterranean conditions (summer drought) must be gradually alleviated, especially just prior to the full glacial stage—as in the late Pliocene.

4) The importance of these topographic, edaphic and climatic changes must be particularly emphasized because significant numbers of new taxa could not have evolved without them. Maps that show the distribution of various taxa demonstrate that they are closely correlated with local differences in topography, soil-rock substrate, and climate which have developed only recently. The relationships are clearly implied by data for *Pasarella* (MILLER, 1956, figs. 1–8), *Horkelia*, *Penstemon*, *Potentilla*, *Zauschneria* (CLAUSEN, KECK and HEISEY, 1940), *Achillea* (CLAUSEN, KECK and HEISEY, 1948, fig. 1), *Potentilla* (CLAUSEN and HEISEY, 1958, fig. 1), *Ceanothus* (MCMINN, 1942, maps 1–7; NOBS, 1963, figs. 35, 103, 109), *Layia* (CLAUSEN, 1951, figs. 56, 57), *Viola* (CLAUSEN, 1951, figs. 67, 68), *Holocarpha* (CLAUSEN, 1951, fig. 43), *Gilia* (GRANT and GRANT,

1954, 1956) and many others. By comparison, taxa of similar rank are much less numerous in areas of low environmental diversity, as on the Great Plains. Clearly without its great variation in relief, substrate and climate, the mediterranean climatic region of California would be much poorer biotically, and especially in herbaceous plants. Opportunities for evolution here increased dramatically during the Quaternary, and the herbaceous flora responded (AXELROD, 1957, p. 38–43).

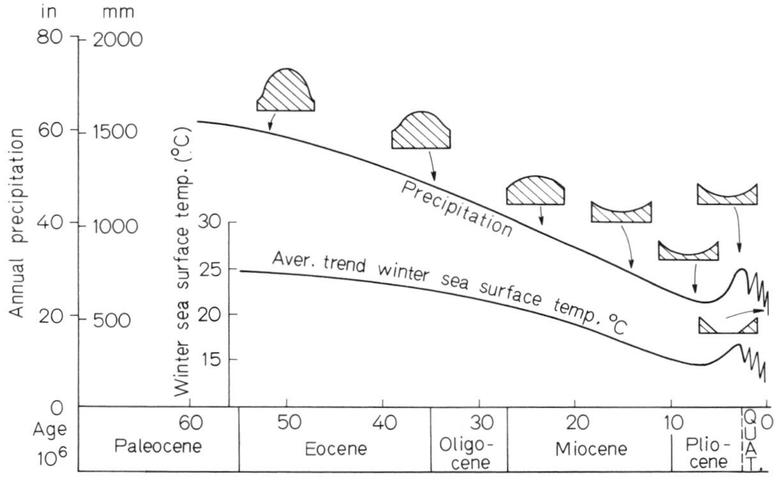

Fig. 2. Showing general average precipitation curve for California, and the inferred seasonal distribution of rainfall. For comparison, the general average minimum winter sea-surface temperature, as estimated chiefly from molluscan faunas (DURHAM, 1954), is also charted

As noted earlier, dry middle Pliocene climate was followed by moister upper Pliocene climate that heralded the start of the first major continental glaciation. Thus, the first mediterranean climate appeared only after the first glacial–which agrees with fossil evidence in the Mediterranean basin (see below). During the Quaternary, therefore, there have been four periods of mediterranean climate separated by pluvial (glacial) ages when the mediterranean condition was greatly ameliorated and restricted – or possibly absent over wide areas. As mountain ranges were elevated during the Quaternary, and interior valleys came into existence, each successive period of mediterranean climate was more severe (more drought) and more numerous discontinuous dry areas developed over the lowlands. During the recurring pluvials dry areas were restricted (or eliminated), and probably disappeared from the windward slopes of the ranges for forests entered the lowlands and spread widely there during the early Pleistocene (AXELROD, 1966). Clearly, there has been a series of rising and falling tempos, with the intensity of mediterranean climate increasing with each stage as the topographic-climatic factor increased and as the successive pluvial phases decreased in strength. During the interglacial (mediterranean) ages, therefore, selection has been for populations able to withstand increasingly longer and more severe periods of summer drought, and greater extremes of summer heat.

Rapid evolution of herbaceous taxa in many families (e.g. Boraginaceae, Caryophyllaceae, Cruciferae, Compositae, Labiatae, Hydrophyllaceae, Leguminosae *(Lupinus,*

Trifolium, Lotus, Astragulus), Polemoniaceae and Scrophulariaceae) was favored because many different selective forces of the environment made it possible for local interbreeding populations, each of variable genetic structure and origin, to reach new adaptive peaks. Furthermore, these selective forces were becoming more diverse and intense as mountain building progressed during the Quaternary. Local dry areas were increasing in number, area, and intensity; diverse soil and rock types were being exposed; new microclimatic areas (canyons with cold air drainage; north-south slopes, dry rocky canyon walls, etc.) were coming into existence as erosion carved the rising ranges; vegetation formed increasingly more complex overlapping mosaics of interfingering and overlapping associations as mountains were elevated, thus enabling divergent populations to live in proximity. Mutation, hybridization, introgression, and other factors could thus give rise to new populations of different adaptive mode, and their success would virtually be insured by selection for any one of the numerous new microenvironments that were coming into existence throughout the entire region.

Alternating cool-moist (pluvial) and hot-dry (mediterranean) climates during the Quaternary resulted in repeated large spatial shifts in populations. As an interglacial age commenced, the trend to drier summer climate would favor isolation of populations in the more humid forest and woodland areas, but those in the more arid grassland, sage, chaparral and savanna regions would spread. On the other hand, a trend to cooler, moister climate would favor the spread of mesic forest and woodland and bring together populations formerly isolated, whereas those in drier areas would become more isolated; and many would become extinct as their areas disappeared.

This dynamic background—increasing topographic relief with its varied new microclimatic areas, diverse relief permitting the juxtaposition of divergent populations, progressively greater edaphic diversity, alternation of cool-moist and warm-dry climates causing spatial shifts of populations—appears to readily account for rapid evolution under mediterranean climate. As a warm, summer-dry mediterranean climate commences to expand following a pluvial (glacial) period, numerous relatively local dry areas would appear initially, each separated by areas of moister climate supporting forest. As the dry areas gradually increase in size, some populations previously isolated by abrupt changes in relief, substrate and microclimate would meet. As a result, rapid shifts in adaptive mode would be expected because of the highly selective forces operating. Selection would particularly favor herbs because their short annual cycle would be attuned to the gradually lengthening period of summer drought. Near the margins of the expanding dry areas where there were steep gradients to new environments (subzones), populations would be subject to strong selection. Under such conditions merging with other partially isolated populations would favor the formation of transgressive populations, and also the emergence of new founder populations. In this manner, many novelties could arise rapidly in ecologically different, new habitats as mediterranean climate expanded. However, most of them would be ephemeral—appearing during a dry cycle and then becoming extinct as moister climate returned with the next pluvial stage and their unique local area disappeared.

As a summer-dry interglacial cycle wanes and dry-adapted taxa become restricted, formerly isolated forest populations would spread and some probably would become established more widely. The new taxa would meet related ones in nearby forested areas that earlier were isolated, and also more distantly related—but still compatible—populations in marginal woodland and savanna areas, to produce new combinations through hybridization, introgression, etc. Taxa near the dry margins of forest might well cross with savanna-border populations, especially if they were juxtaposed in areas of favorable relief and climate. By a process of fusion and refusion of variable populations from semiarid, subhumid and more humid environments, numerous new herbs (and some shrubs) could evolve very quickly in response to the fluctuating Quaternary climate.

On this basis, it is understandable that the distribution of endemics falls into a consistent pattern. As shown by STEBBINS and MAJOR (1965), patroendemics (restricted diploids related to widespread polyploids) occur in the highest numbers in the equable central Coast Range region near the coast, with secondary centers at middle altitudes in Sierra Nevada and Siskyou Mountains where climates are also equable, or along the west margin of the Colorado Desert. The apoendemics (polypoloids related to more widespread diploids, or high polyploids related to more widespread species with lower degrees of polyploidy) are most frequently in the inner central Coast Ranges, on the east slopes of the Sierra, and in the mountains of interior southern California, where they are in ecotones between more mesic and xeric biota. The data also show that the distribution of regional endemics in larger genera and in those of intermediate size are concentrated in local areas characterized by diversity of topography, soil and climate, and where they are ecotonal with respect to relatively mesic conditions and those of marked drought and higher summer temperatures, notably in the central Coast Ranges.

The richness of the California (mediterranean) flora is the result of numerous factors, some of which can now be discerned. It is a reservoir for many ancient relicts which have persisted here under equable climate, whether moist *(Sequoiadendron giganteum, Picea breweriana)* or semiarid *(Carpinteria californica, Lyonothamnus floribundus)*. It also has been an active theatre of recent evolution for the herbaceous component of the flora. This has been most active in the ecotone between derivatives of the Arcto-Tertiary and Madro-Tertiary Geofloras, where a diversity of moisture, temperature and edaphic conditions occur that owe their existence to the diverse topography that came into existence during the Quaternary. Not only are populations in these ecotonal areas subject to fluctuations in response to minor climatic change from year to year, or decade to decade, but the climatic trends of the Quarternary also played a significant role in the origin of new herbaceous taxa by bringing divergent populations into contact and promoting further diversification.

Vegetation

As discussed earlier, the rich sclerophyllous vegetation types gradually lost a number of taxa as summer rainfall was reduced, and as temperatures were lowered over the far West during Neogene time. Woodland species that are now in the southwestern United States and northern Mexico, distributed in *Arbutus, Bumelia, Clethra, Dodonaea, Ilex, Mahonia, Persea, Quercus, Populus, Robinia, Sabal* and *Sapindus,* gradually disappeared from California during the Pliocene. Their elimination left an impoverished woodland over the region, and the surviving woodland species have a more restricted distribution. Furthermore, the floras show that fossil species whose modern descendants now occur in different areas under different climates in California lived together during the Miocene and Pliocene. Clearly, the modern associations must be more recent segregates in response to more extreme climates of the Quaternary. Some examples are reviewed briefly because they indicate how new communities have developed in response to selection for the more localized new environments in which evolution of herbaceous taxa has been proceeding rapidly.

Woodland

The insular woodland flora of southern California is rich in woody endemics, many of which have a fossil record on the mainland (AXELROD, 1939, p. 67; 1958b, p. 479; 1967a). For example, *Lyonothamnus* is recorded in Nevada (Stewart Spring, Buffalo Canyon, Middlegate, Aldrich Station, Purple Mountain, Truckee floras), central California (Mulholland flora), and interior southern California (Tehachapi, Mint Canyon floras). *Laurocerasus (= Prunus) lyonii* has a close fossil relative in interior southern California (Piru Gorge, Tehachapi, Mount Eden floras). The insular *Quercus tomentella* occurs as a fossil in central California (Mulholland, Sonoma floras). These were gradually confined to the insular region as colder climates developed over the mainland area during the Quaternary, as demonstrated by the presence of yellow pine forest in the lowlands of interior southern California during the early Pleistocene (AXELROD, 1967a), and by the occurrence of Douglas fir-redwood forest in the coastal strip near Carpinteria during the later Pleistocene (CHANEY and MASON, 1933; AXELROD, 1967a, p. 295).

A walnut-oak woodland, dominated chiefly by *Juglans californica* and *Quercus agrifolia,* inhabits coastal southern California from Carpinteria to the Santa Ana Mountains, reaching up the major river valleys where there are moderating sea breezes in summer. The present community is new in much of its present area, for the coastal strip was covered with closed-cone pine forest during the pluvial (glacial) phases of the Pleistocene. During the later Miocene the coastal area supported a rich oak-laurel forest of which walnut was a part: the Miocene fossil nuts can scarcely be separated from those of the living species. During the Pliocene, walnut lived in interior southern California with woodland vegetation quite unlike that in the areas of its present occurrence. Among the plants associated with it were members of digger pine-blue oak woodland and an upland oak-laurel-madrone forest that is no longer in the region. Avocado was then a common stream-border tree in the lowlands where walnut lived. Clearly the present walnut woodland is relictual and impoverished, and survives near the coast under equable climate.

The coastal region of California from Sonoma County southward into northern Baja California is marked by woodland and savanna vegetation composed solely of *Quercus agrifolia.* The community occurs chiefly in areas of mild climate, and attains optimum development on northerly-facing slopes where evaporation is reduced. Nowhere in the fossil record is there any evidence of a pure *Quercus agrifolia* association. The fossil species *(Q. lakevillensis)* is regularly recorded with other oaks and woodland species. The modern association appears to be a relatively recent segregate, adapted primarily to mild equable climate of moderate rainfall.

Quercus engelmannii forms a distinctive oak woodland-savanna, reaching from Pasadena southward in the interior valleys to just across the Mexican border near Campo and Tecate. It forms pure stands on drier, more exposed slopes, but is mixed with *Q. agrifolia* in mesic sites and is replaced by it in moister canyons and on cooler north-facing slopes. A fossil species very similar to *Q. engelmannii* is recorded in central California (Mulholland, Sonoma floras), southern California (Mount Eden flora), and in western Nevada (Verdi flora). In central California it occurs with taxa whose nearest descendants are now only in coastal southern California *(Malosma laurina, Ceanothus spinosus, Lyonothamnus floribundus, Quercus tomentella)*; in Nevada it is with an oak similar to *Q. lobata* which is now found in central California chiefly; and in southern California it occurs with fossil members of the digger pine woodland, which is now in central California, with oak-walnut woodland which is now in the coastal region, and with an oak-laurel-madrone forest which is no longer in California. The modern association appears to be a more narrowly-adapted community, of relatively recent derivation.

Pinyon pine-juniper woodland now occurs in the mountains bordering the southern Great Valley of California. It appears to be a relict of the Pliocene conifer-woodland that extended widely over the Great Basin area, and was also in central California during the middle of the epoch. Juniper is recorded at several localities in central Nevada (Esmeralda, Truckee, Chloropagus, Stewart Spring floras), and pinyon pine is in the Esmeralda flora (undescribed cone). Live oaks similar to those now in California (e.g. *Q. chrysolepis, Q. lobata, Q. wislizenii*) regularly contributed to the Pliocene community in Nevada. Also present were other taxa that are no longer in that region but occur now only in California (i.e. *Cercocarpus, Heteromeles, Schmaltzia*), or in the southwestern United States *(Robinia, Sapindus)*. These taxa that were associated of the pinyon-juniper woodland disappeared from Nevada as colder climates developed over the interior, and as summer rainfall was reduced. The surviving pinyon-juniper woodland of Nevada is therefore new: it is impoverished as compared with the richer surviving community in the San Rafael Mountain-Mount Pinos-Tehachapi Mountain region. Whether the differences between the junipers (*osteosperma* of the Great Basin; *californica* of California) developed in response to recent regional climatic differences is not known.

Digger pine-oak woodland, composed of *Pinus sabiniana, Quercus agrifolia, Q. wislizenii, Q. lobata* and *Q. douglasii*, now ranges around the Valley of California, occupying the lower slopes of the Sierra Nevada and the inner Coast Ranges, and it reaches to the margins of southern California in the Liebre Mountain area. During the Pliocene it has been recorded in interior southern California, where fossil species similar to *P. sabiniana, Q. agrifolia*, and *Q. douglasii*, lived with taxa related to those now only in the insular woodland, or the southern California Engelmann oak woodland, or the woodland of Arizona and border areas, notably species of *Arbutus, Persea* and *Sapindus, Robinia*, and others. Clearly, the modern association is impoverished, and has been restricted northward to a region with colder winters.

Chaparral

Chaparral has formed an important sclerophyllous vegetation in southern California since the middle Miocene at least. At that time it was more diverse in composition than it is at present. Not only did it include a mixture of species whose modern analogues are now representative of coastal and interior communities, it also included evergreen and some deciduous shrubs that are now found south of the desert, where they contribute to the chaparral and also to desert-border and thorn scrub vegetation. In central California, chaparral was more diverse in composition in the middle Pliocene – the driest part of the Tertiary – than it is today in this region. Furthermore, it was present in Nevada from later Miocene into the middle Pliocene, yet sclerophyllous chaparral is not represented in that region today because of the colder climate.

Three very different factors appear to account for the development of the present areas of chaparral. First, some of them are due in part to the contraction of range of formerly more widespread species. For instance, dominants of the chaparral of southern California (e.g. *Ceanothus spinosus, Quercus palmeri, Schmaltzia ovata*) have close counterparts in the middle Pliocene of central California (Mulholland flora), where they were associated with taxa that are now typical of the chaparral of that region. They apparently disappeared from central California in response to colder Quaternary climates. Species that are now typical of the chaparral of the mountains of coastal southern California have Miocene and Pliocene equivalents in interior southern California *(Ceanothus spinosus, Malosma laurina)* where they occur with species similar to those now typical of the interior chaparral *(Quercus turbinella, Ceanothus crassifolius, C. leucodermis)*. Also, some species that are now typical of California chaparral are represented by equivalents in the Miocene and Pliocene floras of Nevada (*Ceanothus cuneatus* in the Verdi flora; *Schmaltzia ovata* in Esmeralda; *Heteromeles, Cercocarpus betuloides, Styrax californica* in Middlegate flora). They gradually were eliminated from the interior as colder climates developed following the middle Pliocene. In each of the examples noted, the Neogene communities lived under warm equable climates, and hence had a wide distribution; their present more narrow ranges result from climatic change which has confined taxa to their present areas.

The second factor that has resulted in the diversity of chaparral is speciation in the genera *Ceanothus* and *Arctostaphylos*. A number of relatively new endemic species of these genera occur in California. For instance, the series *cuneatus-ramulosus-sonomensis-masonii-gloriosus* in the Coast Ranges north of San Francisco Bay appears to express gradation in response to climatic change from hotter, drier interior *(cuneatus)* to the coastal region of high equability and adequate rainfall *(gloriosus)*. The series *purpureus-divergens-confusus-prostratus* var. *occidentalis* from the same region is confined to the Sonoma volcanics of late Pliocene age. The Coast Ranges composed of Sonoma volcanics were uplifted only during the Quaternary, and each taxon has a very restricted distribution which coincides with changes in habitats and notably with soil conditions (see Nobs, 1963, figs. 103, 109, 110). It is apparent that in this area chaparral has increased in diversity of taxa in response to Quaternary changes in climate and in substrate.

A third factor that accounts for the diverse communities in chaparral seems to be the result of recent, aggressive expansion by one species at the margin of the range of the formation in drier areas. This is implied by the distribution of *Adenostoma fasciculatum* which ranges throughout California at the lower margins of chaparral vegetation with the exception of parts of the coastal strip where climate appears to be too moist for it. It regularly forms a single dominant over hundreds of square miles in the lower part of the chaparral zone. At moderately higher levels it is associated with *Ceanothus cuneatus*, and then with *Quercus dumosa, Cercocarpus betuloides* and *Arctostaphylos glauca* or other species at higher, moister levels. At still higher levels *Adenostoma* is replaced by a chaparral composed of species of other genera entirely, in *Arctostaphylos, Ceanothus, Cercocarpus, Prunus (Laurocerasus), Quercus, Rhamnus,* and others. *Adenostoma* has not been found in the fossil record, which is not surprising in view of the small, needle-like leaves which could easily pass unrecognized. The pure stands of *Adenostoma* seem to be recent in much the same way that yellow pine and Douglas fir may form single dominants over wide areas, and much as *Artemisia* also does in Nevada, and juniper at levels above it. It is suggested that the areas dominated by chamise may have expanded and reached maximum distribution during the Xerothermic period, which is consistent with the occurrence of relict stands on poor sites on the floor of the northern Great Valley, and also in similar areas on south slopes close to the coast in areas not far removed from other taxa that are now relict there, and which appear to have also expanded their distribution from the drier interior at the same time (AXELROD, 1966, p. 42–54).

Finally, one further point must be made with respect to the area now covered by chaparral. The acreage of chaparral in California has increased greatly over that of the Pliocene. This has been made possible by the uplift of new mountains which provided new areas of poor, shallow soil for its expansion.

Summary

The rich Neogene sclerophyllous vegetation of California and border areas lost many taxa during the period as summer rainfall decreased and as winter temperatures were lowered. Since many of these were deciduous, the surviving vegetation types are more sclerophyllous. As equability and precipitation decreased over the interior, the surviving sclerophyllous vegetation was restricted to the seaward slopes of the Sierra Nevada-Peninsular Range axis chiefly. Changing climate during the Quaternary accounts for segregation of new woodland and chaparral communities from the more widely distributed and more diverse sclerophyllous vegetation types. These new associations have narrower ranges and fewer taxa, and are adapted to more localized subclimates within the mediterranean regime which only appeared after the first glacial.

The California flora increased in richness of herbaceous plants in response to two Quarternary events. Rapidly rising mountains provided a progressively greater diversity of terrain, local climate, and rock-soil type. Under fluctuating global climate, the mediterranean condition was present only intermittently, and became more severe after each succeeding glacial age.

Selection was chiefly for populations of herbaceous plants which shifted rapidly and readily in response to changing climate, and adapted quickly to the varied new selective forces of the rapidly changing environment. Divergent populations lived in proximity not only because of diverse terrain and microclimate. Formerly isolated populations (either from humid or semiarid areas) were alternately brought into juxtaposition by fluctuating wet and dry climatic cycles. During the pluvial stages when the mediterranean phase was greatly weakened and restricted (or eliminated), there must have been widespread extinction of minor taxa as local dry areas to which they were adapted disappeared.

Comparison with the Mediterranean Region

There are many similarities between the history of sclerophyllous vegetation in California and the Mediterranean region. To provide a basis for comparison, certain general aspects of the Cretaceous and Tertiary vegetation history of the Mediterranean region must first be reviewed briefly.

Cretaceous-Paleogene[5] Setting

Much of the region, including the Saharan area to the south, was flooded by the mid-Cretaceous (Cenomanian) transgression. As the sea retreated later in the period non-marine deposition commenced over a wide area. In the north African sector, the non-marine sediments were derived chiefly from erosion of highlands of Precambrian crystalline basement rocks, remnants of which now form the Hoggar, Air and Tibesti massifs, parts of which reach altitudes of 8,000 to 10,000 feet (2440–3050 m). These mountains were covered with forests that provided the source for much of the silificied woods and logs which occur widely over the region. The Cretaceous woods represent conifers (Araucariaceae, Podocarpaceae) as well as angiosperms (Lauraceae, Annonaceae), and indicate moist, warm climate (see FURON, 1968). Fossil leaves from the middle Cretaceous of Egypt occur near Aswan (BARTHOUX and FRITEL, 1925; SEWARD, 1935). Ranging from notophyll to mesophyll in size, and assigned to *Annona, Cinnamomum, Laurus, Magnolia, Dipterocarpophyllum,* Cycadaceae, Proteaceae, *Sabalites,* they indicate a well-watered lowland region with forests living under at least subtropical climate. The existence of such conditions here is consistent with the nature of the Cretaceous floras of southern Europe, which reveal broad-leaved evergreen forests thriving under generally similar conditions (see lists in BERRY, 1916; DEPAPE, 1959).

Nonmarine deposition continued widely over northern Africa down to the close of the Oligocene, as dated by the fossil mammal faunas recovered at many localities. Remains of fossil woods are common in the deposits; references to literature are noted by FURON (1968). The Paleogene forests from northern Africa and the Saharan region include taxa that represent families that are preponderantly tropical to subtropical in their requirements – Annonaceae, Celastraceae, Euphorbiaceae, Guttiferae, Lauraceae, Leguminosae, Malvaceae, Monimiaceae, Moraceae, Myristicaceae, Myrtaceae, Proteaceae, Rutaceae, Sapindaceae, Sterculiaceae, Tamaricaceae. Many of them are not now represented in northern Africa, or in the Sahara, but are in the moister African, Indo-Australasian, and American tropical and subtropical regions. The evidence of environment provided by the fossil woods is consistent with the nature of the few leaf (SEWARD, 1935) and seed floras (CHANDLER, 1954) recovered from the region. The data are further strengthened by the Paleogene floras of southern Europe which contain numerous broadleaved evergreen subtropical plants (ETTINGSHAUSEN, 1854; UNGER, 1869; LAURENT, 1899).

It is apparent that the present Saharan and Mediterranean regions were occupied by tropical to subtropical evergreen forests in the later Cretaceous and early Tertiary – a relation also shown by vegetation in the California region and areas to the south (AXELROD, 1950e, 1958b). In each area, oak-laurel forest covered warm, equable well-watered slopes in the uplands, and rocky drier sites with shallow, poor soils presumably supported small hardy sclerophyllous trees and shrubs. Inasmuch as sea-surface temperatures were sufficiently high to support shallow-water molluscan faunas that indicate tropical and subtropical marine climates, summer was the season of maximum rainfall on the western part of the continent where savanna forests were widespread over the lowlands. The evi-

5 The terms Paleogene and Neogene refer to the Paleocene, Eocene and Oligocene Epochs, and the Miocene and Pliocene Epochs, respectively. The terms are more precise than early Tertiary and late Tertiary.

dence suggests that the Madro-Tertiary Geoflora of southwestern North America and the Mediterrano-Tertiary Geoflora of the Mediterranean region both had their origins over the southwestern parts of the continents by late Cretaceous (and earlier) time (AXELROD, 1960, p. 273–277). Both evolved in regions that are presently desert, but were then well watered though there was a dry season in winter.

As a result of continental uplift in transitional Oligo-Miocene time, deposition became more restricted over the African-Saharan sector of the Mediterranean region (FURON, 1968). The great upwarp which initiated development of the rift valley system of Africa, and affected much of the continent, amounts to fully 9,000 feet (2740 m) since Oligocene time over central Africa (BELOUSSOV, 1969). As the continent was gradually elevated, deposition ceased in the Paleogene basins which were now subject to erosion. Since most of the continent was being drained, sedimentary basins were restricted and the nonmarine Neogene record is much less complete than that of Paleogene time. In response to general uplift, greater extremes of temperature began to develop and precipitation was reduced over the interior. Hence, rainforest and savanna became more restricted and drought-resistant plants began to spread from local dry areas which earlier had been confined to the lee of mountains and to intramontane valleys. It was at about this time that taxa representing species very similar to those of the present Mediterranean sclerophyllous vegetation first appear in moderate number in the floras of southern Europe, and they increased in diversity during the Miocene and Pliocene; the absence of an adequate Neogene plant record in north Africa militates against outlining the history of broadleaved sclerophyllous vegetation there.

Neogene Vegetation-Climate

The Neogene sclerophyllous vegetation was far richer in taxa than that which has survived in the area – a relation also shown by the California region. Among the numerous plants that now contribute to sclerophyllous Mediterranean vegetation, the following have similar species in Miocene and Pliocene floras of southern Europe[6]: *Arbutus unedo, Buxus sempervirens, Chamaerops humilis, Ilex canariensis, Jasminum heterophyllum, Laurus nobilis, Myrsine africana, Nerium oleander, Olea europaea, Phillyrea angustifolia, Quercus ilex, mirbeckii, suber, Staphylea pinnata.* Members of the laurel forest of the Macaronesian Islands that were associated with them included fossil plants similar to *Apollonias canariensis, Laurus canariensis, Ocotea foetens,* and *Persea indica,* as well as *Celastrus cassinoides, Ilex canariensis, Myrsine canariensis,* and *Viburnum rugosum.* The laurel forest of the Canary and Madeira islands attains best development on the steep north and northeast sides of the island which are regularly fog-shrouded in summer and have a typically summer-dry mediterranean climate. The record not only shows that fossil plants similar to those now in the Canary and Madeira islands lived in southern Europe and north Africa together with typical members of the sclerophyllous vegetation of that region, but all of them lived under ample summer rainfall. From the Miocene into the late Pliocene they are recorded with numerous plants, for instance *Annona, Berchemia, Carpinus,*

6 European investigators have regularly referred fossil plants to modern species if they appear indistinguishable – a practice usually not followed by American paleobotanists for various reasons (see CHANEY, 1954: AXELROD, 1967 c. p. 111; BECKER, 1969, p. 10).

Cinnamonum, Diospyros, Glyptostrobus, Juglans, Liquidambar, Paliurus, Paulownia, Pterocarya, Robinia, Sapindus, Sassafras, Sterculia, Taxodium, Ulmus and *Zelkova,* that now live only in areas of ample summer rainfall. They indicate an ecotonal relation with the European sector of the Arcto-Tertiary Geoflora, a relation displayed also in the Miocene and Pliocene from central California and Nevada northward into Oregon.

Some of the Pliocene floras that provide evidence of this relationship along the northern shores of the western Mediterranean basin must be noted briefly because they clearly suggest the age of mediterranean climate. The middle Pliocene (Plasiancian) flora of the lower Rhone Valley (Boulay, 1922) shows that the littoral area was dominated by stands of *Sabal* on sandy tracts near the marine embayment. The lower and middle exposed mountain slopes supported an oak-laurel forest that included plants whose nearest relatives are now in the Canary and Madeira islands *(Laurus canariensis, Persea indica, Ocotea foetens, Notelaea excelsa, Celastrus cassinoides),* as well as oaks that are now typically Mediterranean *(Quercus ilex, mirbeckii, sessiliflora, suber).* The oaks presumably attained dominance over the laurels on somewhat drier slopes where they formed dense woodlands, and were associated there with *Castanea (vesca), Carpinus (orientalis), Acer (opulifolium), Ulmus (parviflora)* and other hardwoods of xeric requirements. On the drier outcrops were thickets of *Quercus (coccifera),* and some *Quercus (ilex),* and also *Buxus (sempervirens), Viburnum (tinus),* and *Phillyrea (angustifolia).* Cooler slopes in the valleys and at higher levels supported a mixed deciduous hardwood forest. Some of the taxa are similar to those now in the eastern United States *(Berchemia volubilis, Carya alba, Diospyros virginiana, Juglans cinerea, Myrica cerifera, Robinia pseudoacacia, Sassafras officinale),* others occur now only in the far East *(Acer palmatum, A. orientale, Cinnamomum camphora, Ginkgo biloba, Liquidambar formosanum, Quercus serrata, Sapindus mukorossi, Torreya nucifera),* and others now live chiefly in the more temperate parts of Europe *(Acer pseudoplatanus, Alnus glutinosa, Carpinus betulus, Fagus silvatica, Pyrus communis, Populus tremula,* and *Ulmus campestris),* though some of them occur in the Mediterranean basin where they occupy cool, moist, northerlyfacing slopes surrounded by sclerophyllous vegetation.

The succeeding rich Meximeux flora of late Pliocene (Astian) age represents vegetation of the lower Rhone Valley just after the middle Pliocene (Plasiancian) sea had retreated (Saporta and Marion, 1876). Members of the Canary Island Element dominate, notably fossil species similar to the living *Apollonias canariensis, Ilex canariensis, Laurus canariensis, Persea indica, Ocotea foetens,* and *Viburnum rugosum.* Associated with them were taxa similar to those that comprise the vegetation of the Mediterranean area *(Acer opulifolium, Cercis siliquastrum, Daphne pontica, Fraxinus ornus, Nerium oleander, Punica granatum, Quercus ilex, Viburnum tinus),* and together they formed a mixed sclerophyllous forest and scrub far richer than any that has survived. The cooler moister slopes near at hand and at higher levels supported a mixed deciduous hardwood forest composed of species related to those now in eastern United States *(Carya alba, Liquidambar styraciflua, Liriodendron tulipifera, Magnolia grandiflora)* and in eastern Asia *(Bambusa nitis, Diospyros lotus, Glyptostrobus pensilis, Paulownia imperalis, Torreya nucifera).* Some of the more xeric hardwoods no doubt entered the oak-laurel forest and contributed regularly to it, especially in intermediate sites.

To the west, the Barcelona flora of late Pliocene (Astian) age gives us a glimpse of the vegetation in the valley of the Llobregat River, where forest zonation was similar to that in the lower Rhone (ALMERA, 1907). The lower forest zone was a rich laurel-leafed forest composed of several species similar to those in the Canary Islands, notably *Laurus canariensis, Persea indica, Ocotea foetens, Ilex canariensis* and *Celastrus cassinoides*. The forest also included fossils scarcely separable from plants in the Mediterranean, notably *Buxus sempervirens, Chamaerops humilis, Fraxinus ornus, Nerium oleander*, and *Quercus ilex*. Others that are in the Mediterranean also range north into Europe or east into Asia Minor in areas where there is summer rain *(Acer pseudoplatanus, Alnus glutinosa, Carpinus betulus, Fagus silvatica, Populus alba, Salix alba, S. incana)*. On bordering cooler slopes was a mixed deciduous forest composed of species similar to those now found in widely separated areas. Some have their nearest relatives in the eastern United States *(Berchemia scandens, Diospyros virginiana, Liquidambar stryciflua, Magnolia grandiflora, Sassafras officinale)*, others are found chiefly in the far eastern Mediterranean region where there is summer rainfall – as in the Caucasus *(Acer orientale, Carpinus orientalis, Juglans regia, Zelkova crenata, Platanus orientalis)*, some now occur only in eastern Asia *(Cinnamomum camphora, Glyptostrobus pensilis, Liquidambar formosana, Sapindus mukurosii)*. It is evident that a number of plants that are now exotic to the Mediterranean mingled with the evergreens that are now distinctive of the area, and hence formed richer, more complex communities than those that have survived.

Quaternary Vegetation-Climate

Particularly significant to the problem of the age of mediterranean climate is the early Pleistocene (Villafranchian) flora from Lake Ichkeul in northern Tunisia which is well dated by a large mammalian fauna (ARAMBOURG, ARENES and DEPAPE, 1953). The flora is composed of numerous plants that still occur in Tunisia – *Ceratonia siliqua, Populus alba, Olea europaea, Quercus afarea, Q. coccifera, Q. ilex, Q. mirbeckii, Q. suber, Rhus coriaria*. In addition, temperate species, several of which are no longer native to the African continent, are also recorded – *Fagus silvatica, Juglans regia, Pterocarya* sp., *Ulmus scabra*. Furthermore, taxa of generally subtropical requirements that are now absent from north Africa are also present – *Cassia, Pittosporum, Salix canariensis, Sapindus*. This assemblage represents diverse vegetation types. A lowland *Olea-Ceratonia* community occupied warm sites together with a few subtropical relicts *(Cassia, Pittosporum, Sapindus)*. On the bordering more mesic slopes was a sclerophyllous oak woodland and oak savanna was on the drier slopes. Occupying the cooler, moister valleys and reaching up to higher levels was a temperate mixed hardwood forest with *Fagus, Juglans, Pterocarya* and *Ulmus*. Climate was mild temperate, with winter and also summer rainfall. Clearly, mediterranean climate must be post-Villafranchian (Glacial I) in age – which agrees with evidence discussed above for California.

That similar vegetation ranged far to the south of the Mediterranean shore in the early Pleistocene, is indicated by a pollen flora recovered from a small diatomite deposit at Ilamane in the Hoggar Mountains in the central Sahara. As described by VAN CAMPO *et al.* (1964), the flora comes from a site near 6,000 feet (1830 m) in a region now very dry. As these authors point out, the high percentage of trees indicates a more humid climate and higher rainfall. There are 31 woody and 40 herbaceous plants in the flora. The Ilamane flora includes plants not previously recorded from the Sahara, notably *Zelkova, Pterocarya fraxinifolia, Picea* cf. *orientalis, Taxus baccata, Ostrya*. There are other pollen taxa that represent species that are not now in Europe, for instance *Tilia* (cf. *rubra*), *Alnus*, several species of *Quercus*, and *Juglans* (not *regia*). The flora has a high Irano-Caucasian group of plants, as represented by *Picea, Pinus, Pterocarya, Platanus, Zelkova, Ulmus, Ostrya, Juglans, Tilia, Alnus, Corylus, Quercus* and *Fraxinus*. The Mediterranean taxa that occur in the flora include species of *Pinus, Cupressus, Olea, Phillyrea*, and others. Of the remainder, *Entada* (Mimoseae), *Antidesma* (Euphorbiaceae) and Sapotaceae give a "tropical" element to the flora. Their most northern representatives are now in the west African savannas, and VAN CAMPO *et al.* (1964) suggest that they may have connected that area with the lower flanks of the Hoggar as gallery forest along drainage lines during the early Quaternary.

The Ilamane flora is generally like that from Lake Ichkeul in Tunisia, though there are differences. The Ichkeul flora has its "tropical" taxa *(Cassia, Ficus*, Lauraceae, *Sapindus)*, and also those of the Caucasus *(Juglans, Pterocarya, Ulmus)*, and the Mediterranean *(Ceratonia, Olea, Quercus* ssp., *Rhus)*. The absence of conifers at Lake Ichkeul reflects its lower altitude, and also the nature of the fossils. The leaves at Lake Ichkeul were derived from plants near at hand, and any conifers in the area must have been in the Atlas Mountains to the south; the pollen flora at Ilamane in the Hoggar not only represents plants that lived near the lake, but also in more distant areas. The flora of Ilamane recalls the present flora of Greece-Turkey, where summer rainfall supports a mixed temperate forest in the mountains, above the sclerophyllous Mediterranean flora of the summer-dry lowlands.

These records of forest and woodland in the central Sahara in the earliest Pleistocene, and presumably also in the later Pliocene, agree with the nature of the pollen floras recovered from the Plateau Tufa at Kurkur Oasis (Lat. 24° N), about 70 miles west of Aswan Dam in southern Egypt (VAN CAMPO, QUINET and COHEN, 1968). Of later (?) Pliocene age, they strongly support evidence from the Hoggar Mountains that taxa of the Arcto-Tertiary Geoflora, which still live in the mountains of the eastern Mediterranean basin, were present in the central Sahara at the end of the Tertiary and that they were associated with taxa that represent both Mediterranean and subtropical African elements. The two samples described from the Plateau Tufa show a change in the relative proportions of phyto-geographic elements which indicate a trend to drier climate. The older sample (U-8) has pollen representing Arcto-Tertiary taxa that are now in the mountains of the eastern Mediterranean *(Alnus, Corylus, Salix, Ostrya*, cf. *Betula, Tilia, Aesculus, Carpinus*, cf. *Quercus, Platanus*, cf. *Ulmus)*, others are assigned to a Mediterranean element *(Pinus:* including *P. halepensis*, Cupressaceae, Caryophyllaceae, Urticaceae, Oleaceae: *Olea?*, Rhamnaceae, Cistaceae, Umbelliferae, *Pistacia, Celtis)*, and some represent a subtropical group *(Podocarpus*, Ericaceae: non-mediterranean, Sapindaceae, Loranthoideae, *Phyllanthus, Phoenix, Hyphaene*, Capparidaceae, *Ficus)*. In the somewhat younger sample (U-4) that they also consider Pliocene, Arcto-Tertiary plants decrease in importance (cf. *Carya, Alnus, Salix, Carpinus)*, the Mediterranean element shows a slight increase *(Pinus:* including *P. halepensis*, Cupressaceae, Caryophyllaceae, Urticaceae, *Olea?*, Oleaceae, Rhamnaceae, Cistaceae, Umbelliferae, *Celtis, Euphorbia* cf. *amygdaloides)*, and the subtropical and tropical mesic-hydric taxa increase appreciably (Ericaceae: non-mediterranean, Combretaceae, Melastomaceae, *Ilex*, Celastrales, Sapotaceae, *Hymenocardia, Phoenix* including *Chamaerops, Hyphaene, Acacia: etbaica-raddiana* group, Capparidaceae). This is accompanied by a rise in the importance of halophytic and xerophytic taxa (Compositae: including *Artemisia*, Chenopodiaceae, Amaranthaceae,

Cruciferae, Leguminosae, Resedaceae, Tamaricaceae, *Trichodesma, Zygophyllum, Fagonia*, Zygophyllaceae, *Ephedra).*

The succeeding flora from Wadi Tufa I (U-16), which is considered middle Pleistocene on the basis of regional stratigraphy, indicates further desiccation in the region. The Arcto-Tertiary temperate plants are fewer in number *(Alnus, Salix,* cf. *Quercus, Platanus)* and chiefly represent riparian plants, the Mediterranean element is moderately represented *(Pinus,* including *P. halepensis,* Cupressaceae, Caryophyllaceae, *Olea?,* Rhamnaceae, Cistaceae, Labiatae), the subtropical xeric and mesic elements are reduced (Ericaceae: non-mediterranean, *Hymenocardia, Phoenix* including *Chamaerops, Acacia: ethaica-raddiana* group, Capparidaceae), and the xeric and halophytic elements are well represented (Compositae: including *Artemisia,* Chenopodiaceae, Cruciferae, Leguminosae, *Echium, Zygophyllum*).

Analyses of late Pleistocene fossil pollen floras in the Saharan region have revealed the nature of conditions there during the past 20,000 years. It is evident that Mediterranean vegetation inhabited the present Saharan region during the pluvial stages. For the central Sahara, MONOD (1963) and QUEZEL (1965) review evidence of pollen studies and radiocarbon dating of sediments in terraces, fossil soils and guano. In the period from approximately 10,000 to 8,000 years ago the region was occupied by Mediterranean vegetation: a desert flora is not recorded. Among the trees, pollen of the following are represented: *Pinus halepensis, Cedrus atlantica, Cupressus dupreziana, Juniperus, Quercus ilex,* as well as *Alnus glutinosa, Acer* aff. *monspeliensis, Celtis australis, Daphne* aff. *gnidium, Pistacia lentiscus, P. terebinthus, Erica arborea, Fraxinus xanthoxyloides, Sambucus, Phillyrea* aff. *media, Salix, Zizyphus, Tilia platyphylla,* and *Juglans regia.* The assemblage represents vegetation like that now in the semiarid Saharan Atlas, though a few (*Tilia, Juglans*) taxa that are not represented there indicate some summer rainfall. Farther south in the Tchadian Sahara, the diatomites of Borkou (QUEZEL and MARTINEZ, 1958) reveal pollen of Mediterranean plants and others – *Cupressus dupreziana, Juniperus oxycedrus, Cedrus atlantica, Pinus halepensis, Prunus* sp., *Ephedra tilhoana, Quercus* cf. *coccifera* or *ilex, Ficus, Acacia flava, A. raddiana, Myrtus nivellei, Olea europaea,* and *Erica arborea.* There is evidence of summer rainfall, to judge from *Acacia* and *Ficus.*

The pollen floras provide evidence of increasing aridity (see MONOD, 1963), and the gradual restriction of the mediterranean sclerophyllous forest, woodland and chaparral to their present areas. The mesic flora of deciduous oaks, maple, linden, alder and cedar is approximately 10,000 years, but the younger beds (~6,000 to 4,000 years) show Neolithic vegetation was dominated by plants of drier requirements – oaks, aleppo pine, juniper. The terminal part of the Neolithic (~2,700 years) reveals cypress, olive, hackberry and pine. As precipitation continued to decrease, there was a replacement of the Mediterranean flora by the Sahelian which disappeared several thousand years ago as the present Saharan Desert flora assumed dominance over wide areas.

Modern Relicts

Finally, it is recalled that relicts of the Mediterranean flora persist in favorable sites in the higher mountains of the Sahara. The diverse geographic elements of the flora noted earlier by PELLEGRIN (1926) have been reviewed by QUEZEL (1965) on the basis of additional work on the region. The coexistence of Mediterranean and subtropical African species, many of which are represented by endemic and vicarious species, reflects evolutionary and climatic changes of later Cenozoic and Quaternary times. The endemics and vicariants among the woody plants are almost certainly of middle to late Tertiary age. Some of the phytogeographic relations, however, reflect climatic changes of the Quaternary which permitted the invasion alternately, of a northern or a southern flora, of which remnants are now found in central Saharan mountains in relict sites. In the Hoggar and Tibesti massifs, the Mediterranean and montane African taxa are concentrated in areas where local climate approaches that of the respective regions from which they have come. In the Hoggar, Mediterranean and European taxa comprise grasses and forbs, as well as local thickets of shrubs and shrubby trees that are restricted to relatively mild sites. Of the woody plants represented in the Mediterranean region, identical species include *Erica arborea, Nerium oleander* and *Pistacia atlantica,* whereas vicariants are illustrated by *Olea europaea-laperrini, Myrtus communis-nivellei, Cupressus sempervirens-dupreziana.* In the Tibesti, the African taxa occur chiefly in moist drainageways on the southwest slopes at moderate altitudes where winters are warmer. Here are species of *Acacia, Capparis, Abutilon, Grewia, Commicarpus, Cordia, Ehretia* and others that indicate mild winter climate, and also summer rainfall that formerly was more widespread to the north.

These occurrences are paralleled by relict distributions in the mountains of the Mohave and Sonoran deserts today. In the eastern Mohave Desert, higher mountains support species that are in cismontane southern California, and provide links with the sclerophyllous vegetation of Arizona–for instance *Cercocarpus betuloides, Ceanothus greggii, Garrya flavescens, Quercus turbinella, Q. chrysolepis, Rhamnus californica, R. ilicifolia,* and *Schmaltzia trilobata.* They occur in sites not far from shrubs of subtropical alliance that reach their northern limit in the low Danby-Cadiz-Bristol-Troy lake-basin trough where species of *Acacia, Agave, Cercidium, Chilopsis, Crossosoma, Dalea, Fouquieria, Holocantha, Lycium, Nolina, Olneya,* and *Prosopis* live under milder winter climate. Comparable relations exist in the mountains of central Baja California, where sclerophyllous vegetation in the mountains is surrounded by the Sonoran Desert flora. As listed elsewhere, the flora of San Julio Canyon near Commondu includes woodland and chaparral plants that are typically Californian and Arizonan, as well as Mexican (BRANDEGEE, 1889; AXELROD, 1958b, p. 496). On the higher summits of the region (Sierra San Borja, Cerro La Higuera, Volcan las Tres Virgines), which are at the general latitude of Cedros Island (27° 30′ N to 28° 25′ N), Dr. REID MORAN reports the following (written communication, Feb. 1971):

Pinus monophylla	*Schmaltzia kearneyi*
Juniperus california	*Malosma laurina*
Nolina bigelovii	*Schmaltzia ovata*
Quercus cedrosensis	*Schmaltzia trilobata*
Quercus peninsularis	*Ceanothus perplexans*
Quercus turbinella	*Rhamnus insula*
Mahonia higginsae	*Arctostaphylos peninsularis*
Ribes quercetorum	*Garrya grisea*
Cneoridium dumosum	*Garrya veatchii*
Adenostoma fasciculatum	*Arctostaphylos glauca*

Cercocarpus betuloides	*Xylococcus bicolor*
Heteromeles arbutifolia	*Forestiera neomexicana*
Prunus ilicifolia	*Eriodictyon angustifolium*
Toxicodendron diversiloba	

In this region, summer rains are sufficiently frequent over the mountains so that there is a 2-period blooming for annuals. A number of woody plants of more southerly distribution reach their northern limit about 100–150 miles farther south, in the mountains southwest of Loreto. These include oaks *(Q. idonea, Q. reticulata)*, cottonwood *(Populus brandegeana)* and others that contribute to sclerophyllous vegetation farther south, and in the Cape Region (see NELSON, 1920). The broadleaved sclerophyllous vegetation of the southern peninsular region is confined to the mountains which have regular summer rain. Present here are a number of endemic broadleaved sclerophylls, for instance *Arbutus peninsularis, Garrya salicifolia, Quercus brandegeei,* and *Q. devia*. Also present are unique species that are known elsewhere only from the mountains of southern Sonora and adjacent Sinoloa, notably *Ilex brandegeei, Quercus albocincta* and *Schmaltzia hartmannii.*

In both the California and Mediterranean regions, therefore, relict sclerophyllous vegetation that now inhabits the higher mountains over the desert includes a mixture of taxa whose close Neogene relatives occupied much of the area of the present desert, and which are now largely confined to regimes of either summer or winter rainfall, though some occur in both. Restricted to favorable sites as aridity increased, they provide living links to areas of summer precipitation south of the desert were vegetation and climate are similar to that which characterized the southern half of the area of sclerophyllous vegetation during Miocene and Pliocene times—the northern half being typified by both winter and summer precipitation.

Discussion

The preceding summary review of some of the fossil evidence available for analysis indicates that there are important points of similarity in the development of sclerophyllous vegetation in the Mediterranean and California regions.

In each area, sclerophyllous vegetation originated over the present region of the tropical deserts and border areas during the Cretaceous and early Tertiary (AXELROD, 1950 e; 1958 b; 1960; 1967 b). The taxa occupied local, scattered dry sites where they were rapidly differentiating from dry tropical and subtropical alliances that lived under warm equable climate. By the early Tertiary sclerophyllous vegetation was well established over the region. It graded north to the rich mixed deciduous hardwood forests of the Arcto-Tertiary Geoflora, and to the south it merged into broadleaved evergreen forest, savanna and thorn scrub of the Neotropical- and Paleotropical-Tertiary Geofloras, respectively. To the north, rainfall was evenly distributed through the year, but to the south it was limited to the warm months. There is no evidence for protracted summer drought during the Miocene or early Pliocene in either area, apart from local rainshadows in more southerly regions. In view of its position between temperate and tropical geofloras, and the regular occurrence of rainfall during the warm season, the Neogene sclerophyllous vegetation (laurel-oak forest, oak woodland-savanna, chaparral) was richer in taxa than the derivative vegetation which has survived in the mediterranean areas of summer drought. In each region, the different kinds of sclerophyllous vegetation gradually lost taxa during Neogene time as summer rainfall gradually decreased and as winter temperatures were progressively lowered.

In southwestern North America and in the Mediterranean region, sclerophyllous vegetation shifted into the present area of the desert during the major pluvials. It was in ecotone with grassland, thorn forest and savanna vegetation at these times when arid sites were highly localized. The progressive restriction of mediterranean vegetation to its present area has resulted primarily from the expansion of the desert climate, and from the recent development of a regime of summer drought on the north and northwest margin of the deserts. Both appear to be related causally to chilling of the ocean early during deglaciation. This results in a strengthening of the major cold-water currents (California, Canary), in a decrease in sea-surface temperature, and hence in the development of more stable air-masses over the region during summer–or all year in the case of the Atacama and Namib deserts which are bordered by the colder Benguela and Humboldt currents.

Segregation of the mediterranean vegetation into simpler communities of more restricted distribution occurred in response to the development of more localized Quaternary subclimates which selected species of differing tolerance ranges with respect to temperature and rainfall. In each area, the richer woody vegetation has survived in insular isolation under equable marine climate, the poorer in the more extreme, less equable ones. Although a new regional climate and vegetation–the desert–displaced wide areas occupied by sclerophyllous vegetation during Quaternary time, relict scrub and woodland vegetation still occur locally in the desert mountains where there is sufficient rainfall, whether in the Saharan or Sonoran regions. To the south, where there is reliable summer rainfall, sclerophyllous scrub, woodland and forest reappear and exhibit an aspect and composition more nearly like that of the Neogene.

More numerous genera of woody plants link the Mediterranean and Californian regions today than any other areas of mediterranean climate. They represent two divergent groups whose origins largely have been separate. First are those of Arcto-Tertiary derivation which ranged widely across the temperate parts of Holarctica during the Tertiary, and are represented by different species (or sections) in the different temperate regions where they now live, whether in Europe, eastern Asia, western America or eastern America. Representative of this group are conifers *(Abies, Picea, Pinus)* and angiosperms *(Acer, Aesculus, Alnus, Betula, Cercis, Clematis, Cornus, Crataegus, Platanus, Populus, Quercus, Rhamnus, Sumac, Rosa, Salix, Smilax, Viburnum, Vitis)*. In California and in the Mediterranean region, they occur chiefly in the mountains where there is higher rainfall[7]. If they descend into the lowlands, they occupy streambanks and cool, moist sites in areas otherwise characterized by sclerophyllous vegetation. The second group is composed of species which appear to have evolved from subtropical alliances in lower middle latitudes as parts of the Madro-Tertiary and Mediterrano-Tertiary Geofloras, respectively. Included in this group are broadly related species of *Pinus, Juniperus, Cupressus, Arbutus, Helianthemum, Laurocerasus, Myrica, Quercus* and *Styrax*. That there are fewer taxa in common to the sclerophyllous vegetation of these areas than of the

7 In California, *Aesculus californica* and *Cercis occidentalis* form exceptions, for they are regularly in oak woodland and chaparral vegetation. In northern Baja California, the unique *Aesculus parryi* is a shrub in the chaparral.

bordering forests is understandable in view of the areal relations of the Tertiary climates under which they lived, and which resulted in their essential isolation. In assessing their history, it is significant that the links between these areas were more numerous in the past. The Miocene and Pliocene floras of each region also had species of *Clethra, Ficus, Ilex, Pistacia, Ocotea, Persea,* and *Sabal.* Although this latter group disappeared from California during the Pliocene, they are represented today by essentially identical species in the oak-laurel, oak woodland, and short tree vegetation of northern Mexico, where there is summer rainfall. By contrast, all of them survived in the Mediterranean region (Canary and Madeira Islands) except for *Sabal.* They (and others) might have persisted in California if islands of sufficient altitude to provide adequate precipitation under mild marine climate had been present in the area from Catalina Island southward to Cedros Island, and if sea-surface temperatures were somewhat higher—as they are in the Mediterranean region.

Also relevant to an interpretation of the historical relations which are implied by these present and past ties, is the occurrence of other taxa in the Mediterranean region that represent subtropical relicts. *Capparis spinosa* and *Myrsine canariensis* are evergreen species of pantropical genera that indicate ancient transtropic connections. *Notelaea* (Oleaceae) of the Canaries is known elsewhere only from New Caledonia and Australia; *Pittosporum coriaceum* of the Madeira Islands belongs to a tropical and subtropical genus common to Africa, Australia, New Guinea and the islands of the Southwest Pacific. *Phoenix canariensis* is a member of a genus that is widels distributed in Africa and southwest Asia, and *Dracaena* occurs in tropical Africa and also on Mauritius. Inasmuch as tropical and subtropical taxa were associated with sclerophyllous vegetation in Neogene time and earlier, explanations that invoke migration around the northern parts of Holarctica to account for the present links between these areas of mediterranean climate raises difficult problems: the very different photoperiod to the north; passage of migrants through or near the geographic cold pole in northeast Siberia; absence of arid corridors to the north due to cyclonic storms which were regularly present; inadequate warmth; problems of long distance dispersal whether by air, wind, water, or animals.

In searching for an alternate explanation, it is recalled that there were land connections across the Atlantic into the late Cretaceous (e.g. DIETZ and HOLDEN, 1970; TARLING, 1971). Migration was no doubt possible somewhat later because islands were scattered along the mid-Atlantic Ridge during the later Cretaceous and into Paleogene times; they have since been transported laterally and to subsea positions by ocean-floor spreading. The general similarities between the floras of the rainforest and bordering savannas of America and Africa have been discussed previously (e.g. AUBREVILLE, 1959, 1969; AXELROD, 1952, 1970; BOUGHEY, 1957; CAMP, 1947, 1952; ENGLER, 1905; HUTCHINSON, 1946, 1949). There also are links between the evergreen montane rainforests of Africa and America (BOUGHEY, 1965; MIRANDA, 1959). Among the genera common to the montane flora of Africa and Mexico are *Aralia, Dalbergia, Diospyros, Eugenia, Ficus, Ilex, Myrica, Podocarpus, Rhamnus, Sapindus, Sterculia, Zizyphus,* and others. Apart from the genera common to the mesophytic evergreen lowland and montane rainforests and savannas that range from the inner to outer tropics, plants adapted to open drier regions also provide links across the tropics. These are in the genera *Acacia, Bursera, Cardiospermum, Celtis, Commiphora, Lycium, Maytenus, Prosopis,* and others (see ENGLER, 1914). On this basis, it seems likely that from the middle Cretaceous into the Paleogene migration across lower-middle latitudes via subhumid to semiarid corridors was also possible for sclerophyllous plants, notably those of oak-laurel forest and oak woodland and their associated shrubs.

Additional evidence is provided by the similarities between the Cretaceous floras of the eastern United States and those of southern Europe (southern France, Spain). Many of the resemblances between them, noted by BERRY (1916), are confirmed by more recent comparative studies (DEPAPE, 1959, 1963; TEIXEIRA, 1952). The links between the broadleaved temperate rain forests of these areas are provided by similar conifers, ferns and numerous angiosperms, notably *Bauhinia, Laurus, Liriodendron, Magnolia,* Palmae, *Platanus,* Myrtaceae, *Quercus, Sassafras* and many others. Not only are the genera similar, many of the species appear to be indistinguishable. Of especial interest is the presence in the Raritan flora (HOLLICK, 1895, pl. 21), and possibly in the Dakota flora as well, of leaves that appear to be very similar to those of *Arbutus canariensis,* which is now confined to the Canary and Madeira islands. The significance of these Cenomanian links becomes apparent when it is recalled that angiosperms first entered the record in middle latitudes, and only appeared at higher latitudes later in the Cretaceous (AXELROD, 1959, 1970). Clearly, the close similarities between the mid-Cretaceous floras of southern Europe and the central United States must have been due to east-west connections which have since been removed by sea-floor spreading. In sum, the recurring links between tropical rainforest, montane rainforest, savanna and thorn scrub vegetation on opposite sides of the Atlantic today, and their similarities in the past, suggest that ties between the small, now-restricted areas of mediterranean climate may have a similar explanation since trans-tropic links were more numerous between these types of vegetation than they are today.

Conclusion

Woody plants of divergent phyletic relations in southwestern North America and the Mediterranean region have responded to a sequence of similar changing physical conditions during Cretaceous and Cenozoic times to produce structurally and functionally similar vegetation. Early in the sequence, evergreen forest and savanna occupied the present areas of mediterranean and desert climate which then had ample rainfall in the warm season. Trees and shrubs in local sites of greater temperature extremes and lower effective rainfall were perfecting various structural and functional features (deep root systems, sclerophyllous leaves, evergreen) which adapted them to increasing drought following Oligocene time. The record shows that varied adaptive types were already established by the middle Eocene, and that by Miocene time most sclerophylls were scarcely separable from living species that now dominate the oak-laurel forest, oak woodland savanna, and chaparral of each area. They then lived under a climate with adequate summer rainfall in the south, and precipitation well distributed through the year to the north.

There were greater similarities between sclerophyllous vegetation of the Mediterranean and California regions during the Neogene than there are today. Furthermore, similarities were also greater between tropical rainforest, savanna, and montane rainforest vegetation during late Cretaceous-Paleogene time than at present. All these relations appear explicable by connections across lower and low-middle latitudes between areas of humid and dry climates prior to the broad separation of these areas by sea-floor spreading.

Commencing in mid-Oligocene time, the ocean became progressively colder and the subtropical high pressure system gradually gained strength offshore and became more stable in position in summer. As a result of increased aridity over lower-middle latitudes, sclerophyllous vegetation gradually was confined to more coastal areas of adequate moisture, to mountains over the developing dry regions, and to areas south of the drought zone where there was adequate precipitation. Although summer rainfall gradually decreased in present areas of mediterranean climate during the Neogene, it was sufficient into the late Pliocene and Glacial I to support plants that now live only in areas of summer rainfall.

The surviving sclerophylls in areas of mediterranean climate contribute to vegetation of lower diversity than that of the Tertiary, or that which persists in areas of summer rain where similar vegetation occurs (i.e. Southwest USA, Mexico). Inasmuch as mediterranean climate did not appear until after the Tertiary, the similarities in structure and function displayed by taxa of divergent origins in areas of mediterranean climate were not shaped by it. Since the sclerophylls are older than the mediterranean climate in which they have survived, they must have been preadapted to summer drought.

During the Quarternary, environments fluctuated in response to glacial and non-glacial climates and many new local environments came into existence in response to rapid mountain building, the appearance of new edaphic areas, and the development of new climatic subzones. Increased diversification of local climates segregated sclerophylls into new communities that are less diverse in taxa than those of the Tertiary, and have more restricted distributions that correspond with local moisture-temperature provinces. The appearance of new environments provided annuals especially, but also some perennials and a few shrubs, with opportunities to evolve numerous new taxa adapted to ever-narrower ways of living. They are at a low taxonomic level, and they are ephemeral. Most will disappear as the ice caps melt further, for as the oceans become warmer and the cold-water currents off the mediterranean coasts wane, summer rainfall will return and mediterranean climates will disappear.

Abstract

The fossil record indicates that species similar to most woody plants that now make up the sclerophyllous mixed evergreen forest, oak woodland-savanna, and chaparral vegetation of California were in existence during Miocene time, and that the Neogene vegetation types were more diverse in taxa and adaptive types than the modern descendant vegetation. The plants lived under warm to mild temperate climates with ample summer rainfall.

Commencing in the later Miocene, the subtropical high pressure system strengthened as the oceans became colder and the lands hotter. As the thermal contrast increased, summer precipitation was reduced and many taxa were eliminated from the woody flora which decreased in diversity. Hence, most of the surviving woody plants, and the adaptive types they represent, are much older than mediterranean climate which only appeared after the early Pleistocene.

It was primarily the herbaceous flora that evolved under mediterranean climate, though a few woody genera also responded. Rapid proliferation of species resulted because intense, continuing mountain building established many new topographic-climatic-edaphic-biotic sub-zones in a new summer-dry climate. However, since summer-dry alternates with moist pluvial climate, many of these new taxa are transients for they (or their descendants) will disappear when moister climates return.

A similar vegetation history is indicated for the Mediterranean region; there appear to have been links between it and the drier parts of North America in Cretaceous and later times.

The sclerophyllous flora and vegetation now in the mediterranean climates of the Southern Hemisphere largely evolved independently, but in response to similar factors.

References

ADAM, D. P.: Ice ages and the thermal equilibrium of the earth. Interm. Research Rept. 15. Dept. Geochronology, Univ. Ariz. 1969.

ALMERA, D. J.: Flora pliocénica de los alrededores de Barcelona. Mem. Real. Acad. Ciencias Artes Barcelona **3**, 321–335 (1907).

ARAMBOURG, C., ARENES, J., DEPAPE, G.: Contribution à l'étude des floras fossiles Quarternaires de l'Afrique du Nord. Arch. Mus. Nat. Hist. nat. **7**, ser. T 2, 1–85 (1953).

AUBREVILLE, A.: Etude comparée de la famille des Legumineuses dans la flore de la forêt équatoriale africaine et dans la flore de la forêt amazonienne. Comp. Rend. Soc. Biogeogr. **36**, 43–57 (1959).

AUBREVILLE, A.: Essais sur la distribution et l'histoire des angiospermes tropicales dans le monde. Adansonia, ser. 2, **9**, 198–247 (1969).

AXELROD, D. I.: A Pliocene flora from the Mount Eden beds, southern California. Carnegie Inst. Wash. Pub. **476**, 125–183 (1937).

AXELROD, D. I.: A Miocene flora from the western border of the Mohave Desert. Carnegie Inst. Wash. Pub. **516**, 1–128 (1939).

AXELROD, D. I.: The Pliocene Esmeralda flora of west-central Nevada. Wash. Acad. Sci. Jour. **30**, 163–174 (1940a).

AXELROD, D. I.: The Mint Canyon flora of southern California: a preliminary statement. Amer. Jour. Sci. **238**, 577–585 (1940b).

AXELROD, D. I.: The Black Hawk flora (California). Carnegie Inst. Wash. Pub. **553**, 91–102 (1944a).

AXELROD, D. I.: The Mulholland flora (California). Carnegie Inst. Wash. Pub. **553**, 103–146 (1944b).

AXELROD, D. I.: The Oakdale flora (California). Carnegie Inst. Wash. Pub. **553**, 147–166 (1944c).

AXELROD, D. I.: The Sonoma flora (California). Carnegie Inst. Wash. Pub. **553**, 167–206 (1944d).

AXELROD, D. I.: The Pliocene sequence in central California. Carnegie Inst. Wash. Pub. **553**, 207–224 (1944e).

AXELROD, D. I.: The Alvord Creek flora (Oregon). Carnegie Inst. Wash. Pub. **553**, 225–262 (1944f).

AXELROD, D. I.: The Alturas flora (California). Carnegie Inst. Wash. Pub. **553**, 263–284 (1944g).

AXELROD, D. I.: A Sonoma florule from Napa, California. Carnegie Inst. Wash. Pub. **590**, 23–71 (1950a).

AXELROD, D. I.: Further studies of the Mount Eden flora, southern California. Carnegie Inst. Wash. Pub. **590**, 73–117 (1950b).

AXELROD, D. I.: The Anaverde flora of southern California. Carnegie Inst. Wash. Pub. **590**, 119–158 (1950c).

AXELROD, D. I.: The Piru Gorge flora of southern California. Carnegie Inst. Wash. Pub. **590**, 159–214 (1950d).

AXELROD, D. I.: Evolution of desert vegetation in western North America. Carnegie Inst. Wash. Pub. **590**, 215–306 (1950e).

AXELROD, D. I.: A theory of angiosperm evolution. Evolution **6**, 29–60 (1952).

AXELROD, D. I.: Mio-Pliocene floras from west-central Nevada. Univ. Calif. Publ. Geol. Sci. **33**, 1–316 (1956).

AXELROD, D. I.: Late Tertiary floras and the Sierra Nevadan uplift. Geol. Soc. Amer. Bull. **68**, 19–46 (1957).

AXELROD, D. I.: The Pliocene Verdi flora of western Nevada. Univ. Calif. Publ. Geol. Sci. **34**, 61–160 (1958 a).

AXELROD, D. I.: Evolution of the Madro-Tertiary Geoflora. Bot. Review **24**, 433–509 (1958 b).

AXELROD, D. I.: Poleward migration of early angiosperm flora. Science **130**, 203–207 (1959).

AXELROD. D. I.: The evolution of flowering plants. 227–305. In: Evolution after Darwin, vol. I: TAX, S. Ed.). Chicago: Univ. Chicago Press 1960.

AXELROD, D. I.: A Pliocene Sequoiadendron forest from western Nevada. Univ. Calif. Publ. Geol. Sci. **39**, 195–268 (1962).

AXELROD, D. I.: The early Pleistocene Soboba flora of southern California. Univ. Calif. Publ. Geol. Sci. **60**, 1–109 (1966).

AXELROD, D. I.: Geologic history of the California insular flora. 267–316. In: Proc. Symposium on the Biology of the California Islands (R. N. PHILBRICK, Ed.). Santa Barbara: Santa Barbara Bot. Garden 1967 a.

AXELROD, D. I.: Drought, diastrophism, and quantum evolution. Evolution **21**, 201–209 (1967 b).

AXELROD. D. I.: Evolution of the Californian closed-cone pine forest. 93–149. In: Proc. Symposium on the Biology of the California Islands (R. N. PHILBRICK, Ed.). Santa Barbara: Santa Barbara Bot. Garden 1967 c.

AXELROD, D. I.: Mesozoic paleogeography and early angiosperm history. Bot. Review **36**, 277–319 (1970).

AXELROD, D. I., TING, W. S.: Late Pliocene floras east of the Sierra Nevada. Univ. Calif. Pub. Geol. Sci. **39**, 1–118 (1960).

BARTHOUX, J. C., FRITEL, P. H.: Flore crétacée du grès Nubie. Mem. Inst. Egypte Cairo, **7**, 65–119 (1925).

BECKER, H. F.: Fossil plants of the Tertiary Beaverhead basins in southern Montana. Paleontographica Abt. B, **127**, 1–142 (1969).

BELOUSSOV, V. V.: Continental rifts. In: The earth's crust and upper mantle. Amer. Geophys. Union, Geophys. Monogr. **13**, 539–544 (1969).

BERRY, E. W.: The upper Cretaceous flora of the world. In: Maryland Geol. Surv., Upper Cretaceous 183–314 (1916).

BERRY, E. W.: A petrified walnut from the Miocene of Nevada. Wash. Acad. Sci. Jour. **18**, 158–160 (1928).

BOUGHEY, A. S.: The origin of the African flora. Oxford: Oxford Press 1957.

BOUGHEY, A. S.: Comparisons between the montane forest floras of North America, Africa and Asia. (Proc. 5th Meetg., Assoc. pour l'étude taxonomique de la flora d'Afrique tropicale). Webbia **19**, 507–517 (1965).

BOULAY, N.: La flore pliocène de la Vallée du Rhône. Ann. Sci. Natur. Bot. ser. 10, **4**, 73–366 (1922).

BRANDEGEE, T. S.: A collection of plants from Baja California. Proc. Calif. Acad. Sci. II, **2**, 117–216 (1889).

CAMP, W. J.: Distribution patterns in modern plants and the problems of ancient dispersal. Ecol. Monogr. **17**, 159–183 (1947).

CAMP, W. J.: Phytophyletic patterns on lands bordering the south Atlantic basin. Amer. Mus. Nat. Hist. Bull. **99**, 205–216 (1952).

CHANDLER, M. E. J.: Some upper Cretaceous and Eocene fruits from Egypt. Bull. British Mus. (Nat. Hist.), Geology **2**, 149–187 (1954).

CHANEY, R. W.: The Deschutes flora of eastern Oregon. Carnegie Inst. Wash. Publ. **476**, 185–216 (1938).

CHANEY, R. W.: A new pine from the Cretaceous of Minnesota and its paleoecological significance. Ecology **35**, 145–151 (1954).

CHANEY, R. W., AXELROD, D. I.: Miocene floras from the Columbia Plateau. Carnegie Inst. Wash. Pub. **617**, 1–237 (1959).

CHANEY, R. W., MASON, H. L.: A Pleistocene flora from the asphalt deposits at Carpinteria, California. Carnegie Inst. Wash. Pub. **415**, 45–79 (1933).

CLAUSEN, J.: Stages in the evolution of plants species. New York: Cornell Univ. Press 1951.

CLAUSEN, J., KECK, D. D., HEISEY, W. H.: Experimental studies on the nature of species. I. Effect of varied environments on western North American plants. Carnegie Inst. Wash. Pub. **520**, 1940.

CLAUSEN, J., KECK, D. D., HEISEY, W. H.: Experimental studies on the nature of species. III. Environmental response of climatic races of *Achillea*. Carnegie Inst. Wash. Pub. **581**, 1948.

CLAUSEN, J., HEISEY, W. D.: Experimental studies on the nature of species. IV. Genetic structure of ecological races. Carnegie Inst. Wash. Pub. **615**, 1958.

CONDIT, C.: The San Pablo flora of west-central California. Carnegie Inst. Wash. Pub. **476**, 217–268 (1938).

CONDIT, C.: The Remington Hill flora (California). Carnegie Inst. Wash. Pub. **553**, 21–55 (1944a).

CONDIT, C.: The Table Mountain flora (California). Carnegie Inst. Wash. Pub. **553**, 57–90 (1944b).

COOPER, W. S.: The broad-sclerophyll vegetation of California: an ecological study of the chaparral and its related communities. Carnegie Inst. Wash. Pub. **319**, 1922.

DEPAPE, G.: Les floras fossiles du Crétacé supérieur en France. 84th Congr. des Sociétées savantes, 61–94 (1959).

DEPAPE, G.: Colloque sur le Crétacé inférieur. Mem. Bureau recherches géologiques et minières, **34**, 349–371 (1963).

DIETZ, R. S., HOLDEN, J. C.: Reconstruction of Pangea: Breakup and dispersion of continents, Permian to Present. Jour. Geophys. Res. **75**, 4939–4956 (1970).

DORF, E.: A late Tertiary flora from southwestern Idaho. Carnegie Inst. Wash. Pub. **476**, 73–124 (1936).

DURHAM, J. W.: The marine Cenozoic of southern California. In: Geology of southern California, Calif. Div. Mines Bull. 180, Chap. III, 23–31 (1954).

EHLIG, P. L., EHLERT, K. W.: Offset of Miocene Mint Canyon Formation from volcanic source along San Andreas fault, southern California. Geol. Soc. Amer., Abstracts with Programs **4**, 154 (1972).

ENGLER, A.: Über floristische Verwandtschaft zwischen dem tropischen Afrika und Amerika, sowie über die Annahme eines versunkenen brasilianisch-äthiopischen Kontinents. Sitzungsber. K. Preuss, Akad. Wissen. **6**, 180–231 (1905).

ENGLER, A.: Über Herkunft, Alter und Verbreitung extremer xerothermer Pflanzen. Sitzungsber. K. Preuss. Akad. Wissen. **20**, 564–621 (1914).

ETTINGSHAUSEN, C. VON: Die Eocene floras des Monte Promina. Denkschr. Kais. Akad. Wiss. Wein, Math.-Naturwiss. Classe **8**, 19–41 (1854).

FURON, R.: Géologie de l'Afrique. 3rd ed. Paris: Payot 1968.

GRANT, A., GRANT, V.: Genetic and taxonomic studies in Gilia. VIII. The Cobwebby Gilias. El Aliso, **3**, 203–287 (1956).

GRANT, V., GRANT, A.: Genetic and taxonomic studies in Gilia. VII. The woodland Gilias. El Aliso, **3**, 59–91 (1954).

HALL, C. A.: Displaced Miocene molluscan provinces along the San Andreas fault, California. Univ. Calif. Publ. Geol. Sci. **34**, 291–308 (1960).

HOLLICK, A.: The flora of the Amboy clays. U. S. Geol. Surv. Monogr. **26**, 1895.

HUTCHINSON, J.: A botanist in South Africa. London: P. R. Gawthorn Ltd. 1946.

HUTCHINSON, J.: The families of flowering plants. 2 vols. 2nd edit. Oxford: Oxford University Press 1959.

LAURENT, L.: Flore des calcaires de Celas. Ann. Musée Hist. natur. Marseille, ser. II, **1**, 1–154 (1899).

MCMINN, H. E.: *Ceanothus*. Santa Barbara: Santa Barbara Botanic Garden 1942.

MILLER, A. H.: Ecologic factors that accelerate formation of races and species of terrestrial vertebrates. Evolution **10**, 262–277 (1956).

MIRANDA, F.: Posible significación del porcentaje de géneros bicontinentales en América tropical. Ann. Inst. Biol. Mex., **30**, 117–150 (1959).

MONOD, T.: The late Tertiary and Pleistocene in the Sahara. 115–229. In: African ecology and human evolution. (F. C. HOWELL and F. BOURLIERE, Eds.). Viking Fund Publ. in Anthropology 36. New York: Wrenner Gren Foundation 1963.

NELSON, E. W.: Lower California and its natural resources. Mem. Nat. Acad. Sci. **16**, 1920.

NOBS, M. A.: Experimental studies on species relationships in *Ceanothus*. Carnegie Inst. Wash. Pub. **623**, 1963.

PAGE, V. M.: *Lyonothamnoxylon* from the lower Pliocene of western Nevada. Madrono **17**, 258–266 (1964).

PELLEGRIN, F.: Les affinités de la flore des sommets volcaniques du Tibesti (Afrique centrale). Comp. Rend. Acad. Sci. **182**, 337–338 (1926).

QUEZEL, P.: La végétation du Sahara. Stuttgart: Gustav Fischer Verlag 1965.

QUEZEL, P., MARTINEZ, C.: Etude palynologique de deux diatomites du Borkou (Territoire du Tchad, A.E.F.). Bull. Soc. Hist. Nat. Afr. Nord. **49**, 230–244 (1958).
SAPORTA, G. DE, MARION, A. F.: Recherches sur les végétaux fossiles de Meximeux. Mus. Nat. Hist. Natur. Lyon, **1**, 131–335 (1876).
SEWARD, A. C.: Leaves of dicotyledons from the Nubian sandstone of Egypt. Geol. Surv. Egypt. 1935.
SMITH, H. V.: Notes on fossil plants from Hog Creek in southwestern Idaho. Papers Mich. Acad. Sci., Arts & Letters, **23**, 223–231 (1938).
SMITH, H. V.: A Miocene flora from Thorn Creek, Idaho. Amer. Midl. Naturl. **25**, 473–522 (1941).
STEBBINS, G. L.: Aridity as a stimulus to evolution. Amer. Natur. **86**, 33–44 (1952).
STEBBINS, G. L., MAJOR, J.: Endemism and speciation in the California flora. Ecol. Monogr. **35**, 1–35 (1965).
TARLING, D. H.: Gondwanaland, paleomagnetism and continental drift. Nature **229**, 17–21 (1971).
TEIXEIRA, C.: Notes sur quelques gisements de végétaux fossiles du Crétacé des environs de Leiria. Revista Faculdade Ciencias Lisboa, 2 a ser. C, **2**, 133–154 (1952).
UNGER, F.: Die fossile flora von Radoboj. Denkschr. Akad. Wiss. Wein. **39**, 125–170 (1869).
VAN CAMPO, M., AYMONIN, G., GUINET, P., ROGNON, P.: Contribution à l'étude du peuplement végétal Quaternaire des montagnes Sahariennes: l'Atakor. Pollen et Spores, **6**, 169–194 (1964).
VAN CAMPO, M., GUINET, P., COHEN, J.: Fossil pollen from late Tertiary and middle Pleistocene deposits of the Kurkur Oasis. 515–520. In: BUTZER, K. W., and HANSEN, C. H., Desert and River in Nubia: Geomorphology and prehistoric environments at the Aswan Reservoir. Appendix I. Madison: Univ. Wisconsin Press 1968.
WEBBER, I. E.: Woods from the Ricardo Pliocene of Last Chance Gulch, California. Carnegie Inst. Wash. Pub. **412**, 113–134 (1933).
WOLFE, J. A.: Miocene floras from Fingerrock Wash, southwestern Nevada. U. S. Geol. Survey. Prof. Paper 454–N. 1–36 (1964).

Section VI: Animal Biogeography and Ecological Niche

This introduction has two main intentions: firstly, to link this section with the previous one, through discussion of how far statements founded on considerations of plant biogeography can be generalized to the biogeography of the faunas of the same regions; and secondly, to outline the different approaches adopted in the five chapters of this section.

In this preliminary comparison of plant and animal biogeography in mediterranean regions, consideration is given only to groups of terrestrial invertebrates, certain of which have been treated in detail in the chapters of SÁIZ and VITALI-DI CASTRI, as well as in that of DI CASTRI on soil animals in Section IV. In regard to other animal groups, terrestrial invertebrates are particularly suitable for comparison with plants from a biogeographical standpoint, because they are in general more directly dependent on the vegetation, are settled in more restricted areas and have fewer possibilities of active dispersal.

The fundamental statement of AXELROD's chapter, which refers to the history of mediterranean ecosystems in California but which is probably also valid for other regions having this climatic type, is that the mediterranean climate is not old but is, in fact, very young in geological terms. AXELROD argues that modern woody mediterranean formations are very recent as regards the present relative abundance among taxa, but that these taxa are far more ancient than the occurrence of the mediterranean climate. These taxa are probably survivors of a richer flora which have been able to overcome the new conditions of summer drought in a scarcely modified form. The woody floras did not evolve in response to the mediterranean climate; rather they were – to a greater or lesser extent – preadapted to it.

Most of these considerations might be applied *mutatis mutandi* to the biogeographical situation of terrestrial invertebrates in mediterranean climate regions. Lack of fossil evidence, of course, precludes the pinpointing of precise ages in this case. Nevertheless, it would hardly be tenable to claim that the major adaptations of terrestrial invertebrates be considered as strictly specific to the mediterranean conditions. The peculiarity of the invertebrate fauna in mediterranean regions is the coexistence of so many different kinds of adaptation in different coexisting taxa. However, each of these adaptations can be found in other climatic zones, and most of these taxa are present elsewhere in scarcely modified forms. For instance, nocturnal habits, strongly chitinized cuticle or periods of diapause are common adaptive features of desert invertebrates, though these features are sometimes modified for the mediterranean conditions, as illustrated by the aestivation of some gastropods. On the other hand, the hygrophilous animals of mediterranean soils are living just in that stratum of the ecosystem which is the least influenced by the daily or seasonal extremes of the mediterranean climate.

In fact, edaphic specialization (life in deep soil) is often considered as a specific response to the mediterranean climate. It is true that no other soils are more populated at depth by peculiar arthropod species than mediterranean soils. However, the relativity of the concept of "edaphism" has already been noted (see DI CASTRI's chapter on soil animals). Thus, these soil animal taxa, sometimes at the level of common species or genus, exist also in other zones having a non-mediterranean climate, even though they may live

The introduction to this section was prepared by FRANCESCO DI CASTRI.

there in more superficial layers or on the ground surface. Also, most of these taxa are very ancient and conservative phylogenetic lines, their adaptations being inconsistent with the recent origin of the mediterranean climate. Probably only a very limited number of groups – related to species still living in litter – have acquired specific edaphic adaptations as an "escape" response to the mediterranean climate; in other words, they have not fitted their biology to the seasonality of the mediterranean climate, but have maintained their previous strict hygric requirements and have avoided the new conditions of seasonality by living in deep soil.

RAVEN comments on the existence of so many species of plants in the mediterranean climate regions, which can be considered as ecotones having accumulated derivatives from tropical and temperate floras. It is also stressed in the introduction to Section V that the mediterranean climate has promoted speciation in a few genera of woody plants but mainly in the herbaceous flora, which is rich and largely endemic in these areas. This general observation is also applicable to the invertebrate fauna of these regions which is composed of elements of distinct biogeographic origin (temperate, tropical, desertic), and which presents phenomena of "pulverization" of phyletic lines in numerous species which have a very restricted distribution. However, as noted by AXELROD in relation to the evolution of herbaceous plants in California, it is unlikely that the main evolutionary factor has been the appearance of the mediterranean climate alone. The coincident large changes in topography, with physiographical fragmentation and development of many different local subclimates and soils, have probably played a fundamental role in the segregation of populations of terrestrial animals in mediterranean areas and in the subsequent processes of speciation.

A further point of comparison arises from the statement of RAVEN that groups with a circum-Antarctic distribution have not contributed significantly to the vegetation of the mediterranean regions of the southern hemisphere (Central Chile, South Africa and South Australia); the woody vegetation of these regions, at least, has been differentiated from tropical ancestors. While Antarctica was probably covered in Cretaceous times by warm temperate rain forests, semiarid or subhumid plant formations, that is, with species compatible to a mediterranean-like climate, were lacking. This is not the case for the soil-related communities of invertebrates in the mediterranean climate regions. Not only are paleoantarctic elements dominant in the Chilean soils, but they are also well represented in other areas – mediterranean and non-mediterranean – of the southern hemisphere. In addition, they constitute a not negligible component of soil communities in California and Mediterranean Europe. In fact, because of the microclimatic stability of soil, these animals are now living under conditions resembling those of a temperate-oceanic habitat with constant high humidity, which do not appear to be very different from the climatic conditions of Antarctica in Cretaceous times.

In summary, it seems that mediterranean ecosystems are, as such, very young, that is, in regard to the community relations among and between species. They are, however, integrated by taxa of different ages: very old for most of the woody floras and the soil invertebrates, much younger for the herbaceous plants and the epigeous animals. Community patterns in the structure of the soil subsystem are probably more ancient than in the above-ground subsystem, because of the great climatic stability of soil and the persistence of so many lines of gondwanian and paleoantarctic origin. Whatever the exact ages of these elements might be, it can fairly be postulated that the historical heterogeneity of the mediterranean climate ecosystems, when different taxa or different subsystems are compared, is greater than in any other terrestrial ecosystem type of the world. It remains to be demonstrated whether couplings between such heterogeneous parts, with contemporary or relict attributes, have implications to the functioning of the entire ecosystem.

In this specific section two distinct sets of chapters can be distinguished. These focus on two different facets of animal biogeography. The chapters by SÁIZ and VITALI-DI CASTRI deal with *historical biogeography,* in a classical way inspired by the works of JEANNEL (see for instance, JEANNEL 1942, 1967), while adopting a more quantitative approach based on principles of numerical taxonomy. In spite of an unavoidable ingredient of speculation attributable to the lack of related fossils, these chapters provide partial explanations of the *origins* and possible pathways of distribution of arthropods inhabiting regions which at present have a mediterranean type of climate.

The last three chapters, by CODY, SAGE, and HURTUBIA and DI CASTRI, approach the problem of animal distribution from the viewpoint of *ecological biogeography,* taking as a basis the concept of ecological niche. Thus, they address their attention to other kinds of questions, such as, "how many species can coexist in a given territory?" or "in which way are available resources divided?" Information of this nature is particularly valuable for understanding the *structure* of these communities. Other illuminating examples of this type of approach are given by MACARTHUR (1972).

SÁIZ and VITALI-DI CASTRI discuss the distribution of two groups of terrestrial invertebrates which are particularly well represented in mediterranean climate zones: the beetles populating the deep strata of soil and the arachnid pseudoscorpions living near the soil surface. Both authors stress the fact that the acceptance of the continental drift hypothesis is an essential requirement for explaining the present distribution of these groups. The chapter of SÁIZ emphasises the role of paleoantarctic lines in colonising soil biotopes even in the mediterranean regions of the northern hemisphere, while VITALI-DI CASTRI illustrates the existence of true biogeographical connections among the three austral continental masses, an existence which is probably not directly linked to the presence of a mediterranean type of climate.

A general biogeographic conclusion of these two chapters, not specifically stressed in them, is that there is no apparent relation between the area where the greatest number of species occurs at the present time and the originary point of these taxa. Both chapters show that the highest species concentration of groups such as *Leptotyphlinae* (Coleoptera, Staphylinidae) or *Gymnobisiinae* (Pseudoscorpionida, Vachoniidae) is found in mediterranean regions. However, the authors do not at all postulate that these groups have originated in these regions. Rather, they advocate a paleoantarctic origin. It is suggested that these groups are remnants of different phylogenetic lines whose rate of speciation was radically increased because of physiographic and climatic factors which favoured their segregation. Reference is made to segregation which has occurred in caves, in deep soil, in small fragmented islands, in humid valleys widely separated by arid territories, and at the top of mountains where there is permanent condensation of marine fogs. However, most of these types of segregation apply to hygrophilous elements; it might be worthwhile to extend this kind of biogeographic analysis to another group of terrestrial invertebrates – the tenebrionid beetles – which is also well distributed in mediterranean regions, but which is mainly represented by xerophilous elements. The fact that all the five mediterranean regions are bordered by desertic zones, where tenebrionids constitute a dominant group, would provide further relevance to such a study.

A final point, which has been mentioned in most of the chapters of this volume but which should be particularly stressed in this introduction, is that man has become the main factor influencing species distribution in mediterranean regions. This action can be considered to take two main forms. On the one hand, there is the intentional or accidental introduction of species which have a broad distribution and a large range of ecological tolerance – up to cosmopolitan ubiquitous species – in ecosystems that originally were integrated by species having a more restricted distribution and requirements. On the other hand, there are actions facilitating the invasion or the spreading out of xerophilous species – of autochthonous nature or coming from the neighbouring deserts – following the degradation of marginal areas through overgrazing and erosion. At the present time, many mediterranean zones can be envisaged as being compound by sporadic spots of relatively humid habitats dotted in a stroma of arid environments whose extension and continuity are progressingly increasing. The existence of gradients of different types and degrees of human impact and intervention makes these areas particularly suitable for study – with a dynamic diachronic view – of the process of desertification and the biogeographic mechanisms involved in it.

The final group of chapters in this section focusses on the concept of ecological niche. These chapters deal, implicitly or explicitly, with the question of whether different phyletic lines of two disjunct ecosystems that arose from a different basic situation *must* adopt the same strategy of adaptation when confronted with a similar kind of physical environment; in other words, whether these phyletic lines are converging toward a similar type, performing the same function in the ecosystem and occupying essentially the same ecological niche. The chapters of CODY and SAGE, through their comparisons of Californian and Chilean faunas, present some evidence that the adaptive possibilities of different species in a given type of environment are fairly narrowly limited. In fact, these multispecies communities seem to be organised in a similar way in the two areas considered, according to similar niche characteristics. However, both CODY and SAGE also describe peculiar differences between Chile and California in niche occupation. The chapter of HURTUBIA and DI CASTRI refers only to Chilean fauna. It illustrates some evolutionary strategies, such as feeding and reproductive tactics, which provide for the existence of so many co-generic sympatric species in mediterranean regions. The taxa considered in these three chapters, both birds and lizards, have already proved to be excellent groups for studies on ecological niche.

CODY's chapter on the similarities of bird niches in response to parallel selective forces places in a more general context the comparative analysis between California and Chile, since it presents conclusions from other intercontinental biome comparisons of bird faunas. CODY argues that the most interesting difference between the bird niches of California and Chile concerns the rate at which species are replaced between adjacent habitats. In fact, while bird species in California divide the state by habitats, in Chile division is mainly by geographic area and to a much lesser extent by habitat. This sort of behaviour might well be referable to the more isolated conditions of Central Chile.

The two chapters of SAGE and HURTUBIA and DI CASTRI deal with the same lizard group. These chapters reach similar conclusions in certain aspects, particularly as regards the great morphological overlap in Chilean species. There might seem to be a contradiction between these two chapters. SAGE postulates the absence of specialised species of lizards in Central Chile, while HURTUBIA and DI CASTRI frequently report different kinds of specialisation for this faunal group. This apparent contradiction is attributable to the different approaches of these two chapters and to the relativity of the concept of "generalist" species. In this respect, and integrating the conclusions of the two chapters, it can be affirmed that, although the Chilean fauna of lizards is in general less specialised than the analogous Californian fauna, different degrees of increasing specialisation (or decreasing generalization) nevertheless exist among the various species of Chilean lizards as regards the delimitation of their niches.

References

JEANNEL, R.: La genèse des faunes terrestres. Eléments de biogéographie. Paris: Presses Universitaires 1942.

JEANNEL, R.: Biogéographie de l'Amérique Australe. 401–460. In: Biologie de l'Amérique Australe. C. DELAMARE DEBOUTTEVILLE and E. RAPOPORT (Eds.), Vol. 3. Paris: C.N.R.S. 1967.

MACARTHUR, R. H.: Geographical Ecology. Patterns in the distribution of species. New York: Harper and Row 1972.

1

Biogeography of Soil Beetles in Mediterranean Regions

Francisco Sáiz

Edaphic or endogeous forms are characteristic biological components of regions with a mediterranean climate. These forms have maintained the constancy of their ecological requirements by means of increased penetration into the soil (edaphic environment).

Ecological-Evolutionary Aspects

Penetration has been attained either by colonisation of caves (e.g., Trechinae and Bathysciinae (Carabidae), of the fissures of the soil (sometimes associated with buried stones), or of the soil itself. The adaptive strategy in each case is different, as illustrated by the body flattening of the fissures inhabitants (e.g., members of the subgenus *Phacomorphus* of the genus *Speonomus*, Staphylinidae), the false physogastria, enlarged legs and strong development of chaetotaxy in cave species (e. g., *Aphaenops*, Pselaphidae), and all the many adaptive features which characterize the soil beetles, the main subject of this chapter.

Table 1. Relation between habitat of *Nothoesthetus* and climate in Chile

Species	Locality	Latitude	Habitat	Climate[a]
coiffaiti	Palmas Cocalán	34°	soil	Mediterranean semiarid
scitulus	Concepción	37°	humus	Mediterranean humid
obesus	Valdivia	40°	humus	Oceanic with mediterranean influence
montanus	Valdivia	40°	humus	Oceanic with mediterranean influence
australis	Bahía Orange	55°	moss	Oceanic sub-antarctic

[a] Bioclimatic regions of Chile according to di Castri (1968).

Edaphism, or more precisely euedaphism (strict adaptation to the life in deep soil), is not a characteristic of a particular morphological or higher taxonomic group, but is rather the adaptive response of epigeous or humicolous forms of different groups (Protura, Diplura, Coleoptera, etc.) to survival in mediterranean climates. These forms originated from regions or from geological ages with a more constant climate, especially from an hydric viewpoint. Within the Coleoptera, members of the genus *Nothoesthetus* (Staphylinidae, Euaesthetinae), comprising 5 species recorded only in Chile (Sáiz, 1969) shift their habitat to deeper soil levels as the climate becomes more arid (Table 1). The Protura are traditionally considered as strictly euedaphic forms and are found as such in Central Chile (34° S with a mediterranean climate). In the Valdivian forests of Chile (40° S with a fundamentally oceanic climate) they live, however, in humus, in litter and in moss, and can even be found in the fronds of ferns (see also DI CASTRI's chapter on soil animals). This is indicative of the fundamental importance of the hydric aspect of the environment in the phenomena of edaphism.

In this respect, several groups of Coleoptera should be mentioned. Within the Staphylinidae family, the *Osoriini* and the *Phloecharis,* among others, mainly show their endogeous forms within regions having a mediterranean climate, and their epigeous forms outside of these areas. Similar observations have been made for several groups within the Trychopterigidae (genus *Ptinella),* Carabidae and Pselaphidae. The more arid the mediterranean climate, the greater the degree of edaphism adopted by the hypogeous forms living

Fig. 1. *Macrotyphlus politus* of Chile

there. Euedaphism is therefore the result of historical (climatic and vegetational changes) and contemporary factors (mediterranean climate, topography). This is particularly true for taxa that have secondarily colonized the soil, as in the case of the Coleoptera.

The endogeous Coleoptera, which include members of such families as the Carabidae,

Colydiidae, Trychopterigidae (= Ptiliidae), Curculionidae, Scydmaenidae, Pselaphidae and Staphylinidae, can be subdivided in two groups. One group is formed by taxa which have very close relatives (at the level of tribe, genus, subgenus) living in an epigeous or in a cave environment. The other group is composed of taxa (e. g. *Mayetia* (Pselaphidae), Leptotyphlinae Staphylinidae) which are integrally adapted to the edaphic environment and which lack close ties with epigeous forms.

The Leptotyphlinae is a very old predatory tribe of edaphic microcoleoptera (Fig. 1). Members of the group have a very weak capacity for active dispersal; this lends particular interest to considerations of the biogeographic status of the group. The problems of survival of the Leptotyphlinae in the mediterranean climate have been solved by adoption of an edaphic way of life. This has entailed a number of morphological and behavioral adaptations and consequences:

1) Strong reduction of size; individuals range from 0,7 to 1,7 mm in length and are thus the smallest representatives of the family Staphylinidae.

2) Subcylindrical and filiform body, associated with the homodinamy of the segments (i.e., the different body segments have approximately the same length).

3) Constriction between the prothorax and the mesothorax; this enables the insect to make large circular movements and facilitates its movement in soil interstices or in the microgalleries of dead rootlets.

4) Anophtalmy and body depigmentation.

5) Apterism, which is accompanied, according to the degree of edaphism, by the shortening and scapular welding of the elytra. Only the more primitive tribe *(Neotyphlini)*, that with a greater diversity of genera and a distribution which extends somewhat beyond the mediterranean climate area, does not have complete scapular welding.

6) Strong head, which is as well developed as, or more so, than the prothorax; strong mandibular development; anterior border of the forehead made stronger by increased chitinization; shortening of the antennae. These characteristics (plus the absence of digging modifications in the front legs, the general shortening of the legs, the reduction of the number of tarsal segments, and the small general development of the chaetotaxy) put this group within the "cephalodiggers", in contrast to the "leg-diggers", an adaptive strategy little used by the edaphic Coleoptera.

7) Finally, the combination of "mediterranean adaptations" in this taxonomic group has resulted in an external morphological convergence and in a very high specific divergence of internal structures, such as the male aedeagus. These internal structures show ultraevolutions which are much more developed than in the epigeous forms, and which must be considered as signs of phylogenetic antiquity. The absence of common species and genera in the different regions with a mediterranean climate is an additional indication of group antiquity. Groups such as the *Osoriini* (Staphylinidae) which have only recently colonised the soil and for which a greater percentage of species live in an epigeous environment, have a primitive type of aedeagus, lacking ultraevolutions and having similarities to the aedeagus of non-edaphic forms. This group, as a whole, maintains a high plasticity and a strong capacity for dispersal, and thus contains a number of potential soil colonizers.

Examples of morphological variation with climate are the external morphological characteristics such as the secondary sexual characters of the male, which are lost on adoption of euedaphism. These characteristics are frequently recorded in the more primitive tribe, the one least adapted to edaphic life *(Neotyphlini)*, and are occasional or absent in the tribes that are better adapted to edaphic life. In Chile, where 6 genera and 23 species have been reported, there is increase in frequency of secondary sexual characteristics with decreasing climatic aridity (Table 2).

The presence of a single species with secondary sexual characters in zones with an arid mediterranean climate is referable to the occurrence of this species in a relict Valdivian type of forest (cloud forest). The presence of two Leptotyphlinae in this type of forest (at Fray Jorge) substantiates the idea of a colonizing influence of the austral fauna of Chile.

Table 2. Number of species of *Neotyphlini* with or without secondary male sexual characters in relation to the climate in Chile

	Mediterranean Climate					Oceanic climate with mediterranean influence	
Species	arid	semiarid	subhumid	humid	perhumid		Total
Without	2	9	1	–	–	–	12
With	1	1	1	2	3	3	11
Total	3	10	2	2	3	3	23

The vegetational diversity and the climatic fluctuations typical of the mediterranean climatic regions, in addition to the antiquity and the poorly developed capacity for active dispersal of the taxa, have resulted in a proliferation of species of Leptotyphlinae, as well as of *Mayetia* (Pselaphidae). The greater the severity of the mediterranean climate, the greater the diversification in different species.

Biogeographical Aspects

The present taxonomic status of the subfamily Leptotyphlinae is summarized in Table 3. Further information is given by COIFFAIT (1959, 1961, 1962, 1963, 1964) and COIFFAIT and SÁIZ (1965). Its present distribution is:

Cephalotyphlini: a monospecific and aberrant tribe, found only in Corsica (Fig. 2). This tribe has strong affinities to the *Entomoculini.*

Table 3. Present systematic status of the subfamily Leptotyphlinae

Tribe	No. of genera	No. of species
Cephalotyphlini	1	1
Entomoculini	5	162
Leptotyphlini	8	165
Metrotyphlini	5	9
Neotyphlini	14	34
Total	33	371

Metrotyphlini: a tribe apparently in the process of extinction, with very few species, completely euedaphic; 7 species (1 of which parthenogenetic) are found in the Mediterranean Basin, 2 species in parts of Chile having a mediterranean climate (Fig. 2).

Entomoculini: a very homogeneous tribe, euedaphic, with a distribution confined to the Mediterranean Basin (Fig. 3).

Leptotyphlini: a tribe centered in the Mediterranean Basin (97.5% of total species), with a small nucleus in central East Africa (2.5% of total species). Three geographic centers are evident (Fig. 4): a) Portugal, Tunisia, Corsica, Sardinia and the northern part of

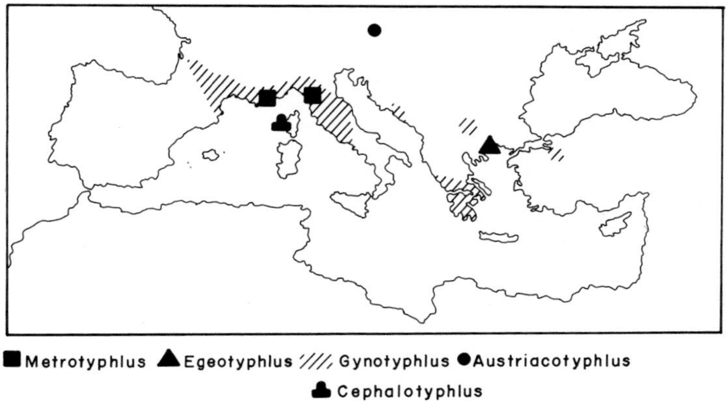

Fig. 2. Distribution of genera of the *Cephalotyphlini* and *Metrotyphlini* tribes. The *Metrotyphlini* tribe has also a genus *(Apotyphlus)* with two species in Chile

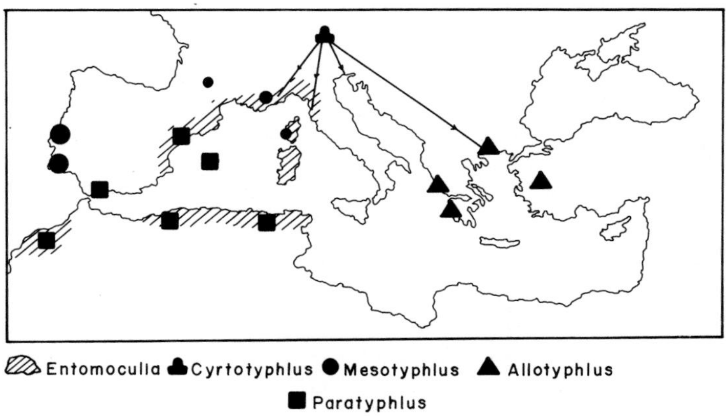

Fig. 3. Distribution of genera of the *Entomoculini* tribe

the western Mediterranean region *(Leptotyphlus, Hesperotyphlus* and *Epalxotyphlus);* b) Turkey, Lebanon *(Kenotyphlus, Eotyphlus);* and c) Africa *(Kilimatyphlus, Sekotyphlus, Afrotyphlus).* The tribe is almost totally euedaphic.

Neotyphlini: a tribe less adapted to the edaphic habitat, having preserved certain characteristics of humus-living species, such as greater size and incomplete welding of the scapular articulation. The tribe has a high number of genera and a low number of species. It is found in Chile (6 genera and 23 species), California (7 genera and 10 species), South Africa (1 genus and 1 species), and the north coast of the Adriatic Sea (1 genus and 2 species) (Fig. 5).

According to the present state of knowledge about the distribution of the subfamily, the following origins and dispersal can be postulated. The group has a Gondwanian origin (Fig. 6), and was divided into two nuclei when the first two big continental blocks were separated. One nucleus, that produced the *Leptotyphlini* line, remained in the continental block that includes the present Africa. The line gave rise to *Kilimatyphlus, Afrotyphlus*

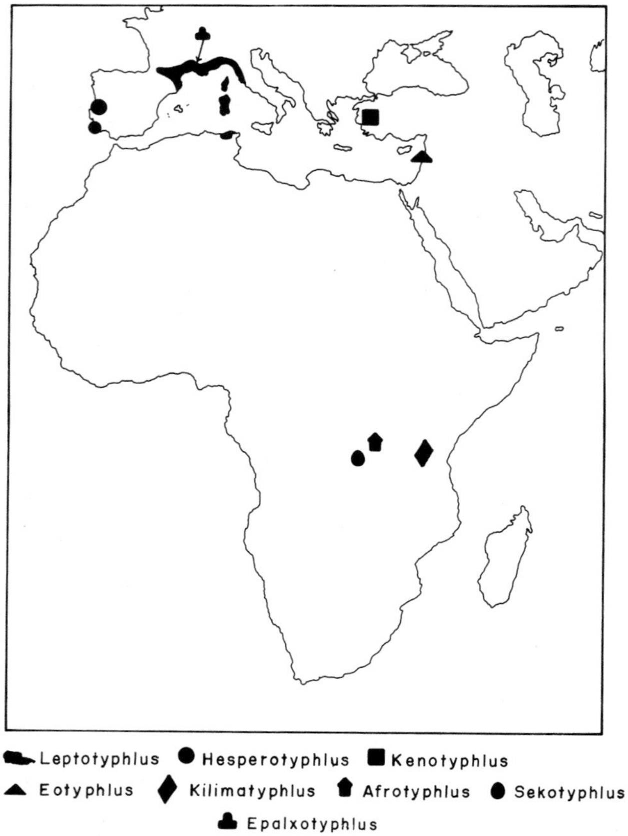

Fig. 4. Distribution of genera of the *Leptotyphlini* tribe

and *Sekotyphlus* in Africa. It then crossed through Ethiopia and Arabia and reached the eastern *(Kenotyphlus* and *Eotyphlus)* and western Mediterranean region *(Leptotyphlus, Hesperotyphlus* and *Epalxotyphlus).* The *Entomoculini,* probably forming part of a previous population wave, have followed a similar path. Due to its greater antiquity, the lack of survival of this group in Africa is understandable. The *Cephalotyphlini* have apparently declined, in the evolutionary sense, and have been supplanted by the *Entomoculini.*

The other nucleus of the Leptotyphlinae remained in the paleantarctic block. This nucleus gave rise to a first distributional wave (*Metrotyphlini*), which, passing through Chile *(Apotyphlus),* must have reached the Californian region, and then later to Europe, where there are at present 4 genera with 7 species. Its apparent present-day absence in

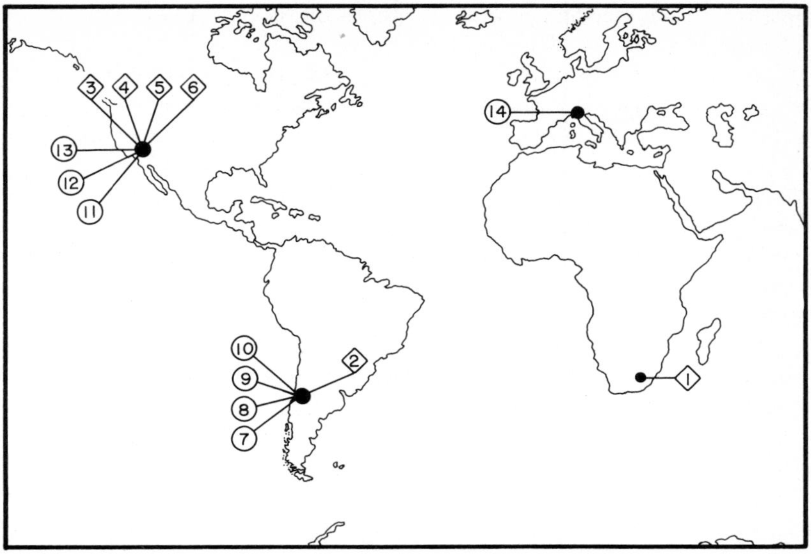

Fig. 5. Distribution of genera of the *Neotyphlini* tribe. Subline A: 1. *Cafrotyphlus*, 2. *Oreotyphlus*, 3. *Homeotyphlus*, 4. *Xenotyphlus*, 5. *Cainotyphlus*, 6. *Telotyphlus;* Subline B: 7. *Chiliotyphlus*, 8. *Eutyphlus*, 9. *Macrotyphlus*, 10. *Paramacrotyphlus*, 11. *Neotyphlus*, 12. *Heterotyphlus*, 13. *Prototyphlus*, 14. *Megatyphlus*

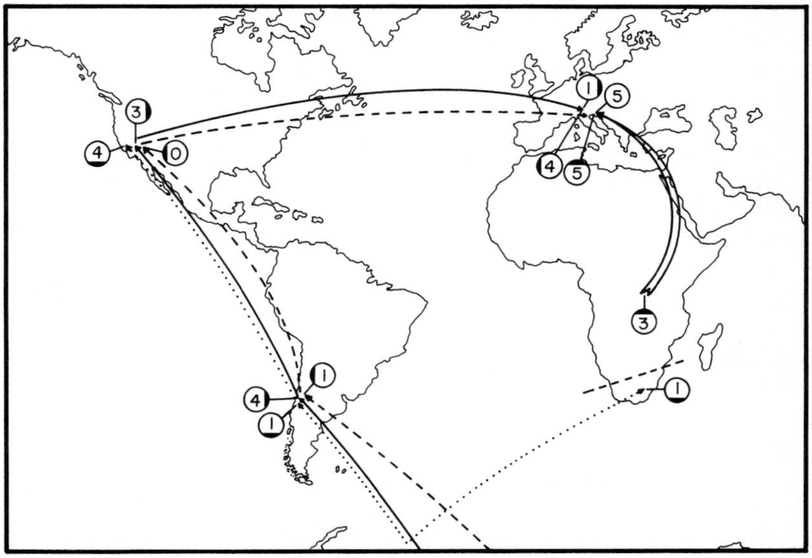

Fig. 6. Map showing the possible origin and the direction of dispersal of the different phylogenetic lines of the Leptotyphlinae. In a circle the number of genera by region and tribe. Solid line: *Neotyphlini* B. Dotted line: *Neotyphlini* A. Dashed line: *Metrotyphlini*. Double line: *Entomoculini* and *Leptotyphlini*

California is probably referable to the lack of relevant faunistic study in the soils of this region. Nevertheless, as an evolutionary line, this group appears to be declining.

A second dispersal wave originated in the Paleantarctic, corresponding to the *Neotyphlini* tribe. It is the most recent one, and the one less adapted to the life in the soil, having the greatest vitality and plasticity. It is possible to observe two sublines in this group; subline A has separated gular sutures and is represented by one genus in South Africa, one in Chile and four in California; subline B has joined gular sutures, and has four genera in Chile, three in California and one in Europe.

Considerations of the diversity and degree of association among genera of this group tend to confirm the conclusions given above, which were based on morphology and distribution.

The greater diversities (Brillouin's index) correspond to the tribe with the largest degree of euryapty *(Neotyphlini)* and therefore to the largest geographical distribution, and to the regions with the most extensive areas of mediterranean climate, where a greater species proliferation has been possible (Table 4). The Mediterranean Basin seems to be the convergent point of several different phylogenetic lines.

Table 4. Geographic distribution of the tribes of the subfamily Leptotyphlinae, and regional and tribal diversity

	Central Africa	Mediterranean region	Chile	California	South Africa	Diversity bits
Cephalotyphlini		X				0.00
Entomoculini		X				1.59
Leptotyphlini	X	X				0.78
Metrotyphlini		X	X			1.43
Neotyphlini		X	X	X	X	2.81
Diversity bits	0.90	2.25	2.54	1.88	0.00	

As an attempt towards a phylogenetic organization of the subfamily, a dendrogram of affinities among genera is given in Fig. 7. It is based on 41 generic characteristics analysed through the "simple matching coefficient" and by the clustering method of SOKAL and SNEATH (1963). Further taxonomic study in California, South Africa and Australia should provide data for establishing, in a more precise way, these relationships.

Meanwhile, our current work is aimed at elucidating the following points:

1) High homogeneity of the *Neotyphlini* tribe, which shows strong relationships among genera distributed in different paleantarctic fragments. The isolation of the European genus *(Megatyphlus)* and its great affinities with the genera of the *Metrotyphlini* tribe point to the strong adaptive convergence recorded in regions with a mediterranean climate.

2) Connection of the *Neotyphlini* and *Metrotyphlini* tribes, through study of forms found in Chile *(Apotyphlus)* and Europe *(Megatyphlus* and *Metrotyphlus)*. This will probably show that the existence of an associated genus in California should be expected, and that a genus with the basic characteristics of *Metrotyphlus* could have been the basis of the differentiation of the tribe in Europe.

3) Connection between the *Metrotyphlini* and *Leptotyphlini* tribes, which is considered to be a product of a convergence through euedaphic adaptation.

4) Connection between *Leptotyphlini* and *Entomoculini* tribes, derived from a com-

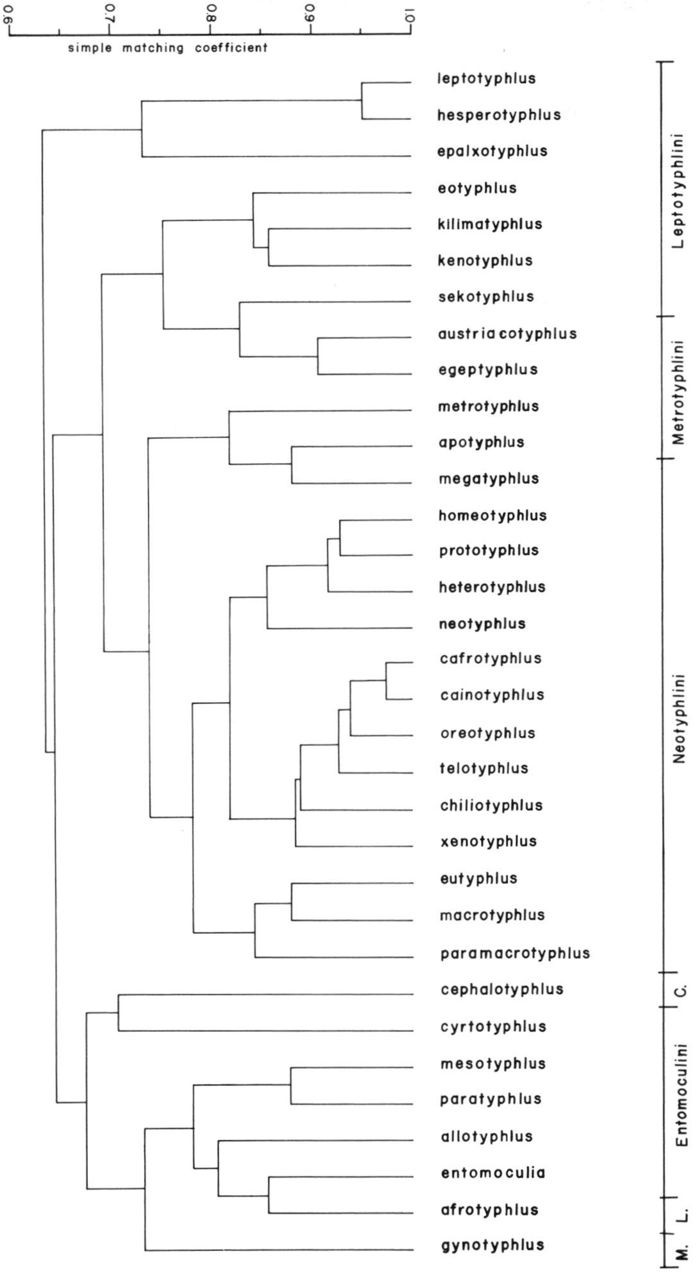

Fig. 7. Dendrogram of affinities among genera of the subfamily Leptotyphlinae

mon origin, from an edaphic adaptation and from a common geographic distribution. A similar relationship can be postulated for the *Cephalotyphlini* and *Entomoculini.*

5) Relations between genera of central East Africa and the group centered in Turkey and Lebanon *(Leptotyphlini),* reaffirming the theory of a migration route through these regions during dispersal from Africa to Europe.

6) Isolation of the *Leptotyphlini* genera located in the western Mediterranean region, showing the termination point of the phylogenetic line.

In these considerations, the genus *Gynotyphlus* is not taken into account, since it is a monospecific and parthenogenetic genus; the number of non-computable elements therefore is very high, due to the absence of male characters.

References

CASTRI, F. DI: Esquisse écologique du Chili. 7–52. In: DELAMARE DEBOUTTEVILLE, C., RAPOPORT, E. (Eds.), Biologie de l'Amérique Australe, Vol. **4.** Paris: C.N.R.S. 1968.

COIFFAIT, H.: Monographie des Leptotyphlites (Col. Staphylinidae). Revue fr. Ent. **26,** 237–437 (1959).

COIFFAIT, H.: Etat actuel de nos connaissances sur la biogéographie de quelques microstaphylinidae (Coleoptera) édaphobies. XI Int. Kongress für Entomologie Wien 1960, **1,** 512–516 (1961).

COIFFAIT, H.: Les Leptotyphlitae (Col. Staphylinidae) de Californie. Revue fr. Ent. **29,** 154–166 (1962).

COIFFAIT, H.: Les Leptotyphlitae (Col. Staphylinidae) du Chili. Systématique et Biogéographie de la sous-famille. 371–383. In: DELAMARE DEBOUTTEVILLE, C., RAPOPORT, E. (Eds.), Biologie de l'Amérique Australe, Vol. **2.** Paris: C.N.R.S. 1963.

COIFFAIT, H.: Un nouveau Leptotyphlite de l'Afrique du Sud. Revue fr. Ent. **31,** 13–17 (1964).

COIFFAIT, H., SÁIZ, F.: Nouveaux Leptotyphlinae (Col. Staphylinidae) du Chili. Rev. Ecol. Biol. Sol.**2,** 129–136 (1965).

SÁIZ, F.: *Nothoesthetus,* nouveau genre humicole et endogé des Euaesthetinae chiliens (Col. Staph.). Bull. Soc. Hist. Nat., Toulouse **105,** 295–310 (1969).

SOKAL, R., SNEATH, P.: Principles of numerical taxonomy. San Francisco-London: Freeman 1963.

2

Biogeography of Pseudoscorpions in the Mediterranean Regions of the World

VALERIA VITALI-DI CASTRI

Introduction

The Pseudoscorpionida (Arachnida) constitute, from a biogeographical point of view, a very interesting faunal group in mediterranean climate regions, namely the Mediterranean Basin, a part of western North America, central Chile and the south of South Africa and Australia. These regions of the world are the areas where the greatest number of species of Pseudoscorpions is concentrated. Though there is lack of data on the abundance of Pseudoscorpion species in the humid tropics, our preliminary observations on tropical materials tend to confirm that the number of species of this Order is greatest in the mediterranean zones, even when compared with the tropical rain forests.

The diversity of habitats occupied by Pseudoscorpions in these zones is remarkably high. They live in the humus layers of the soil, in litter, in mosses and lichens, in the epigeous stratum, in caves, under bark and under stones; they also colonise rodent and bird nests, and show phoresis in flies and in the hair of animals of different groups; finally they are frequently observed in houses. In addition, there are species with very different ecological tolerance, ranging from hygrophilous forms such as the Chthoniidae and some Chernetidae to very xerophilous forms such as the Olpiidae and Menthidae.

Abundance of Species in the Mediterranean Regions

For comparing the number of species in the different mediterranean climate areas, we made a complete bibliographic revision. It is impossible to quote here all the publications dealing with Pseudoscorpion faunas in the mediterranean regions, but particular attention should be drawn, in addition to the general works of BEIER (1932a, 1932b, 1963) and CHAMBERLIN (1931), to the publications of BENEDICT and MALCOM (1970), CHAMBERLIN (1962), HOFF (1958), MUCHMORE (1967, 1968, 1969a, 1969b) and SCHUSTER (1962, 1966a, 1966b, 1966c) for the western part of North America; BEIER (1948, 1954, 1966b, 1967, 1968, 1969) for Australia; BEIER (1947, 1964a, 1966a) for South Africa; and BEIER (1964b) and VITALI-DI CASTRI (1962, 1963, 1965, 1969a, 1969b, 1969c) for Chile. In order to limit conventionally the extension of the mediterranean climate regions, the principles of the UNESCO-FAO bioclimatic map (1963) were adopted.

Fig. 1 shows that of 441 species found in Europe, 269 are present in the Mediterranean zone, from which more than 80% are endemic in this region. Most of them are cave-dwelling species.

In the United States, California has the greatest number of species. On the entire western coast of the United States, 70 species have been described. Only 17 of these are present also in the States with a non-mediterranean climate.

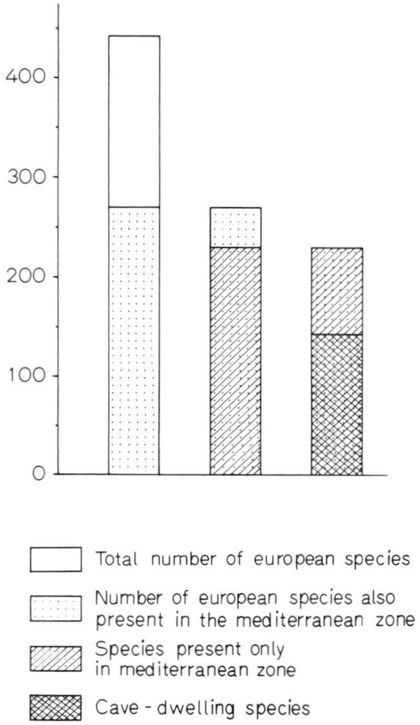

Fig. 1. Number of species of Pseudoscorpions in Europe and in the mediterranean climate zones of Europe

In relation to Chile, Fig. 2 presents the latitudinal sequence of the number of species of Pseudoscorpions from the "Norte Grande" down to the Austral region. It can be clearly seen that their highest concentration is in the Central Zone of Chile, with a true mediterranean climate, followed by the "Norte Chico", with an arid mediterranean climate. These two zones are inhabited by many more species than the Austral Zone or the "Norte Grande", even though these former zones are less large in area. Of the 40 species that have been found to date in the Chilean mediterranean zone, only 14 live also in zones with a different climate.

A somewhat similar situation is also seen in South Africa and Australia. In the first region there are 20 mediterranean species among a total of 46; in the second one, 28 among 73.

To explain this mediterranean concentration, as well as the accentuated endemism of the mediterranean species of Pseudoscorpions, hypotheses based on present or historical factors can be advanced.

The mediterranean regions show a high topographical and vegetational heterogeneity, so that different species with diverse ecological tolerances might coexist in these regions. For example, in "Quebrada de la Plata" (near to Santiago), in a very reduced area, there are hygrophilous, mesophilous, and even very xerophilous plant formations (see SCHLEGEL, 1966). The same phenomenon occurs in almost all the small valleys of the Central Zone and "Norte Chico".

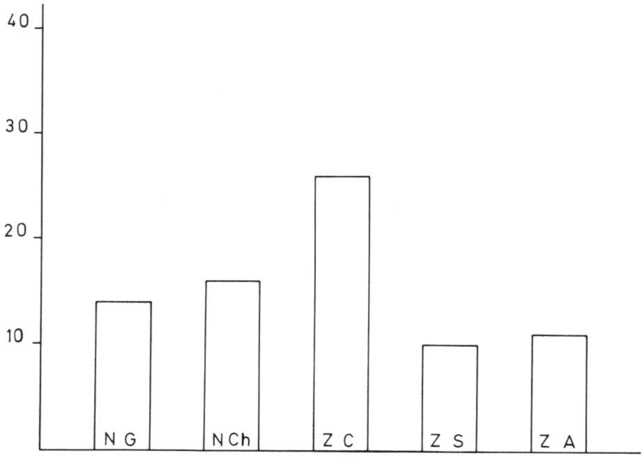

Fig. 2. Number of species of Pseudoscorpions in different zones of Chile, from North to South. NG: Great North ("Norte Grande"). NCh: Little North ("Norte Chico"). ZC: Central Zone. ZS: Southern Zone. ZA: Austral Zone

Historically, the paleoclimatic changes and the cliseric shifts of vegetation facilitated the advance of a series of hygrophilous lines up to the central zone of Chile. During the withdrawal of plant formations toward the south, due to other climatic changes, hygrophilous relicts remained in all the small valleys, widely separated one from each other, and in the heights of the Coastal Range. The geographical isolation and the effect of small populations have permitted the fragmentation, in many species, of previously homogeneous phylogenetic groups (VITALI-DI CASTRI, 1963).

This same evolutionary phenomenon, but with even more marked effects, can be observed in the European Mediterranean zone, where segregation occurred particularly in caves. Many elements had to search for the cave environment, because of its greater thermic and hygric stability. These groups have afterwards remained completely separated from one another.

Another occurrence frequently observed in the mediterranean zones of Chile, an occurrence which according to HOFF (1959) is very unusual in other regions, is the high intrageneric sympatry. For example, in Fray Jorge (Norte Chico) there are numerous species of the *Austrochthonius* genus living together. In Paposo (perarid mediterranean zone of the Norte Grande according to DI CASTRI, 1968), nearly all the species of the subfamily *Cheiridiinae* are coexisting, while in the southern part of the country these species are isolated, and in general are not sympatric. This finding could also be explained by historical reasons; probably the *Cheiridiinae* had a continual distribution from the southern Valdivian region up to Fray Jorge – Paposo. When the climate became more arid, these hygrophilous elements, already differentiated into diverse species, progressively concentrated in a few refuges (the forests or close bushes still hygrophilous because of coastal fogs, as in Fray Jorge and Paposo, or because of superficial phreatic bands as in Til-Til).

As a final example, attention is drawn to the subfamiliy *Gymnobisiinae*, a group that in spite of its paleoantarctic origin, shows in the mediterranean zone of Chile its greatest fragmentation in species and a superspecialization, with progressive increase of abnormal characters of sexual dimorphism (VITALI-DI CASTRI and DI CASTRI, 1970).

It is evident that for a better understanding of the processes of this sympatry, more detailed research on the ecological niche of these species and their strategy for niche delimitation in the different mediterranean climate zones should be undertaken.

In the light of preliminary observations, some possible evolutionary strategies are:

– Vertical migrations from the under-bark environment to the soil and vice versa, in order to find the most suitable microclimatic conditions in different seasons. This can be done in different development phases (for instance, larvae in soil, adults in surface);

– Habitat turnover or specialization to a narrow habitat;

– Food specialization or plasticity in relation to prey;

– Well defined altitudinal preferences, as in some *Gymnobisiinae;*

– Strict specialization to the under-bark environment, as in *Gigantochernes;*

– Strict specialization to the under-stone environment, as in Menthidae and Ideoroncidae;

– Cave specialization, this adaptation being very common in Europe, present also in other çontinents (North America, Australia) but not yet found in South America;

– Specialization to very arid habitats, as in Olpiidae.

Faunistic Relationships of the Five Regions with a Mediterranean Climate

A first biogeographical comparison has been made between the five regions of the world where zones of mediterranean climate exist. These are the Mediterranean Sea Basin (including also Morocco and Portugal), the western coast of North America, the south of South Africa, the southern strip of Australia and the Chilean mediterranean zone.

In these five mediterranean regions analysis is made of:

– the number of families, subfamilies or tribes, genera and species present;

– the "mediterranean endemism", that is the percentage of species that at present are found only in mediterranean climate conditions, and not under any other climate;

– the affinities among the Pseudoscorpion faunas of these five regions at the level of families, subfamilies, genera and species.

Fig. 3 shows the number of families, subfamilies, genera and species for each mediterranean region. A first consideration is that in general the regions of the austral hemisphere have a smaller number of families and subfamilies relative to the boreal zones, thus suggesting a very ancient isolation between the boreal and austral faunas.

A large number of genera and species exists in the Mediterranean Sea Basin regions. It might be argued that this is referable to the more intensive research that has been undertaken on the group in Europe. However, it seems that this difference is due mainly to the fundamental role of caves in promoting species segregation in Europe.

Also the greatest "mediterranean endemism" corresponds to Europe (80% of endemic species), while the smallest is observed in Chile (60%). This fact is easy to explain in Europe, evoking previous considerations about the cave life. In Chile, the fact that many species present in mediterranean zones are also found in the more humid southern regions is probably due to the paleoantarctic origin of most of these species.

In order to elucidate the affinities between the five regional faunas, use was made of the association coefficient "simple matched coefficient", derived from methods of numerical taxonomy (SOKAL and MICHENER, 1958; SOKAL and SNEATH, 1963). The affinities between the five regions were calculated at different taxa levels: family, sub-

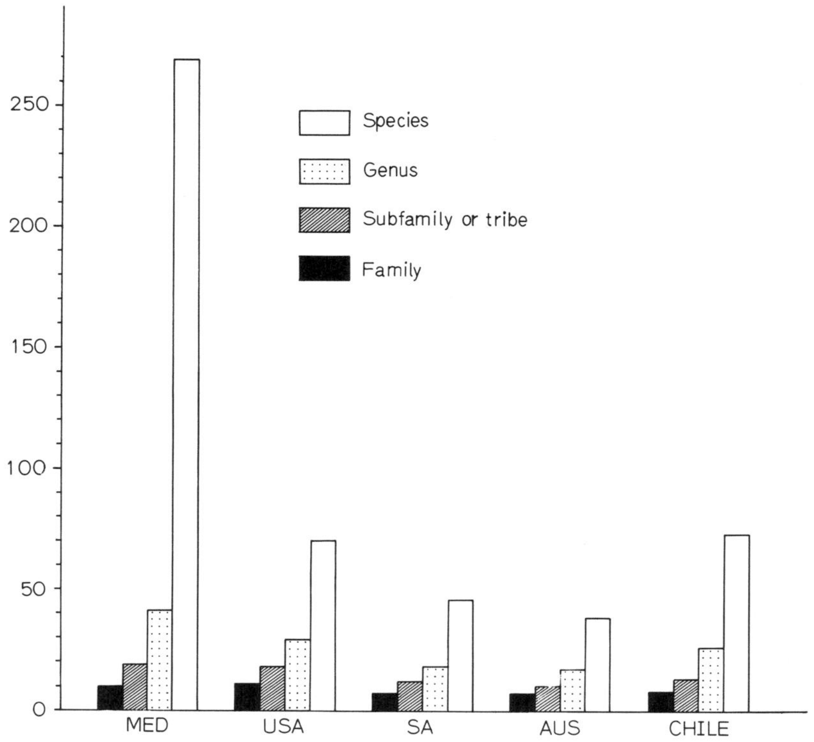

Fig. 3. Number of different taxa of Pseudoscorpions in the five regions of the world with mediterranean climate. MED: regions around the Mediterranean Sea, plus Morocco and Portugal. USA: western coast of United States of America. SA: southern part of South Africa. AUS: southern Australia. CHILE: mediterranean climate zones of Chile

family or tribe, genus and species. The degree of affinity shown in Fig. 4 and 5 is directly proportional to the thickness of the intercontinental lines. The lowest affinity values are not shown in these figures.

Affinity between regions decreases from family to subfamily or tribe and to genus; it disappears completely at the species level. In fact, there is no species common to these regions, except for some cosmopolitan species such as *Cheiridium museorum, Chelifer cancroides* and *Withius subruber,* species which were probably carried by man.

Analysing affinities at the family level, Fig. 4 shows that there are two areas of high affinity: the regions of the Mediterranean Sea Basin and the United States on the one hand, the austral parts of Africa, South America and Australia on the other. Only four families are present in all the five mediterranean climate regions: Chthoniidae, Olpiidae, Garypidae and Cheiridiidae.

Affinities at the genus, but not the subfamily, level are shown here (Fig. 5). Very high affinities exist between Australia, South Africa and Chile, thus confirming the paleo-antarctic nature of these links. The lowest affinity at this taxa level is between the Mediterranean Sea region (including North Africa) and South Africa, in spite of the present territorial continuity. In California the genera of holarctic and paleoantarctic origin seem to be mixed in a relatively balanced way.

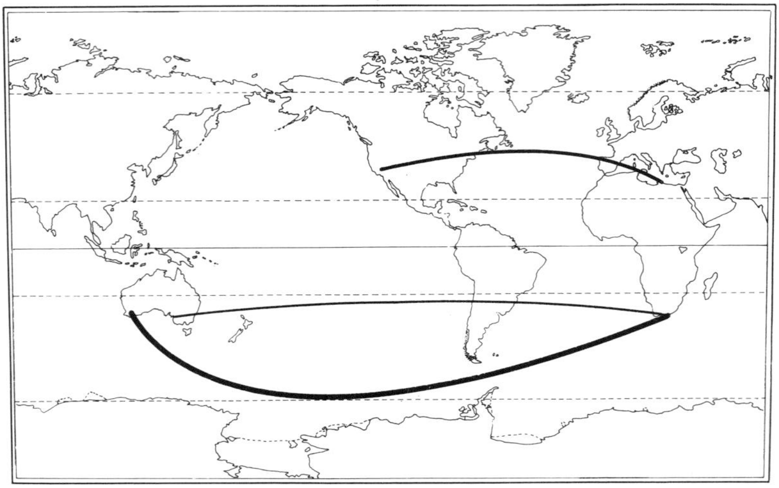

Fig. 4. Principal affinities among Pseudoscorpion families in regions with mediterranean climate. The degree of affinity is directly proportional to the thickness of lines

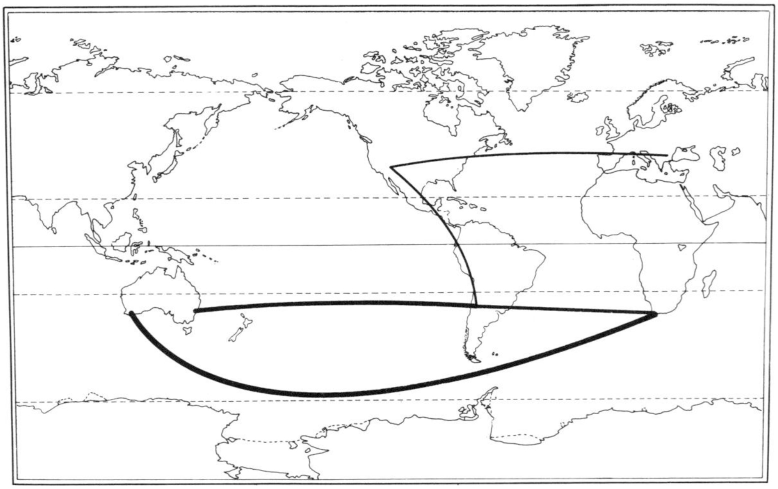

Fig. 5. Principal affinities among Pseudoscorpion genera in regions with mediterranean climate. The degree of affinity is directly proportional to the thickness of lines

Biogeographic Heterogeneity of the Mediterranean Faunas

As a general conclusion to this chapter, attention might be drawn to the heterogeneity, both from a biogeographical and an ecological viewpoint, of the different elements integrating the Pseudoscorpion faunas in the mediterranean regions. To further illustrate this point, the probable origin and biogeographic characteristics of some different lines of Pseudoscorpions living in the mediterranean zone of Chile are given below. It must be admitted that a certain degree of speculation is required to infer the history of these lines from their present known distribution.

a) a most ancient group, such as the Menthidae (Fig. 6), exhibiting now a very fragmentary relict distribution restricted to regions which at present have a mediterranean climate (California, Chile, Israel) and to their more adjacent desertic areas (VITALI-DI CASTRI, 1969 a).

b) probably pantropical elements, such as the *Cheiridiinae,* perhaps originated from the tropical nucleus of *Pseudochiridiinae* (Fig. 7) through an evolutionary process of progressive neoteny (VITALI-DI CASTRI, 1966, 1970 a).

c) many elements of the family Olpiidae, which is dominant in all the desertic zones of the world, are largely represented as well in the five regions with a mediterranean climate, mainly in correspondence with xerophilous formations (arid areas and sun-exposed slopes).

d) lines of a paleoantarctic origin, dominant in all the hygrophilous plant formations

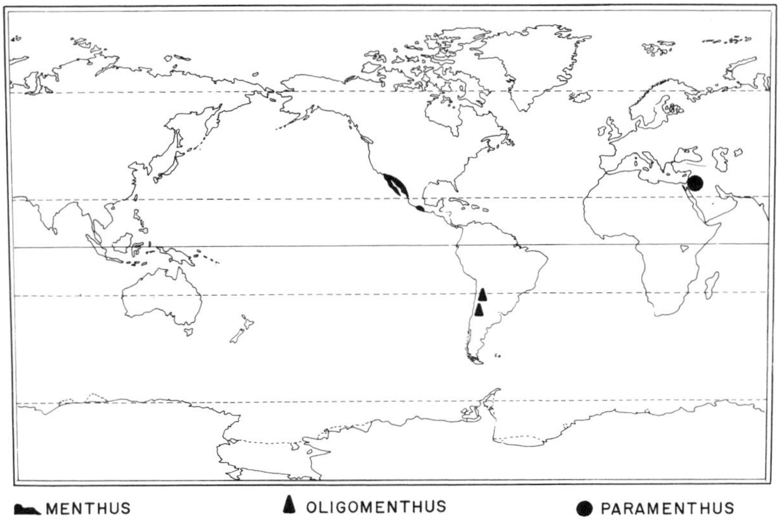

Fig. 6. World distribution of the family Menthidae

of southern Chile, but very abundant also in the soils of the sclerophyllous forests and even of the open mediterranean woodlands of central Chile. The present patterns of geographic distribution of these different lines fit well with the modalities and timing of repartition proposed by JEANNEL (1967). In this sense, the genus *Austrochthonius* is present in all the austral regions (Fig. 8): Australia and New Zealand, Chile and southern Argentina, some subantarctic islands such as the Crozet archipelago (VITALI-DI CASTRI, 1968) and South

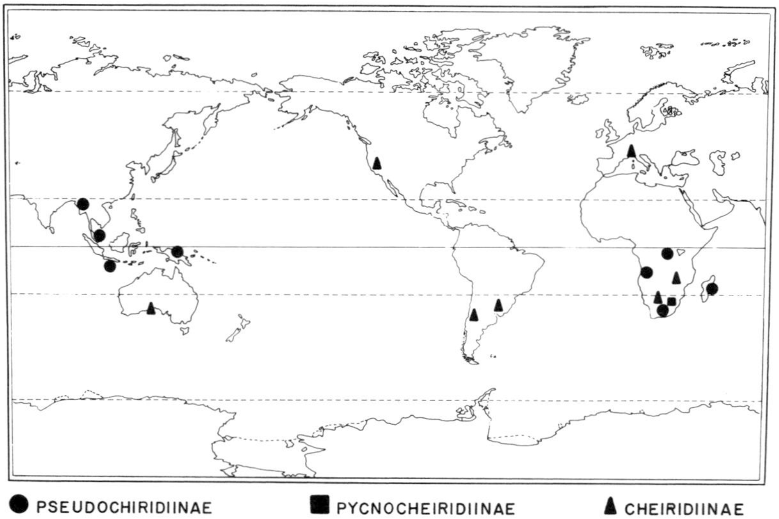

Fig. 7. World distribution of the three subfamilies of Cheiridiidae

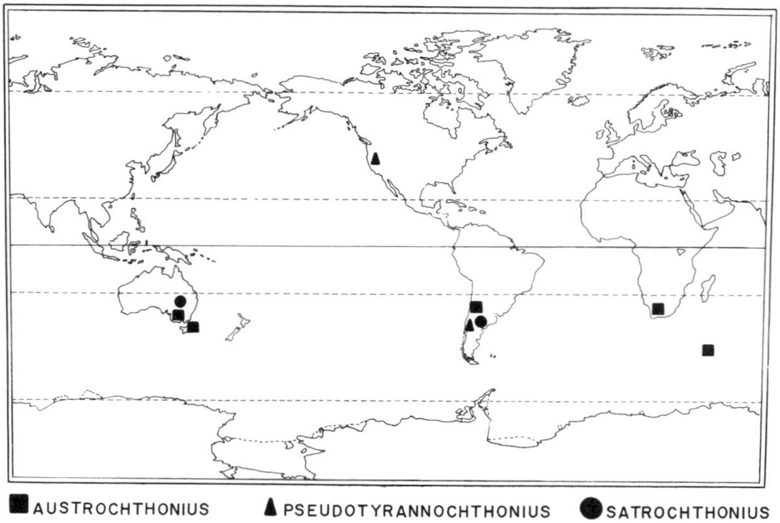

Fig. 8. World distribution of three genera of Chthoniidae

Africa (in fact, the *Paraustrochthonius* of South Africa is likely to be a true *Austrochthonius* as indicated by VITALI-DI CASTRI, 1968); this kind of distribution allows the postulation of a very ancient cretaceous origin (JEANNEL, *op. cit.*); our discovery of *Austrochtho-*

nius in southern Brazil makes the distribution of this genus comparable to that of the *Araucaria* (including its fossil findings). On the other hand, the genus *Satrochthonius*, known from the australian region and found also in Chile, is absent in South Africa (Fig. 8),

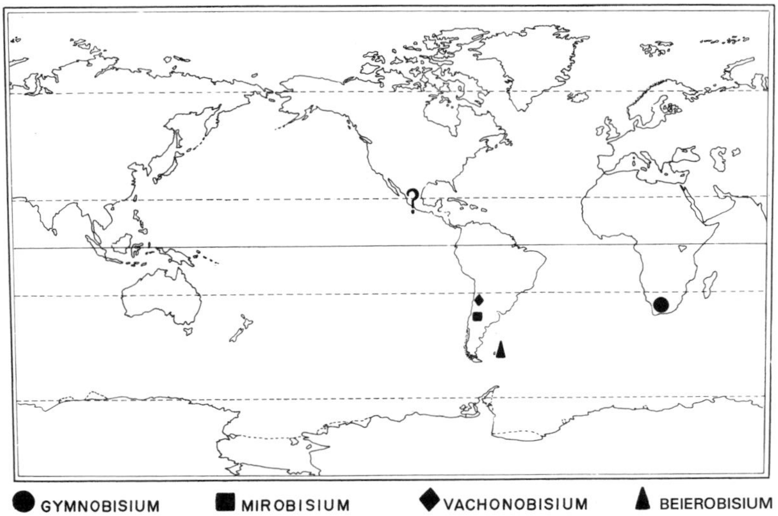

Fig. 9. World distribution of the four genera of the subfamily *Gymnobisiinae* (Vachoniidae). The question mark signifies the presence in Mexico of two genera *(Vachonium* and *Paravachonium)* of the subfamily *Vachoniinae*, whose degree of systematic affinity with the *Gymnobisiinae* is not clear

thus indicating a more recent cretaceous origin; its present distribution seems to follow that of *Nothofagus*. Finally, the subfamily *Gymnobisiinae* is distributed in South Africa, the Falkland Islands (VITALI-DI CASTRI, 1970b), southern Argentina and Chile (VITALI-DI CASTRI, 1963, 1969c) but is lacking in Australia and New Zealand (Fig. 9); this should correspond to a repartition of this line during the Tertiary through an hypotetical subatlantic bridge between austral Africa and austral South America (JEANNEL, *op.cit.)*; while presenting the highest number of species in the mediterranean region of Chile, morphological and phylogenetic considerations exclude that the *Gymnobisiinae* have had their origin in this region (VITALI-DI CASTRI and DI CASTRI, 1970).

e) elements of amphitropical distribution (Chile and western North America), such as *Pseudotyrannochthonius* (Fig. 8); this genus could represent a relict of a paleoantarctic line having migrated toward the north along the Andes.

f) some neotropical elements, such as *Gigantochernes*, found only in the sclerophyllous forests of central Chile and in the subtropical woodlands of northern Argentina, Paraguay and southern Brazil (VITALI-DI CASTRI, 1972).

g) endemics restricted to the mediterranean zone of Chile, for instance *Drepanochthonius* and several species of Chernetidae, whose affinities are not clear.

h) cosmopolitan species carried by man, in Chile represented only by *Withius subruber*.

References

BEIER, M.: Pseudoscorpionidea I. Subord. Chthoniinea et Neobisiinea. 1–258. In: Das Tierreich. Berlin–Leipzig: Walter de Gruyter & Co. 1932 a.

BEIER, M.: Pseudoscorpionidea II. Subord. C. Cheliferinea. 1–294. In: Das Tierreich. Berlin-Leipzig: Walter de Gruyter & Co. 1932 b.

BEIER, M.: Zur Kenntnis der Pseudoscorpionidenfauna des südlichen Afrika, insbesondere der süd-west- und südafrikanischen Trockengebiete. "EOS", Rev. Esp. Ent. **23**, 285–339 (1947).

BEIER, M.: Über Pseudoscorpione der australischen Region. "EOS", Rev. Esp. Ent. **24**, 525–562 (1948).

BEIER, M.: Report from Prof. T. GISLÉN's expedition to Australia in 1951–1952. 7. Pseudoscorpionidea. Lund Univ. Ars. **50** (3), Kungl. Fysiogr. Sällsk. Handl. **65** (3), 3–26 (1954).

BEIER, M.: Ordnung Pseudoscorpionidea (Afterscorpione). 1–313, Lief. 1. In: Bestimmungsbücher zur Bodenfauna Europas. Berlin: Akademie-Verlag 1963.

BEIER, M.: Weiteres zur Kenntnis der Pseudoscorpioniden-Fauna des südlichen Afrika. Ann. Natal Mus. **16**, 30–90 (1964 a).

BEIER, M.: Die Pseudoscorpioniden-Fauna Chiles. Ann. Naturhistor. Mus. Wien, **67**, 307–375 (1964 b).

BEIER, M.: Ergänzungen zur Pseudoscorpioniden-Fauna des südlichen Afrika. Ann. Natal Mus. **18**, 455–470 (1966 a).

BEIER, M.: On the Pseudoscorpionidea of Australia. Aust. J. Zool. **14**, 275–303 (1966 b).

BEIER, M.: Some Pseudoscorpionidea from Australia, chiefly from caves. Austr. Zool. **14**, 199–205 (1967).

BEIER, M.: Some cave-dwelling Pseudoscorpionidea from Australia and New Caledonia. Rec. South Austr. Mus. **15**, 757–765 (1968).

BEIER, M.: Neue Pseudoskorpione aus Australien. Ann. Naturhistor. Mus. Wien, **73**, 171–187 (1969).

BENEDICT, E. M., MALCOLM, D. R.: Some pseudotyrannochthoniine false scorpions from western North America (Chelonethida: Chthoniidae). J. New York Entom. Soc. **78**, 38–51 (1970).

CASTRI, F. DI: Esquisse écologique du Chili. 7–52. In: DELAMARE DEBOUTTEVILLE, C. and RAPOPORT, E. (Eds.), Biologie de l'Amérique Australe. Vol. IV. Paris: C. N. R. S. 1968.

CHAMBERLIN, J. C.: The Arachnid Order Chelonethida. Stanford: Stanford University Press 1931.

CHAMBERLIN, J. C.: New and little-known false scorpions, principally from caves, belonging to the families Chthoniidae and Neobisiidae (Arachnida, Chelonethida). Bull. Amer. Mus. Nat. Hist. **123**, 299–352 (1962).

HOFF, C. C.: List of the pseudoscorpions of North America North of Mexico. Amer. Mus. Novitates No. 1875, 1–50 (1958).

HOFF, C. C.: The ecology and distribution of the pseudoscorpions of north-central New Mexico. Univ. New Mex. Publ. Biol. (8), 1–68 (1959).

JEANNEL, R.: Biogéographie de l'Amérique Australe. 401–460. In: DELAMARE DEBOUTTEVILLE, C. and RAPOPORT, E. (Eds.), Biologie de l'Amérique Australe. Vol. III. Paris: C.N.R.S. 1967.

MUCHMORE, W. B.: Pseudotyrannochthoniine pseudoscorpions from the western United States. Trans. Amer. Microsc. Soc. **86**, 132–139 (1967).

MUCHMORE, W. B.: Two new species of chthoniid pseudoscorpions from the western United States (Arachnida: Chelonethida: Chthoniidae). Pan-Pacific Entom. **44**, 51–57 (1968).

MUCHMORE, W. B.: The pseudoscorpion genus *Neochthonius* CHAMBERLIN (Arachnida, Chelonethida, Chthoniidae) with description of a cavernicolous species. Amer. Midl. Natur. **81**, 387–394 (1969 a).

MUCHMORE, W. B.: New species and records of cavernicolous pseudoscorpions of the genus *Microcreagris* (Arachnida, Chelonethida, Neobisiidae, Ideobisiinae). Amer. Mus. Novitates No. 2392, 1–21 (1969 b).

SCHLEGEL, F.: Pflanzensoziologische und floristische Untersuchungen über Hartlaubgehölze im La Plata-Tal bei Santiago de Chile. Ber. d. Oberhess. Ges. f. Natur- und Heilkunde Giessen, Neue Folge **34**, 183–204 (1966).

SCHUSTER, R. O.: New species of *Kewochthonius* CHAMBERLIN from California (Arachnida: Chelonethida). Proc. Biol. Soc. Wash. **75**, 223–226 (1962).

SCHUSTER, R. O.: A new species of *Allochthonius* from the pacific northwest of North America (Arachnida: Chelonethida). Pan-Pacific Entom. **42**, 172–175 (1966a).

SCHUSTER, R. O.: New species of *Apochthonius* from western North America (Arachnida: Chelonethida). Pan-Pacific Entom. **42**, 178–183 (1966b).

SCHUSTER, R. O.: New species of *Parobisium* CHAMBERLIN (Arachnida: Chelonethida). Pan-Pacific Entom. **42**, 223–228 (1966c).

SOKAL, R. R., MICHENER, C. D.: A statistical method for evaluating systematic relationships. Univ. Kansas Sci. Bull. **38**, 1409–1438 (1958).

SOKAL, R. R., SNEATH, P. H. A.: The principles of numerical taxonomy. San Francisco and London: W. H. Freeman Co. 1963.

UNESCO-FAO: Carte bioclimatique de la zone méditerranéenne: notice explicative. Recherches sur la zone aride XXI. Paris: UNESCO 1963.

VITALI-DI CASTRI, V.: La familia Cheiridiidae (Pseudoscorpionida) en Chile. Inv. Zool. Chilenas **8**, 119–142 (1962).

VITALI-DI CASTRI, V.: La familia Vachoniidae (= Gymnobisiidae) en Chile (Arachnidea, Pseudoscorpionida). Inv. Zool. Chilenas **10**, 27–82 (1963).

VITALI-DI CASTRI, V.: *Cheiridium danconai* n. sp. (Pseudoscorpionida) con consideraciones sobre su desarrollo postembrionario. Inv. Zool. Chilenas **12**, 67–92 (1965).

VITALI-DI CASTRI, V.: Observaciones biogeograficas y filogenéticas sobre la familia Cheiridiidae (Pseudoscorpionida). 379–386. In: Progresos en biología del suelo. Montevideo: UNESCO 1966.

VITALI-DI CASTRI, V.: *Austrochthonius insularis,* nouvelle espèce de Pseudoscorpions de l'archipel de Crozet (Heterosphyronida, Chthoniidae). Bull. Mus. Hist. nat. **40**, 141–148 (1968).

VITALI-DI CASTRI, V.: Remarques sur la famille des Menthidae (Arachnida Pseudoscorpionida) à propos de la présence au Chili d'une nouvelle espèce, *Oligomenthus chilensis*. Bull. Mus. Hist. nat. **41**, 498–506 (1969a).

VITALI-DI CASTRI, V.: Tercera nota sobre los Cheiridiidae de Chile (Pseudoscorpionida) con descripción de *Apocheiridium (Chiliocheiridium) serenense* n. subgen., n. sp. Bol. Soc. Biol. Concepción **41**, 265–280 (1969b).

VITALI-DI CASTRI, V.: Revisión de la sistemática y distribución de los *Gymnobisiinae* (Pseudoscorpionida, Vachoniidae). Bol. Soc. Biol. Concepción **42**, 123–135 (1969c).

VITALI-DI CASTRI, V.: *Pseudochiridiinae* (Pseudoscorpionida) du Muséum National d'Histoire Naturelle. Remarques sur la sous-famille et description de deux nouvelles espèces de Madagascar et d'Angola. Bull. Mus. Hist. nat. **41**, 1175–1199 (1970a).

VITALI-DI CASTRI, V.: Un nuevo género de *Gymnobisiinae* (Pseudoscorpionida) de las islas Malvinas. Revisión taxonómica de la subfamilia. Physis **30**, 1–9 (1970b).

VITALI-DI CASTRI, V.: El género sudamericano *Gigantochernes* (Pseudoscorpionida, Chernetidae) con descripción de dos nuevas especies. Physis **31**, 23–38 (1972).

VITALI-DI CASTRI, V., DI CASTRI, F.: L'évolution du dimorphisme sexuel dans une lignée de Pseudoscorpions. Bull. Mus. Hist. nat. **42**, 382–391 (1970).

3

Parallel Evolution and Bird Niches

Martin L. Cody

Introduction

The chief aim of this chapter is to examine the extent to which bird niches show similarities in response to parallel selective forces. The problem can be approached at various levels, for parallel evolution can be demonstrated within a broad range of sample sizes or units on each of two variables. The first is a gradient of decreasing proximity of habitat sites, so that initially comparisons may be made within habitats, then between adjacent habitats, and extended to comparisons on a continental and finally an intercontinental scale. A second axis ranks various sample sizes of numbers of bird species used in these comparisons. The comparisons may be made at the level of the single species, of groups of related or coexisting species, of whole communities or of the complete avifauna. The particular emphasis of this review is the comparison of communities, a grouping of species at an intermediate to higher level of organization, between continents, the broadest possible scaling of habitat isolation. At these levels parallel evolution is at its most dramatic, for the former choice extends the matching of individual species niches to include the relative juxtaposition of these niches in the community, their number and shape, and the latter condition assures us that the species in these communities so compared will bear minimal taxonomic affiliation to each other and thus minimal resemblance due to factors other than parallel selection. While comparisons at different levels of species organization and geographic separation will be treated in turn as available information allows, the intercontinental community comparisons will be given special attention. I have studied several sorts of avian communities, especially in North and South America; this information will be presented and the community organization contrasted. In particular, the Californian and Chilean bird communities of broad-leaf sclerophyll scrub will be discussed.

Convergence at the Level of the Species

I. Replacement of Species on a Local Scale

1. Within-habitat Replacements

First mention will be given the few instances in which species replacement within habitats show apparent effects of convergent evolution. However, such cases are speculative, as most such replacement pairs are closely related, and we seek to explain not so much a convergence in appearance and behavior as an exceptional lack of divergence. In grasslands of North America, the two sibling species of meadowlarks *Sturnella magna* and *S. neglecta* occupy geographic ranges with only marginal overlap (stasipatry), the former in the east and the latter in the west. The two co-occur over a narrow strip of territory from Texas to Ontario; no hybridization occurs, for the species are reproductively isolated (LANYON, 1957). However, little or no divergence has occurred in feeding behavior and morphology. The lack of divergence seems tantamount to parallel evolution, a less convincing example of the same phenomenon, especially when the same appearance and behavior has been evolved in unrelated grassland species elsewhere (see below; the meadowlark niche will be a *leitmotiv* of the chapter). Likewise the grasshopper sparrow *Ammodramus savannarum* is replaced by its congener Baird's sparrow, *A. bairdii,* in the northwestern part of its range in middle North America. The two are virtually identical in ecology and behavior, and very similar in appearance (as are many grassland sparrows). These cases are further complicated, beyond the taxonomic factor, by the fact that the two species pairs are interspecifically territorial where they meet (discussed at length in CODY, 1969). Thus there are three selective forces promoting similarity in appearance: a) common ancestry, b) common niche characteristics and c) selection for a common signal (appearance and/or voice) used in territory defense. In other finch genera, such as *Junco, Aimophila, Amphispiza* and *Pipilo,* the same factors are probably operating, and the Chilean species of *Spinus* and *Phrygilus* may also be included.

Below discussion will be restricted to species pairs in which the second factor in phenotype convergence, that of niche coincidence or near-coincidence, can be singled out. But it should first be stressed that there are many taxonomically related species pairs and groups which fail to show divergence in appearance although they have become reproductively isolated (including the so-called sibling species). The genera *Empidonax* and *Myiarchus* are represented in North America by usually not more than one species in any one habitat. Within each of the genera the species pose problems in field identification, and in the former genus at least species may be interspecifically territorial (JOHNSON, 1963); the problem in isolating the various selective forces complicates such examples.

2. Species Replacements between Habitats

The phenomenon of parallel or even convergent evolution can be seen locally in the replacement of species among habitats in the same geographic area. Thus the role of digging into the ground or leaf litter for insects is occupied in North America by thrashers (Aves: Mimidae). Although classified into several species and two genera *(Toxostoma, Oreoscoptes),* the thrashers are similar in body size and build with sturdy legs and long tails; with the exception of two species which occupy drier country with perhaps harder ground (*O. montanus* and *T. bendirei*), all possess long, decurved bills to dig for their food. The brown thrasher *T. rufum* of the east is replaced in California chaparral by *T. redivivum,* in the Mojave Desert by *T. lecontei,* in the Lower Sonoran Desert by *T. curvirostre* and in the mesquite washes of the southwest by *T. dorsale.* The long-billed thrasher *T. longirostre* is found only in the lower Rio Grande valley. None occurs in forests, but all are typical of brushy habitats. Apparently the adaptations to the feeding niche described are common to several variations of brushy habitat, and largely override the differential habitat-specific selection to which each species is exposed. Woodpeckers (Aves: Picidae) and hummingbirds (Aves: Trochilidae), amongst many other possibilities, provide similar examples of species groups in which a turnover in species names but not in species ecological role occurs with a change in habitat. We might say of such examples that the niche is not confined to a single habitat type, but remains relatively unaltered with minor or even considerable changes in habitat structure and geographic location.

II. Intercontinental Species Replacements

1. Historical Aspects

When replacement or turnover amongst ecologically equivalent species takes place between taxonomically unrelated species in habitats which themselves are floristically unrelated and geographically isolated, the results are considerably more convincing. We are convinced, that is to say, that the component of natural selection which is comprised by habitat is dominant in producing the convergence, and that genealogy and history play but minor roles.

The first indication that species on different continents could become very similar in appearance and gross morphology in response to selection for parallel niches came from the descriptions of the old systematists. When biological exploration took place in new worlds, the material was received by taxonomists whose basis of comparison was their own civilized and familiar species quota. What could be more natural and, indeed, what a fine *ex tempore* test of the convergent evolution concept, than that they should name the new forms, on the basis of gross morphology, after the closest local types? Of course, it is just this gross morphology which so readily becomes adapted by natural selection to a particular niche, and structurally similar habitats provide similar ingredients for specific niches. For the sake of science and to the detriment, perhaps, of an ecologically-oriented traveller, the true taxonomic affiliations of these exotic species were eventually revealed; feathers were counted, bones numbered and muscles displayed, and new and different families and genera were discovered to have come to resemble and replace in distant lands the local fauna. Thus LINNAEUS (1758) first classified the North American warblers after the Old World warblers he knew from Sweden (Sylviidae), rather than a distinct family (Parulidae). Likewise the North American mimic thrushes Mimidae were classified with the Old World thrushes Turdidae, its vultures (Carthartidae) with the Old World vultures (Aegypiidae) and its blackbirds (Icteridae) with the Old World Blackbirds (also Turdidae). In this way LINNAEUS inadvertently and unknowingly came across one off the great cases of convergence. A species from the North American prairies was classified as a lark *Alauda magna* (Alaudidae) by CATESBY, and LINNAEUS concurred. It is found, said LINNAEUS, not only in America but also in Africa! BUFFON (1812) was sceptical of such a range: "LINNAEUS asserts that it also occurs in Africa". In fact two species were involved, and neither are larks. The American bird is the meadowlark *Sturnella magna* (Icteridae), and we recognize the other as the African pipit *Macronyx croceus* (Motacillidae), whose similarity to *Sturnella* is now legendary. Taxonomists of that time had much difficulty with the meadowlark, as BUFFON *(op. cit.)* discusses. LAWSON and CATESBY, followed by LINNAEUS, called it a lark, BRISSON thought it a blackbird *(Turdus:* Turdidae), and BUFFON thought it was obviously neither. PENNANT called it a "stare" *(Sturnus:* Sturnidae), and LATHAM a "Louisiana stare", which may be intended, by the description, to refer to the bobolink *(Dolichonyx:* Icteridae) and hence comes closest to the true affinities.

Similar confusion, albeit entertaining and even educational confusion, reigned when the systematists began naming South American birds. To one familiar with the habits and appearance of European and New World species, all the "right" mistakes were made. In Chile, for example, many of the small passerine niches are occupied by flycatchers Tyrannidae and ovenbirds Furnariidae, families restricted to the New World and dominant in South America. Many tyrannids were named *Muscicapa* after the Old World flycatchers Muscicapidae, open country forms of both families (*Lessonia, Geositta, Cinclodes)* named *Alauda, Motacilla* or *Anthus* after common open country birds in the lark and pipit families in the Old World, and foliage-hunting forms *(Tachuris, Phleocryptes)* called *Sylvia* after the Old World warblers. The designation of *Eugralla* (in the endemic South American family Rhinocryptidae) as *Troglodytes* (a wren) is particularly apt. These examples could be multiplied at great length, but the point has been made. Works calling attention to these overall similarities between intercontinental species and familial replacements are numerous, and include FRIEDMAN (1946), MOREAU (1966), LACK (1968, 1969), DUBOST (1968) and FRY (1970).

2. *Species of Simple Habitats*

Many examples of convergent evolution may be drawn from species which occupy simple, like habitats. This is because a) such habitats may be compared and matched with accuracy, b) numbers of species, and hence the possibilities for niche rearrangements and alternative juxtapositions, are few. Grasslands, deserts and tundra are examples of habitats which from the avian ecology viewpoint may be regarded as simple. Many of the classic cases of convergent evolution are found in these habitats: the mammalian example of *Dipodomys* (Heteromyidae, North American deserts)–*Antechinomys* (a marsupial in Australian deserts)–*Gerbillus* and various Dipodidae (Asian and African deserts); the reptile example of *Phrynosoma-Moloch* (North American and Australian deserts respectively); the grassland birds *Sturnella-Macronyx,* plus their South American replacements *Pezites* and *Leistes* (both Icteridae). This bird complex is quite remarkable, for all species are terrestrial and walk around in grasslands using a pickaxelike bill to dig and poke for insects. The *Sturnella* (2 species) are streaked brown above, yellow below with a black breastband, a pattern exactly duplicated by the pipit *Macronyx croceus.* The South American *Pezites* repeats this, except that the yellow is replaced by red. Curiously enough, a more southerly African species, *Macronyx ameliae,* is also red below rather than yellow; this may simply be part of the association of a red coloration with wetter and yellow with drier habitat which is general in birds. All species are alike in size, show similarities in voice, build domed nests on the ground and lay 3–4 eggs at latitude 30°. (*Pezites* also has an enlightening taxonomic history, for LINNAEUS called it a *Sturnus,* which designation was changed to *Sturnella* by FILIPPI in 1847). Grassland birds are discussed below with respect to community patterns.

For mountain birds, we could cite the resemblance of the furnariid *Chilia melanura* which lives in rocky places in the Chilean Andes to the European wall creeper *Tichodroma muraria* (Certhiidae) of the Alps. Again, these examples are legion; anyone wishing for such entertainment could spend some time with WALKER's (1964) Mammals of the World or other similar works of wide coverage.

3. *General North American-South American Comparisons*

Marshes qualify as simple habitats for birds. In North America their characteristic resident birds are the icterids *Agelaius phoeniceus, A. tricolor* and *Xanthocephala xanthocephala,* the red-winged, tricolored and yellow-headed blackbirds. The first two species are all black, with respectively buff-bordered and white-bordered red shoulder patches ("epaulets"). In Chilean marshes lives a single icterid, the yellow-shouldered marsh bird *A. thilia,* in which the epaulets are yellow rather than red; this color covers the head of the third North American species, *Xanthocephala.* As discussed in LACK (1968) and below, the weaverbirds Ploceidae fill this niche in African marshes; some species (e.g. *Diatropura*) closely duplicate in appearance, behavior and ecology these icterid species. No convergence has occurred in the other marsh-dwelling passerines, which in North America include two wren species (Troglodytidae) and several Emberizidae *(Ammospiza, Melospiza).* In Chile one of the same two wrens occurs, together with a multi-colored tyrannid *Tachuris* and a warbler-like ovenbird *Phleocryptes.*

The chats *Oenanthe* and *Saxicola* (Aves: Turdidae) are palaearctic in distribution; a single species *Oenanthe oenanthe* has crossed the Bering Sea and breeds across northern North America while migrating south-west back to the Old World. The chats are charac-

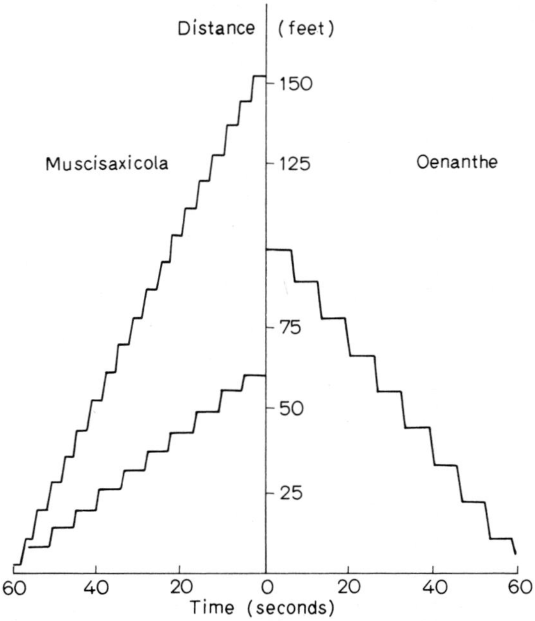

Fig. 1. Comparison of feeding behavior in the north temperate wheatear *Oenanthe oenanthe* and the south temperate *Muscisaxicola maculirostris* and *M. alpina* (Families Turdidae and Tyrannidae respectively)

teristic of open habitats, stony heaths and commons, tundra and barren hillsides. They use elevated perches on rocks and sally for or pounce on flies and other moving insects. In South American the aptly-named genus *Muscisaxicola* (Tyrannidae) provides for north temperate bird watchers extremely convincing "chats", in feeding behavior, habitat preference and in breeding biology. In central Chile six species can be found on the Andean slopes from 1200 to 3750 m elevation (CODY, 1970). Feeding behavior is perhaps the most basic and critical part of bird species' biology, upon which depend and to which are related most other characteristics (habitat choice, morphology, seasonality, etc.). Feeding behavior in birds can be quantified in a simple way, by timing feeding activity with stop-watches; then a saw-tooth feeding behavior curve is drawn by plotting distance moved by the bird against time, in an average feeding sequence. Most species feed by stops and starts which when averaged show little intra-specific variation within populations. Three parameters are required to characterize the behavior curve: average speed *s,* duration of average stop in the sequence *d* and per cent time the bird is stationary, *t* (CODY 1968). In this way the wheatear *O. oenanthe* and the two *Muscisaxicola* species studied by CODY (1970) closest to it in size are compared in Fig. 1. As is to be expected, the former exhibits a feeding behavior bracketed by the tyrannids. *Muscisaxicola* spp. at any one place are interspecifically territorial; apparently food resources cannot be subdivided by species to result in territory overlap. This also corresponds to what has been observed in chats, for STRESEMANN (1950) described the non-overlapping territories of three wintering *Oenanthe* species in Egypt.

Another north temperate genus which does not reach South America and for which a very convincing replacement genus can be named is *Toxostoma,* the thrashers whose general biology was mentioned earlier. In Chile the furnariids *Upucerthia* occupy the thrashers niche, and have come to resemble thrashers in many ways. They are large terrestrial birds of open country, and are equipped with the same decurved bill which they use to dig for food. I am familiar with *Upucerthia dumetaria,* which in habitat most closely resembles *T. lecontei* or *T. dorsale.* The color plates in PETERSON (1961) and JOHNSON, GOODALL and PHILIPPI (1957) may be compared, and similarities in body proportions deduced from Table 1.

Table 1. Body measurements (mm) in two north temperate thrashers (Mimidae) and two south temperate ovenbirds (Furnariidae) which are ecologically homologous

Species	Body size	Tail	Wing	Bill
Upucerthia dumetaria	210	84	96	30.7
Upucerthia vallidirostris	210	76	88	34.5
Toxostoma lecontei	230	122	98	32.8
Toxostoma redivivum	250	130	103	36.3

A niche which must be present and constant over those parts of the world that are densely populated by man is that associated with his urban buildings and refuse, the "house sparrow niche". In Europe, and in North America since their introduction to New York in the 1850's and subsequent spread to virtually all parts of the country, house sparrows (*Passer domesticus;* Aves, Ploceidae) are a sight familiar to everyone. They must be described as "generalists", and perhaps a part of their success story (from humble, probably African, origins, to cosmopolitan ubiquity) is their willingness to nest almost anywhere (buildings, tree hollows or outer branches, pylons, rocks) and to eat almost anything (insect, fruit, seeds, garbage, feces). In California they have apparently displaced house finches *(Carpodacus mexicanus)* from heavily urban centers, and coexist with the finches in suburbs where there is sufficient greenery for the latter to survive. The native "house sparrow" of much of Central and South America is the rufous-naped sparrow, *Zonotrichia capensis,* whose extraordinary geographic range extends from Mexico to Tierra del Fuego. House sparrows were first seen in Chile in 1904, and have since colonized the whole country. In city centers the native "chincoles" are rarely seen, but as in the house sparrow-house finch situation, they still occupy gardens and city parks where they can apparently hold their own. In downtown Lima, Peru, I saw no *Zonotrichia,* but the plaza at Arequipa was about 50 : 50 *Passer-Zonotrichia,* and downtown Cuzco was occupied solely by *Zonotrichia* in 1965. House sparrows are evidently without competitive superiors in urban centers, but coexistence by some resource division is achieved at some point along an urban-suburban-rural gradient with the tree sparrow *Passer montanus* in Britain, the house finch in California and the rufous-naped sparrow in South America. This is a fascinating, ongoing process, and merits close attention.

Convergence at the Level of the Community

I. Habitats of Simple Structure

The most precise comparisons between communities which occupy similar habitats in different geographic regions can be made where the habitat involved is structurally simple. This is because habitat structure influences not only the number of species present (MACARTHUR, 1965; PIANKA, 1966; SHELDON, 1968), but also the methods available to the occupant species to divide resources (CODY, 1968; PIANKA, 1967). Thus if habitat structure, by which must be understood its diversity, spatial complexity and associated distribution of physical and non-physical resources, is kept at a low level, sources of error due to inaccurate matching are minimized.

Many striking examples of parallel communities exist among structurally simple habitats. PATRICK (1961) was early to emphasize that the number of species of producers may be very similar among simple habitat types. Working with fresh-water diatomaceous algae, she demonstrated that the rivers of northeastern North America supported about the same number of species per substrate area as rivers in the southeast; even tributaries of the Amazon in Peru provided a close match in numbers of species, even though their names are quite different (PATRICK, 1964). From an algal viewpoint a river which supports over 100 species may not be so simple in structure, but the point still stands. Intertidal communities, especially those of sandy beaches, show close parallels among coasts of different oceans. The facts that many of the animal genera are cosmopolitan in distribution (e.g. *Donax, Emerita)* and that no strict area controlled parallels from actual censuses have been drawn lessens the value of this example. The work on sea-bottom communities, developed by THORSON (1957, 1958, 1960) and to which KUZNETSOV (1970) is a recent contributor, has been approached strictly from the point of view of parallel community studies. The results illustrate, perhaps more convincingly than any other data, that given a habitat (sea-bottom) simple and relatively constant in structure over large geographic areas, community structure in terms of numbers and types of species also remains constant. Some infaunal communities, such as the shallow water mixed-bottom *Macoma* association, show quite precise parallels in the numbers of species and in their niche relationships among coasts as dissociate as those of Greenland, Denmark, Washington (northwest United States) and Japan.

Terrestrial communities of almost the simplicity of sandy beaches are provided by dry, sandy deserts. In such deserts remarkable parallels at both the level of the species and of the community may be found. The parallels between the lizard species of the Australian and North American deserts have been described by PIANKA (1969a, 1969b, 1970, 1971), and SAGE (elsewhere in this volume) discusses those found in his work on North and South American parallels. Another species group intimately associated with the substrate in deserts is the ants. KUSNEZOV (1953, 1956) has discussed parallelism as a general phenomenon and with particular reference to desert ants. Despite wide difference in taxonomic origins, the niches of "hunting ants" and of "graminivorous ants" show similar occupancy and adaptations between Central Asia (genera *Cataglyphis* and *Messor* respectively) and Argentina (genera *Dorymyrmex* and *Pogonomyrmex*). Apparently selection must await chance for the Old World development of a third niche group, however, for the fungus-farming ants occur only in the New World. In the Mohave Desert of southeastern California 8–12 species of ants may be found in a 3-acre patch of uniform desert at 1200 m elevation (CODY and BERNSTEIN, unpublished); this number dimishes steadily to 6–8 species at 600 m (BERNSTEIN, 1971). In Utah, 1 acre of substructurally similar Great Basin Desert supports an average 12 species of ants (communities 2, 4, 5, 6, 7 of INGHAM, 1963), whereas an acre of Mohave Desert supports an average 6 species there (communities 10–17, INGHAM, *op. cit.*). In the central Sahara Desert BERNARD (1961) finds that 14 species regularly occur in the area of Tassili n'Ajjer. Further, the relative abundances of these ant species show similarities among desert locations, as do their size distributions.

Bird species are more intimately associated with the vegetation which grows in these deserts, and this varies much more in structure than the substrate on which it grows. Thus the Great Basin Desert vegetation grows to a height of about 1 m, the Mohave up to 2 m, and the Lower Sonoran vegetation up to 4–5 m; correspondingly the small-bird populations, with counts of 6, 13 and 17 species/10 acres respectively, show less parallelism than lizards and ants.

II. Marine Birds

The habitat exploited by marine birds might be expected to show little variation from coast to coast, at least in the structural variables which influence the composition of terrestrial communities. Six species of the bird family Alcidae breed on the Olympic Peninsula in the northwest corner of the United States (CODY, 1973). Five of these species dive for fish, and can be ranked according to the distances from the breeding rock at which they feed. The sixth feeds on plankton at greater distances offshore (Fig. 2).

Grimsey Island, northern Iceland, also supports six alcid species, five of which are different in name from the Pacific set. A strong correspondence exists, however, in the organization of their feeding zones (Fig. 2). The distances over which the birds have to travel for food is the predominant influence on morphological, ecological and behavioral traits, and hence the correspondence between pairs of species which feed at the same points in the feeding zone sequence indicated extends to these additional characters. In each set of species the sequence of body sizes and bill morphologies is similar. Only the species feeding closest to the breeding cliffs rear two young, the others but one. Chick growth rates and precocity are in both sequences inversely and directly, respectively, parallel to the distance the parents travel for food. These trends are summarized in Table 2.

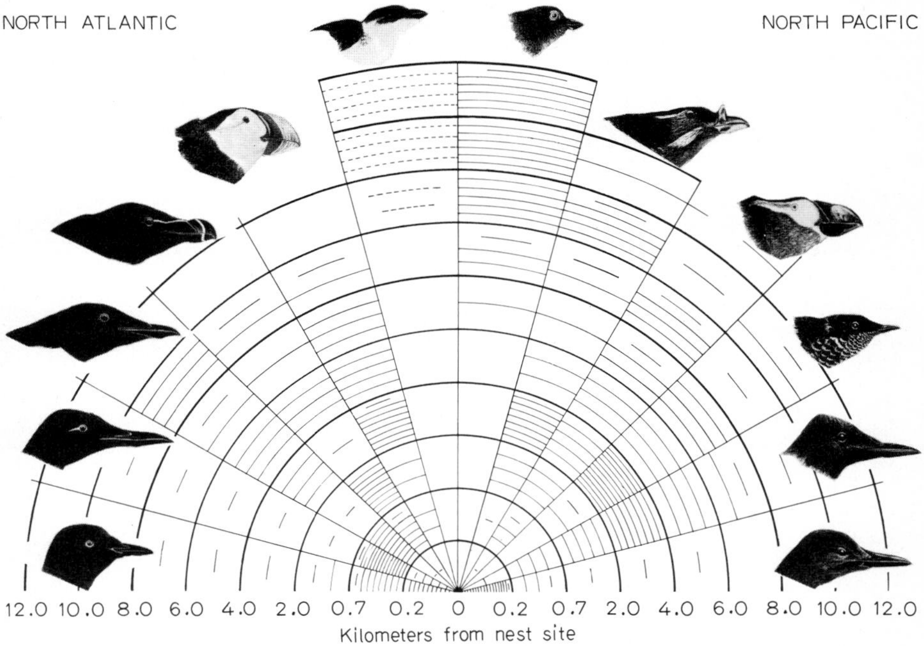

Fig. 2. Comparison of the feeding zonation of marine birds, Family Alcidae, in eastern Pacific and northern Atlantic communities. Locations are the Olympic Peninsula, northwestern United States, and Grimsey Island, northern Iceland. All bird profiles are to the same scale. Feeding activity is represented by solid arcs, each of which amounts to 5% of the total feeding activity

The Alcidae are restricted in their geographic distribution to the Northern Hemisphere. In the Southern Hemisphere, and in particular in Chilean waters, their closest ecological counterparts are the diving petrels Pelecanoididae and the penguins Spheniscidae. The former, in particular, bear an extremely close resemblance in their morphology, ecology and behavior to certain of the smaller alcids, at least as far as existing information goes. The morphology of the common Diving Petrel *Pelecanoides urinatrix* has been investigated in detail by KURODA (1967), who finds striking parallels in proportions, osteology and musculature between it and the slightly larger Ancient Auklet *Synthliboramphus antiquus*. Unfortunately for a U.S.-Iceland-Chile comparison, almost nothing is known of the habits and breeding biology of these birds. The only life history study is that of RICHDALE (1943) on *P. urinatrix*. In common with the Ancient Murrelet and with most of the smaller murrelets and auklets, *urinatrix* rears one chick in a burrow, which is visited at night and fed small fish and the larger crustaceans of the zooplankton, and is molested by the local larid predator. Judging by the chick growth rate of 5 gm/day and its attainment of a maximum weight in excess of that of the adult, the parents feed at considerable distances offshore.

The penguins Spheniscidae also show many characteristics which are convergently similar to those of alcids. Indeed, following a period when Alcidae was grouped with Colymbidae (divers, loons), from 1840–1893 the former were classified with Spheniscidae;

Table 2. Feeding and breeding biology of two six-species communities of marine birds (Alcidae), in the northeastern Pacific Ocean (Olympic Peninsula) and north Atlantic Ocean[a]

	Species	Body size (inches)	Distance to food (km)	Time in nest (days)	Chick rate (gm/day)	Precocity[b]	Nest site	Daily rhythm
North Pacific	*Cepphus columba*	14	0.27	35	10.7	9/10	crevice	diurnal
	Uria aalge	$16^1/_2$	3.10	21	8.5	1/3	ledge	diurnal
	Lunda cirrhata	$14^1/_2$	4.67	55	7.7	3/4	burrow	diurnal
	Brachyrhamphus marmoratus	9	?	(42?)	(5–6)	(4/5)	(burrow)	nocturnal
	Cerorhinca monocerata	$13^1/_2$	5.34	61	5.0	4/5	burrow	crepuscular
	Ptychorhamphus aleuticus	8	$20^{\pm}$	45	3.5	7/6	burrow	nocturnal
North Atlantic	*Cepphus grylle*	13	0.47	35	9.1	7/8	crevice	diurnal
	Uria aalge	$16^1/_2$	0.59	23	6–9	1/3	ledge	diurnal
	Uria lomvia	18	3.28	23	6–9	1/3	ledge	diurnal
	Alca torda	17	2.62	17	6.8	$^{\pm}1/4$	niche	diurnal
	Fratercula arctica	$12^1/_2$	3.97	42	5.0	10/13	burrow	diurnal
	Plautus alle	8	10	?	?	?	burrow	(nocturnal)

[a] References given in CODY (1973)
[b] Maximum proportion of adult weight gained in nest.

the extinct great auk still bears the name *Pinguinus impennis*. GYSELS and RABAEY (1964) discuss the systematic position of the Alcidae as evidenced by an electrophoretic study of lens proteins. Several genera showed divergent patterns, and the murre *Uria aalge* in particular appeared to be similar to the Spheniscidae with respect to this character. These authors cite figures which indicate that the Alcidae and the Spheniscidae hold 54% anatomical characters in common, while another 27% are variable between the groups, and conclude that the families show common origin and parallel evolution.

In spite of the evident morphological and adaptive similarities between the Alcidae and southern hemisphere replacements, the possibility of ordering a set of southern species into a feeding zone sequence as above does not look good. In the absence of the appropriate data, MURPHY (1936) is perhaps the best source of information. And it appears that on most of the Chilean coastline only one *Pelecanoides* and one *Spheniscus* coexist. In the extreme south perhaps two of the latter cooccur, but this is still far from a six-species group. Apparently the vacancies are filled by numerous Procellariformes, but pending field studies no conclusions can be drawn.

III. Grasslands

Grasslands are simple habitats which can be duplicated in structure with high fidelity on different continents. Different families of birds predominate in different regions. In North America these are Icteridae and Emberizidae, with very few Motacillidae and

Alaudidae. In South America the grassland passerines belong to Emberizidae, Furnariidae and Tyrannidae, with an occasional Motacillid. In Asia Emberizids, Motacillidae and Sturnidae are encountered, and in Africa there are many grassland Alaudidae and Sylviidae.

Many grasslands, including those of widely scattered north and south temperate zones, the neotropics, and even arctic and antarctic heathlands, show striking parallels in bird species number and composition. In homogeneous areas around 4 ha in size, three of four passerines are present; a large vegetarian "grouse-type" species, both long- and short-billed "wader-type" species, and two or three raptorial species (including a "pursuer", often *Falco*, and a "searcher", often *Buteo*) are usual (CODY, 1966). Other grassland bird censuses confirm this picture (ALM *et al.*, 1966; KURODA, 1968; NAKAMURA, 1963; NAKAMURA *et al.*, 1968; OGASAWARA, 1966; SOIKKELI, 1965; WILLIAMSON, 1967), and those which on the surface appear to contradict it, such as the South African grasslands and their avifauna, have not yet been censused at the 4 ha resolution required to test the apparent discrepancies. Of course, larger area with greater habitat heterogeneity can produce higher species counts. Thus WIENS (1969) found 7 regularly breeding species (passerines) in 32 ha of Wisconsin grassland, and SMITH (1966) found the same number in 16 ha of prairie in southern Saskatchewan. In 8 ha of grassland near Buffalo Pound Provincial Park, Saskatchewan, located to include native prairie, mowed prairie, sowed hayfield and taller grassy borders to prairie ponds, CODY (unpublished) counted 10 passerines of strictly grassland affinities; thus the need for control of area and vegetation type is obvious.

Similarities between Chilean and North American grassland bird communities extend much further than parallels between numbers of species. The following summary is from CODY (1968). In grasslands of different heights the resources available to avian exploiters are allocated in different ways. In short grasslands 4 passerines usually occur; they exhibit very different feeding behaviors but little or no habitat separation within the study area. In tall grasslands, however, feeding behaviors are very similar, apparently due to the restrictions imposed on the bird's feeding options by the dense vegetation. But here bird species have distinct habitat preferences which result in extensive interspecific separation.

Grasslands of intermediate heights are also intermediate in the methods of resource division employed by the resident species, with partial separation of territories among species and some differences in feeding behavior. Thus a mixed grassland of intermediate height (mean = 0.97 ft) in Kansas supports grasshopper sparrows *(Ammodramus savannarum)* in the tallest and densest vegetation (mean = 1.02 ft), horned larks *(Eremophila alpestris)* in the shortest, open areas (0.76 ft) and eastern meadowlark *(Sturnella magna)* in intermediate vegetation (0.96 ft) with considerable overlap between adjacent pairs. Their feeding behaviors, following the notation described above, are (0.382 ft/sec; 2.26 sec; 25.9%) for the horned lark, meadowlark – (0.165; 1.81; 22.4) and sparrow – (0.010; 65.0; 99.2). In a field of comparable structure in central Chile (mean ht = 0.89 ft), one also finds three passerines. *Pezites militaris* duplicates the meadowlark niche in those sections of the field intermediate in grass height and density (ht 0.86 ft; dens 6.07), and displays a corresponding feeding behavior (0.107; 1.23; 15.8). *Sicalis luteola* (0.012; 65.00; 97.7) occurs only in the tallest parts of the field (mean ht 0.93), and therefore replaces the grasshopper sparrow. In the most open parts of the field (ht 0.86; dens 5.90) lives *Anthus correndera,* doing just what the horned lark does in Kansas (0.378; 0.97;

18.1). These matches in feeding behaviors are remarkably close (Fig. 3), as are those of the species' body proportions on which the behavior is dependent (Table 3).

Thus the Chile-Kansas convergence extends from number of species and their relative abundances to the precise way in which habitats are subdivided and food harvested, and the resultant behavioral and morphological characters. The net outcome is that the

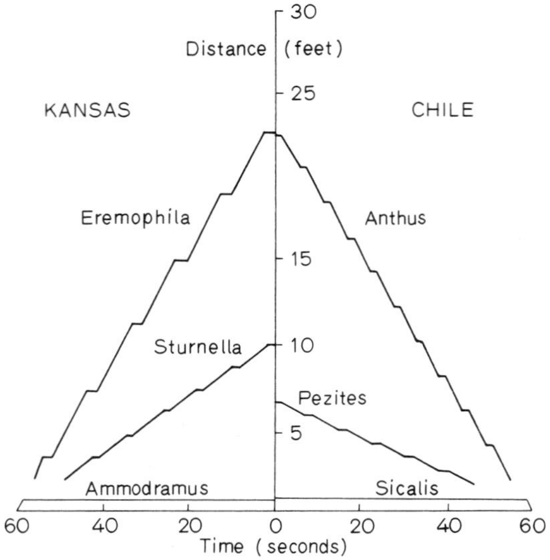

Fig. 3. Comparison of the feeding behaviors of the three bird species resident in fields of medium grass height in Kansas, central United States, and central Chile

Table 3. Comparison of grassland bird phenotypes between Kansas and Chile[a]

Species	Total length	Wing	Tail	Bill length	Bill depth	Bill depth / Bill length	Tarsus
E. alpestris	157	104	69	11.2	5.6	0.50	22
A. correndera	153	79	60	13.0	5.5	0.42	23
S. magna	236	122	79	32.1	11.6	0.36	42
P. militaris	264	117	98	33.3	13.3	0.40	34
A. savannarum	118	62	47	10.8	6.5	0.60	19
S. luteola	125	70	49	9.7	7.1	0.73	16

[a] All lengths are in mm.

ecological overlaps among species pairs within each study area are very similar; mean habitat overlap is 63% in Kansas and 60% in Chile, the extent of vertical overlap in feeding heights (78% in Kansas *vs* 89% in Chile) and of overlap in feeding behavior (18% *vs* 21%) is likewise close.

The structure of the field is the causative factor in the convergences, for as variables such as grass height and grass density change, the methods of resource division change in parallel in Chile and in North America. Such parallels exist despite the greatly divergent histories of the two areas and, as in the case of irrigated fields in Chile, such as that just described, the relatively recent beginnings of some of the habitats involved. NAKAMURA's field *(op. cit.)* had a variety of grass heights from medium to tall, and there also the species showed partial habitat separation. *Acrocephalus bistrigiceps* occupied the tall reeds, *Emberiza furcata* and *E. yessoënsis* the shorter grasses bordering water and elsewhere, and *Saxicola torquata* was found on the more barren hillsides. *E. furcata,* for example, shows a mean habitat overlap with the other species present of around 50% or a little less, just what we would expect in grasslands of the range and diversity of vegetation heights such as this.

IV. Beech Forests

The north temperate forest association of beech-maple, *Fagus-Acer,* is widespread over the Palaearctic and Nearctic regions, and has a close parallel in south temperate *Nothofagus* forests. The southern beech is also widely distributed (DARLINGTON, 1965), and can presently be found in New Guinea, Australia and Tasmania, New Zealand and a few smaller southwest Pacific islands, and in South America in Chile and Argentina.

The bird populations of three widely separate northern sites have been censused: Japan (URAMOTO, 1961), Denmark (JOENSEN, 1965) and Ohio, U.S.A. (WILLIAMS, 1936). The relative abundances of the resident species are also known, and information on the feeding habits of the birds is included. Information which is similarly detailed is available from two south temperate locations, Chile (CODY, 1970)[1] and New Zealand (KIKKAWA, 1966, to which I have added pertinent data of J. DIAMOND, pers. comm.). Two further localities, the *Nothofagus* forests of Australia and Tasmania, have also been studied by ornithologists, but only species lists are available (KIKKAWA, HORE-LACY and BRERETON, 1965, and RIDPATH and MOREAU, 1966, respectively). All census areas are alike in that the predominant tree species are beech *(Fagus* or *Nothofagus),* other broad-leaf deciduous trees are present (e.g. *Acer),* and a dense understory of bush-type vegetation slows progress through the forest for the observer (bamboo-grass *Sasa* in Japan, the bamboo *Chusquea* in Chile, cutting grass *Gahnia* (Cyperaceae) in Tasmania, and so on). The areas studied in the north temperate sites are all similar (around 28 ha); the New Zealand site is 20 ha, but the Chilean area only 6.6 ha.

Species can be compared among censuses for one-to-one correspondence of niches, and in the abundances of the occupants of those niches. Niches can then be grouped (see Table 4) in such a way that ecologically related species are lumped; species within such a grouping are supposed to be strong competitors, such that within these groups a reduced density or absence of one species is most likely to be compensated by an increased density of another species within the group. Thus a correspondence in the densities of species within groups is expected among sites, even though the number of species in the groups does not exactly correspond. Finally the sites may be expected to correspond in numbers of species and overall population densities; the extent to which they do not can be attributed variously to a) chance effects, b) productivity differences, c) non-correspondence of habitat, d) historical factors (man's influence, "island effects", introduced predators or competitors, etc).

1 VEILLEUMIER (1972) has produced an almost identical census from the Argentinian side of the Andes in similar *Nothofagus*.

Table 4. Bird species and niches in north temperate *Fagus-Acer* forests in comparison to those of south temperate *Nothofagus* forests

Niche	Japan[a] (75 acres)		Denmark[b] (77 acres)		Ohio[c] (65 acres)		New Zealand[d] (50 acres)		Australia[e]	Tasmania[f]		Chile[g] (16.5 acres)	
Sallying	*Muscicapa*	(76)	*Phoenicurus*	(78)	*Sayornis*	(87)	*Rhipidura*	(74)	*Rhipidura*	*Rhipidura*	(74)	*Elaenia*	(77)
flycatchers	*narcissina*	0.150	*phoenicurus*	0.140	*phoebe*	0.015	*fuliginosa*	0.080	*rufifrons*	*fuliginosa*		*albiceps* $^1/_2 \times$	0.57
low	*Muscicapa*	(72)	*Muscicapa*	(85)	*Empidonax*	(74)			*Rhipidura*				
medium	*latirostris*	0.030	*striata*	0.029	*virescens*	0.054			*fuliginosa*				
high	*Muscicapa*	(92)	*Muscicapa*	(83)	*Contopus*	(83)			*Petroica*				
	cyanomelana	0.030	*hypoleuca*	0.078	*virens*	0.108			*rosea* (part)				
					Myiarchus	(106)							
					crinitus	0.023							
		0.210		0.247		0.200		0.080					0.288
Aerial	(*Delichon*	(107)			(*Progne*	(142)						(*Tachycineta*	(103)
flycatchers	*urbica*)				*subis*)							*leucopyga*)	
Foliage	*Phylloscopus*	(62)	*Phylloscopus*	(76)	*Vireo*	(77)	*Mohoua*	(82)	*Sericornis*	*Zosterops*	(60)	*Spinus*	(73)
insecti-	*occipitalis*	0.180	*sybilatrix*	0.075	*flavifrons*	0.069	*ochrocephala*		*magnirostris*	*lateralis*		*barbatus*	0.273
vores	*Phylloscopus*	(62)	*Phylloscopus*	(64)	*Dendroica*	(66)	$^1/_2 \times$	0.188	*Pardalotus*	*Pachycephala*	(100)	*Elaenia*	(77)
canopy	*tenellipes*	0.055	*colybita*	0.123	*cerulea*	0.019	*Zosterops*	(63)	*punctatus*	*pectoralis*		*albiceps* $^1/_2 \times$	0.575
high	*Aegithalos*	(59)	*Phylloscopus*	(67)	*Piranga*	(96)	*lateralis*	0.143	*Zosterops*			*Sylviornitho-*	(51)
medium	*caudatus*	0.095	*trochilus*	0.169	*erythromela*	0.119			*lateralis*			*rhynchos*	
low			*Regulus*	(53)	*Dendroica*	(64)			*Gerygone*			*desmurii*	0.121
			regulus	0.007	*virens*	0.054			*richmondi*				
					Vireo	(81)							
					olivacea	0.493							
		0.330		0.374		0.754		0.237					0.682
understory	*Urospena*	(54)	*Hippolais*	(79)	*Setophaga*	(61)	*Gerygone*	(55)	*Pachycephala*	*Pachycephala*	(99)	*Aphrastura*	(59)
	squamaiceps	0.160	*icterina*	0.039	*ruticilla*	0.261	*igata*	0.088	*pectoralis*	*olivacea*		*spinicauda*	0.545
	Cettia	(68)	*Sylvia*	(78)	*Wilsonia*	(68)	(*Xenicus*	(55)	(*Pachycephala*	*Sericornis*	(62)		
	diphone	0.165	*borin*	0.315	*citrina*	0.215	*longipes*)[j]		*rufiventris*)	*humilis*			

			Sylvia atricapilla *Sylvia communis*	(74) 0.117 (72) 0.026					*Sericornis frontalis*				
		0.325		0.497		0.476		0.088					0.545
		0.655		0.871		1.230		0.325					1.515
Insecti-vores twigs and branches	*Parus ater* *Parus atricapillus* *Parus varius* *Parus major*	(59) 0.015 (65) 0.120 (76) 0.120 (68) 0.200	*Parus caeruleus* *Parus palustris* *Parus major*	(67) 0.221 (62) 0.045 (76) 0.351	*Parus atricapillus* *Parus bicolor*	(66) 0.062 (80) 0.108	*Finschia novarsee-landiae* *Mohoua ochro-cephala* $^1/_2$ ×	(65) 0.050 (82) 0.188	*Acanthiza pusilla* *(Acanthiza lineata)* *Eopsaltria australis*	*Acanthiza ewingii*	(48)	*Anaeretes parulus*	(48) 0.182
		0.455		0.617		0.170		0.144					0.182
trunk surface	*Sitta europaeus* *Troglodytes troglodytes*	(80) 0.085 (50) 0.095	*Sitta europaeus* *Certhia familiaris* + *C. brachy-dactyla* *Troglodytes troglodytes*	(86) 0.140 (65) 0.159 (48) 0.221	*Sitta carolinensis*	(88) 0.054	*Acanthisitta chloris*	(49) 0.280	*Climacteris leucophaea* *Climacteris eryhtrops*	*Acanthiza magnus*	(51)	*Pygarrhichas albogularis* *Troglodytes aedon*	(80) 0.061 (52) 0.182

[a] Data from URAMOTO (1961)
[b] JOENSEN (1965)
[c] WILLIAMS (1936)
[d] KIKKAWA (1966) and J. DIAMOND, pers. comm.
[e] KIKKAWA *et al.* (1965)
[f] RIDPATH and MOREAU (1966)
[g] CODY (1970)
[h] Introduced
[i] Brood parasite
[j] Rare; nearly extinct
[k] Extinct
() Not in census, but known to be present in area

Niche	Japan[a] (75 acres)		Denmark[b] (77 acres)		Ohio[c] (65 acres)		New Zealand[d] (50 acres)		Australia[e]		Tasmania[f]		Chile[g] (16.5 acres)	
trunks high low	*Dendrocopus kizuki*	(86) 0.100	*Dendrocopus major*	(141) 0.039	*Dryobates pubescens*	(89) 0.046	(*Philesturnus carunculatus*)[j]	(100)	(*Calypto-rhynchus funereus*)	(375)	*Calypto-rhynchus funereus*	(375)	(*Dendrocopus lignarius*)	(94)
	Dendrocopus major	(130) 0.015			*Dryobates villosus*	(120) 0.038	*Nestor meridionalis*	(295) 0.010					(*Campephilus magellanicus*)[j]	(215)
	Dendrocopus leucotus	(150) 0.015			*Centurus carolinus*	(131) 0.004							*Colaptes pitius*	(159) 0.030
	Picus awokera	(143) 0.015			*Dryocopus pileatus*	(229) 0.004								
					(*Colaptes auratus*)	(150) 0.004								
		0.145		0.039		0.096		0.010						0.030
		0.780		1.176		0.320		0.434						0.455
Ground feeders	*Phasianus soemmerringii*	(218) 0.040	*Phasianus colchicus*	(248) 0.013	(*Bonasa umbellus*)	(175)	(*Apteryx australis*)[j]		*Menura superba*		*Menura superba*[h]		*Turdus falklandii*	(129) 0.121
	Turdus dauma	(158) 0.015	*Turdus merula*	(128) 0.426	*Turdus migratorius*	(134) 0.015	*Gallirallus australis*	(188) 0.060	*Oreoconcla lunulata*	(144)	*Oreocincla lunulata*	(144)	*Pteroptochos tarnii*	(106) 0.121
	Turdus sibiricus	(124) 0.020	*Turdus philomelus*	(116) 0.218	*Hylocichla mustelina*	(109) 0.300	*Turdus merula*[h]	(128) 0.055	*Colluricincla harmonica*	(124)	*Colluricincla harmonica*	(124)	(*Scelorchilus rubecula*)	(71)
	Turdus chrysolaus	(121) 0.020	*Fringilla coelebs*	(90) 0.513	*Pipilo erythrophthalmus*	(89) 0.027	*Turdus philomelos*[h]	(116) 0.038	*Psophodes olivaceus*		*Turdus merula*[h]	(128)	*Scytalopus magellanicus*	(49) 0.121
	Erithacus cyane	(75) 0.290	*Emberiza citrinella*	(88) 0.045	*Seiurus aurocapillus*	(73) 0.200	*Miro australis*	(98) 0.107)	*Orthonyx teminckii*					
	Erithacus akahige	(75) 0.005	*Erithacus rubeca*	(73) 0.286	*Seiurus motacilla*	(81) 0.011	*Fringilla coelebs*[h]	(90) 0.075	*Petroica rosea* (part)					
			Prunella modularis	(68) 0.107			*Prunella modularis*[h]	(68) 0.015						
							Petroica macrocephala	(71) 0.183						
		0.390		1.608		0.553		0.533						0.363

Raptors nocturnal	*Otus scops*	(143)	*Strix aluco*	(255) 0.010	*Strix varia*	(333) 0.015	*Ninox novaeseelandiae*	(193) 0.02	*Ninox novaeseelandiae*	(200)	*Ninox novaeseelandiae*	(200)	*Bubo virginianus*	(340)
diurnal	*Spiaëtus nipalensis*	(497)	*(Buteo buteo)*	(383)	*(Buteo jamaicensis)*	(380)	*(Falco novaeseelandiae)*	(285)	*(Accipiter novaeseelandiae)*	(302)	*Podargus strigoides*	(260)	*Milvago chimango*	(288) 0.061
			(Accipiter nisus)	(198)					*(Aquila audax)*	(598)				
Seeds and fruit Omnivores	*Cucculus saturatus*[i]	(200)	*Chloris chloris*	(88) 0.065	*Richmondena cardinalis*	(94) 0.073	*Cyanorhamphus auriceps*	(119) 0.160	*Aprosmictus scapularis*		*Platycercus caledonicus*	(183)	*Microsittace ferruginea*	(179) 0.061
	Sphenurus sieboldii	(185) 0.010	*Carduelis cannabina*	(80) 0.013	*Hedymeles ludovicianus*	(101) 0.008	*Carduelis flammea*	0.075	*Platycercus elegans*		*Kakatoë galerita*	(370)	*Columba araucana*	(210) 0.061
	Garrulus glandarius	(170) 0.065	*Coccothraustes coccothraustes*	(103) 0.016	*(Coccyzus americanus)*	(144)	*Chalcites lucidus*[h]	(103) 0.040	*(Cacomantis pyrrhophanus)*[i]		*Phaps elegans*	(160)	*Curaeus curaeus*	(128) 0.303
			Columba palumbus	(245) 0.169	*Ectopistes migratorius*[k]	(205)	*Eudynamis taitensis*	(192) 0.027	*(Chalcites basilis)*[i]		*Strepera fuliginosa*	(255)		
			Columba oenas	(223) 0.049	*(Cyanositta cristatus)*	(132)	*(Turnagra capensis)*[j]	(131)	*Acanthochaera corunculata*					
			Oriolus oriolus	(156) 0.055			*Hemiphaga novaeseelandiae*	(259) 0.02	*(Leurcosarcia melanoleuca)*					
			Garrulus glandarius	(179) 0.003			*Prosthemadera novaeseelandiae*	(147) 0.075	*Strepera graculina*					
		0.085		0.370		0.081		0.322						0.425
Nectivores							*Anthornis melanura*	(88) 0.163	*(Acanthorhynchus tenuirostris)*				*Sephanoides sephanoides*	(64) 0.121
									(Meliphaga chrysops)					
Scavengers			*Corvus corone*	(332) 0.013	*(Cathartes aura)*	(535)							*Coragyps atratus*	(538) 0.091

The results of the 5 censuses and the two species lists are summarized in Table 4. Species densities and body sizes (total length in mm) are also included. Each species is aligned as closely as possible with its counterpart(s) in the other forest habitats, although in a good number of cases this is not possible. Some "split appointments" are made for more accurate niche representations. The niche associations are listed from those I consider most distinct—the sallying flycatchers— to those in which associations are made only tentatively and at considerable risk of error. The last major category, "Seeds & Fruit, Omnivores", is to some extent a catch-all grouping, numbering as it does the true seed and fruit eaters of the canopy, the omnivorous jays and their homologues, the insectivorous cuckoos, the pigeons and doves and finally the oriole-equivalents.

There are obvious qualitative similarities among the species lists and likewise there are some anomalies which might at first sight shake our faith in natural selection. To what extent do fewer species per niche grouping compensate by increased density per species for the absense of competitors? For each of the ten pairs of habitats I obtained the correlation coefficients r_{sp-sp} of the densities of species in one census with the densities of their replacements, the species which provide the best possible one-to-one matches, in the second locality. These correlations coefficients averaged 0.356. A second set of correlation coefficients r_{gr-gr} was obtained by comparing densities between niche groupings rather than species to species. Higher values were scored (mean = 0.512). The ratio r_{gr-gr}/r_{sp-sp} is a measure of the degree to which species compensate within niche groupings for more of fewer species, and is independent of productivity considerations; this ratio exceeds unity in 8 out of 10 pair-wise comparisons, in some cases by a great margin. The extent to which this compensation takes place is in part due to the

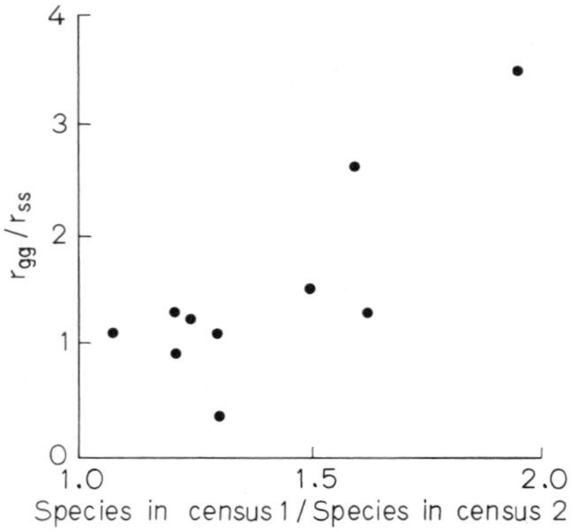

Fig. 4. In beech forest censuses, the densities of ecological homologues are correlated between species (r_{ss}) and between groups of ecologically similar species (r_{gg}). The extent to which the between-census correlation is improved by grouping species into associations of strong competitors (r_{gg}/r_{ss}) is seen to be proportional to the extent that one census is species-depauperate with respect to the other

discrepancies in the numbers of species available to fill the niche space. The more similar are species counts per locality, the less opportunity to exercise density compensations by the resident species. Accordingly, when the ratio of higher to lower species counts per census is plotted against the ratio of grouped to single niche density correlation coefficients, a positive relation is obtained (Fig. 4).

The beech forest data, whilst crude, serve to illustrate various factors that influence species density. Historical or chance effects must be mainly responsible for the differences in species numbers, among the niche groupings and in the census totals. The total density of pairs/unit area can be taken as a rough measure of productivity which varies, then, from a low in New Zealand of 1.857 pr/ac to a high in Denmark of 4.485 pr/ac and appears, happily, to be largely independent of species numbers. How important is this productivity estimate in the absolute densities of niche groupings? The sallying flycatchers appear remarkably constant in density among 4 habitats (the exception is a low count from New Zealand), indicating, perhaps, that an upper density limit has been reached at each site, regardless of the number of species (1–4) which compose this density figure. Other niche associations, such as the canopy and understory insectivores, are variable in density from site to site; in the former but not the latter density increases with the number of species per niche association (Fig. 5, a and c). When a correction is made for productivity by expressing group density as a percentage of the total plot density, the scatter remains for the understory species, and increases for the canopy groups (Fig. 5, b and d). Contrary to this, the insectivores of twigs and branches show a some-

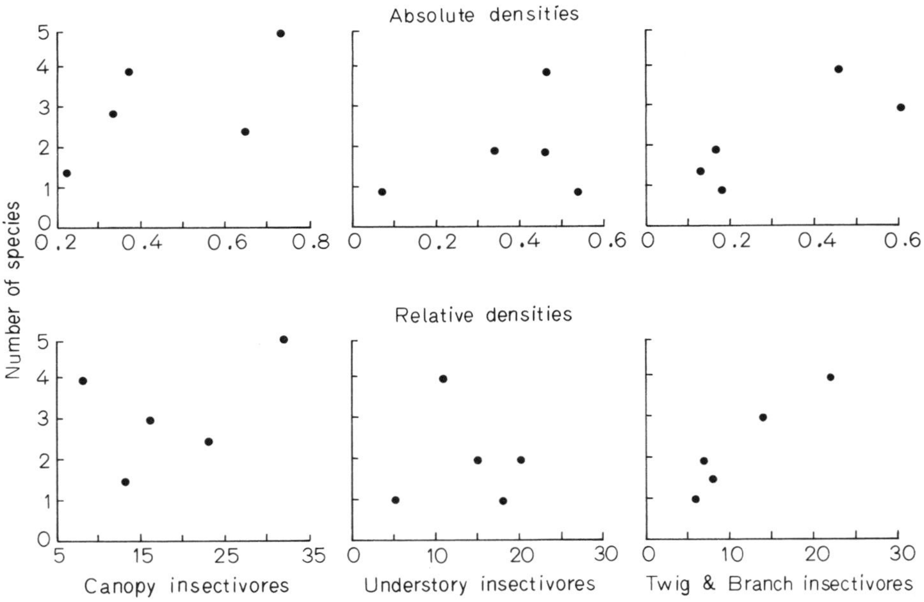

Fig. 5. Comparison between five intercontinental beech forest censuses. Three niche groupings are considered: canopy insectivores, understory insectivores and twig-and-branch insectivores. Numbers of species/niche group may or may not be correlated with absolute and relative group densities

what orderly relation between numbers of species and absolute group density, which after correction for productivity resolves into a positive linear relation (Fig. 5, e and f). Thus the density of the niche grouping is directly proportional to both the number of occupant species and the so-called productivity: $D_{gr} = 0.05\ nP$ in this case. Little of more sophistication is worth attempting with these data; suffice it to say that the accidents of history and chance may be partially compensated by niche expansion, which is reflected by increased densities per species when fewer species divide resources, and may or may not be influenced by overall productivity.

V. Mediterranean Scrub

1. The Habitat

From earlier parts of this volume the reader will now be familiar with the vegetation types found in coastal regions that enjoy a mediterranean climate. In particular, the chaparral association of California, composed predominantly of fire-adapted broad-leafed sclerophylls around 2–3 m in height, parallels in many aspects of physiognomy, morphology, flowering seasons and many other adaptive features the matorral of central Chile (MOONEY and DUNN, 1970).

Since 1966 the writer has studied a 3 ha patch of chaparral at 600 m elevation in the Santa Monica Mountains of coastal southern California. The area was (for it burned in the Malibu fire of September 1970) a mixture of broad-leaf sclerophylls *(Rhus, Quercus dumosa, Arctostaphylos, Ceanothus),* with some narrow-leaf species *(Adenostoma).* Until 1970 the area had not been burned for 38 years. Since 0.24 ha of this site included taller trees of *Quercus agrifolia,* that section will be excluded for the purposes of comparison with a Chilean area.

In 1968 a Chilean matorral study area was selected on the basis of its physiognomic similarity to that in the Santa Monica Mountains. The site chosen is 1.84 ha at 270 m elevation on the coastal range near Puchuncavi, 75 km north of Santiago. The dominant

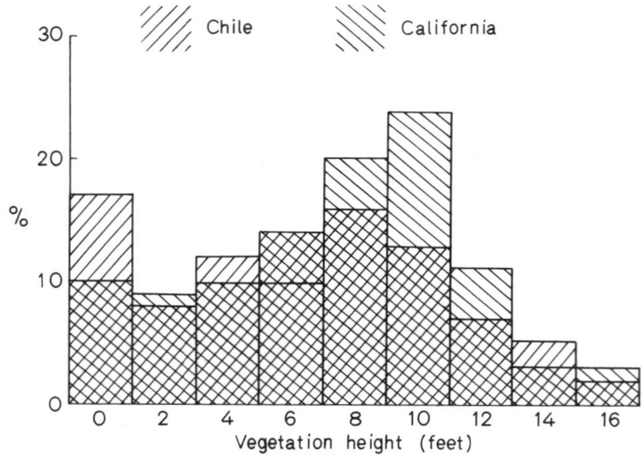

Fig. 6. Frequency distribution of foliage in California chaparral and Chilean matorral

plant species there are of *Lithraea* and *Quillaja,* and their convergent similarities with Californian shrubs are described elsewhere in this volume. In both plots the vegetation is virtually continuous, a close-canopy shrub cover of variable height but of constant high density. Vegetation height distributions for the sites are given in Fig. 6. The Chilean plot was censused throughout the 1968 breeding season, and comparable information on its use by bird species to that of the California site was obtained. This information includes: a) numbers and kinds of species, b) their distribution and abundance within the study areas, c) their feeding height distributions within the vegetation and d) quantitative measurements of their feeding behaviors.

2. *Niche Dimensions*

The resident bird species must be aligned between plots according to their niche similarities, but there are several dimensions or resource variables which serve to separate competitors. First is habitat. In California chaparral is intermediate on a moisture gradient between the drier coastal sage association *(Artemisia-Salvia-Eriogonum)* and the moist riparian or oak woodlands *(Quercus agrifolia, Umbellularia, Juglans)*. In Chile a similar ordering of habitats occurs, with matorral intermediate on a habitat (moisture) gradient between coastal *Bahia-Haplopappus* and inland *Cryptocarya* woodlands. Central Chile has another habitat, "savanna", with no counterpart in California. This is a low, open savannah-like habitat chiefly of *Acacia-Prosopis,* and may have been derived from matorral-like vegetation by the influence of man and his goats. For the purposes of this work I shall not distinguish the two habitats, as their avifaunas are extremely similar.

Secondly, species differ in the foods they eat, the places they seek that food (height above the ground; site within the vegetation–trunks, foliage) and the way in which the food is secured (flycatching by sallying, plucking, or searching over trunks, etc.). Thus a pictoral representation of niches described in this way would need at least twice as many dimensions as a sheet of paper. Table 5 is therefore a compromise; it gives the census results, and shows specific niches grouped according to simple combinations of the above-mentioned variables.

3. *Species Distributions over the Habitat Gradient*

In Table 5 are listed only those species which may be seen in the primary habitat, chaparral or matorral, in the breeding season; names are spaced horizontally according to where each species reaches its maximal abundance in the habitat gradient.

Very few species are restricted to the primary habitat type. The wrentit *Chamaea fasciata* and the California thrasher *Toxostoma redivivum* occur only in chaparral, and *Asthenes humicola* (the wrentit homologue) plus perhaps *Phytotoma* only in matorral (the Chilean positions are somewhat tentative, pending more extensive field work there). About the same number of the species listed in each locale reach maximal abundance to the woodland side of scrub in the gradient. However, rather more of the remaining species reach higher abundances in chaparral than in drier, lower vegetation than is the case in Chile; the matorral contains relatively more birds which are commoner in drier habitats. To some extent this is a reflection of the greater proportion of open ground (17% *vs* 10%) and of lower mean height of vegetation (6.3 ft *vs* 8.7 ft) in the matorral as compared to the chaparral site. On average, the Chilean species are more widely distributed over the habitat gradient (+10–20%). Thus there are 22 unlisted species which occur in oaks or coastal sage but not in chaparral, and only 8–10 such Chilean species, in spite of the fact that the number of species per habitat is at least as great in Chile. Californian species-area curves are therefore much steeper.

Table 5. Census results and species correspondence between two mediterranean scrub sites

No.	Coastal sage	California: Chaparral (6.9ac)	Oak woodland	Chile: Woodland	Matorral (4.6ac)	*Bahia-Haplopappus*
1	118/0.71[a]	*Thryomanes bewickii* ⟶			⟵ *Troglodytes aedon* ⟶	120/0.65
2	151/1.03	*Chamaea fasciata*			⟵ *Asthenes humicola*	165/0.33
3	102/0.65	⟵ *Psaltriparus minimus* ⟶			⟵ *Leptasthenura aegithaloides*	160/0.22
3a	108	⟵ *Polioptila caerulea* ⟶			*Aphrastura spinicauda* ⟶	140
4	127/0.16	⟵ *Parus inornatus* ⟶			⟵ *Anaeretes parulus* ⟶	110/1.09
5	146/0.03	⟵ *Empidonax difficilis* ⟶			⟵ *Elaenia albiceps* ⟶	150/0.33
6	197/0.13	⟵ *Myiarchus cinerascens* ⟶			⟵ *Pyrope pyrope* ⟶	210/0.27
7	96/0.45	⟵ *Calypte anna* ⟶			⟵ *Patagona gigas* ⟶	212/0.45
7a	86	⟵ *Archilochus alexandri* ⟶			⟵ *Sephanoides seph.*	110
					⟵ *Nothoprocta perdicaria*	305/0.04
8	250/0.39	⟵ *Lophortyx californica* ⟶			⟵ *Lophortyx californica* ⟶	250/0.22
9	205/0.32	⟵ *Pipilo fuscus* ⟶			⟵ *Diuca diuca* ⟶	175/0.33
10	138	*Carpodacus mexic.*			⟵ *Zonotrichia capensis* ⟶	147/0.82
11	540	*Geococcyx* ⟶			⟵ *Pteroptochos megapodius*	232/0.22
12	298	*Zenaidura macr.* ⟶			⟵ *Zenaidura auriculata*	265/0.33
13	289/0.45	⟵ *Aphelocoma coerulescens* ⟶			⟵ *Agriornis livida* ⟶	275/0.11
14	199/0.52	⟵ *Pipilo erythrophthalmus* ⟶			⟵ *Turdus falklandii* ⟶	280/0.33
15	284/0.45	*Toxostoma redivivum* ⟶			⟵ *Mimus tenca* ⟶	290/0.44
16	181/0.19	*Pheuticus melanocephalus* ⟶			⟵ *Phytotoma rara* ⟶	195/0.33
16a	103	⟵ *Spinus lawrencei* ⟶			⟵ *Spinus barbatus* ⟶	135
17	184	⟵ *Dendrocopus nuttallii* ⟶			⟵ *Dendrocopus lign.* ⟶	182
18	280/0.10	⟵ *Colaptes cafer* ⟶			⟵ *Colaptes pitius* ⟶	320/0.02
19	147	⟵ *Aeronautes saxatalis* ⟶			⟵ *Tachycineta leucopygia* ⟶	135/0.44
20	197	*Phalaenoptilus nuttallii* ⟶			⟵ *Chordeiles longipennis* ⟶	243
		⟵ *Bubo virginianus* ⟶			⟵ *Bubo virginianus* ⟶	
		⟵ *Cathartes aura* ⟶			⟵ *Cathartes aura* ⟶	
		⟵ *Buteo jamaicensis* ⟶			⟵ *Parabuteo unicinctus* ⟶	
		Accipiter cooperii ⟶			⟵ *Milvago chimango* ⟶	

[a] Body size/density

4. *Status in the Censuses: Full Members and Marginal Species*

Not all the species listed in the table were resident within the census areas ("full members"). Some full members of one census have obvious counterparts in the other locale which were recorded as "marginal" there (part of a territory within census area; breeds adjacent to census area). Such marginal species are also listed in the table. Thus in chaparral the western flycatcher *Empidonax difficilis*, common in oak woodlands, occurs on the edge of the study plot and occasionally feeds within it. Its Chilean counterpart,

Elaenia albiceps, is a full member of the matorral census. The reverse is true for the woodpeckers *Dendrocopus.*

5. Ecological Counterparts

At once it can be seen that the two bird communities parallel each other with remarkable precision. Almost every full member species in each census can be matched with one-to-one correspondence to species in the parallel habitat, and 20 pairs of suggested ecological counterparts are numbered. Of these 15 involve 2 full member species, and 5 associate a full member species (4 Chilean, 1 Californian) with a marginal species. In each census 4 foliage insectivores, 2 flycatchers, 2 woodpeckers, 1 nectivore, 1 seed-fruit eater and both diurnal and crepuscular aerial feeders occur. The one-to-one matching inspires less confidence with the ground-feeders, where some pairings are more anomalous than homologous. Only one full member species is not matched up, the Chilean tinamou *Nothoprocta perdicaria.* It is displaced in the matching process by the quail *Lophortyx californica,* which was introduced into Chile from California in 1870. Three numbered pairs are of matched marginal species, and each site also supports a large owl, a scavenger, and both searching and pursuing hawks.

Some of the species pairings provide near-perfect matches, the more interesting of which are developed below. Others, especially among ground feeders, are not so perfect. When the taxonomic differences between censures are considered, the two communities duplicate each other with impressive fidelity. The 23 numbered matches include, after all, 9 between the members of different families and 8 between confamilials, with only 5 between congeners plus 1 conspecific, the Chile-California quail. If the paired species

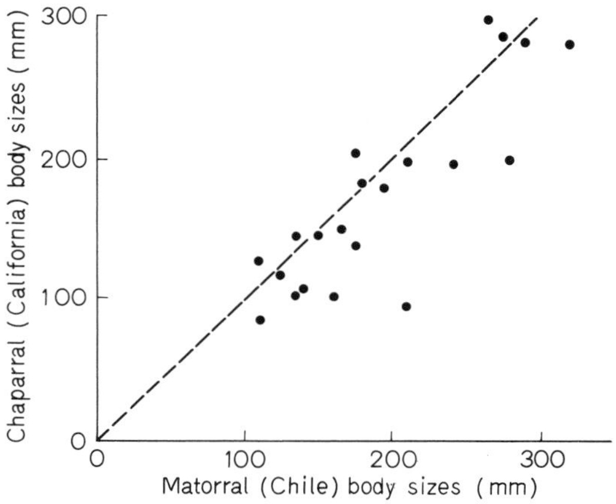

Fig. 7. Correspondence between chaparral and matorral ecological equivalents in body size

in Table 5 are ecological counterparts, then we would expect their morphologies and their general ecologies to coincide, almost by definition. This can be tested in a simple way by plotting Chilean and Californian values of various indices against each other. This is done in Fig. 7 for body size. The correlation is good ($r = 0.66$) and the slope of the regression line is very close to unity ($b = 1.05$). An interesting across-the-board increase in body size Chilean over California species shows up (a = –12.6 mm), which can probably be attributed to the differences in lineage between the occupants of the two sites, and which has not quite been erased by natural selection. To my surprise, the correlation coefficient for a similar plot of bill lengths rather than body sizes indicates a slightly poorer correspondence (r = 0.61); *a priori* bill size might be (and has been shown from eg. island-mainland comparisons) expected to show more plasticity than body size.

General ecological requirements are reflected in average territory sizes or in densites (pr/ac). The matched species show a considerable scatter in one-to-one density correspondence (Fig. 8). Part of this scatter is resolved when niches are grouped and densities summed, but the Chilean site maintains a greater productivity with 6.5 pr/ac *vs* 5.6 pr/ac for the chaparral (17 measured niche densities). Mean feeding heights between matched pairs are very similar (n = 12; r = 0.85), in spite of a poor match between the height of the selected habitat (r = 0.17).

It is informative to look at the similarities between these two bird communities in the way species coexist or niches arranged with respect to each other. Given that species co-occur in a certain habitat type, then coexistence is achieved by means of a subdivision of the food supplies. As discussed above, food is sought and niches segregated by different feeding behaviors, at different heights above ground and with different bill apparatus. These factors may be combined into a single figure of ecological overlap, akin to the competition coefficient of laboratory workers. Two 20×20 symmetrical matrices of the

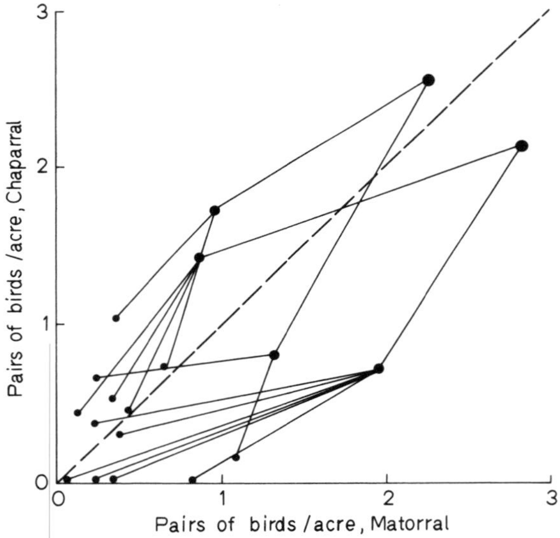

Fig. 8. Correspondence between chaparral and matorral ecological equivalents, species, niche groups and census subtotals, in density of pairs of species

Californian and Chilean bird communities containing all pair-wise ecological overlaps are given in Table 6, and these are then used to construct dendrograms of ecological affinities within the communities Figs. 9 and 10. This shows at a glance the ecological relations among the resident species; although the rough groupings of Table 5 are helpful, the picture is now concise, quantitative and accurate. Inspection of the matrices

Table 6. Two symmetrical 20 × 20 matrices of ecological overlaps. Upper right: California chaparral; lower left: Chilean matorral

	Thryomanes bewickii	*Chamaea fuscata*	*Psaltriparus minimus*	*Parus inornatus*	*Empidonax difficilis*	*Myiarchus cinerascens*	*Calypte anna*	*Lophortyx californica*	*Pipilo fuscus*	*Carpodacus mexicanus*	*Geococcyx californianus*	*Zenaidura macroura*	*Aphelocoma coerulescens*	*Pipilo erythrophthalmus*	*Toxostoma redivivum*	*Pheuticus melanocephalus*	*Dendrocopus nuttallii*	*Colaptes cafer*	*Aeronautes saxatalis*	*Phalaenoptilus nuttallii*
Troglodytes aedon	1	0.71	0.54	0.40	0.45	0.43	0.63	0.39	0.31	0.31	0.16	0.37	0.36	0.28	0.23	0.42	0.36	0.27	0.14	0.46
Asthenes humicola	0.75	1	0.67	0.53	0.76	0.47	0.51	0.44	0.41	0.45	0.18	0.39	0.43	0.38	0.24	0.53	0.36	0.31	0.27	0.45
Leptasthenura aegithaloides	0.72	0.55	1	0.65	0.67	0.56	0.34	0.39	0.30	0.38	0.10	0.30	0.29	0.27	0.14	0.46	0.40	0.17	0.19	0.53
Anaeretes parulus	0.73	0.69	0.70	1	0.46	0.40	0.31	0.35	0.43	0.32	0.08	0.25	0.34	0.41	0.16	0.37	0.54	0.15	0.15	0.50
Elaenia albiceps	0.65	0.68	0.68	0.85	1	0.65	0.35	0.25	0.29	0.38	0.10	0.28	0.34	0.28	0.26	0.49	0.39	0.16	0.11	0.19
Pyrope pyrope	0.60	0.52	0.64	0.57	0.60	1	0.34	0.17	0.21	0.33	0.11	0.17	0.30	0.20	0.17	0.53	0.39	0.19	0.05	0.35
Patagona gigas	0.33	0.18	0.43	0.28	0.25	0.41	1	0.23	0.21	0.14	0.12	0.20	0.29	0.17	0.18	0.27	0.35	0.21	0.13	0.40
Lophortyx californica	0.43	0.48	0.36	0.35	0.36	0.31	0.12	1	0.71	0.24	0.60	0.84	0.66	0.57	0.38	0.25	0.29	0.62	0.21	0.22
Diuca diuca	0.42	0.54	0.34	0.33	0.32	0.27	0.09	0.68	1	0.52	0.43	0.57	0.66	0.74	0.54	0.36	0.35	0.51	0.13	0.14
Zonotrichia capensis	0.61	0.72	0.51	0.50	0.51	0.42	0.14	0.65	0.72	1	0.31	0.35	0.49	0.50	0.24	0.58	0.23	0.33	0.01	0.14
Pteroptochos megapodius	0.29	0.42	0.29	0.28	0.32	0.39	0.15	0.50	0.41	0.39	1	0.62	0.52	0.31	0.37	0.12	0.11	0.59	0.09	0.11
Zenaidura auriculata	0.36	0.48	0.24	0.31	0.27	0.23	0.15	0.59	0.68	0.45	0.45	1	0.59	0.47	0.35	0.16	0.19	0.55	0.24	0.29
Agriornis livida	0.17	0.29	0.14	0.16	0.18	0.30	0.12	0.36	0.45	0.26	0.47	0.56	1	0.50	0.55	0.38	0.46	0.73	0.08	0.19
Turdus falklandii	0.48	0.58	0.33	0.35	0.33	0.32	0.18	0.66	0.69	0.60	0.62	0.62	0.57	1	0.58	0.37	0.28	0.36	0.12	0.18
Mimus tenca	0.47	0.71	0.44	0.56	0.58	0.45	0.16	0.53	0.47	0.58	0.65	0.56	0.38	0.72	1	0.20	0.29	0.63	0.07	0.08
Phytotoma rara	0.55	0.56	0.56	0.62	0.74	0.53	0.26	0.26	0.36	0.47	0.24	0.27	0.12	0.31	0.44	1	0.37	0.24	0.01	0.28
Dendrocopus lignarius	0.73	0.58	0.59	0.68	0.61	0.53	0.36	0.28	0.31	0.41	0.18	0.34	0.14	0.34	0.57	0.57	1	0.27	0.19	0.36
Colaptes pitius	0.35	0.45	0.23	0.28	0.20	0.21	0.24	0.54	0.55	0.37	0.34	0.73	0.52	0.58	0.42	0.34	0.30	1	0.07	0.15
Tachycineta leocopyga	0.17	0.14	0.19	0.28	0.17	0.18	0.10	0.21	0.14	0.20	0.12	0.22	0.04	0.14	0.14	0.12	0.11	0.18	1	0.36
Chordeiles longipennis	0.19	0.18	0.25	0.21	0.22	0.25	0.17	0.22	0.16	0.20	0.13	0.25	0.02	0.17	0.16	0.18	0.12	0.18	0.70	1

shows that overlaps generally increase horizontally and vertically away from the diagonals, supporting the original grouping in Table 5. Following standard methods (see, eg., SOKAL and SNEATH, 1963), species are grouped into similar arrangements in the two communities, with major associations foliage-level and ground-level feeders. An extra group is formed in chaparral by combination ground-, foliage- and aerial feeders, but the rank correlation remains very high. In summary, the average matorral Chilean species differs from the average chaparral Californian species in a) broader occupancy of

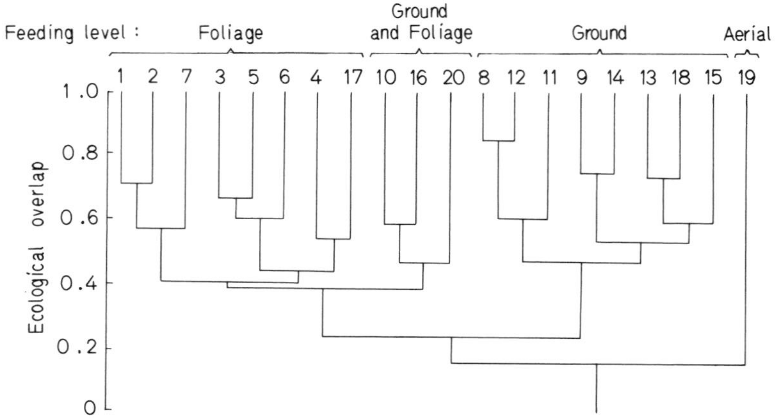

Fig. 9. Dendrogram of ecological relations among 20 chaparral bird species of California. Species numbers are as in Table 5

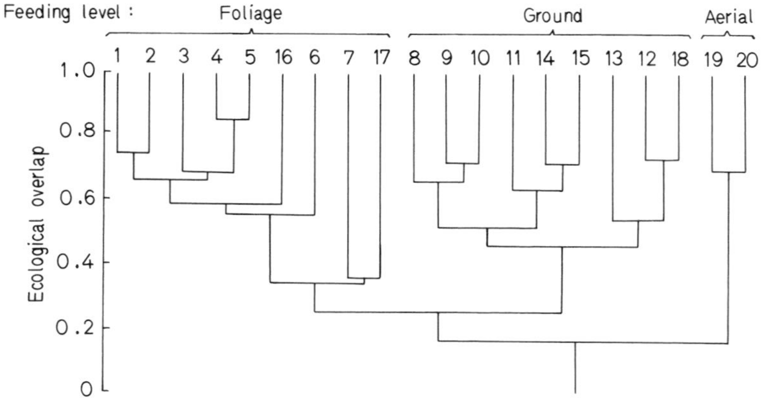

Fig. 10. Dendrogram of ecological relations among 20 matorral bird species of Chile. Species numbers are as in Table 5

the "coastal-sage"-woodland habitat gradient (Table 5), b) broader range of vertical feeding zones (Fig. 11); the average matorral species pair differs from the average chaparral species pair in a) 6% less habitat overlap within the study site (89% *vs* 95%), b) 9% less overlap in vertical foraging zones (26% *vs* 35%), and c) 12% more overlap in food feeding behavior (50.5% *vs* 38.5%). These differences may be regarded as minor in comparison with the methods of resource division in other bird communities (CODY, unpublished); the net outcome is that bird species tolerate a 12.7% total ecological overlap in chaparral (the product of the above three factors) *vs* 11.7% in matorral, figures which

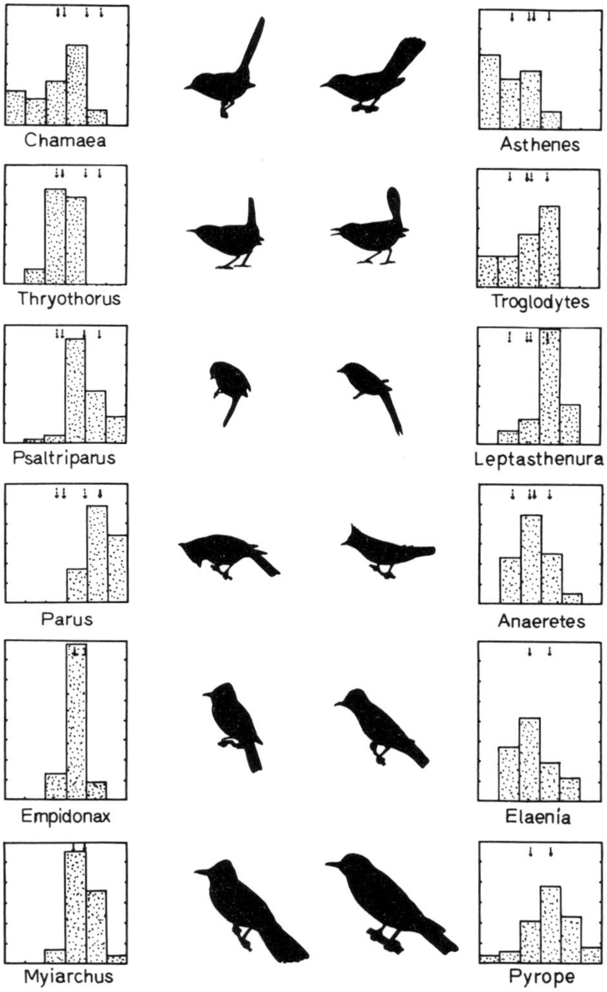

Fig. 11. Frequency distributions of feeding heights of four foliage insectivores and two sallying flycatchers in chaparral and of the six ecologically homologous species in matorral. Silhouettes are all to the same scale. Arrows above each distribution indicate its mean (in terms of the six equivalent units of the abscissa) and those of the frequency distribution of neighboring species

differ by only 1%. For perspective and to appreciate the similarity of the overlap figures, bird communities have been found to vary in percentage ecological overlap of pairs of resident species from less than 10% to greater than 25%.

6. *The Foliage Insectivores and Sallying Flycatchers*

The canopy insectivores may be treated in further detail. These species, grouped into the first two categories of Table 5, number six at each site and correspond well in total density (2.71 *vs* 2.89). All are full members except *Empidonax difficilis,* which is marginal at the chaparral site. The four foliage insectivores are mainly separable on the basis of foraging height. In chaparral, the wrentit *Chamaea fuscata* feeds low in the vegetation (average 3.09′) and to some extent on the ground. Above it may be ranked Bewick wren *Thryothorus bewickii* (average 3.30′), bushtit *Psaltriparus minimus* (average 4.47′) and plain titmouse *Parus inornatus* (average 8.0′). In Chile the homologues are similarly stacked in feeding zones, with *Asthenes* in the lowest position (average 1.71′), *Troglodytes* above it (average 3.88′), and *Leptasthenura* and *Anaeretes* in the highest positions. The latter two occupy reversed position with respect to the chaparral species (averages 7.76′ and 4.20′ respectively), and show the effect of decreased vegetation height at the matorral site. In each plot the larger flycatcher feeds at higher levels than does its smaller competitor, and in each country is characteristic of drier and more open vegetation while the smaller flycatcher in more abundant in forests. Fig. 11 gives the feeding height histograms using six height categories; chaparral species have a reduced spread of

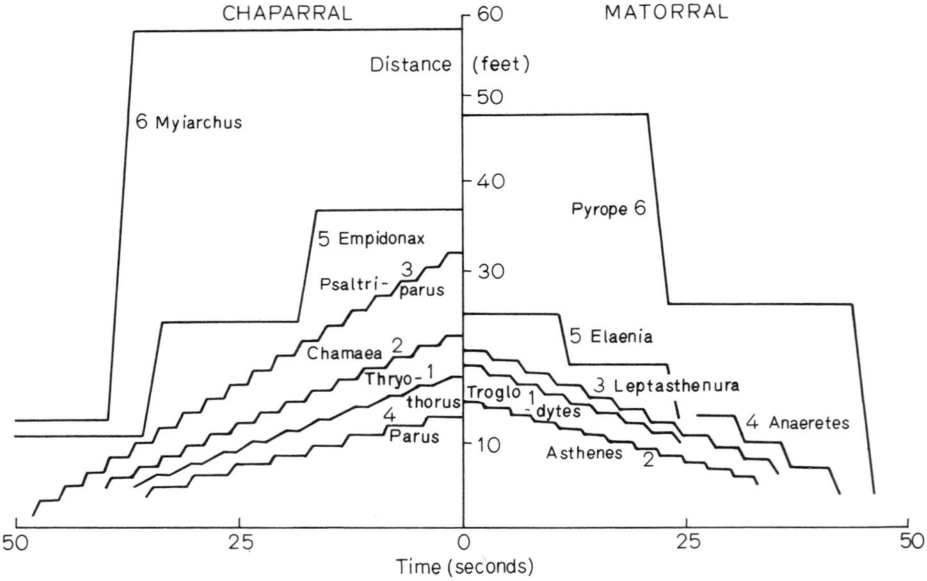

Fig. 12. Feeding behaviors of four foliage insectivores and two sallying flycatchers in chaparral and in matorral

activity among these categories as compared to the matorral species, for a diversity index of their coverage of vegetation height averages 2.55, only 78% of the 3.25 figure for their Chilean counterparts.

Silhouettes of the 12 species are included in Fig. 11 and are drawn to scale; the phenotypic correspondence in size and proportions is very striking. Lastly, the feeding behaviors of the species are compared in Fig. 12. The matorral Chilean species are rather less diversified than the chaparral Californian counterparts, which is perhaps associated with their broader foraging height niches. The flycatchers provide distinctive behavior graphs, and correspond well, but the foliage insectivores are not as easily matched. There are evidently, at this level of detail, a variety of efficient ways bracketed by the titmouse and bushtit feeding behaviors in which birds can move through scrub vegetation and harvest insects.

7. *Chaparral Elsewhere*

It would be extremely interesting to extend the Chile-California comparison to the bird communities of the same vegetation type elsewhere. Unfortunately almost nothing has been published on the southwestern Australian equivalent, and only bird species lists from South Africa (WINTERBOTTOM, 1966; BROEKHUYSEN, 1966). Some comprehensive studies have been conducted in *garrigue* in southern France by BLONDEL (1965, 1969), but his site appears to support a rather lower and more patchy vegetation, and is extremely low in bird species. Further work with careful habitat matching should prove to be very valuable.

Convergence between Avifaunas

The final question to be considered in this chapter is: Does the whole avifauna of a particular region show any correspondence to those of other, comparable regions? The answers to this question are obscure for several reasons. If the question is to be at all meaningful, the way in which "regions" should be "comparable" is in habitat types and perhaps also in the relative proportions of each habitat type. Such matches between large geographic areas are correspondingly difficult to make and conclusions that are drawn less valuable.

CODY (1970) attempted this sort of comparison between the non-marine bird faunas of California and Chile, although in a superficial way. These are two of the very few regions worth comparing, perhaps, for California and Chile are similar in size and latitudinal range, both face an ocean and are backed by tall mountains, and both possess a range of habitats from dry deserts at low latitudes through mediterranean habitats to wet forests at higher latitudes. Each region also possesses a coastal mountain range and prominent north-south valleys.

The total number of species in each region is extremely similar, with 235 in California and 230 in Chile. However, the distribution of these species among habitats within each region differs, as the Californian species tend to be habitat-specific and occur over the whole state (inasmuch as do their specific habitats). In Chile, on the other hand, species are much less habitat specific but are more restricted in geographic distribution; a species turnover occurs with shifts on a north-south axis rather than between habitats (CODY, 1970). The Chilean distribution type is characteristic of island avifaunas, and is thought to reflect the more isolated nature of central Chile from sources of potential colonists and the subsequent reduced rate of immigration.

The number of bird species which can be found in a patch of habitat is related to the structural complexity of the habitat (MACARTHUR, 1965). Both North American and Australian habitats follow the same linear relation between bird species diversity and diversity of habitat structure (RECHER, 1969), but the Chilean habitats, although within the same range of temperate latitudes, support species-packing levels closer to those of tropical habitats (MACARTHUR, RECHER and CODY, 1966; CODY, 1970).

A different approach to avifaunal equivalence was taken by LEIN (1972), who examined the occupancy in different zoogeographic areas (Sclater regions) of various trophic levels in terms of number of species. Correspondence was high (90%) between regions of similar habitat types (eg. Nearctic and Palaearctic) and of close proximity and historical association (eg. Nearctic and Neotropics), and reduced to around 70% to other pair-wise comparisons. Several anomalies showed up, such as a low incidence of nectivory in the Palaearctic, and a scarcity of ground-foraging birds in Australia, but the overall similarities in the number of species which exploit different sets of food resources remain impressive.

Less comprehensive comparisons, for instance between homologous families, have been drawn and some have already been mentioned in this chapter (see also FRY, 1970). LACK (1968, Ch. 5) has shown that many parallel species can be identified in the Icteridae (North and Central America) and the Ploceidae (Africa), and discusses many more instances of convergent evolution at lower taxonomic levels. These comparisons, and indeed most of those made between families and avifaunas are necessarily of a qualitative nature, and lie more in the realm of comparative and historical zoogeography than in community ecology. Pending more quantitative and detailed studies of these higher groups, I consider them to be of reduced value in furthering the conclusions already made.

Acknowledgements

The original field work reported in this chapter was financially supported by the National Science Foundation (GB-6150, GB-13651), and the Ford Foundation through the University of California-Universidad de Chile exchange program.

References

ALM, B., MYHRBERG, H., NYHOLM, E., SVENSSON, S.: Densities of birds in Alpine heaths. Vår Fågelv. **25**, 193–201 (1966).

BERNARD, F.: Biotopes habituels des fourmis sahariennes de plaine, d'après l'abondance de leurs nids en 60 stations très diverses. Bull. Soc. Hist. Nat. Afrique du Nord **52**, 21-40 (1961).

BERNSTEIN, R. A.: The ecology of ants in the Mohave Desert: Their interspecific relationships, resource utilization and diversity. Ph. D. Thesis, Los Angeles: University of California 1971.

BLONDEL, J.: Etude des populations d'oiseaux dans une garrigue Méditerranéenne: description du milieu, de la méthode de travail et exposé des premiers résultats obtenus à la période de reproduction. Terre et Vie **112**, 311-341 (1965).

BLONDEL, J.: Synécologie des Passeraux résidents et migrateurs dans le Midi Méditerranéen français. Marseille: Centre Régional de Documentation Pédagogique 1969.

BROEKHUYSEN, G. J.: The avifauna of the Cape "Protea-Heath Macchia" habitat in South Africa. Ostr. Suppl. N. 6, 323–334 (1966).

BUFFON, G. L.: Histoire Naturelle XI. History of Birds v. 1. Trans. W. Smellie. London: Cadell and Davies 1812.

CODY, M. L.: The consistency of inter- and intra-continental bird species counts. Amer. Natur. **100**, 371–376 (1966).

CODY, M. L.: On the methods of resource division in grassland bird communities. Amer. Natur. **102**, 107–147 (1968).
CODY, M. L.: Convergent characteristics among sympatric populations: A possible relation to interspecific territoriality. Condor **71**, 222–239 (1969).
CODY, M. L.: Chilean bird distribution. Ecology **51**, 454–464 (1970).
CODY, M. L.: Coexistence, coevolution and convergent evolution in seabird communities. Ecology **54**, (1973).
DARLINGTON, P. J.: The Biogeography of the Southern End of the World. Cambridge, Mass: Harvard Univ. Press 1965.
DUBOST, G.: Les niches écologiques des forêts tropicales sud-américaines et africaines, sources de convergences remarquables entre rongeurs et artiodactyls. Terre et Vie **22**, 3–28 (1968).
FILIPPI, G.: (1847) Mus. Mediol., Anim. Vertebr., cl. 2, Aves, pp. 15, 32. (Cited in JOHNSON, GOODALL and PHILIPPI, 1957).
FRIEDMANN, H.: Ecological counterparts in birds. Sci. Monthly **43**, 395–398 (1946).
FRY, C. H.: Convergence between jacamars and bee-eaters. Ibis **112**, 257–259 (1970).
GYSELS, H., RABAEY, M.: Taxonomic relationships of *Alca torda, Fratercula arctica* and *Uria aalge* as revealed by biochemical methods. Ibis **106**, 536–540 (1964).
INGHAM, C. O.: An ecological and taxonomic study of the ants of the Great Basin and Mohave Desert regions of southwestern Utah. Dissert. Abstr. **24**, 1759–1760 (1963).
JOENSEN, A. H.: An investigation on bird populations in four deciduous forest areas on Als in 1962 and 1963. Dansk Ornith. Foren. Tidsskr. **59**, 111–186 (1965).
JOHNSON, A., GOODALL, J. D., PHILIPPI, R. A.: Las Aves de Chile. Buenos Aires: Platt Establ. Graf. S. A. 1957.
JOHNSON, N. K.: Biosystematics of sibling species of flycatchers in the *Empidonax hammondii-oberholseri-wrightii* complex. Univ. Calif. Publ. in Zool. **66**, 79–238 (1963).
KIKKAWA, J.: Population distribution of land birds in temperate rain forest of southern New Zealand. Trans. Royal Soc. New Zealand **7**, 215–277 (1966).
KIKKAWA, J., HORE-LACY, I., BRERETON, J. LEGAY: A preliminary report on the birds of the New England National Park. Emu **65**, 139–143 (1965).
KURODA, N. H.: Morpho-anatomical analysis of parallel evolution between Diving Petrel and Ancient Auk, with comparative osteological data of other species. Misc. Rep. Yamashina Inst. Ornith. Zool. **5**, 111–137 (1967).
KURODA, N. H.: Avifaunal survey of Mt. in Iwate, N. Honshu. Misc. Rep. Yamashina Inst. Ornith. **5**, 214–240 (1968).
KUSNEZOV, N.: Formas de vida especializadas y su desarollo en diferentes partes del mundo. Dusenia (Curitiba, Brasil) **4**, 85–102 (1953).
KUSNEZOV, N.: A comparative study of ants in desert regions of central Asia and of South America. Amer. Natur. **90**, 349–360 (1956).
KUZNETSOV, A. P.: Ecology and distribution of the sea bottom fauna and flora. Trans. Shirshov Inst. Oceanol. **88**, 1–112 (1970).
LACK, D.: Ecological Adaptations for Breeding in Birds. London: Methuen 1968.
LACK, D.: Tit niches in two worlds, or, Homage to G. E. Hutchinson. Amer. Natur. **103**, 43–49 (1969).
LANYON, W. E.: The comparative biology of the meadowlarks *(Sturnella)* in Wisconsin. Nuttall Ornith. Club Publ. **1**, 1–67 (1957).
LEIN, H. R.: A trophic comparison of avifaunas. Syst. Zool. **21**, 135–150 (1972).
LINNAEUS, C.: Systema Naturae. Regnum Animale. Tenth Edition. Leipzig: Engelmann 1758.
MACARTHUR, R. H.: Patterns of species diversity. Biol. Rev. **40**, 510–533 (1965).
MACARTHUR, R. H., RECHER, H., CODY, M. L.: On the relation between habitat selection and bird species diversity. Amer. Natur. **100**, 319–332 (1966).
MOONEY, H., DUNN, E. L.: Convergent evolution of mediterranean-climate evergreen sclerophyll shrubs. Evolution **24**, 292–303 (1970).
MOREAU, R. E.: The Bird Faunas of Africa and its Islands. London: Academic Press 1966.
MURPHY, R. C.: The Oceanic Birds of South America. New York: American Museum of Natural History 1936.
NAKAMURA, T.: A survey of an upland grassland bird community during the breeding season. Misc. Rep. Yamashina Inst. Ornith. Zool. **3**, 334–357 (1963).

NAKAMURA, T., YAMAGUCHI, S., IIJIMA, K., KAGAWA, T.: A comparative study on the habitat preference and home range of four species of the genus *Emberiza* on peat grassland. Misc. Rep. Yamashina Inst. Ornith. Zool. **5**, 313–336 (1968).

OGASAWARA, K.: Bird survey of Mt. Kurikoma and its surrounding area, northern Honshu, with ecological notes. Misc. Rep. Yamashina Inst. Ornith. Zool. **4**, 371–377 (1966).

PATRICK, R.: A study of the numbers and kinds of species found in rivers in Eastern United States. Proc. Acad. Nat. Sci. Philadelphia **113**, 215–258 (1961).

PATRICK, R.: A discussion of the results of the Catherwood Expedition to the Peruvian headwaters of the Amazon. Proc. Internat. Assoc. Theor. Appl. Limnol. **15**, 1084–1090 (1964).

PETERSON, R. T.: A Field Guide to the Birds of the Western United States. New York: Houghton-Mifflin 1961.

PIANKA, E.: Convexity, desert lizards and spatial heterogeneity. Ecology **47**, 1055–1059 (1966).

PIANKA, E.: On lizards species diversity: North American flatland deserts. Ecology **48**, 333–351 (1967).

PIANKA, E.: Habitat specificity, speciation and species diversity in Australian desert lizards. Ecology **50**, 498–502 (1969 a).

PIANKA, E.: Sympatry of desert lizards *(Ctenotus)* in Western Australia. Ecology **50**, 1012–1030 (1969 b).

PIANKA, E.: The ecology of *Moloch horridus* (Lacertilia: Agamidae) in Western Australia. Copeia **1970**, 90–103 (1970).

PIANKA, E.: Comparative ecology of two lizards. Copeia **1971**, 129–138 (1971).

RECHER, H.: Bird species diversity and habitat diversity in Australia and North America. Amer. Natur. **103**, 75–79 (1969).

RICHDALE, L. E.: The Kuaka or Diving Petrel, *Pelecanoides urinatrix* (Gmelin). Emu **43**, 24–48 (1943).

RIDPATH, M. G., MOREAU, R. E.: The birds of Tasmania: Ecology and evolution. Ibis **108**, 348–393 (1966).

SHELDON, A. L.: Species diversity and longitudinal succession in stream fishes. Ecology **49**, 193–198 (1968).

SMITH, H. J.: A breeding bird survey of an uncultivated grassland at Regina. Blue-Jay **24**, 129–131 (1966).

SOIKKELI, M.: On the structure of the bird fauna on some coastal meadows in western Finland. Orn. Fenn. **42**, 101–112 (1965).

SOKAL, R., SNEATH, P.: Principles of Numerical Taxonomy. San Francisco: Freeman & Co 1963.

STRESEMANN, E.: Interspecific competition in chats. Ibis **92**, 148 (1950).

THORSON, G.: Bottom communities (Sublittoral or Shallow Shelf). In: Treatise on Marine Ecology and Palaeoecology, Ed. J.W. HEDGPETH, Ch. 17. Mem. Geol. Soc. America **67**, 461–534 (1957).

THORSON, G.: Parallel level bottom communities, their temperature adaptation and their "balance" between predators and food animals. Persp. Marine Biol., Los Angeles, Calif. (1958, 1960).

URAMOTO, M.: Ecological study of the bird community of the broad-leaved deciduous forest of central Japan. Misc. Rep. Yamashina Inst. Ornith. Zool. **3**, 1–32 (1961).

VUILLEUMIER, F.: Bird species diversity in Patagonia (temperate South America). Amer. Natur. **106**, 266–271 (1972).

WALKER, E.: The Mammals of the World. v. I-III. Baltimore: The Johns Hopkins Press 1964.

WIENS, J.: An approach to the study of ecological relationships among grassland birds. Ornith. Monogr. **8**, (1969).

WILLIAMS, A. B.: The composition and dynamics of a beech-maple climax community. Ecol. Monogr. **6**, 318–408 (1936).

WILLIAMSON, K.: A bird community of accreting sand dunes and salt marsh. Brit. Birds **60**, 145-157 (1967).

WINTERBOTTOM, J. M.: Ecological distribution of birds in the indigenous vegetation of the southwest Cape. Ostrich **37**, 76–91 (1966).

4

Ecological Convergence of the Lizard Faunas of the Chaparral Communities in Chile and California

RICHARD D. SAGE

Introduction

This chapter is an attempt at documenting some impressions of ecological similarities in the lizard faunas of the chaparral habitats of central Chile and southern California. These impressions are based on many years of residence in the California region, and from a series of short visits to Chile during the years 1968–70. The ecologies of the species will be compared in a very simple manner. The morphological description of the species and faunas is compared against the ecological patterns. A discussion of the origins of the present-day ecological configuration of these faunas in the two habitats is related to the idea of a similar set of selection factors in operation in the two environments, resulting in ecological convergence.

The first section will deal with the taxonomic entities involved, in order to familiarize the reader with the general diversity patterns in these temperate regions of North and South America.

The Faunal Diversity

Faunal diversity, as used here, is simply the number of taxa which occur in the habitat. It is a measurement of the number of evolutionary units present in the environment, and as such may be a reflection on the complexity of the ecological system. Table 1 is a listing of the species which occur in the chaparral habitats in the respective areas.

The species list for the Chilean habitat is based on my collections made at one specific site: Puente de las Dehesas, 820 m elevation, Lo Barnecha, Santiago province. This locality is a disturbed habitat at the base of the mountains. The area includes a dry river bed, the adjacent riverine habitat and a hillside. Introduced, weedy-species of plants are prominent members of this community, and include the genera *Ribes, Pinus* and *Eucalyptus*. A total of six days were spent in the field at this locality during March of 1968 and 1969. Field observations on the species *L. chiliensis, lemniscatus* and *tenuis* made in other parts of Chile are in accord with those made at this locality. The conclusions in this discussion are based on limited samples and should be considered in this light. Other species of *Liolaemus* do occur within only kilometers of this site, and of course may be considered as members of this chaparral community. However, it is my experience with this genus in Chile and Argentina that there is rapid taxa replacement over short distances, although ecological diversity remains about the same. Thus in this genus many species seem to have very similar ecologies and simply replace one another geographically.

The list of lizard species selected as members of the California chaparral fauna is based on personal experience in the area during many years. The list is composed of basic, widespread elements found in the chaparral. It does not include intruders from other habitats, such as the fauna of rock-dwellers and desert species, which do occur in this formation in extreme southern California. This selection of species would be found over the south-central part of the California chaparral, from Orange to San Luis Obispo counties.

Table 1. The lizard species found in the chaparral habitat

Chile		California	
Callopistes maculatus	(Teiidae)	*Cnemidophorus tigris*	(Teiidae)
Liolaemus chiliensis	(Iguanidae)	*Gerrhonotus multicarinatus*	(Anguidae)
Liolaemus lemniscatus	(Iguanidae)	*Eumeces skiltonianus*	(Scincidae)
Liolaemus tenuis	(Iguanidae)	*Sceloporus occidentalis*	(Iguanidae)
Liolaemus monticola	(Iguanidae)	*Uta stansburiana*	(Iguanidae)
		Phrynosoma coronatum	(Iguanidae)
		Anniella pulchra	(Anniellidae)

The two points to consider are the absolute numbers of species per habitat, and the taxonomic diversity of the two areas. A total of five species were found in Chile, while seven species would be expected to occur on a similar site in California. This is a difference of 28.5% between the two habitats. Such a difference does not seem very great.

The higher taxonomic structure of the two faunas differs notably. The Chilean fauna is simple, composed of two genera in two families. One genus, *Liolaemus*, dominates. In California the structure is quite distinct. Here there are seven genera belonging to five families. There is no case of two or more congeneric species.

The Ecologic Organization

HUTCHINSON'S (1959) conceptualization of a series of volumes which combine to describe the niche of an animal species is a convenient way of dividing this complex state into more manageable parts. Three important sub-components of the niche are considered: the time, place and food exploitation patterns of the species. These will be divided into very simple types of categories, in keeping with the preliminary nature of the research.

The time exploitation pattern can be broken down into two major, non-overlapping categories, i.e., a nocturnal or diurnal activity period. The groups in both habitats are all diurnal in this pattern[1].

1 In making the list for the California fauna, the nocturnal gecko, *Coleonyx variegatus*, was omitted, since it enters only the southernmost part of the chaparral. It is a more typical member of the desert fauna to the south and east of the coastal chaparral. In Chile a similar situation exists in the presence of gecko species of the genus *Garthia* towards the driest, northernmost limits of the chaparral formation.

The place exploitation pattern considers the three-dimensional spatial distribution of the species during their activity periods. The divisions here are very basic ones, but with some sub-divisions. The terms are largely self-explanatory except for those in the subdivisional areas. The "Open" and "Closed" parts of the Terrestrial space are respectively the exposed areas without vegetation and the area covered by vegetation. In Table 2 a comparison is made of the faunas with respect to the place exploitation patterns. The major difference between the two faunas is the absence of a subterranean element in the Chilean region. Otherwise, similarities are great between the two faunas in the number of places inhabited.

Table 2. The place exploitation pattern in the lizards from chaparral communities

Place	Chile	California
Subterranean	none present	*A. pulchra*
Terrestrial:		
Open	*C. maculatus*	*C. tigris*
	L. monticola	*U. stansburiana*
Closed	*L. lemniscatus*	*E. skiltonianus*
	L. chiliensis	*G. multicarinatus*
	(partly arboreal)	(partly arboreal)
Arboreal	*L. tenuis*	*S. occidentalis*

The food exploitation pattern consists of two portions: the diet and the predation strategy patterns. The first deals with what foods are eaten, i.e. a trophic position in the community's energy chain. The second involves the manner of selection of prey items and in what proportions the food species are taken. Table 3 shows the dietary patterns for the two faunas and Table 4 compares their predation strategies. A comparison of the diet patterns for the faunas shows much similarity. The one major difference is the absence of an ant-eating specialist in Chile, while it is represented in California by a species of *Phrynosoma*. HURTUBIA and DI CASTRI (in this volume) indicate that the species *Liolaemus fuscus* seems to be a specialist on ants. It is a widespread and common species in Chile, but was not found in the area where the observations were made. The presence of the secondary predator in the Chilean fauna, as represented by *C. maculatus* is based on the statement by

Table 3. The diet patterns of the chaparral lizard faunas

Diet	Chile	California
Herbivorous	none	none
Insectivorous	*C. maculatus* (part.)	*C. tigris*
(small arthropods)	*L. lemniscatus*	*E. skiltonianus*
	L. chiliensis	*G. multicarinatus*
	L. monticola	*U. stansburiana*
	L. tenuis	*S. occidentalis*
		A. pulchra
(ants)		*P. coronatum*
Secondary Carnivore	*C. maculatus* (part.)	

DONOSO-BARROS(1966) that this species includes small lizards in its diet. This is not considered to be a major departure from the general similarity of the dietary patterns. Such a tendency to eat other lizards, and thus place themselves in this other category, is largely the direct consequence of a prey-size versus lizard-size relationship in this generalist class of predators. When the predatory lizards reach the large sizes, as *Callopistes* does, smaller lizards become suitable food items.

Table 4. The predation strategies in the lizards of the chaparral communities

Predation Strategy	Chile[a]	California
Generalist		
"Sit and Wait"	*L. monticola*	*U. stansburiana*
	L. tenuis	*S. occidentalis*
"Active Foraging"	*C. maculatus* (presumably)	*C. tigris*
		A. pulchra
		E. skiltonianus
		G. multicarinatus
Specialist		
"Sit and Wait"	none	*P. coronatum*
"Active Foraging"	none	none

[a] No observations on the prey capture methods of *L. chiliensis* and *L. lemniscatus* are available. From their other ecological characteristics it is predicted that they will be more of the "Active Foraging" type than are *L. monticola* or *L. tenuis*.

Table 5. Ecological counterparts in the chaparral communities

Chile	California
C. maculatus	*C. tigris*
L. chiliensis	*G. multicarinatus*
L. lemniscatus	*E. skiltonianus*
L. tenuis	*S. occidentalis*
L. monticola	*U. stansburiana*
none	*P. coronatum*
none	*A. pulchra*

A survey of simple measures of important ecological parameters suggests similarity between the two faunas. A consideration of what species-members in the two faunas are most similar is attempted here. By simply matching up species which share most of the measures in common, pairs of species which are ecological equivalents are established. These species-pairs are those whose adaptive responses have been most similar and may be considered as convergent upon one another. These ecological equivalents are presented in Table 5. It is concluded that two California species have no ecological counterparts in the Chilean habitat studied.

One of these is a subterranean form, *A. pulchra;* and the other one is a dietary specialist on ants, *P. coronatum*. The other pairs share their time, space and food exploitation patterns in common manners.

The Morphological Space

The niche hypervolume concept is useful in determining similarities and differences in ecological relationships in multi-species studies. A similar type of concept for the morphological entities which are the species populations is used here. This *morphological space* is a volume describing the topological characteristics of the species. For use as an adjunct in ecological studies the morphological parameters used should be those which demonstrate a functional relationship to important ecological behavioral patterns.

The operational basis for such a comparative work is the widely held assumption that form has function in behavioral patterns and is subject to adaptive modification. This means that form does to some degree reflect behavioral pattern. If the chosen characters are important to behavioral patterns of ecological relevance, then we are describing the ecologically important, morphological-species. Such a description provides the material for comparison to the related behavioral patterns if such data are available, or it may serve in comparisons of morphologies with other species, from the same or different habitats. The latter use will be employed here.

In this analysis only two sets of characters were used to create a planar space, instead of a volume of greater dimensionality. The two characters are involved in locomotory and prey-capture behaviors. A ratio between fore and hindleg lengths describes the relative importance in movement that these limbs contribute. Legs of more equal length would contribute more equally in movement. Longer hindlegs indicate that the forelegs are contributing less during forward propulsion. This condition is found in fast funning and hopping forms. Fast, or bipedal running is commonly found in species of open habitats, where obstructions to high velocities are minimal. Legs of more equal lengths are found in forms living in more closed types of habitats, or on vertical surfaces. Here the forelegs are important in pulling the animal forward or upwards.

The jaws are important in the prey-capture act. The relative conformation of the head and jaws will affect the manner in which the predator can carry out its prey-capture behavior. A long, narrow head will provide the organism with the ability to swing the snout faster and snap the jaws shut more quickly than a form tending towards the opposite condition (LUNDELIUS, 1957). The more narrow snout also produces a pincer-like structure which can be used for poking into crevices.

Measurements of these characters were made on series of specimens from both habitats. The results are displayed in Fig. 1 (a and b). For the California plot the legless *Anniella* has been omitted because the loss of legs results in its loss of planar dimensionality. In the plot of the California fauna, the species sort out into almost completely non-overlapping clouds of points. Some of the known behavioral patterns correlate with the relative positions of the species in this morphological space. With the exception of *Phrynosoma*, the species of the open habitats (*Uta* and *Cnemidophorus*) have relatively longer hind legs than the species of closed habitats (*Eumeces* and *Gerrhonotus*) or arboreal sites *(Sceloporus)*. The explanation of the aberrant position of *Phrynosoma* on this locomotory axis almost certainly relates to the vagility of its prey items (ants), and its escape response (crypticity and immobility). In regard to the relationship between the morphology of the head and the diet, the distributions cannot be so easily explained at present. This is because the correct types of analyses have not been done. The specialization of *Phrynosoma* on the slow ants is in accord with its position in the space.

Whether the other species take relatively increasing percentages of more vagile prey items is unknown, but the prediction is that such a trend would emerge.

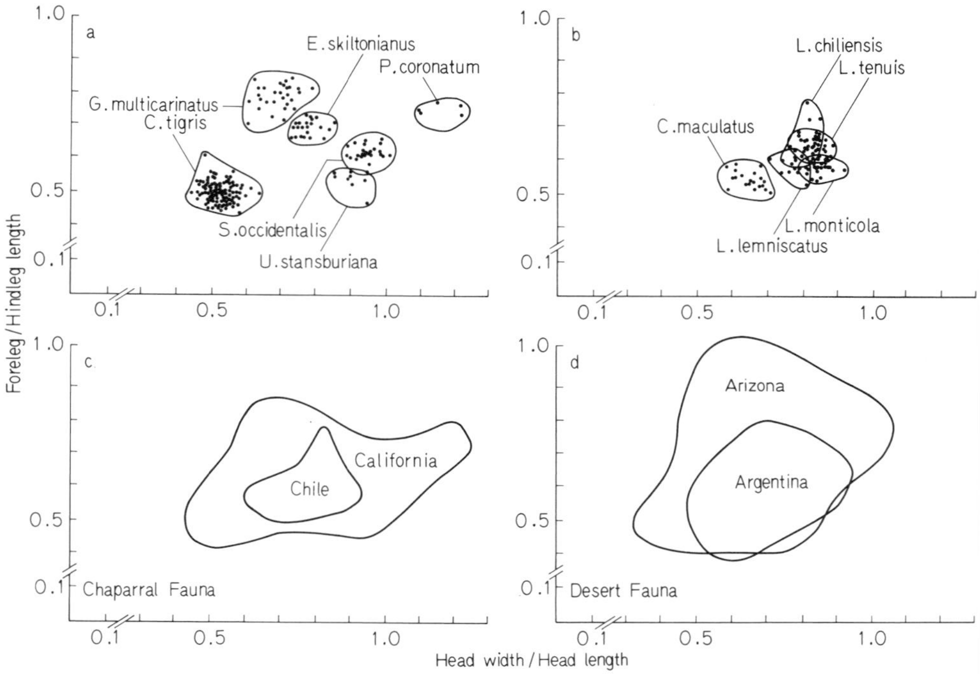

Fig. 1. The Morphological Spaces of the species in the California (a) and the Chilean (b) Chaparral faunas, and the Faunal Spaces of the Chaparral (c) and Flatland Desert (d) communities

The picture presented by the Chilean fauna (Fig. 1 b) is quite different from that for California. In this case there is much overlap in the space occupied by the different species. Nevertheless, the amount of area occupied by each species is about the same as that found in the California species. Although broadly overlapping, the species of *Liolaemus* do show some separation. The tendency toward separation in the same directions as their California equivalents is in part present. The two more arboreal forms, *L. chiliensis* and *L. tenuis* have legs of more equal length than do the terrestrial forms. However, there appears to be no separation between the terrestrial forms from the closed habitat, *L. lemniscatus* and the open habitat, *L. monticola*. The relationship between head form and prey type can only be speculated upon at this time. The single sample for food habits that I have looked at revealed that *L. monticola* took many more ants than the other three species of *Liolaemus*. If this is typical it at least conforms to the prediction in relation to *L. chiliensis* and *L. lemniscatus*.

The inclusive area covered by the species of the habitat may be called the *faunal space*. This describes all of the morphological area swept out by that fauna. It is a collective description of the morphologies of the sympatric species. Knowing the exact areas swept out by the individual species, a sum value can be obtained. If the faunal space is divided by this sum value a measure of the overlap in morphological space results. A value of 1 for this quotient indicates that there is no overlap and no blank spaces. Values of less than 1 could indicate that overlap exists. These calculations were not made in this instance, due to the inadequate numbers of measurements for some of the species. However, visual inspection shows that for the California fauna a value greater than 1 would be obtained due to the empty spaces between the species. A value of considerably less than 1 would result for the Chilean fauna due to the large amount of overlap exhibited by the species of *Liolaemus*.

Discussion

The diversity figures for the two habitats show that two more species are found in the California chaparral than in a comparable habitat in Chile. The analysis of the ecological organization of the two faunas demonstrates that one of the place exploitation patterns is not found in any of the Chilean species, and that one of the food exploitation patterns is perhaps absent (see earlier comment by HURTUBIA and DI CASTRI). Tests of the consequences of such absences on the prey species, either on the ant fauna or the subterranean fauna would be of value, but feasible experiments do not come to mind.

Of the remaining five species, ecologically equivalent pairs seem to be present in the two regions. They span a fairly broad range of the space exploitation patterns, from arboreal to terrestrial and in open and closed habitats. Prey attack patterns range from the "Sit and Wait" to the "Active Foraging" methods in both areas. One consequence of such conclusions is that if placed in sympatry with one another (as in an experimental situation) the paired species would be in closer competition with one another than they would be with their resident faunal members. It would seem that with the two previously mentioned exceptions the breadth of the environmental occupation by the lizard faunas is about the same in both of these North and South American chaparral communities.

While the ecological organization is about the same, the morphological pattern is different. These discordancies provide for a number of speculations. For the sake of an argument let us assume that two of the earlier statements are correct. The first is that the five species-pairs are ecologically equivalent; secondly, that the measurements used in defining the morphological space and the faunal space do measure functionally adaptive anatomies in areas of prey capture and locomotion, in the same degree for all of the species. If these conditions are granted, then it is necessary to conclude that one species in each pair is not as well adapted for its ecological role as its equivalent in the other fauna. If the morphological characters are functionally important ones then they will have differing efficiencies of operation, depending on their degree of adaptive perfection. One state will be more efficient than others (and for all others if efficiency is a unimodal function). If selection will favor reduction in niche overlap in coexisting species, it is not unreasonable to suppose a concomitant trend for divergence in the functionally related anatomical structures. On the bases of these last two premises it is concluded that some members of the Chilean fauna, which are more similar among themselves, will be the ones which are less efficiently adapted to their ecological roles. In particular the two species from the closed habitats, *L. chiliensis* and *L. lemniscatus* would appear to be the least adapted for their niches. In absolute measurements these two are the most distantly separated species from their North American equivalents. Relative comparisons have been made here as it cannot be assumed that the North American forms are necessarily the most perfectly adapted types for their ways of life. Tests of this hypothesis might be done with physiological experiments on the climbing or locomotory efficiencies of the different species. A corollary would be that less efficient species would have to devote proportionally more of their energy intake to their own maintenance budgets, rather than in the reproductive efforts. This should be tested by calculating amounts of energy devoted to egg clutches in the different species.

If the hypothesis is correct it is an interesting consideration that some species are less well adapted to their niches. In this case two alternatives for this situation might be invoked. The first is that the evolution of niche differences has proceeded more rapidly than morphological modifications. The other alternative might be that anatomical changes have reached limits set by the genetic constitution, particularly in the *Liolaemus* species. Unfortunately, neither alternative is amenable to experimental testing.

As mentioned earlier, the faunal space can be used for comparative purposes. In the case of the California-Chile situation (Fig. 1 c), the larger space (50 square units) of the California fauna, versus a smaller space (9.9 square units) for the Chilean one gives an *a priori* indication of a larger ecological space. This is in accord with the prediction based on faunal diversity, but it is not a linear relationship. For another system we can compare a set of desert faunas. The faunas from the flatland desert communities of Arizona and central Argentina were depicted in a graphical manner (Fig. 1 d). In these two habitats 10 species are expected in Arizona and 9 in Argentina. These faunas have more species than those of the chaparral communities, and larger faunal spaces (70.3 and 29.4 square units for the Arizona and Argentina faunas respectively). The increase in areas is roughly proportional to species diversities, if the North and South American faunas are treated as belonging to different sets.

The Ecological Convergence

Ecological convergence is suggested for the systems under discussion. Convergence requires that similar selective forces have been acting in the past on two systems, which originally differed from one another, to produce a currently similar condition. Examination and demonstration of current similarities is a *post facto* study, and while a positive conclusion may be logically sound it cannot be considered a proven fact. Proof is impossible to obtain for the long-term evolutionary phenomena under discussion. We are left with comparative studies to provide tentative interpretations of the past historical sequence. This study of the chaparral faunas suggests one way such a sequence may take place.

The taxonomic diversity patterns in the two habitats indicate that two different methods of occupation of the chaparral have taken place. Information on the geographic origins of the genera and species is difficult to obtain. If the present day areas of maximum diversity of congeneric species is a good estimator of the general place of origin, then we can assume that the chaparral areas in California did not serve as such a site for most of the resident species. GREER (1970) says that *Eumeces* is of an Old World origin. The greatest diversities of the *Gerrhonotus, Sceloporus* and *Cnemidophorus* are in parts of Mexico and Central America. The group to which *Uta* belongs is most diverse in the desertic regions of western United States and northern Mexico. Only *Anniella* appears restricted to this region. The suggested sequence of occupation in California has been that of additions of ecologically fully "formed" invaders to the area, followed by minor adjustments to the immediate situation.

In Chile the pattern of two genera and families is different. *Callopistes* is presently restricted to the lowland regions of coastal western South America, from Ecuador to Chile. The four species of *Liolaemus* are all species restricted to central Chile. The diversification pattern of this latter genus is unknown, but this central region has as great a diversity of species as almost any other area within the distribution of the genus. So it is not too unreasonable to assume that these species had their origins, at least at the species level, in this general region. It is in this same region that they probably underwent their principal ecological modifications. This fauna was built up by autochthonous resident elements, which after initial genetic isolation, underwent ecological changes in the presence of one another (or closely related ancestral stocks), and arrived at a faunal structure similar to the one in California. The production of this ecological arrangement in the two habitats by different methods fits well with the assumption of the presence of similar sets of selective forces producing a set of similar adaptive responses.

Summary

The ecological organization of the lizard species in the chaparral habitats in Chile and California is arranged in a very similar way. The difference in species diversities is related to the absence of species with particular exploitation patterns, rather than an enlargement of the behavioral patterns of the other resident species to include these areas. In both sites there is utilization of both of the arboreal and terrestrial environmental space. In the terrestrial space both the open and closed portions are occupied. In California the subterranean space is occupied by one species, while in Chile this region is unoccupied. Feeding behaviors in California include both generalist species and a single specialist. In Chile only the generalist type of behavior seems to be present.

Granted a roughly equivalent similarity in the ecological space filled by the two faunas, the morphological spaces for the faunas are quite different. Morphological divergence is great between the species of the California fauna, while comparatively little separation and much overlap occurs among the members of the Chilean fauna. A suggested hypothesis to explain the smaller amount of divergence for the Chilean fauna might be the more rapid adoption of separate behavioral patterns than the appropriate morphological states, or that inherent genetic limitations have stopped the extent of divergence along the appropriate lines. The situation suggests that mechanical efficiencies of the organisms during their ecologically important behavioral patterns are less for one member of each of the equivalent species pairs. This hypothesis can be tested in the laboratory. One of the ecological consequences of such a difference may be a shift in energy appropriations away from reproduction to self-maintenance activities.

The faunas though ecologically similar demonstrate two different patterns of establishment. The California fauna is composed of taxa whose major ecological adaptations occurred outside this area. The fauna was built up by additions of well formed ecological types. In Chile the resident fauna is autochthonous, having undergone its ecological development *in situ,* and in the presence of other current residents. The case suggests an example of convergence of ecological structure for the lizard faunas, presumably due to similar selective forces acting in the two environments.

Acknowledgements

I would like to express my sincere thanks to a number of people who have helped in various ways in the development of this research. Mr. KLAUS BUSSE and his parents, who provided living facilities and kind hospitality, made my stays in Santiago during the field work very enjoyable. Mr. RAYMOND HUEY provided valuable discussion and ideas during the development of the work. Dr. WILLIAM MILSTEAD presented this contribution at the symposium conference in Valdivia, Chile, and made comments on the manuscript. Mrs. JEAN GRUBB typed it. To Dr. W. F. BLAIR I am particularly grateful for the opportunities to carry out studies of reptiles in temperate South America. Financial support for this work came from grants to Dr. BLAIR: AF-ASOSR 1327-67, NSF-GB 8265 and NSF-GB 15768. These three grants are in support of the United States effort in the International Biological Programme.

References

DONOSO-BARROS, R.: Reptiles de Chile, 382. Santiago: Ediciones de la Universidad de Chile 1966.

GREER, A. E.: A subfamilial classification of Scincid lizards. Bulletin Museum of Comparative Zoology **139,** 151–184 (1970).

HUTCHINSON, G. E.: Concluding remarks. Cold Spring Harbor Symposium of Quantitative Biology **22,** 415–427 (1957).

LUNDELIUS, E. L.: Skeletal adaptations in two species of *Sceloporus.* Evolution **11,** 65–83 (1957).

5

Segregation of Lizard Niches in the Mediterranean Region of Chile

JAIME HURTUBIA and FRANCESCO DI CASTRI

Introduction

Quantitative studies on ecological and morphological characteristics of congeneric sympatric species have recently had a strong influence on the development of evolutionary theories concerning niche breadth as an inverse measure of specialization, and to niche overlap or distance between the niche as a measurement of competition (HUTCHINSON, 1965; KLOPFER and MACARTHUR, 1960; MACARTHUR, 1968; LEVINS, 1968 a, b; SCHOENER, 1965; SCHOENER and GORMAN, 1968). The definition of these measurements has made it possible to enunciate several questions about the way in which the species segregate their niches, the size of the niche and the consequences of different niche breadths.

In addition, the concept of character displacement (BROWN and WILSON, 1956), together with the approach of LEVINS (1968 b) and MACARTHUR and WILSON (1967) on adaptive strategies in different environments, have given a new perspective to the study of sympatric species.

The purpose of this chapter is to discuss the type of strategies developed by species of the lizard genus *Liolaemus* (Sauria: Iguanidae) in order to segregate their niches in mediterranean-climate ecosystems of Chile. Briefly, the main questions posed are: What are the strategies segregating the niches among these species? How similar can these strategies be? How have the various strategies evolved?

The Genus Liolaemus in the Mediterranean Climate Region of Chile

The genus *Liolaemus* is a strictly neotropical group presenting a great variety of forms and is one of the most complex genera within the family Iguanidae. It has a continuous distribution in South America from the southern part of Ecuador and the coast of Rio de Janeiro in Brazil, south to Tierra del Fuego.

The physical geography of Chile has contributed greatly to making the Chilean species of *Liolaemus* a very isolated group from the rest of South American congeners. Furthermore, the intermountain valleys, broad zones of lowlands, the slopes of the volcanos in the southern region and a great part of the Atacama Desert in the north, are zones that have, in geological recent times, opened up for colonization by these lizards (HELLMICH, 1951).

In the mediterranean-climate region of Chile a great variety of habitats are found, referable to the range of climatic gradients and topographical features characterizing the zone. In this region the genus *Liolaemus* reaches its highest species diversity and highest degree of sympatry.

Between the parallels 25° and 41° S, which include the Chilean zone with a mediterranean climate, 30 species of *Liolaemus* occur. In this study, the zone between latitudes 32° and 36° S, is considered as the most representative of the mediterranean climate type. In this area there are 13 species of *Liolaemus* (DONOSO-BARROS, 1966). From this complex, nine of the ecologically better known and more common species have been chosen for study: *chiliensis, nitidus, schröderi, monticola monticola, lemniscatus, fuscus, tenuis tenuis, gravenhorsti* and *nigroviridis campanae.* The distribution of these species is given in Fig. 1.

Data Analysed

The data presented in this chapter were collected during investigations on altitudinal distribution (PINTO, HERMOSILLA, DI CASTRI and ASTUDILLO), on trophic and species diversity (HURTUBIA, DI CASTRI and PINO) and on herbivory under experimental conditions (DI CASTRI and FUENTES), still unpublished. In addition to these observations, information on morphological aspects of the various species was assembled from the literature, particularly from PFLAUMER (1944), HELLMICH (1952) and DONOSO-BARROS (1966).

Table 1. Distribution of character states in *Liolaemus* species (see text for coding)

Character	Morphological														
	1	2	3	4	5	6	7	8	9	10	11	12	13	14	15
chiliensis	1	0	0	0	0	0	0	0	0	0	1	0	0	0	0
gravenhorsti	0	1	1	1	1	0	1	0	1	0	0	0	0	0	0
tenuis tenuis	0	1	0	0	0	1	0	1	0	0	0	0	1	0	0
lemniscatus	0	0	1	1	0	1	0	0	1	0	0	0	0	0	1
fuscus	0	1	1	1	0	1	0	0	1	1	0	0	0	1	1
monticola monticola	0	1	0	1	1	1	0	1	1	1	1	0	0	0	0
nitidus	1	1	0	1	1	0	0	0	1	0	1	1	0	0	0
schröderi	0	0	1	1	1	1	1	0	1	1	1	1	0	0	0
nigroviridis campanae	1	0	0	0	0	1	1	1	1	1	0	1	0	0	0

In this analysis 15 ecological and 15 morphological characters were arbitrarily considered for each species. These characters are: A) **morphological.** 1) Size superior to 70 mm; 2) Snout-vent length/hindleg length; 3) Coloration pattern in bands; 4) Brown-sand pigmentation; 5) Coloration variability; 6) Side of the neck with folds; 7) Viviparous reproduction system; 8) Sexual dimorphism; 9) Prominent auditive scales; 10) Rostral at least three times wider than high; 11) Males with larger size; 12) Snout-vent length/tail length; 13) Size inferior to 50 mm; 14) Hindleg length/hindfoot length less than 1,8; 15) Hindleg length/hindfoot length superior to 2,2. B) **ecological.** 16) Constant herbivory; 17) Occasional herbivory; 18) High trophic diversity; 19) Arboreality; 20) High habitat diversity; 21) Altitudinal preference; 22) Xerophilous preference; 23) High density; 24)

Tendency to sympatry; 25) Altitudinal segregation; 26) Wide geographical distribution; 27) Colonisation of human settlements; 28) Southern displacement; 29) Northern displacement; 30) High frequency.

The distribution of these characters is summarized in Table 1. The characters have been subdivided into only two character states or attributes, by "absence-presence" and they are coded "0" and "1". For some characters it was necessary to have a double entry in this distribution; for instance, body size (characters 1 and 13): hindleg length/hindfoot length (characters 14 and 15), and geographical displacement (characters 28 and 29).

From Table 1 the similarity matrices shown in Table 2 were computed, according to the "simple matched coefficient" (CATTELL, 1949; SOKAL and MICHENER, 1958). Briefly, these values express a measure of the number of matches or agreements of characters out of the total number of such agreements possible in the study. Phenetic similarity (in standard type) and ecological similarity (in boldface) are presented. These similarity matrices correspond to a Q-Type Similarity Matrix (SOKAL and SNEATH, 1963).

Phenetic and Ecological Relationships in Liolaemus

In order to make more evident the inter-species similarities, two dendrograms have been constructed (Fig. 2). One is based on morphological characters (Fig. 2a) and the other on ecological attributes (Fig. 2b). These dendrograms have been elaborated by the

Table 1 (continued)

Ecological														
16	17	18	19	20	21	22	23	24	25	26	27	28	29	30
0	0	0	1	0	0	0	0	0	0	1	0	1	0	0
0	0	1	1	1	0	0	0	1	0	0	0	0	0	0
0	0	1	1	1	0	0	1	1	1	1	1	1	1	0
0	0	1	0	1	0	0	1	1	0	1	1	1	1	0
0	0	0	0	0	0	1	1	0	0	1	0	0	1	0
0	1	0	0	0	0	1	0	0	0	1	0	0	1	1
0	1	0	0	0	1	1	0	1	1	0	0	0	1	1
0	1	0	0	1	1	1	0	0	1	0	0	0	0	0
1	1	0	0	0	1	1	1	0	0	0	0	0	0	0

"weighted variable group" clustering method of numerical taxonomy (SOKAL and SNEATH, 1963), using SPEARMAN's "sums of variable" method for recomputing the correlation coefficients.

The classical type of dendrogram has been modified by introducing a logarithmic scale horizontal to the dendrogram, in order to highlight the different degree of similarity between the neighbouring species according to matrix 1 (see Table 2).

It is interesting to note that the phenetic relationships drawn from Fig. 2a group the species of *Liolaemus* in a similar way to that established by classical taxonomists. DONOSO-BARROS (1966) divides the Chilean species of the genus *Liolaemus* into 5 groups according to their sizes, dorsal patterns, coloration, type of scales, body shape, etc., thus

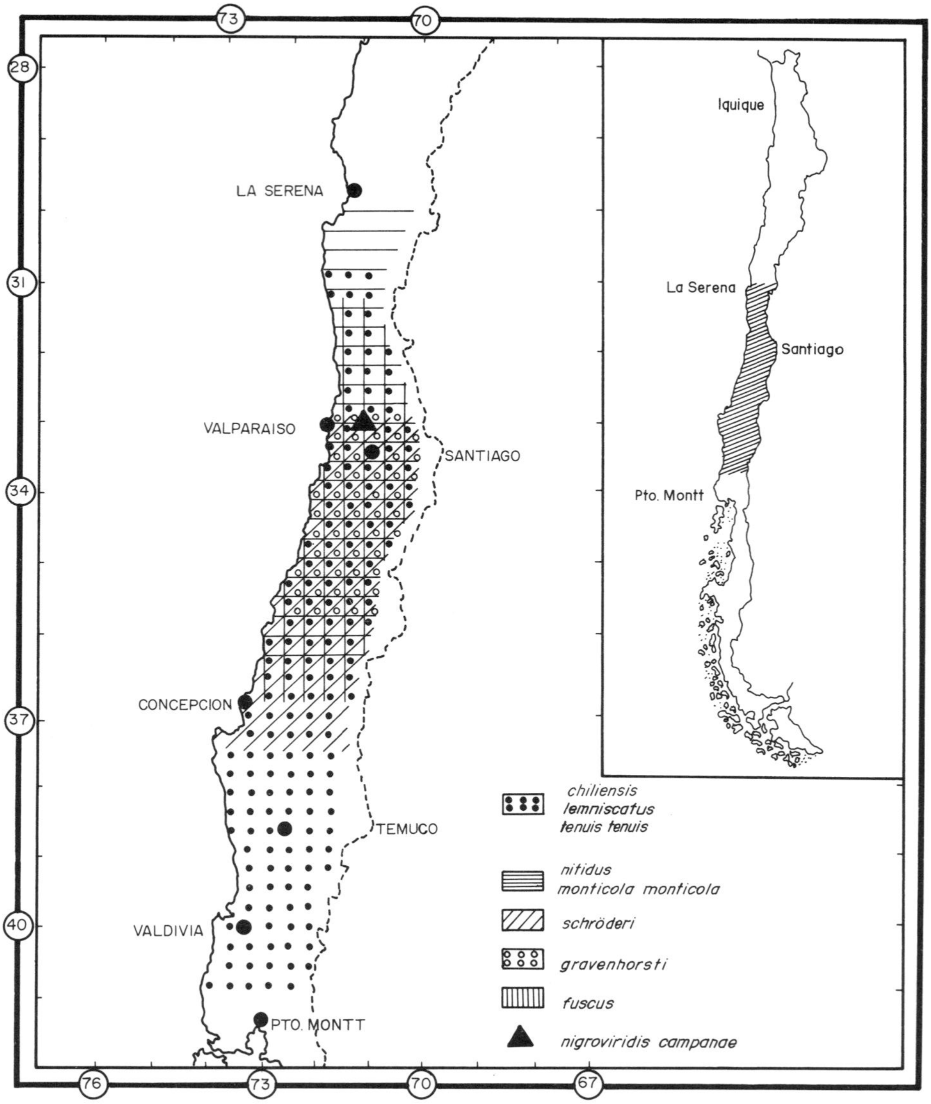

Fig. 1. Geographic distributions for the 9 *Liolaemus* species, showing high concentration in the typical mediterranean region of Chile

associating the species studied here in the following way: Group A: *lemniscatus, schröderi, gravenhorsti* and *fuscus;* Group B: *tenuis tenuis, monticola monticola* and *nigroviridis campanae;* and Group C: *chiliensis* and *nitidus.* In Fig. 2a it can also be seen that *fuscus* and *lemniscatus* are morphologically more similar (87%) than the other pairs of species.

The values of similarity for the ecological relationships of Fig. 2b are greater than for the morphological characters of Fig. 2a. From the dendrogram of Fig. 2b two groups can be identified: Group I. *chiliensis, gravenhorsti, tenuis tenuis* and *lemniscatus;* and

Group II. *fuscus, monticola monticola, nitidus, schröderi* and *nigroviridis campanae*. These two groups show two fairly well defined ecological tendencies. For instance, the order of species in Group II from right to left matches exactly with the altitudinal preference of these species in the coastal Cordillera of Central Chile, especially in those

Table 2. Similarity matrices for nine *Liolaemus* species

	Species 1	2	3	4	5	6	7	8	9
1 *gravenhorsti*	–	67	60	60	47	40	40	53	40
2 *chiliensis*	47	–	53	53	67	60	33	47	47
3 *tenuis tenuis*	47	60	–	87	47	27	27	27	13
4 *lemniscatus*	67	53	53	–	60	40	27	27	27
5 *fuscus*	60	33	47	80	–	80	53	53	67
6 *monticola monticola*	60	47	60	53	60	–	73	60	60
7 *nitidus*	67	67	40	47	40	67	–	73	60
8 *schröderi*	67	40	27	60	53	67	60	–	73
9 *nigroviridis campanae*	40	53	53	47	40	53	47	60	–

Roman standard type: phenetic similarity (per cent). Boldface: similarity (per cent) based on ecological relationships.

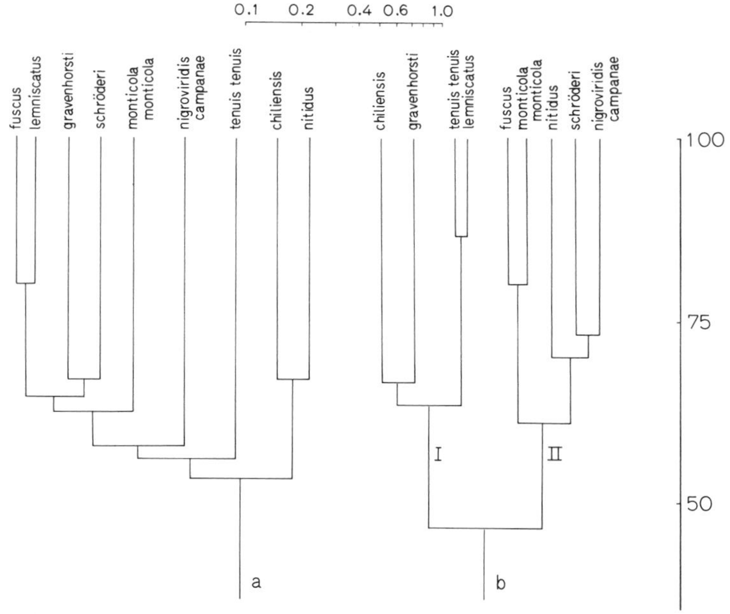

Fig. 2. Relationships of the 9 *Liolaemus* species, a based on morphological characters and b based on ecological attributes

ecosystems where our studies were carried out: Quebrada de la Plata (500 to 1,000 m) and Cerro El Roble (800 to 2,200 m). Species of Group II are those having a greater food specialization and a distribution more limited to arid open habitats.

Group I includes species with the highest degree of plasticity; that is, those species with the broadest distribution and a high trophic diversity, and those that exhibit arboreality. In addition, Fig. 2b shows the highest ecological similarity between the arboreal *tenuis tenuis* and the terrestrial *lemniscatus,* two of the most common species in the mediterranean climate region of Chile.

Strategies Segregating the Ecological Niche

In order to elucidate the main strategies segregating the niches of these species, all characters were first analyzed according to a Q-Type Similarity Matrix (SOKAL and SNEATH, 1963). Subsequently, through several successive analyses with the clustering

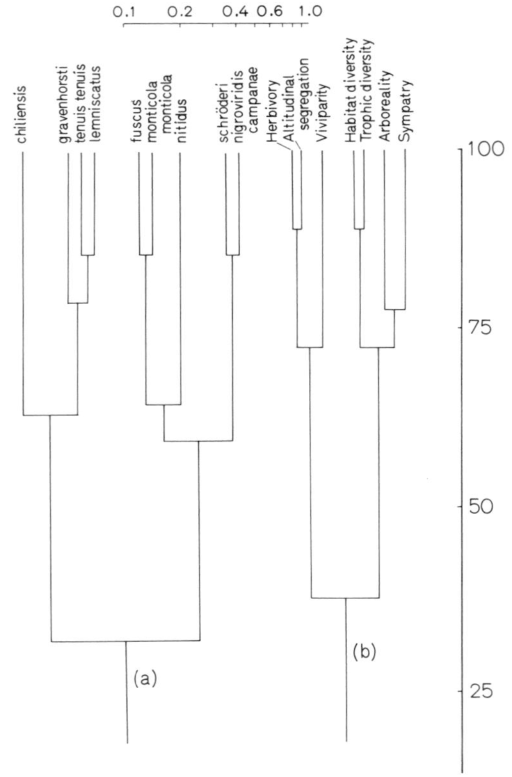

Fig. 3. Relationships of the 9 *Liolaemus* species; (a) dendrogram constructed according to the seven characters considered as main strategies segregating the ecological niche, and (b) dendrogram constructed with R-Type Similarity Matrix data for the association of the seven strategies

method, the number of characters was reduced. In this way seven characters were found which seemed to explain the main strategies for niche segregation. These strategies are: trophic diversity, habitat diversity, arboreality, tendency to sympatry, altitudinal segregation, herbivory and viviparity.

The dendrogram of Fig. 3a shows the relationships among the species when only these seven strategies are considered. In fact, species are associated in a similar way to that shown in Fig. 2b, thus demonstrating the relevance of the seven strategies selected.

The similarity among these strategies (Fig. 3b) was also measured according to the R-Type Similarity Matrix (CATTELL, 1952; SOKAL and SNEATH, 1963). The association between herbivory, altitudinal segregation and viviparity is striking, as well as between trophic diversity and habitat diversity; to a lesser degree, arboreality is associated with the tendency to sympatry.

Altitudinal Segregation and Viviparity

In the Group II of *Liolaemus* (Fig. 2b), each species has an altitudinal preference which performs a strategic role segregating the niche space among the species. *L. fuscus* appears in the lower elevations up to 1,000 m and *nigroviridis campanae* at the opposite end up to 2,200 m. This altitudinal gradient implies a change in the fundamental niche for the species occurring at higher altitude. They tend to be viviparous, presumably because eggs would not develop due to low air temperature or to the limited duration of the favourable season. The time of egg hatching is inversely proportional to the air temperature (BLANCHARD, 1926). In the mediterranean climate region of Chile, the 1,500 m elevation seems to be the limit above which most of the *Liolaemus* are viviparous *(schröderi* and *nigroviridis campanae*); below this level the oviparous forms are predominant. Several species of *Liolaemus* of the southern Valdivian zones are also viviparous. Thus the genus *Liolaemus* in Chile shows similar reproductive strategies in the mountains as in the southern latitudes. TINCKLE, WILBUR and TYLLY (1969) pointed out that viviparity must be considered a "late-maturing, single brooded" reproductive form, since almost all viviparous species of lizards produce one litter per year and have a later age at first reproduction.

Herbivory

One dimension in the ecological characterization that was previously unknown for several species of *Liolaemus* is the exploitation of the herbivorous niche. In the Group II (Fig. 2b) all the species, with the exception of *fuscus*, eat a certain amount of plant matter. Among these species an herbivory gradient exists, which is very similar to the altitudinal preference; that is, the degree of herbivory increases in the species of higher altitudes. Since most of the species of *Liolaemus* in Southern Chile are also partially herbivorous, this strategy in food niche seems to appear when the lizards are facing strong environmental stresses. The switch of this response is probably due to temporal scarcity of animal food, or extreme poverty of the environment.

There is also a positive correlation between herbivory and the size of lizards of the genus *Liolaemus*. In general, the species of large size show the greatest degree of herbivory *(monticola monticola, nitidus, schröderi* and *nigroviridis campanae,* in order of increasing frequency). SOKOL (1967) and SCHOENER and GORMAN (1968) have found the same correlation in other genera of lizards.

Herbivory has been experimentally demonstrated in *Liolaemus* by DI CASTRI and

FUENTES (unpublished). The degree of experimental herbivory was similar to the gradient observed in the field from the analysis of stomach contents. The experimental herbivory was deduced from the time before the individuals of one given species would eat some plant matter under the pressure of a partial or total lack of animal food.

Not one of the species of *Liolaemus* studied is strictly herbivorous; even species such as *schröderi* and *nigroviridis campanae,* which show a relatively constant herbivory, also take insect food. In other species, herbivory has only a temporal or occasional character, as in *monticola monticola* and *nitidus* for example. The exploitation of the herbivorous niche, that is changing the trophic level in order to receive a greater energy input, has unquestionable evolutionary implications in lizards.

Trophic Diversity and Habitat Diversity

The two morphologically most related species are *fuscus* and *lemniscatus* (Fig. 2a). In relation to the strategies segregating their niches, *lemniscatus* has a much higher habitat and trophic diversity. The occurrence of *fuscus* is associated with strictly xerophilous vegetation, such as the *Puya-Trichocereus* formation. Its food diversity is the lowest among the species studied here.

There are several difficulties in estimating trophic diversity (prey species diversity) from the analysis of stomach contents. For example, each stomach typically contains only a small part of the possible prey species that serve as the food of one predator species. Thus the method derived by PIELOU (1966), which estimated the diversity of a patchy population of sessile organisms from a series of randomly placed quadrats, was adopted. Each of the stomachs of the predator species was treated as equivalent to a quadrat and their prey species diversity (H) estimated first separately according to $H = (1/N)\ (\log_2 N! - \Sigma \log_2 N_i!)$ (BRILLOUIN, 1960), in which N is the total number of prey individuals and N_i is the number of individuals in the i_{th} species. These individual estimates were then pooled using $k = 1, 2, 3, \ldots$ up to z stomachs. When the stomachs are pooled in this way, the accumulated trophic diversity (H_k) resulting from the greater number of prey individuals and prey species increases at a decreasing rate, eventually reaching stabilization at a point t. The stabilization of the curve shows when the parent populations of prey species for one predator species in a single habitat are adequately represented; therefore, the trophic diversity value is representative. Paraphrasing PIELOU (1966), we may say that any number of stomachs in excess of t suffices to "represent" the population of prey species for the predator, in the sense that enlargement of the sample would produce no further increase in trophic diversity.

This stabilized value of the accumulated trophic diversity (H_k) facilitates the comparison of food specialization and niche breadth among sympatric predatory species. In addition, this method permits the estimation of the average trophic diversity per individual (H'_p) with its variance and standard error (PIELOU, 1966). Many data in the literature on prey species diversity in lizards (SCHOENER, 1968; SCHOENER and GORMAN, 1968; PIANKA, 1969), in gastropods (KOHN, 1966) and in mammals (ROSENZWEIG, 1966), have been given without regard to this stabilization point.

Fig. 4 shows the results obtained by the method explained here for sympatric populations of *fuscus, lemniscatus* and *tenuis tenuis* in the mediterranean climate region of Chile (Quebrada La Plata). When comparing the values of H in the histograms of Fig. 4

and the mean trophic diversity ($\overline{H}$) per individual, there is only a slight difference observable between *lemniscatus* and *tenuis tenuis* (1.41 bits vs 1.45 bits) occupying the same habitat (sclerophyllous forest). But when the contents of *k* analyzed stomachs are pooled (curves of Fig. 4), thus obtaining more representative accumulated values of their trophic diversity, a strong difference appears: 3.86 bits per individual for *lemniscatus* compared to 4.96 bits per individual of *tenuis tenuis*. Fig. 4 also shows the minimal trophic diversity

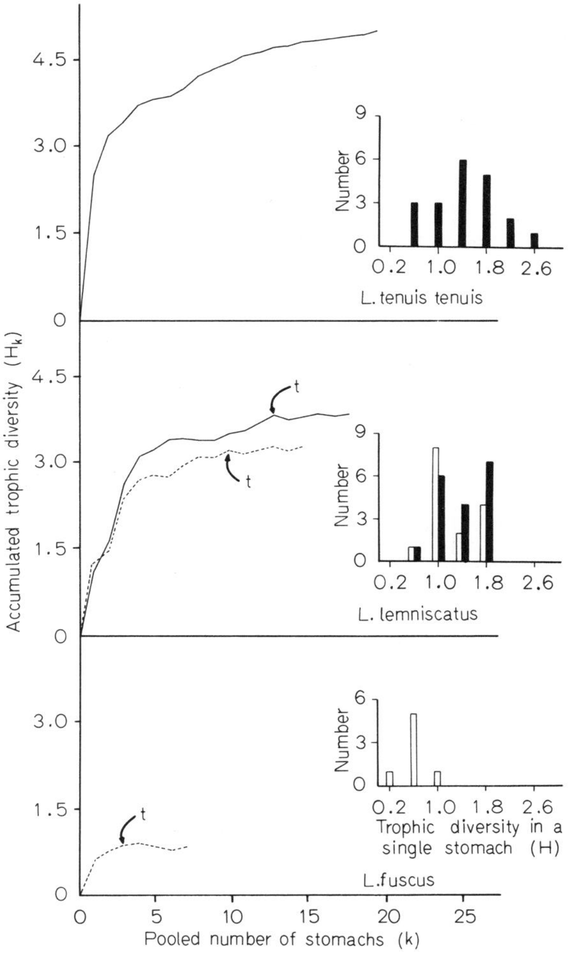

Fig. 4. Curves of the accumulated trophic diversity (H_k) versus the pooled number of stomachs (k), and the frequency distributions of H (trophic diversity per prey individual in separate stomachs) for three species of *Liolaemus* (HURTUBIA, unpublished). Solid lines and shaded columns: *Liolaemus* individuals from sclerophyllous forest. Dashed lines and unshaded columns: *Liolaemus* individuals from low chaparral

of *fuscus;* that is, its strong tendency toward specialization in food (mainly ants), as well as in habitat. Inversely, *lemniscatus* and even more so *tenuis tenuis* are plastic species presenting both high habitat and high trophic diversity. In general, these two parameters seem to be associated within the species of *Liolaemus*. Other general considerations revealed by these studies are that the highest values of trophic diversity have been found in the habitats of greater plant complexity (sclerophyllous forest) and that, on the other hand, the stabilization point of the accumulated trophic diversity was reached earlier in the simplest habitats, even when the same species was concerned (see *lemniscatus* in Fig. 4).

Arboreality

As previously stated, the two species with the largest niche overlap (Fig. 2b) and the highest degree of coexistence in several habitats are *lemniscatus* and *tenuis tenuis*. This is probably due to the strategy of arboreality adopted by *tenuis tenuis*. In those habitats of highest sympatry and highest species diversity of lizards, *tenuis tenuis* lives in the tree stratum. In other zones of lower species diversity, populations of this species do not show the same degree of arboreal preference. This fact, as well as its facility for colonizing new man-modified habitats, is another indication of the plasticity of *tenuis tenuis*.

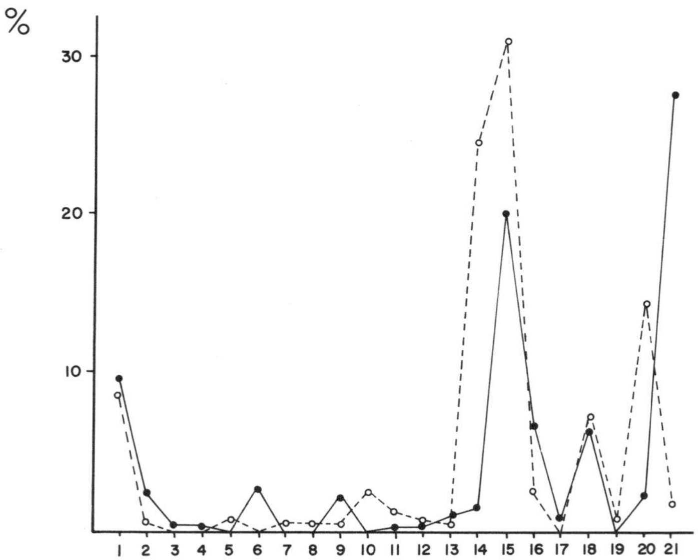

Fig. 5. Percentage of different groups of invertebrates in the food of an arboreal lizard species of *Liolaemus (tenuis tenuis)* compared with the food of a terrestrial one of the same genus *(lemniscatus)*. 1. Araneae 2. Solifugae 3. Pseudoscorpionida 4. Scorpionida 5. Chilopoda 6. Microcorhyphia 7. Ephemeroptera 8. Blattaria 9. Orthoptera 10. Hemiptera 11. Neuroptera 12. Trichoptera 13. Lepidoptera larvae 14. Diptera 15. Coleoptera 16. Coleoptera larvae 17. Thysanura 18. Hymenoptera 19. Psocoptera 20. Homoptera 21. Formicidae. Solid line: *L. lemniscatus*. Dashed line: *L. tenuis tenuis*

The analysis of the stomach contents of the arboreal species *(tenuis tenuis)* and of the terrestrial species *(lemniscatus)* living in the same ecosystem clearly indicates a great difference of food items (Fig. 5). The animals eaten by *tenuis tenuis* were essentially flying insects of large size (Diptera, Hymenoptera, Homoptera, Coleoptera), in contrast to the smaller and frequently apterous prey of *lemniscatus* (Formicidae, Coleoptera, Araneae).

In *gravenhorsti* and *chiliensis* the possibility of sympatry with other *Liolaemus* species seems also to be associated with their arboreality. The main dissimilarity between *gravenhorsti* and *chiliensis* is in their geographical distribution. In fact, with exception of *nigroviridis campanae*, *gravenhorsti* has the most restricted distribution of all the species studied (see Fig. 1), being found now in isolated refuges in the semi-arid parts of Central Chile. From a morphological point of view, *gravenhorsti* is strongly related to *L. cyanogaster cyanogaster*, a typical form of the southern Valdivian humid forest, which is a viviparous and frequently arboreal species. The existence of viviparity in *gravenhorsti*, in spite of its present geographical distribution, suggests that this reproductive strategy is a relict holdover for this species.

Concluding Remarks

The distinctive strategies segregating the niches of the Chilean *Liolaemus* derive from evolutionary adaptations developed by species over a long period of environmental contingencies, mainly paleoclimatic and geomorphological changes, typical of the regions which presently have a mediterranean climate, and more particularly of Chile.

The principal strategies are summarized in Fig. 6. It can be seen, firstly, that differences in trophic diversity and habitat diversity segregate the two species with the highest mor-

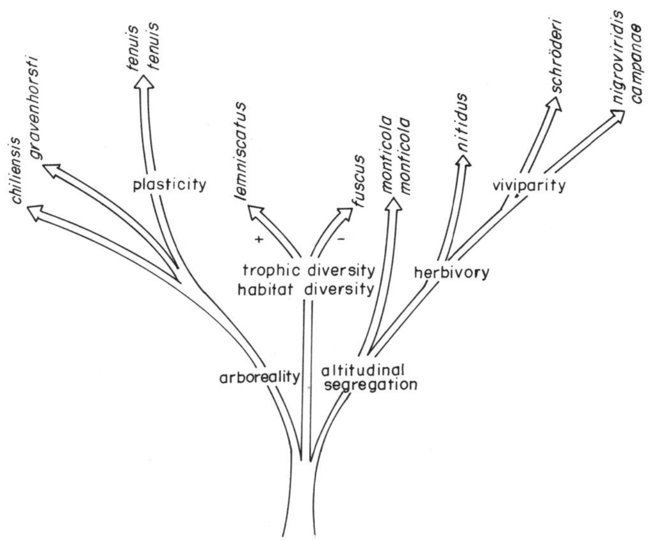

Fig. 6. The major strategies segregating the ecological niche of the mediterranean species of *Liolaemus*

phological similarity *(lemniscatus* and *fuscus)*. The former species has evolved towards plasticity in prey species and habitat, the latter species towards specialization in food (mainly ants) and restriction to a peculiar xeric habitat.

Secondly, arboreality segregates two plastic species, the arboreal *tenuis tenuis* and the terrestrial *lemniscatus*, which otherwise overlap in their ecological characteristics. Furthermore, the arboreal preference allows *gravenhorsti* and *chiliensis* to live in sympatry with other species of *Liolaemus*.

Thirdly, the possibility of different altitudinal preferences, perhaps the major ecological feature of the Chilean landscape, groups a number of species. The species of lower elevations *(monticola monticola, nitidus)* eat animal food in preference, although herbivory occasionally occurs, particularly in *nitidus*. In species of higher altitude *(schröderi* and particularly *nigroviridis campanae)* the quantity and frequency of ingestion of plant matter increases; furthermore, these two species are viviparous.

References

BLANCHARD, F. N.: Eggs and young of the eastern ring-neck snake, *Diadophis punctatus punctatus*. Papers Mich. Acad. Sci. **7**, 279–292 (1926).

BRILLOUIN, L.: Science and information theory. 2nd ed. New York: Academic Press 1960.

BROWN, W. L., WILSON, E. O.: Character displacement. Systematic Zool. **5**, 49–64 (1956).

CATTELL, R. B.: r_p and other coefficient of pattern similarity. Psychometrika **14**, 279–298 (1949).

CATTELL, R. B.: Factor Analysis. New York: Harper 1952.

DONOSO-BARROS, R.: Reptiles de Chile. Santiago: Ediciones de la Universidad de Chile 1966.

HELLMICH, W. G.: On ecotipic and autotipic characters, a contribution to the knowledge of the evolution of the genus *Liolaemus* (Iguanidae). Evolution **5**, 359–369 (1951).

HELLMICH, W. G.: Ensayo de una clave para las especies chilenas del género *Liolaemus*. Inv. Zool. Chilenas **1**, 10–14 (1952).

HUTCHINSON, G. E.: The ecological theatre and the evolutionary play. New Haven: Yale University Press 1965.

KLOPFER, P. H., MACARTHUR, R. H.: Niche size and faunal diversity. Amer. Natur. **94**, 293–300 (1960).

KOHN, A. J.: Food specialization in *Conus* in Hawaii and California. Ecology **47**, 1041–1043 (1966).

LEVINS, R.: Toward an evolutionary theory of the niche. 325–340. In: Evolution and Environment (DRAKE, E. T. Ed.). New Haven and London: Yale University Press 1968 a.

LEVINS, R.: Evolution in changing environments. Monographs in population biology, **2**. Princeton: Princeton University Press 1968 b.

MACARTHUR, R. H.: The theory of the niche. 159–176. In: Population biology and evolution (LEWONTIN, R. C. Ed.). Syracuse: Syracuse University Press 1968.

MACARTHUR, R. H., WILSON, E. O.: The theory of island biogeography. Monographs in population biology, **1**. Princeton: Princeton University Press 1967.

PFLAUMER, K.: Las lagartijas vivíparas del género *Liolaemus* (Fam. Iguanidae). Rev. Univ. Stgo. (An. Acad. Chil. Cienc. Natur.) **29**, 113–116 (1944).

PIANKA, E. R.: Sympatry of desert lizards *(Ctenotus)* in western Australia. Ecology **50**, 1012/1030 (1969).

PIELOU, E. C.: The measurement of diversity in different types of biological collections. J. Theoret. Biol. **13**, 131–144 (1966).

ROSENZWEIG, M. L.: Community structure in sympatric Carnivora. J. Mammal. **47**, 602–612 (1966).

SCHOENER, T. W.: The evolution of bill size differences among sympatric congeneric species of birds. Evolution **19**, 189–213 (1965).

SCHOENER, T. W.: The *Anolis* lizards of Bimini: resource partitioning in a complex fauna. Ecology **49**, 704–726 (1968).

SCHOENER, T. W., GORMAN, G. C.: Some niche differences among three species of lesser antillean anoles. Ecology **49**, 819–830 (1968).

SOKAL, R. R., MICHENER, C. D.: A statistical method for evaluating systematic relationships. Univ. of Kansas Sci. Bull. **38**, 1409–1438 (1958).

SOKAL, R. R., SNEATH, P. H. A.: The principles of numerical taxonomy. San Francisco and London: W. H. Freeman Co. 1963.

SOKOL, O. M.: Herbivory in lizards. Evolution **21**, 192–195 (1967).

TINCKLE, D. W., WILBUR, H. M., TILLY, S. G.: Evolutionary strategies in lizards reproduction. Evolution **24**, 55–74 (1969).

Section VII: Human Activities Affecting Mediterranean Ecosystems

Studies of ecosystems on this planet, except for those of very small areal extent, can be neither complete nor valid unless they take account of the pervasive and sometimes overwhelming role played by man's activities. Unlike other organisms whose role in an ecosystem is a function of physiology, instinctive behavior, and number, all but the latter changing only slowly through the processes of biological evolution, man's impact on his environment varies drastically with his culture which may change rapidly. It may be appropriate to recognize two classes of cultural developments. The first involves technical capability and tends to be progressive. The knowledge of how to control the biota through domestication of plants and animals, construction of irrigation works, knowledge of chemistry, and the application of vast amounts of inanimate energy once gained by a society is retained and added to, though at varying rates. How societies choose to apply their culturally derived capabilities, however, can and has changed in virtually every conceivable direction. Political structures that permitted the organization of labor to perform vast centrally directed tasks have been replaced by local atomistic economic units; the pig, an extraordinarily valuable domesticated animal, has disappeared from the eastern and southern sides of the Mediterranean Basin because of a religious taboo; modern commercial economies can cause the abandonment of cultivation on marginal lands that have been farmed for centuries. Finally, all these sorts of human impacts to the ecosystem are reversible.

The foregoing comments illustrate the complexity of effective inquiry into the part man may play and has played in establishing, giving character to, and sustaining a particular ecosystem. The salient difference between studying man's effects and those of other organisms is that in the former case historical rather than experimental data are central. We must try to learn what the resident society, or societies in sequence, did to their environment, when, and for how long in order to establish the framework in which other organisms might establish themselves and live.

Two chapters introduce some of the questions and provide an idea of the state of our knowledge for this volume. ASCHMANN takes a broad look at the comparative histories of land use in the several areas of the world with mediterranean climates, focussing on the drastically different durations of the kinds of pressures man is now putting on mediterranean type ecosystems. NAVEH and DAN center their attention on a single area, Israel, where intensive, more or less modern, human occupation is of greatest duration and perhaps as well known historically as in any part of the world. In both chapters the abundance of pertinent but presently unanswerable questions will be evident.

The introduction to this section was prepared by HOMER ASCHMANN.

1

Man's Impact on the Several Regions with Mediterranean Climates

HOMER ASCHMANN

There is tremendous variation among the several regions of the world with mediterranean climates in the nature and intensity of the modification of their ecosystems as the result of human activity. The duration of occupation by man is not a significant variable. It exceeds 10,000 years in all cases and may well be longest in South Africa, one of the less modified regions. Through his use of fire in an area that is seasonally inflammable even primitive man's influence on the ecosystem was not negligible, but it was probably comparable in all parts of the world, favoring some organisms and disfavoring or even exterminating others. Areas fully free of fire would undoubtedly carry an heavier forest than they do now. Natural fires would still obtain in some regions, but perhaps with much less effect on the vegetation. The island of Madeira is unique in having the only area of mediterranean climate untouched by man until historic time. The tremendous fires that burned for years shortly after the first settlement early in the fifteenth century altered its vegetational pattern from forest to brush and grass, evidently permanently.

The next great technological innovation that generally affected local ecosystems, the domestication and propagation and protection of some species of plants and animals, however, occurred perhaps 10,000 years ago at the eastern end of the Mediterranean Basin and reached Western Australia less than two centuries ago.

There are good reasons to believe that horticulture, the manipulation of individual plants, commonly vegetatively reproduced clones, antedates broadcast sowing of seed grains (SAUER, 1969). Our earliest archeological records, however, come from the uplands at the eastern end of the Mediterranean Sea and seem to represent a society that planted grass seeds in late fall in fields that had been cleared of brush by burning at the end of the summer dry season. They would sprout in the winter rains, survive the mild winter cold, and be harvested in late spring. HAHN'S (1896) inferences that domestication of the cow, sheep, and goat, respectively the fiercest and wildest of animals in their wild forms, could only be accomplished by a rich and sedentary, i.e. farming, population look ever more probable in the light of new archeological discoveries (ISAAC, 1962). That the motivation for domestication was religious rather than economic is also likely.

Before the dawn of written history, some 5000 years ago, a complex set of variant farming patterns, adapted to distinctive climatic and topographic situations, had been developed in the Middle East, and had begun to spread at varying rates in all directions including westward on both shores of the Mediterranean Sea. Irrigation probably began with small streams easily diverted over alluvial fans. It floresced in the great riverine oases of the deserts to the south and east where evolving societies could direct masses of human labor in water control. Plowing with oxen perhaps also originated in these great oases, but it could spread into dry-farmed areas that had extensive level surfaces. True pastoralism, often accompanied by slash and burn farming with the hoe rather than the plow, led the way westward through the Mediterranean Basin as well as into Central and Northern Europe. Herdsmen could exploit effectively lands too dry, too rough, and too cold to plant, but in the early stages a single community usually carried on both activities, individuals and families only being specialized. This pattern persists to the present in the rougher and less developed parts of Anatolia, Greece, Italy, Iberia, and North Africa. Specialized and somewhat transhumant communities composed entirely of pastoralists are associated with the dry lands and the high mountains rather than areas of mediterranean climate.

These agricultural activities all had profound effects on the native biota. An important element is that farming and grazing permit a great increase in the human population, often by an order of magnitude or more. Horticulture and agriculture involve the more or less complete replacement of the natural vegetation by the cultivated one over limited or extensive areas. The cultivated land may be maintained as such more or less indefinitely or may be abandoned for a number of years after one or a few years of cultivation. In either event the disturbances of cultivation also prove to favor a set of plants called weeds which may actually evolve by selection along with the deliberately bred crops. Pastoralism similarly puts heavy grazing and trampling pressure on the natural vegetation favoring some species but reducing or eliminating others. In the regions of mediterranean climate with their characteristically broken topography only the smoother surfaces are cultivated, at least until there is severe population pressure on the land, while almost any place is accessible to and affected by goats (KOLARS, 1966).

The modification of an ecosystem by cultivation or grazing is a relatively slow and cumulative process, progressing over time measured in millenia. In general human populations and their capacity to affect the landscape tend to increase with time. Disequilibria such as erosion of cleared slopes that tend to decrease a resource will force a concentration of activity and perhaps cause additional disequilibria in other locales. Further, the way in which the general ecosystem is affected is modulated by cultural practices and values that vary in detail among societies. Such features as the preferred grain for breadstuffs, the species of domestic animal herded, or whether agriculture is a commercial enterprise or a subsistence way of life will strongly affect the choice of surfaces for cultivation or abandonment and the intensity of implantation of a cultural landscape in place of the natural one. Because of these particularities it is appropriate to characterize briefly the histories of human occupation of each of the several regions of mediterranean climate over the past 10,000 years, attending especially to varying patterns and intensities of land use.

The Mediterranean Basin

From the beginnings of the neolithic with its domesticated plants and animals, which was present at the eastern end of the Mediterranean Sea about 10,000 years ago, nearly to the beginning of the Christian era, that region or its immediate neighbors recurrently produced both technological and social innovations that spread at varying rates in many directions. A fairly steady spread was to the western end of the basin. The neolithic may have reached the Atlantic by 2000 B. C. or a bit earlier, but by that time highly organized and populous states, specialized cities, and long distance commerce, even in bulky goods, were present around the eastern end of the Mediterranean Sea. The cultivation of fruit trees, especially grapes, figs, and olives, was an eastern introduction, and new fruit crops such as citrus kept coming from the east into Roman times.

The brilliance of the Middle Bronze age around the Aegean Sea in the middle of the second millenium B. C., and its fall into decadence shortly thereafter, as well as the political turmoil that began even earlier in the irrigated deserts to the south and east, suggest that locally populations and land exploitation had peaked and were pressing on resources. The stable and unified control of the Roman Empire had the effect of bringing the western basin up to the technological level of the eastern, with major and modern urban centers developing in Iberia. For a time population growth was rapid in the west, and land use and abuse of the eastern sort developed. Though there were local florescenses in Iberia and Morocco from the 8th to the 12th centuries A.D., evidently associated with specialized crops and irrigation techniques, the Mediterranean Basin had by later Roman times established a land use system that was not significantly modified until the present century.

The political stability or law and order maintained by the Roman Empire seems also to have favored the extension of farming into drier areas than are now cultivated in Syria, Cyrenaica, Libya and Southern Tunisia, areas of ancient civilization even then. Both dry farming and the maintenance of drought-resistant perennials, olives, grapes, and figs were involved, and areas of ancient cultivation can be traced archeologically beyond the modern border between the steppe and the sown. Full famine relief for the populace could be provided by the empire in years of drought. In the politically disordered centuries following the fall of the Roman Empire and extending into the nineteenth, intensive pastoralism, often associated with overgrazing and the obliteration of all shrubs but the toxic oleander, pressed into areas of mediterranean climate, especially in North Africa (MIKESELL, 1961).

Through this long period level land was planted with small grains, sometimes with long fallow periods; in small alluvial basins intensively cultivated and irrigated gardens of fruits and vegetables were developed; the olive and the vine found places on steep lower slopes, and locally where population pressures had developed some artificial terraces on similarly steep slopes were planted to small grains; domestic animals, especially the goat and sheep were pastured in the roughest and poorest country, putting severe pressure on the native vegetation. Native forests, however, never abundant or fast growing, had come to be recognized as a resource that needed care and protection. Pines, oaks, and chestnuts were preserved to yield food and fodder, and selectively their wood. The intensive care and exploitation of native trees, at about the same level as olive cultivation, is a peculiarly mediterranean complex that has spread very little to other parts of the world. The live oaks of the Sierra Morena of Southern Spain are a part of an ancient wild ecosystem that is preserved as a crop. They are pruned, stripped of their cork, and yield a regular acorn crop for fattening hogs.

This system of land exploitation could support stably a large but not infinite population. Pressures of overexploitation were most acute in the upland grazing areas, but erosion of overgrazed slopes could clog drainage systems and create marshes on some of the best alluvial lands. When the social fabric was rent by wars and political instability, burning and cutting of forests and degradation of terraced fields and groves for the expansion of poor pasture tended to destroy resources.

Although the areas of mediterranean climate remain the poorest parts of Europe, a product of rural overpopulation, an atomistic family-oriented social organization, and the retention of an ancient land-tenure system in Iberia and Southern Italy, as well as of man-caused damage to the land-resource base, the past century has witnessed significant changes in agricultural practices. With the improvement of transportation facilities winter vegetables and subtropical fruits can reach markets in Northern Europe as well as local urban centers. Although dependent on risky international luxury markets the producers gain enormously higher returns than from subsistence farming. This specialty production is concentrated in irrigable lowlands, and even paddy rice is cultivated on marshy tracts. Israel and the North African states of the Maghreb are participating in the development. Where they are practiced and extended these intensive cultivations effectively eliminate the native biota, or create a new ecological system. There is a parallel, perhaps indirectly related development in the abused and eroded uplands. Returns are so low that villagers who have raised grain in small patches of terraced or sloping land and grazed sheep or goats where the terrain is more rugged are abandoning their ancestral homes and moving to cities or the irrigated lowlands. Reforestation or rationally restricted grazing has become politically feasible though substantial effects on the long denuded upland slopes may take some generations to appear. In the poorest areas with populations that continue to grow explosively, as in North Africa, the prognosis is for even more destructive exploitation of the uplands.

Chile

When peoples from the western Mediterranean basin set out to explore and exploit the world in the fifteenth century, the land use system of their homeland was what they knew and tried to spread. In Central Chile in the mid-sixteenth century the Spaniards encountered a similar physical environment. Hunting, fishing and gathering Indians had lived in Chile for at least 10,000 years (MONTANE, 1968), but the spread of horticulture to that region from the north probably was much more recent, perhaps starting about the beginning of the Christian era. The local domestication of three grain-bearing plants adapted to winter rains and summer drought, unique to the New World, suggests a considerable antiquity for agriculture in Central Chile. *Teca* disappeared in the eighteenth century and cannot be identified; *mango* was a grass, *Bromus mango; madi* was a composite produced for the oil in its seeds. Llamas and alpacas were herded throughout northern and central Chile at the time of contact, but numbers seem to have been too small to put serious pressure on pastures (KELLER, 1952). Population densities had evidently not been built up to near saturation except in the irrigated valleys from Aconcagua northward. From Santiago south to Chiloé a neolithic shifting cultivation was practiced on gentler slopes. Summer burning for clearing is not readily controllable in a mediterranean climate, and it is likely that the vegetation of the gentler alluvial surfaces of the Central Valley had been greatly modified by Indian practice as far south as the Rio Bio-Bio even though only a small fraction of those surfaces was under cultivation at any one time.

The Spaniards introduced winter sown small grains and the Old World grazing animals. The Old World grains were enough superior to eventually eliminate the native ones. Chile was extraordinarily isolated, and its only realizable wealth for a long time was from gold placers, and the manner of their exploitation had a destructive effect on the native population. When the placers or the labor force to work them were exhausted a colonial society developed that could feed itself with a combination of Indian and European crops, expecially wheat, beans, and potatoes, and European grazing animals. With labor scarce and markets small extensive haciendas exploited the best land, the gentle slopes of the Central Valley, primarily for grazing cattle, each growing enough crops to feed itself. Indian Communities found shelter in little pockets of the coast ranges, joined by Spanish and mestizo small holders. Both the latter groups were more likely to graze goats and sheep than cattle.

The formative period for the modern Chilean landscape is really the mid-nineteenth century. The sudden development of a cash market for wheat to supply the California gold rush, and shortly thereafter one in Australia, led to an expansion of its cultivation into any place that wheat would grow, including steep and vulnerable slopes in the Coast Ranges (SEPULVEDA, 1959). The *fundos* or large estates on better land in the Central Valley produced more and profited more. Wealth from the silver mines of the Norte Chico was invested in irrigation systems as well as in the best land, and an intensification of cultivation, focussing on the vine, proceeded. With European weeds and ornamental trees as well as crops there is probably no more completely exotic vegetation in the world than that which covers irrigable level land in mediterranean Chile.

The slopes of the Andes and those of the Coast Ranges seem to have had consistently different histories of exploitation. Andean slopes were parts of large, lowland-based estates or were government land and were grazed with cattle, on an extensive basis. Large parts of the Coast Ranges were planted in wheat; grazing was by small holders, commonly of goats and sheep. The combination of severe soil erosion and a progressively less favorable ratio between costs and returns continues to cause the abandonment of wheat land. In some cases planted forests have been introduced; in others a wild regrowth that continues to be grazed is the replacement.

The recently instituted (1964) and now accelerated land reform program focusses primarily on breaking up large estates of irrigated and intensively cultivated land. Many such estates also contain even more extensive acreages of rough land or notably poor soil and have been used for grazing or forest plantations. It is too early to tell whether such holdings can be managed as large units on a sustained yield basis or will be subjected to overgrazing as small holders seek to increase their incomes. Intensely political rather than economic pressures are likely to determine government policies. As in Southern Europe, however, rising aspirations of the rural population and migration to the cities have a tendency to reduce grazing pressure and some recovery of the mediterranean scrub vegetation, now thinned or destroyed in extensive areas, may be foreseen.

California

Although discovered by the Spaniards at the same time as Chile, California was not settled by them until the late eighteenth century. The Indian population consisted of hunters, fishers, and gatherers of vegetable stuffs, but despite its lack of horticulture it was numerous and adept at exploiting the wild resources. A conservative culture had achieved a delicate balance with its environment (KROEBER, 1939; ASCHMANN, 1959). The 150,000 humans were a major element of the biota and strongly affected it with fire and collecting pressures on particular species, but they seem to have been in dynamic equilibrium for more than a millenium. They may have developed the most humanized landscape any non-agricultural society ever created.

The establishment of Missions, though humanitarian in aim, caused a catastrophic collapse of the Indian population, especially through introduced epidemic diseases (COOK, 1943). By the time of the accession of California to the United States, and the almost coincident Gold Rush, mediterranean California had an extensive grazing and limited farming economy and perhaps one third the population it had supported with the more primitive system. Beginning before 1850 around San Francisco Bay, and three decades later in Southern California, the level areas shifted rapidly into the production of small grains. On the pastured grasslands Old World weeds rapidly replaced the native grasses and herbs, possibly as early as Mission times. Strongly after 1880, areas subject to irrigation were devoted to specialty fruit and vegetable crops, favored by newly available railroad transportation and free access to the whole United States market. At about the same time a retreat of the dry farming frontier began, associated with rising living standards that made cultivating slopes too steep for mechanized farming uneconomic. Most such abandoned land has reverted to chaparral or scrub or has been maintained as pasture dominated by the weedy Old World grasses and herbs. Grazing has remained extensive and highly commercial with only big ranchers who are able to rationalize their operations and avoid overgrazing participating. Thus much steeply sloping land carries a wild vegetation that is very little abused by grazing. Fire, as in Indian times, has important effects on the extremely vulnerable vegetation. In Southern California where fire protection is assiduously sought fuel accumulation may cause the very intense fires, when they occur, to have an impact on the chaparral, oak grasslands, and coniferous forests different from that of the lighter more frequent burns of earlier times, but the relationships are complex and as yet little understood (SHANTZ, 1947).

The tremendous sprawling urbanization that has afflicted California in the past few decades has been concentrated in areas that had been devoted to irrigated specialized fruit production except in hilly sectors immediately adjacent to the major metropolitan centers. New irrigated plantings have tended to occur in desert rather than mediterranean climates. Thus despite its very large population California retains extensive rough areas with their natural vegetation only moderately modified. These chaparral or brush covered slopes constitute a growing set of management problems to land managers and planners. Can their recreational potential be increased without increasing fire hazard? Could a grass cover be artificially maintained to reduce fire hazard, and would it afford adequate erosion control and slope stabilization? At the present time the direct agricultural or pastoral productivity of these wildlands seems to be their least relevant capability.

South Africa

Only a few centuries before the Dutch brought European land use patterns to the mediterranean part of South Africa in the mid-seventeenth century the pastoralist Hottentots had entered the area, displacing the hunting and gathering Bushmen. The pastoralist tradition had diffused slowly southward through eastern Africa from its ancient hearth near the Eastern end of the Mediterranean Sea. Hottentot grazing activities do not seem to have developed much intensity in the area they had newly occupied. After the Dutch settlers had slowly gained a foothold they spread inland into drier areas with extensive grazing activities. In the Cape District itself small grains and garden produce and fruits found a market in passing ships. These crops were produced on the alluvial lowlands, and with the growth of a modern economy in the Union of South Africa the specialized irrigated crops including vines have tended to displace the small grains. On the steeper slopes neither commercial grazing nor dry farming became important, and the native subsistence grazing has been eliminated within the region of mediterranean climate though it persists in the drier Karoo to the east. There was ample flat land too dry for anything but grazing in the interior for the European commercial grazers to exploit. Consequently the chaparral-like forests on steep slopes in the mediterranean part of South Africa are perhaps the least modified of any in the world.

The Australian Lands with Mediterranean Climates

The aboriginal populations of both Western Australia and South Australia were hunters and gatherers with cultures as conservative as those of the California Indians, but their population density was considerably lower, possibly because of a less food-rich native flora. European settlement came late, 1828 in Western Australia and 1836 in South Australia. The early settlers were few, but they looked on the extensive, relatively level lands of mediterranean climate strictly in terms of their capacity to produce commercial products for export. Local subsistence needs were and remain small. The characteristic chaparral-like forest of other mediterranean areas was largely absent. In the wetter areas there were dense forests of giant eucalyptus trees. In drier locales large shrubs, primarily other species of eucalypts, were scattered over a grassland. With very extensive lands and little labor available sheep grazing became the dominant land use. What labor there was was devoted to eliminating the scrub or forest and upgrading the pasture with European grasses, especially in the wetter areas. On level lands towards the drier interior huge wheat fields were planted, and still further into the drier interior sheep grazing again became dominant. Limited areas of specialized fruit and vine production in the hills behind Adelaide and Perth satisfy local markets (MEINIG, 1962). The extensive mallee scrub of Kangaroo Island and the Ninety Mile Desert southeast of Adelaide remained a meager pasture despite a sufficiency of rainfall until well into this century when it was discovered that adding traces of copper, cobalt, and other trace elements to the soil could both upgrade the pasture and protect the sheep from deficiency diseases (BAUER, 1963).

Despite the fact that the extensive areas of mediterranean climate in Australia support the lowest human population densities of any region of the world with that climate, and even then more than half of the people live in the two large cities, the natural ecosystem may well be the world's most disturbed. The relative lack of relief coupled with uninhibited concentration on extensive commercial grazing and cropping are the determinants.

Interregional Dispersals

It may be appropriate to conclude with a few comments on how certain plants native to one of the world's mediterranean climate regions have successfully established themselves in others after being transported, deliberately or accidentally, by Europeans during the past four centuries. Platyopuntias (*Opuntia* spp.) or prickly pear cactus, possibly from California but more likely from Central Mexico, have been established in all the mediterranean regions. In Southern Italy and Sicily and to a degree in Chile they are cultivated for their fruits, but they also tend to establish themselves on overgrazed, eroded, and rocky slopes where they prevent pasturing and are hard to eradicate. Their clumps offer refuge to small animals and can form exotic bits of enduring wilderness in otherwise heavily grazed terrain.

The Monterey pine, *Pinus radiata,* is endemic to restricted areas along the cool coast of Central California. It has been established in extensive plantations in the moister parts of the Chilean mediterranean area, both in the Central Valley and the Coast Ranges. Its rapid growth there allows it to be the most important source of lumber in the country.

The almost complete replacement of native grasses and herbs in California and Chile by weeds of Old World origin such as wild oats, foxtail millet, and mustard has already been noted. The importation was almost certainly accidental, but the immigrant annuals are generally believed to have been successful because they could tolerate, better than the native plants, pasturing and trampling by the introduced domesticated grazing animals. One is tempted to regard the Old World weeds as having evolved and adapted in their eight to ten thousand year association with man and his domesticated grazing animals (ANDERSON, 1967).

Eucalypts from Australia, many of them species from the parts of that continent with mediterranean climates, have in the past century and a half been given a remarkable worldwide distribution and propagation. They are now abundant in all areas of mediterranean climate, as well as being the most visible tree in the drier parts of the highlands of Western South America and East Africa. Even in humid Southeastern Brazil they are strikingly abundant. Almost invariably they were planted deliberately, and often hopes for their commercial utilization as lumber were unrealized. But they survive without care and withstand repeated cutting for firewood or lumber by stump-sprouting, growing slowly on dry slopes and fast where their roots have access to ground water. I have seen few places where eucalyptus groves are expanding by the dispersion of seedlings. This may take a long time. But I suspect that eucalypts will long be a prominent feature in the man-modified landscapes of all the mediterranean lands.

Reference

ANDERSON, E.: Plants, Man and Life. 3–15. Berkeley: University of California Press 1967.

ASCHMANN, H.: The Evolution of a Wild Landscape, and its Persistence, in Southern California. Annals of the Association of American Geographers **49** Supplement, 34–56 (1959).

BAUER, F. H.: A Pinch of Salt: Trace-Element Agriculture in Australia. California Geographer **4**, 55–62 (1963).

COOK, S. F.: The Conflict between the California Indian and the White Civilization: I. The Indian versus the Spanish Mission. University of California Publications, Ibero-Americana **21**, 1943.

HAHN, E.: Die Haustiere und ihre Beziehungen zur Wirtschaft des Menschen. Leipzig: Drucker und Humblot 1896.

ISAAC, E.: On the Domestication of Cattle. Science **137**, 195–204 (1962).

KELLER, C.: Introducción a Los Aborigines de Chile por José Toribio Medina. 47–59. Santiago: Fondo Histórico y Bibliográfico José Toribio Medina 1952.

KOLARS, J.: Locational Aspects of Cultural Ecology : The Case of the Goat in Non-Western Agriculture. Geographical Review **56**, 577–584 (1966).

KROEBER, A. L.: Cultural and Natural Areas of Native North America. University of California Publications in American Archaeology and Ethnology **38**, 1939.

MEINIG, W.: On the Margins of the Good Earth. Chicago: Rand McNally 1962.

MIKESELL, M. W.: Northern Morocco: A Cultural Geography. 19–31, 95–116. University of California Publications in Geography **14**, 1961.

MONTANE, J.: Paleo-Indian Remains from Laguna de Tagua Tagua, Central Chile. Science **161**, 1137–1138 (1968).

SAUER, C. O.: Agricultural Origins and Dispersals. 19–39. Cambridge: M.I.T. Press 1969.

SEPULVEDA, S.: El Trigo Chileno en el Mercado Mundial. 11–50. Santiago: Editorial Universitaria 1959.

SHANTZ, H. L.: The Use of Fire as a Tool in the Management of the Brush Ranges of California. Sacramento: California Division of Forestry 1947.

2

The Human Degradation of Mediterranean Landscapes in Israel

ZEV NAVEH and JOEL DAN

"Behold, the Lord thy God giveth thee a good land, a land of water, brooks and fountains that spring out of the valley and depths, a land of wheat and barley, of vines, figs and pomegranates, of olive oil and honey, a land in which thou shalt eat bread without scarceness, thou shalt not lack anything in it." (Deuteronomy 8: 7–9.)

Introduction

It is now generally agreed that the Mediterranean region has suffered more than other regions in the world from landscape decay and desiccation, not because of adverse climatical changes, as HUNTINGTON (1924) claimed, but as a result of man's misuse of this landscape (LOWDERMILK, 1944; REIFENBERG, 1955; BUTZER, 1961; WHYTE, 1961).

Nowhere more than in Israel and its neighbouring countries can we find better and more substantiated proof that this desiccation can be traced back to an unfortunate combination of a vulnerable environment, and a long and chiefly destructive history of man-land relationships. Also, nowhere else has it been more convincingly demonstrated that man has not only the power to destroy and deplete his habitat, but also to reclaim and redeem it by virtue of willpower and skill. At the same time, nowhere in this region are the dangers of modern neo-technological landscape despoilation becoming more apparent than in the highly urbanized and industrialized parts of this country.

It seems therefore appropriate to use Israel as a model for the study of man's impact on the mediterranean landscape throughout history in selected ecosystems.

Geographical, Historical and Ecological Background of Man-Land Relationships in Israel

Israel is located on the eastern shores of the Mediterranean Sea and on the equatorward and driest border of the Mediterranean climatic zone and the South West corner of Asia. Its unique position as a meeting ground of plants, animals and people of three continents – Europe, Asia and Africa – and as a land bridge between two oceans – the Mediterranean and Red Sea – and two great hydraulic civilisations – Egypt and Mesopotamia – has determined its great ecological and cultural diversity. It also made it a constant battleground between powerful political centers in the North and South and between "the sown and the desert".

Israel can be subdivided into the following main bioclimatological and phytogeographical regions:

1) The extremely arid zones with Saharo-Arabian desert vegetation in the South.

2) The mildly arid zone with Irano-Turanian steppe vegetation, which is marginal for stable, rain-fed agriculture.

3) The transitional, semi-arid zone on the xeric borders of the mediterranean phytogeographic region.

4) The proper subhumid and humid mediterranean zone in Central and North Israel, to which most of this discussion will be devoted. In this zone – with exception of the narrow belt of coastal sand dunes – mediterranean sclerophyll forest and maquis have been considered as the original climax vegetation by EIG (1933), ZOHARY (1962) and BOYKO (1954).

In a recent, first extensive, palynological study of sediments of Lake Kinneret and the Hule Basin, HOROWITZ (1968) has shown that mediterranean sclerophyll trees, such as *Quercus calliprinos, Q. infectoria, Olea europaea, Pistacia,* as well as *Pinus halepensis,* were already present in this area 60–80,000 years ago. Their relative abundance changed, apparently according to climatical changes, between the dry and warm interpluvial and cooler and wetter pluvial phases, corresponding to the glacial and interstadial phases in Europe. According to HOROWITZ, a slightly drier period occurred between 3000 and 1500 B.C., but no significant climatical changes can be assumed in the last 3000 years.

Three main physiographic-lithological units can be distinguished:

1) The coastal belt with light to medium, chiefly arable soils, derived mainly from aeolian sand sediments and at present densely populated and urbanized.

2) The mountainous belt, rising to about 1000 m with various Cretaceous and Eocene calcareous rocks, including hard limestone and dolomite with terra rossa and dark rendzina soils, and chalk and marls with pale rendzinas.

3) The valleys, plains and plateau region, including the southern and eastern parts of the coastal plains as well as basaltic plateaux in the north. These valleys are usually covered by fine silty clayey aeolian or alluvial material and are all cultivated.

The history of human impact on the mediterranean landscape in Israel can be subdivided into seven major phases of landscape modification:

Phase 1. A very long period – probably lasting 500,000 years – with very slight interference by Lower Paleolithic food gathering and hunting economies in the early and middle Pleistocene.

Phase 2. The beginning of more intensified modification and creation of edge habitats, chiefly by the deliberate use of fire due to more advanced hunting, fishing and foodgathering technologies by the Middle and especially Upper Paleolithic economies in the Late Pleistocene. These economies are documented by the findings of the Carmel Caves of the Lavelloisio-Mousterian *Paleoanthropus palaestinensis* (the "Palestinian Neanderthal man") and the Natufian *Homo sapiens* cultures (GARROD, 1958; GARROD and BATE, 1937).

Phase 3. The beginning of early agricultural and pastoral modification and conversion through domestication of cereal crops and herd animals by Late Mesolithic and Early Neolithic economies, between ca. 9000 to 5000 B.C. For these the Yarmukian culture in the Kinarot Valley is typical. WHYTE (1961), discussing evidence of early agriculture in south-western Asia, concluded that the transitional zones between the arid, steppic Irano-Turanian and the sub-humid open woodland Mediterranean regions in Israel were among the cradles of domestication of cereals and herd animals. They also served as the first centers of successful cereal and stockbreeding farming economies. WHYTE states that the Galilee uplands are centers of distribution of both the wild, domesticable barley, *Hordeum spontaneum,* and the large-grained wild "Emmer" wheat, *Triticum dicoccoides.* He also states that remnants of the ancestors of modern goat breeds, dating back 9000 years, were found in Jericho. This view is verified by recent archeological findings of HIGGS and LEGGE at Gezer and Mount Carmel (LEGGE, 1970, personal communication).

Phase 4. Widespread and intensive agricultural conversion by land clearance, cultivation, terracing and irrigation of arable ecosystems and pastoral modification of non-arable and chiefly upland ecosystems, combined with burning, cutting and lumbering, and coppicing. This phase of intensive land use began with a transitional, prehistorical Chalcolithic period, between 5000–3000 B.C. It lasted from the Bronze and Early Iron Age – the beginning of Israel settlement – (1200 B.C.) throughout the Middle and Late Iron Age, the Persian, Hellenistic, Roman and Byzantian periods. Agricultural expansion reached its first peak during the Israelite period, when the woodlands of the Judean and Samarian mountains were cleared for cultivation and were turned into terraced vineyards, olive plantations and fields. The first, great relapse, to be followed by others, occurred during the Babylonian conquest, 586–232 B.C. A second and even greater peak in land use was reached during the Early Roman period, when the population of Israel (including Transjordanian) reached at least 3 millions and according to Flavius Josephus more than 5 millions, with 2.5 millions in Galilee alone. At that period soil and water conservation measures, as well as crop rotation, organic fertilization and terracing of hillslopes, were widely applied. After a serious relapse during the Roman-Judean wars, the country recovered steadily during Byzantian rule (330–640 A.D.) and reached a final peak in intensive land use before the onset of the darkest phase in the history of Israel.

Phase 5. Increasing agricultural decline and landscape desiccation, starting after the Moslem conquest (640 A.D.) and lasting for more than 1300 years throughout the Arab, the Crusader, the Mameluke and the Turkish rule. The westward expansion of nomad Arab tribes induced a radical change in land use by replacement of the settled, mediterranean agriculture by pastoral nomadism (REIFENBERG, 1955). He – as well as LOWDERMILK (1944) – vividly described the decline of the land and the people through internal warfare and instability. These led to the depopulation of once flourishing cities and villages. The decay of lowland fields, irrigation ditches and terraced hill land, and the destruction of terrace walls, caused catastrophic soil erosion, siltation and creation of badlands ans swamps. This landscape desiccation has its counterpart in other mediterranean countries (SEMPLE, 1931), and especially in North Africa (MURPHEY, 1951). Its impact on geomorphological processes by erosive changes in river channels upstream and by aggradation and sedimentation downstream, has been described recently by VITA-FINZI (1969) and – at least in one case – also in Israel (GUY, 1954). The lowest point of depopulation was reached during the end of the 19th century under the Turkish rule. At that time the total population of Israel amounted to only 200,000–300,000 people (TAYLOR, 1946).

Phase 6. Beginning of Jewish colonization, land reclamation, drainage of swamps, conversion of lowlands and plains into intensive agricultural ecosystems of rain-fed or irrigated fields and plantations, and reconstruction of terraces and afforestation of uplands. This phase began under Turkish rule, but the main land development took place during British Mandatory rule, between the two World Wars and, with accelerated speed, since the foundation of the State of Israel in 1948. During this period the population of Israel grew from less than 1 million to 3 million. Of the total area of Israel of about 2 million ha, only about 550,000 ha is suitable for cultivation. Of the 400,000 ha located in the mediterranean mountainous portion, only about a third is arable land. Of the remainder, around 150,000 ha are actual and potential grazing land, 50,000 ha are gazetted as national parks and reserves and about 75,000 ha as forest land, with 40,000 ha planted – chiefly with Aleppo pines – and 35,000 ha designated as "low value, natural scrub forest".

Phase 7. Intensification of agricultural land use, urbanization and industrialization in recent years. This present phase is characterized by the rapid loss of open space with diversified, "natural" ecosystems and their replacement by urban and industrial wasteland and neotechnological despoilation. This process is aggravated by the biological impoverishment (the functional and visual degradation of the cultivated landscape due to the creation of large "monoculture steppes" of cotton and wheat, and the ever increasing use of chemical fertilizers, herbicides and pesticides) and by the increasing threat of water, air and soil pollution. That the danger for the vanishing of open landscape and its deterioration is very acute can be appreciated from the following facts: rapid population growth (4%/year), the high population density in the coastal region, the increase in electricity demand (doubling within 5 years) and in motor vehicles (300% increase in 9 years).

No figures are available on which the over-all technological burden on the landscape in Israel could be estimated. But if the present rates of increasing pollution of the rivers in the coastal plains are taken as syndromes of environmental deterioration, this burden has already reached a critical level.

At the same time, however, there is a growing public awareness for the need of environmental protection and for nature conservation in Israel. At the end of 1968 there were 32 declared nature reserves covering about 13,000 ha, and more than twice this area has been set aside for additional reserves and parks. These are in some of the most scenic and ecologically significant landscapes, under the control of the Nature Reserves Authority, which also provides effective protection for most of the endangered wild flowers and animals in Israel.

This short account of the main phases of landscape modification can usefully be supplemented by further details on the ecological features and implications of the chief agencies of human intervention on the mediterranean landscape.

Fire has been recognized as the first major agent used deliberately by primitive man to modify his environment for his own benefit, and thereby to become the dominant manipupulator of ecosystems (SAUER, 1957; STEWART, 1957; OAKLEY, 1961; HOWELL, 1962).

In Israel, during September-October, when hot and desiccating "Sharav" winds blow from the desert and the relative air humidity drops below 40%, sclerophyll maquis species and their dry herbaceous understorey are extremely prone to fire. These trees and shrubs, as well as some perennial grasses, regenerate immediately after the fire from roots and dormant buds even before the onset of the winter rains. Thereby they provide highly nutritious, lush and attractive fodder for wild and domestic ungulates in early winter. Later on in the season, germination and vegetative sprouting of edible root, tuber and bulb plants are also encouraged in the newly opened edge habitats (NAVEH, 1960). The rich faunal collections of the Carmel caves, described by GARROD and BATE (1937) included leopard, bear and a great variety of woodland ungulates, especially *Dama mesopotamica*. This points to the existence of advanced and diversified hunting economies during Phase 2.

For similar economies in Greece, it appears that fire has been used in forest and maquis as a means to increase food choice, to facilitate collection of edible plants and to drive and attract browsing game.

The successor of this paleolithic hunterer-gatherer, the mediterranean grazier, has used and is still using fire to open dense maquis tree and shrub thickets in order to increase the amount and quality of pasture for his goats and cattle. Greatest damage is afflicted thereby to the soil of fire-denuded slopes and to resprouting trees, shrubs and perennial grasses in the first year after the burning (NAVEH, 1960). This destructive combination of frequent burning, grazing and cutting has been recognized as the chief cause of the degradation of maquis into poorer and more xeric vegetation types. These are dominated by unpalatable, aromatic and thorny shrubs, such as *Cistus* and *Calycotome villosa,* whose germination is increased after fire, and in lower stages by fire resistant geophytes, such as *Asphodelus* (RICKLI, 1943; KNAPP, 1965; WALTER, 1968).

The cutting and grubbing of woody plants for domestic fuel probably also dates back to Phase 2. It reached, without doubt, a considerable extent in Phase 3, when large amounts of wood and charcoal were required for lime and pottery kilns and for copper and ore smelting. Also the felling of trees for timber and building material for houses, temples and ships had early beginnings, as shown by the first Egyptian and Mesopotamian documents dealing with the export of Phoenician timber from Israel to Pharao Snefru, about 2600 B.C. During the period of King Solomon and his great copper ores in Wadi Araba, around 1000 B.C., the felling of cedars from Mount Lebanon also reached its peak (MIKESELL, 1969). As described by REIFENBERG (1955), the deforestation proceeded with increasing speed, due to growing demands of the population and to frequent warfares. This process has been well documented in the Bible and in classical literature.

The expansion and contraction of scrub vegetation in mediterranean lands corresponded closely to the density of population and their agricultural activities (BARRY and LEROY LADURIE, 1962). Its greatest extent was probably reached during Phase 5 in north and central Israel (TAYLOR, 1946). It was possibly during this period that the present maquis and degradation stages, as described below, were finally established. However, since then, with growing population pressure even in the montainous regions, more and more of these shrublands have again been cleared for recultivation and exploited for fuel, charcoal and limekilns. As described in detail by EIG (1933), the last wave of largescale forest destruction swept Israel during the First World War, when the Turkish rulers needed wood for fuel for their army and railways. Subsequently, the last remnants of the great Tabor oak forest in the Sharon Plains have also been converted into citrus plantations and into urban land.

As already noted, agricultural and pastoral ecosystem modification and conversion began and spread during Phase 3, reaching its peak in Phase 4. After the invasion of the nomadic stock-husbandry men from the desert (Phase 5), pastoral land use became more and more dominant. In this system, the sole source of fodder is derived from year-round grazing of rough natural pastures on land too steep, shallow or rocky for arable farming and from seasonal grazing of stubble fields and crop remnants. On the rugged and shrubby mountain terrain, which constitutes considerable parts of north and central Israel and of other Mediterranean countries (TISDALE, 1967), the black Mamber goat could make best use of the lignified and thorny browse plants. Thus, out of 40 main woody species in Galilee, 30 are browsed regularly by these goats. Cattle, however, utilize only 10 species, chiefly young shoots and fruit pods during the dry season and after fire (NAVEH, 1960). Being able to exist and produce in such extremely poor conditions, where cattle and sheep can no longer survive, goats have been blamed as the main culprits of mediterranean land ruin. However, goats have been only the last link in the vicious cycle of land abuse. In more enlightened systems of rational use of mediterranean wildlands, they could still fulfil a very useful function (FRENCH, 1970). The sole reliance of the mediterranean grazier on grass from natural pastures and stubble fields, without supplementing it even in critical seasonal shortages in early winter, is one of the main reasons for the low level of animal husbandry and of pasture deterioration.

In typical subhumid mediterranean hill pastures, annual fluctuations in rainfall and distribution cause variations of 400% and more in pasture output between wet and dry years. In these dry years not only the livestock suffers from lack of pasture, but ecosystems are subjected to greatest damage (NAVEH, 1967, 1970).

In the modern Jewish hill settlement, these natural pastures have become an integral part of the intensive mixed crop and animal husbandry system. They are grazed chiefly by beef cattle and Havassi milk sheep throughout the green season and for shorter periods in summer. Uncontrolled seasonal grazing of these large herds has caused undesirable changes in pasture composition and has threatened soil stability on non-arable land in the vicinity of the settlements (SELIGMAN *et al.*, 1959). But more recently, intensive agro-ecological methods of pasture ecosystem management have been adopted and most hill pastures are fenced, are rotationally grazed, and are in parts also improved by fertilization, chemical weed and shrub control, and reseeding of annual legumes (NAVEH and RON, 1966).

The separation of arable and livestock farming in the traditional agricultural systems which evolved during Phase 5 did not induce the mediterranean farmer to include any soil improving forage and hay legumes in his crop rotations. These rotations are based solely on soil-exploiting winter grain and on summer crops or bare fallow which leave the soil exposed to water and wind erosion. Soil exploitation and erosion are also favored by the land tenure system, causing the fragmentation of land and its subdivision into narrow strips, ploughed up and down the slope and leaving no incentive for land improvement. The few organic crop residues left on the stubble fields are removed by the livestock which return only a minor part of the soil fertility through their droppings.

Great parts of Israel agricultural land has been converted during the last 30 years into intensively managed and irrigated fields or plantations. The adverse impact of increasingly intensive "chemical" agriculture, coupled with urban and industrial expansion, on the natural and cultivated landscape in Israel has already been mentioned in the description of the present Phase 7.

Very little improvement has been achieved in the exploitation of uplands. As in other Mediterranean countries, the tendency for increased agricultural production by extending cultivation to sub-marginal land and by patch cultivation of rocky and steep slopes is coupled with increasing grazing pressure and is causing accelerated land depletion (TISDALE, 1967).

Thus, in summary, the over-all impact of man on the mediterranean landscape in Israel – as in other countries – has been a destructive one. In spite of many achievements in the distant and recent past, the future of this landscape is threatened both by "traditional" as well as "modern" exploitive land use.

Man-Induced Degradation Pattern in Mediterranean Ecosystems in Israel

Soil-vegetation degradation patterns caused by man's use of land in Israel as well as in other mediterranean countries can be best described as "anthropogenic biofunctions", *sensu* JENNY (1961). In these, man, as the chief controlling agent, has changed the other state factors or variables of the ecosystem in a series of degradation cycles, corresponding roughly to the phases of landscape modification described above.

These "anthropogenic biofunctions" can be written as a generalized ecosystem equation in the following way:

E s,v = f (*h*,cl,o,r,p,t . . .) where

E s,v are the dependent ecosystem properties of soil and vegetation

h is the dominant controlling state factor of human agency

cl and o are external flux potentials of (regional) climate and organisms (plant and animal taxa available)

r and p are the initial state factors of relief (slope and exposure)

t is time, with t_0 at each new cycle after change in state factors and land use

and .... are other undefined controlling state factors.

Our knowledge of earlier cycles can be only speculative and be based on a hypothetical climax and on successional stages. Therefore, in describing earlier degradation cycles we will rely only on pedological and geomorphological evidence and limit ourselves to the description of a few examples of present cycles which commenced during Phase 5.

A useful starting point is the semi-arid or the xero-thermo-mediterranean region of the UNESCO/FAO classification (1963), namely the Philistine Plains and the Southern Judean mountains. These were amongst the earliest centers of intensive agricultural and pastoral activities and they suffered more than the more arid and more humid regions to the south and north. Their pristine vegetation was destroyed apparently at a very early stage and their underlying soils were severly eroded and in some cases "badlands" were created.

The grumosolic dark brown soils were the typical deep soils of this region. These are well developed calcareous and fine textured soils that formed from fine aeolian dust. Due to continuous deposition throughout the Pleistocene and cumulative deposition, several similar paleosols might be found at depth. The overall soil-paleosol complex reaches a depth of 12-14 m in the southern parts which are closest to the desert. Due to the limited rainfall and their high water-holding capacities, moisture penetration is restricted to about 1 m and typical Ca horizons have been formed in deeper layers (DAN and YAALON, 1971).

As a result of severe erosion, the undulating to rolling slopes of the coastal plains and the inland plateaux are now more or less dissected by small gullies. The B horizons and, in time, also the more saline, deeper paleosols, were exposed by erosion of the slopes. In some instances, the whole Pleistocene-paleosol cover has been dissected, revealing the underlying sand and sandstones which have also been partly eroded (Fig. 1).

Towards the east, on the foothills and steeper rocky mountain slopes, soil depth decreases and its characteristics change. Here, due to the steeper topography, most of the dusty aeolic material has been re-eroded and, as a result, the effect of the underlying parent rock material on pedogenic processes is more pronounced. The overall waterholding capacity of these shallow soils on hard rock is lower and therefore water may penetrate deeper into the rock crevices. Carbonates have been leached through the whole profile and as a result noncalcareous terra rossa and brown rendzina soils have been formed on hard carbonate rocks (DAN, 1966). Because of the restricted rainfall in these semi-arid regions, it can be assumed that these soils were never really deep and their present depth is

Fig. 1. Typical erosion and slope pattern in the "badlands" of the semi-arid coastal plains. The fine texture of the B horizon and the underlying, somewhat saline, paleosols (lower part of figure) facilitate the erosion process

a reflection of the initial pedogenic factors described above, and not a result of man-induced erosion as has in general been assumed. On soft rock, on the other hand, soils were deeper and waterholding capacities greater. These soils were cultivated in the past and suffered from erosion. The land use history of these slopes can best be demonstrated on a typical soil-vegetation catena near Beth Guvrin, as shown on two opposing slopes (Fig. 2). The North-facing slope is covered with Nari – an ancient calcrete crust rock – with many outcrops, a non-calcareous brown rendzina soil and a fairly dense canopy of maquis shrubs, dominated by *Quercus calliprinos, Phillyrea media, Pistacia palaestina* and *P. lentiscus*. This slope, because of its rockiness, has apparently never been cultivated and lacks all remnants of terraces. Under the present moderate grazing regime followed by the nearby collective settlement, the slope seems to be well stabilized. On the South-facing slope, however, where the Nari has been stripped off and the underlying soft chalk has been exposed and a highly calcareous pale rendzina has been formed (DAN, 1963), the degradation pattern took a very different course. During Phase 4 this slope – like most, if not all, other pale rendzina slopes – was cultivated. For this purpose the slope was cut to trans-

a

b

Fig. 2. Typical soil profiles from opposing slopes. a Non-calcareous brown rendzina from the Nari slope. The soil is fine and granular-structured, and is well protected by a layer of shrub leaf litter and mulch and by an extensive root system. b Deep, highly calcareous pale rendzina from the soft chalk slope of the non-eroded centre of the terrace

form the circular shape of the hill into a series of rather broad, flat and levelled terraces, bordered by stone wall (RON, 1966). During Phase 5 these terraces were neglected and since then they have been badly eroded. After their abandonment they were covered by dwarfshrub and grass communities, classified by EIG (1927) and ZOHARY (1962) as "batha".

A typical profile diagram of the soil-vegetation degradation pattern on such a terrace is presented in Fig. 3. Where erosion has been most severe, above the ancient terrace wall, the soft chalk rock has been exposed, covered with a scarce vegetation chiefly of *Thymus capitatus*. This is one of the most prominent dwarfshrubs of batha and is a successful invader of erosion pavements, rock outcrops and highly calcareous rendzinas, where heavy competition of quicker growing therophytes is prevented due to the low soil fertility level (LITAV and ORSHAN, 1963).

On the slightly less eroded belt, towards the central part of the terrace, plant cover is more dense and *Asphodelus microcarpus* dominates. This is a tall unpalatable geophyte with a very wide distribution, especially in heavily grazed and degraded open herbaceous communities. On the least eroded and flattest belt, beneath the terrace wall, the rather deep soil includes a definite A and C horizon.

The dense herbaceous sward is dominated by *Asphodelus microcarpus* as well as by *Hyparrhenia hirta* and *Andropogon distachyus,* two drought resistant perennial bunch grasses, prospering on sunny and rocky slopes. These are accompanied by *Avena sterilis,* the most abundant annual grass under light grazing conditions, flowering geophytes such as *Cyclamen persicum, Anemone coronaria* and *Iris palaestina,* and a highly variable mixture of other annual grasses, legumes and herbs.

On the colluvial and gravelly remnants of the broken down terrace wall, dwarfshrubs can be found, while on the stone remnants and the chalky rocks shrubs such as *Pistacia lentiscus* and *Rhamnus palaestina* can be seen. The calcareous, colluvial-alluvial soils of the valley floor, as well as the grumosolic, dark brown clays on the hill plateaux, were cultivated and abandoned only 23 years ago. They are now covered by a dense herbaceous sward. This contains perennial grasses, geophytes and hemicryptophytes, but is dominated by a highly variable therophytic mixture of grasses, legumes, Compositae, Cruciferae, etc. Similar derived annual grasslands can also be found in the more humid mountainous regions and, if not heavily overgrazed and dominated by unpalatable weeds, they constitute the most valuable natural hill pasture in Israel (NAVEH, 1955; SELIGMAN *et al.,* 1959).

With complete cessation of human interference, as implemented in a nearby nature reserve, the maquis shrub vegetation has been transformed into an almost impenetrable thicket, 2–3 m high, uniform in structure and devoid of any herbaceous understorey. A similar trend has also been observed in completely protected maquis reserves in Northern Israel (NAVEH, 1970, 1971).

On the other hand, in the densely populated Hebron mountains, the degradation cycle has been continued and even accelerated as a result of heavy human and livestock pressure: the pale rendzinas which had already suffered in the past from severe erosion are now cultivated again, without reclaiming or reconstructing the broken down terrace walls. Thus their erosion cycle has been renewed. The uneven and poor stands of the barley crop reflects the different erosion belts on these terraces.

On the Nari slopes also, small soil patches between the rock outcrops have now been ploughed and sown with barley. Most of the shrubs have been cut and hand-grubbed for fuel, but a dense grass cover still remains near rock edges and around the sown patches. This cover is dominated by *Poa bulbosa* and *Hordeum bulbosum* which are not only resistant to grazing pressure and drought but serve here as effective soil binders and protectors.

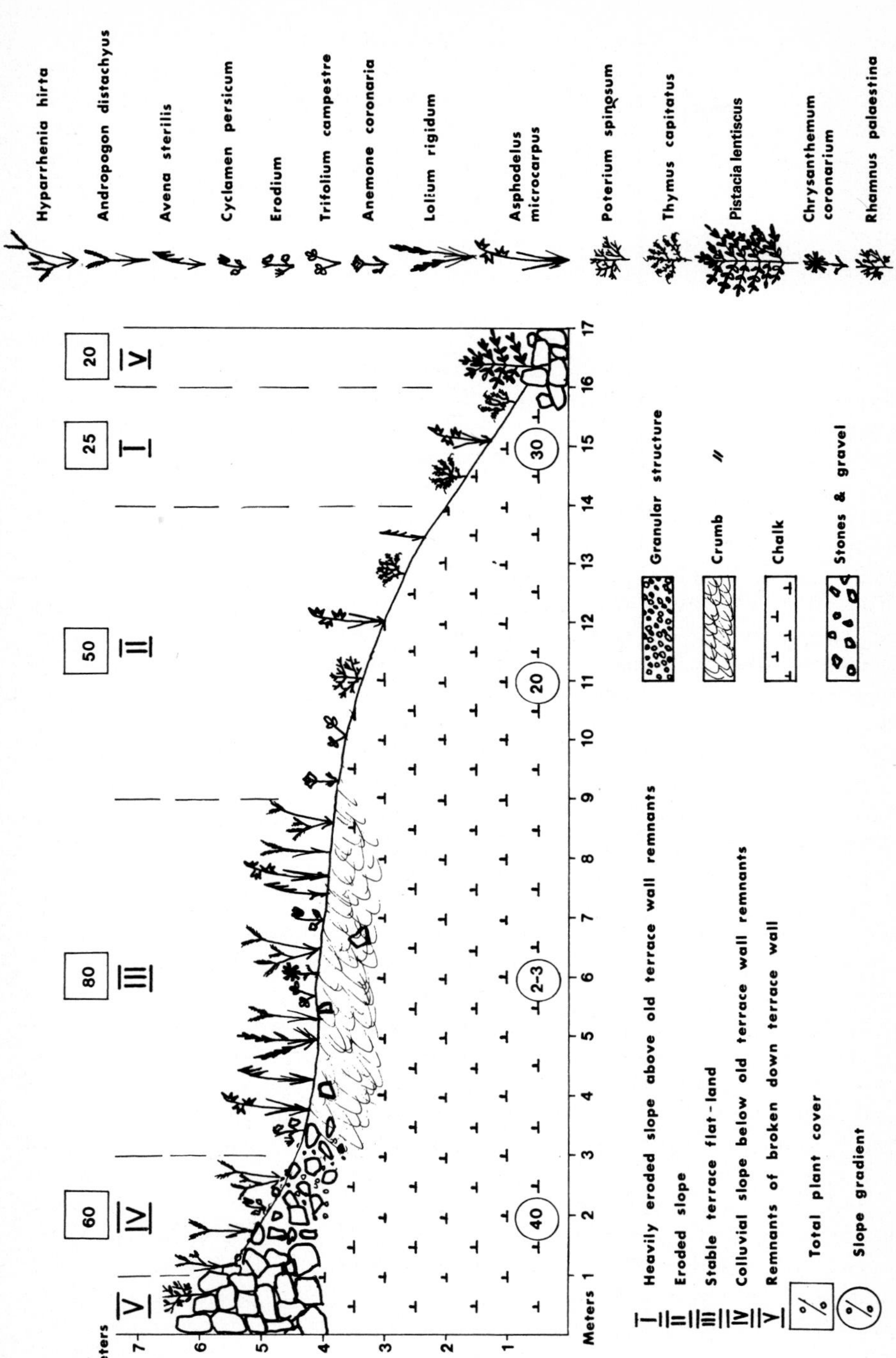

Fig. 3. Soil-vegetation degradation pattern on eroded terraces of pale rendzina on chalk in the southern Judean Hills, near Beit Guvrin

The steeply inclined slopes of hard limestone and dolomite in the Hebron mountains which have never been cultivated, and those with levelled geological bedding whose terraces were abandoned during Phase 5 (RON, 1966), are now also exposed to heavy human and livestock pressure. This pressure, superimposed on the low soil fertility and the harsh climatical conditions, has produced some of the most advanced degradation stages in Israel. They consist of shallow and eroded terra rossa and xeric brown rendzina with a sparse dwarfshrub cover – mainly *Poterium spinosum* on the north-facing slopes – and of bare soil with stunted herbs and geophytes – mainly *Asphodelus microcarpus* on the more exposed and drier sites.

Proceeding further north we reach the subhumid and humid zones of north and central Israel. Here, in the plains and valleys, typical soils are comprised of fine textured, deep vertisols and, on the plateaux of Eastern Galilee and Golan, of comparatively shallow, basaltic vertisols. Like their counterparts in the drier regions, these soils have also been cultivated since Phase 4, and after complete destruction of their pristine vegetation, they have apparently suffered from severe sheet erosion. This is especially true on the basaltic slopes which have been partly stripped of their soil cover, leaving shallow lithosols and many rock outcrops.

In the eastern coastal plains, erosion has caused the exposure of ancient, calcified paleosols (DAN and YAALON, 1966). From the deeper vertisols no direct evidence on erosion can be acquired because their soil profile has the same soil texture all through the soil-paleosol complex. These soils are all intensively cultivated, in most cases with some precautions for soil erosion control. Nevertheless, in years with high rainfall intensities, accelerated erosion is evident, especially on the basaltic slopes sown with winter cereals or clean-fallowed for summer crops. Thus, the amount of soil carried from a 22 km^2 large basaltic watershed into a water reservoir from a single rainstorm in December 1955, was 20.000 m^3 – equivalent to about 1 mm layer of the total surface soil. The average of 6 years erosion amounted to 0.4 mm per year (NIR, 1970).

The mature soil of the sandy and loamy parts of the Sharon Coastal Plains comprise a specific catenary system (DAN, YAALON and KUYUMDJISKY, 1968) with medium textured, typical ABC red mediterranean soils ("Hamra") on the moderate slopes, replaced by impervious pseudogleys ("Nazaz") and hydromorphic vertisols in the depression. The Hamra soils resemble red mediterranean soils on granite and related rocks in California and Chile. They have also suffered accelerated erosion after destruction of the original, pre-settlement vegetation. The upper, porous sandy A horizons, and on steeper slopes even the more impervious B horizons, have been eroded away, leaving the sandy C horizons to form a new soil profile. Most of the eroded material has accumulated at footslope positions, covering the pseudogleys and fine textured soils of the catena. Pottery shreds in the various layers of these footslopes and floodplains reveal their recent origin. In this case, destructive human activity has also to some extent improved the agriculture potentials of these soils which are now used chiefly for citrus plantations, but are also increasingly encroached upon by built-up areas.

The extensive Tabor oak woodlands which covered the Sharon Plains before their destruction by Turks during World War I were considered by EIG (1933) to be remnants of the original climax forests before their early cultivation. However, in contrast to other maquis species, this deciduous and drought resistant tree germinates and develops freely in open fields. It could, therefore, have been an efficient re-invador or survivor of abandoned and eroded fields and foothills (NAVEH, 1969). Herbaceous communities, dominated by

Destmostachya bipinnata and segetalic plants, appear to be a further degradation stage (EIG, 1933; ZOHARY, 1962). They now remain only on few sites not yet used for irrigated cultivation or for built-up areas.

In the mountainous areas of the subhumid zone the soils resemble in general those of the drier regions, although terra rossa and karstic areas are usually much more widespread. This is also true for the maquis scrub vegetation which is usually much higher and better developed and has apparently also better regenerating powers after cutting, burning and grazing. Here also, as long as the rocky, Nari and karstic slopes are covered by more or less dense shrub and grass vegetation, the danger of accelerated erosion is very slight. The shallow terra rossa and brown rendzina soils are well structured, rich in organic matter and have a high infiltration capacity. Thus, even on fire denuded, steep slopes, no signs of run-off and soil movement have been observed after heavy rain storms in Western Galilee. On the other hand, on poorer, more exposed sites and on less fertile, highly calcareous slopes disturbed and compacted by cattle and goat grazing, erosion damage after fire has been considerable (NAVEH, 1960).

In these higher rainfall regions, pale rendzinas which were terraced, cultivated and subsequently neglected, have suffered most from severe erosion. Some are still cultivated and on eroded slopes advantage is taken of the rapid soil formation, due to the easy breakdown of the soft chalky parent rock. Those which were abandoned in 1948, are now densely covered with *Poterium spinosum* "batha".

Mountain plateaux and moderate slopes with considerably deep terra rossa were cultivated during Phase 4 and are still used for olive, fruit and vine plantations. This was accomplished by breaking the long slope into terraces and erecting stone walls on rock outcrops, which collected the soil from the upper part of the terrace and levelled its surface (RON, 1966).

These terraces were apparently abandoned after the Crusader period, and since then they have been covered with a dense canopy of maquis trees, shrubs and dwarfshrubs. On the plateaux, part of the old terraces have been reclaimed and new ones have been constructed with the help of heavy machinery. The soil depth on these plateaux reaches 60 to 80 cm.

In general, trees and shrubs growing between the stone walls on cultivated terraces were not removed by Arabs and older cultures. This may explain why these ancient wall-remnants harbour by far the most developed maquis trees, chiefly *Quercus calliprinos, Phillyrea media* and *Pistacia palaestina*. They reach a height of 6 m on the plateaux, but down the slope, where erosion has removed part of the soil, the maquis vegetation becomes gradually poorer and is replaced by garrigue shrubs, chiefly *Calycotome villosa*, and by batha dwarfshrubs – chiefly *Poterium spinosum*. In certain instances where drainage has been impaired the central and lowest part of the terraces are occupied by a dense herbaceous sward, dominated by perennial and annual grasses, chiefly *Hordeum bulbosum, Dactylis glomerata, Oryzopsis milacea* and *Avena sterilis* (Fig. 4).

However, on similar plateaux and slopes near villages which were not deserted after the Crusader defeat, these terraces were cultivated and grazed for much longer periods and therefore anthropogenic degradation has proceeded much farther. The terrace walls were destroyed and the soil was washed away, leaving only a shallow soil layer of 20 cm and exposing the rock surface. Moreover, stone heaps were left in the center of the terraces from stones which, after being exposed by erosion, were no longer carried away to the terrace walls as in ancient times. The vegetation cover is scarce and consists chiefly of

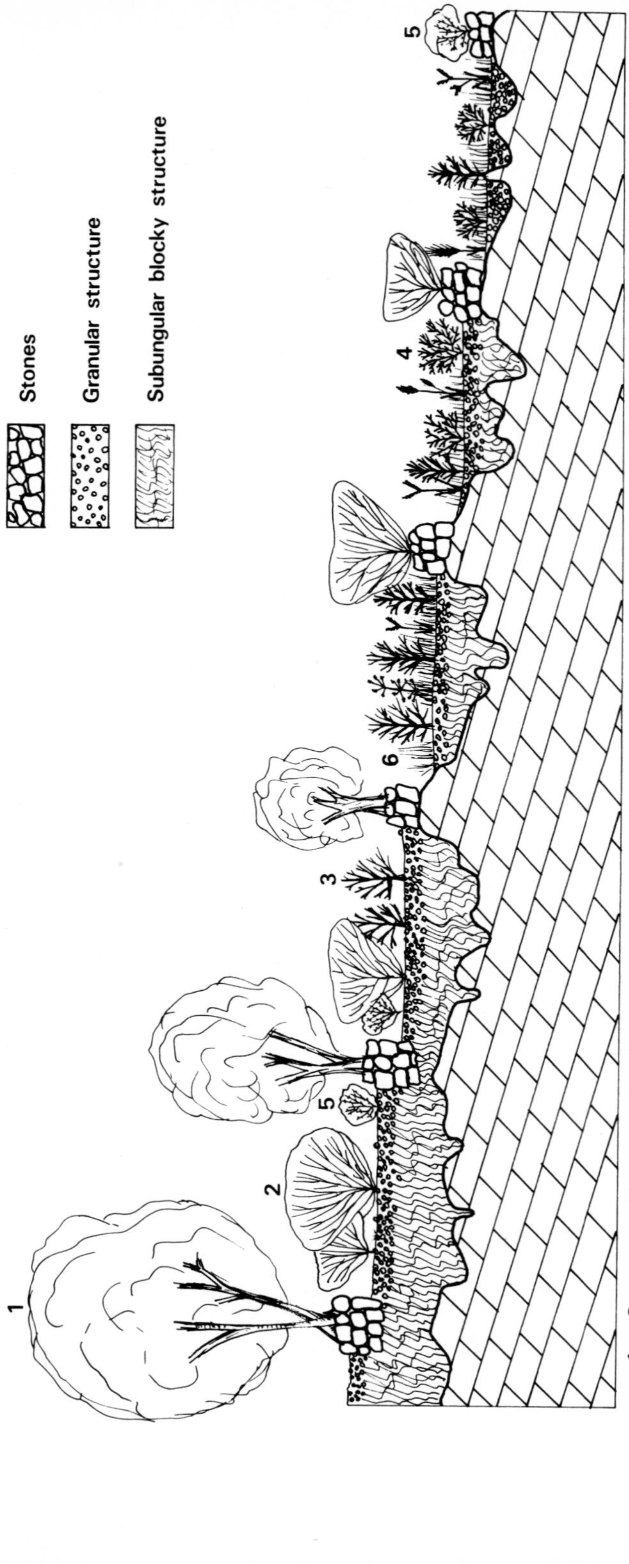

Fig. 4. Schematic downslope degradation pattern of soil-vegetation systems on abandoned terraces of terra rossa on dolomite in Western Galilee near Avdon

Calycotome villosa and stunted *Pistacia lentiscus* bushes near the remnants of the terrace walls and of *Poterium spinosum* with some scattered *Hordeum bulbosum* plants and geophytes, chiefly *Asphodelus microcarpus*.

Discussion and Conclusions

Most of the studies on degradation of mediterranean vegetation types have recently been summarized by WALTER (1968), who concluded that the whole mediterranean region is composed of innumerable variants of different degradation and regeneration stages.

In this chapter we have attempted to describe man-induced landscape degradation as a function of the soil-vegetation system. This descriptive work should be followed by more detailed quantitative ecosystem studies along environmental and anthropogenic gradients in Israel, as well as in other mediterranean countries.

These multivariate "biofunctions" are governed by complex interactions between mode, duration and intensity of human interference, initial site conditions and state factors, climatic and biotic flux potentials and by the resilience and recuperative powers of the ecosystem. It will, therefore, be difficult to quantify these equations and they can – at the most – be presented as crudely simplified models of allelogenic, regressive and progressive successions.

Nevertheless the following generalizations can be made:

1) The harsher and more fragile the environment, the more far-reaching and irreversible will be the man-induced changes in state variables and the slower and more difficult will be the process of recovery. Thus, the complete destruction of the vegetation canopy in the semi-arid zone at a very early date exposed the soil to the direct impact of the harsh climate and to unimpeded rainfall, and created conditions similar to those of more arid regions with a scarce plant cover, climatic extremes and flashfloods. The increase in sediment load from the denuded slopes exceeded in many cases the sediment load capacities of the runoff water, especially at footslope positions and floodplains, and caused aggradational processes which characterize alluvial fans in extremely arid regions. This has also been shown by VITA-FINZI (1969) for other mediterranean countries.

2) These anthropogenic biofunctions were of greatest complexity and most far-reaching on upland arable slopes. Here, several degradation and regeneration cycles have occurred. The first erosion cycle was caused by the clearance of forest and woodlands in the early days of Phase 4. This was followed by an aggradation cycle of terracing – one of the few instances where man's land use improved the factors controlling soil formation and created deeper and more fertile soil. Terracing was undertaken to control erosion through change of original base levels and formation of various smaller secondary slopes with local base levels. However, on negligence of these terraces, during Phase 5, the local base level disintegrated. Moreover, many very steep subslopes were exposed and catastrophic erosion cycles occurred. As illustrated in examples from the Judean hills and Western Galilee, this has been followed by a regeneration cycle, leading to maquis, garrigue, batha and derived grasslands. But under recently increasing human and livestock pressure the erosive degradation cycle has been renewed, leading to bare soil and *Asphodelus* "deserts".

3) The soil recovery process has been slowest on shallow soils covering hard rocks, and faster on soft rock (DAN, 1966). Only in the mountainous region can natural vegetation

recovery still be studied. The dynamics of recovery are apparently determined by the recent "biofunction" and especiallly by soil depth, by slope exposure and by the "initial floristic composition" (EGLER, 1954) of the abandoned site. Thus, the presence of maquis trees and shrubs are chiefly the result of root re-vegetation from coppice, and not a result of directional re-ecesis from seeds into lower seral stages of garrigue, batha or grassland. The complex, highly diversified and variable pattern of vegetation-regeneration of the latter is determined chiefly by past and present biotic history, and by competition for light, moisture and nutrients (LITAV, 1965). Allelopathic interactions may sometimes be involved (NAVEH, 1967).

4) The long duration of these "biofunctions" has also enabled the change in biotic flux potentials, by encouraging not only the invasion of more xeric elements from adjacent more arid zones, but also the selection and evolution of local biotypes with the best adaptive resistance to the constant pressure of defoliation by fire, cutting and grazing. This might explain the remarkable regeneration powers of shrubs such as *Pistacia lentiscus* and trees such as *Quercus calliprinos* and *Phillyrea media*. These regenerative powers, coupled with their capacity to afford effective soil protection on steep and rocky slopes, are the reasons for the relative stability of these shrub ecosystems, in spite of their long history of misuse and degradation (NAVEH, 1960). Similar striking resilience is also shown by mediterranean grasslands, containing hardy species and biotypes of perennial grasses such as *Hordeum bulbosum, Dactylis glomerata judaica, Poa bulbosa* and *Hyperrhenia hirta*.

The history of human degradation of mediterranean ecosystems in California and Chile has probably been too short to enable similar evolutionary processes to take place. Many of the indigenous plants have succumbed to human pressure and have been replaced by invading species from Europe, and especially from the Mediterranean region. Unfortunately, these invaders are mostly weedy annuals and do not include any of the abovementioned perennial grasses (NAVEH, 1967). Thus, in spite of their much shorter duration, these disruptions have, in most cases, been much more catastrophic and far-reaching than in Mediterranean Basin regions.

One of the main conclusions which can be derived from the study of the mountainous region of Israel is that during the long phase of agricultural decay and population decline a new equilibrium has been established on the non-cultivated upland ecosystems which are neither overgrazed and heavily coppiced, nor completely protected. This man-maintained equilibrium between trees, shrubs, herbs, grasses and geophytes, and between dependent and controlling ecosystem state factors, has contributed much to the biological diversity, stability and attractiveness of the mediterranean landscape. It is, without doubt, its most important asset for recreation and tourism. It is now endangered not only by population explosion and by increasing intensity of traditional land use, but also by accelerated speed of urban sprawl and neo-technological despoilation, pollution and erosion.

If these most recent and most threatening trends of landscape degradation proceed unhampered with their present speed and intensity, the few spots of open landscape with natural ecosystems which will remain at all, will be turned into over-crowded recreational slums. This fate has already reached the shores of the Mediterranean Sea and inland waters which rank highest in outdoor recreation demand, but it will very soon also reach mediterranean mountains and hill country in Israel and elsewhere.

References

BARRY, J. R., LEROY LADURIE, E.: Histoire agricole et phytogéographie. Annales Economies, Sociétés Civilisations **17**, 434–477 (1962).

BOYKO, H.: A new plant geographical subdivision of Israel (as an example for South West Asia). Vegetatio **5/6**, 309–318 (1954).

BUTZER, K. W.: Climatic change in arid regions since the Pliocene. 31–56. In: A history of land use in arid regions (STAMP, D., Ed.) Paris: UNESCO 1961.

DAN, J.: The disintegration of Nari lime crust in relation to relief, soil and vegetation. 189–194. In: Trans. Int. Symp. Photo-Interpretation. Delft: ITC 1963.

DAN, J.: The effect of relief on soil formation and distribution in Israel. Pamphlet No. 100. Rehovot: The Volcani Institute of Agric. Res. 1966.

DAN, J., YAALON, D. H.: Trends of soil development with time in the mediterranean environments of Israel. Trans. Int. Conf. Medit. Soils, Madrid. 139–145 (1966).

DAN, J., YAALON, D. H.: On the origin and nature of the paleopedological formations in the central coastal desert fringe areas of Israel. 245–260. In: Paleopedology: origin, nature and dating of paleosols (YAALON, D. H., Ed.) Jerusalem: Israel Universities Press 1971.

DAN, J., YAALON, D. H., KOYUMDJISKY, H.: Catenary soil relationship in Israel. I. The Netanya catena on coastal dunes of the Sharon. Geodezmu **2**, 95–120 (1968).

EGLER, F. E.: Vegetation science concepts. I. Initial floristic composition, a factor in old-field vegetation development. Vegetatio **4**, 412–417 (1954).

EIG, A.: On the vegetation of Palestine. Inst. Agr. and Nat. Hist. Agr. Exp. Sta. Bull. **7**, (1927).

EIG, A.: A historical-phytosociological essay on Palestinian forests of *Quercus aegilops* L. ssp. *ithaburensis* (Desc.) in past and present. Beih. Bot. Cbl. **51**, 225–272 (1933).

FRENCH, M. H.: Observations on the goat. FAO Agricultural Studies No. 80. Rome: FAO 1970.

GARROD, D. A.: The Natufean culture: The life and economy of Mesolithic people in the Near East. Proc. Brit. Acad. **43**, 211–227 (1958).

GARROD, D. A., BATE, D. M. A.: The Stone Age of Mt. Carmel. Oxford: Claredon Press 1937.

GUY, P. L. O.: Archeological evidence of soil erosion and sedimentation in Wadi Musrara. Israel Expl. J. **4**, 77–87 (1954).

HOROWITZ, D.: Upper Pleistocene-Holocen climate and vegetation of the Northern Jordan Valley. Ph. D. thesis. Jerusalem: Hebrew University 1968. (In Hebrew with English summary.)

HOWELL, F. C.: Ambrona/Torralba: Acheulian open-air occupation sites in northern Spain. Annual meeting Amer. Anthrop. Soc. Chicago, No. 15–18 (1962).

HUNTINGTON, E.: Civilization and climate. New Haven: Yale University Press 1924.

JENNY, H. J.: Derivation of state factor equations of soil and ecosystems. Proc. Soil Sci. Soc. Am. **25**, 385–388 (1961).

KNAPP, R.: Die Vegetation von Kaphallinia Griechenland. Königstein 1965.

LITAV, M.: Effects of soil type and competition on the occurrence of *Avena sterilis* L. in the Judean Hills (Israel). Israel J. Bot. **14**, 74–89 (1965).

LITAV, M., ORSHAN, G.: Ecological studies on some sub-lithophytic communities in Israel. Israel J. Bot. **12**, 41–54 (1963).

LOWDERMILK, W. C.: Palestine, Land of Promise. London-New York: Harper 1944.

MIKESELL, M. W.: The deforestation of Mount Lebanon. Geogr. Rev. **59**, 1–28 (1969).

MURPHEY, R.: The decline of North Africa since the Roman occupation: climatic or human? Ann. Ass. Am. Geogr. **41**, 116–132 (1951).

NAVEH, Z.: Some aspects of range improvement in a mediterranean environment. J. Range Mgmt. **8**, 265–270 (1955).

NAVEH, Z.: Agro-ecological aspects of brush range improvement in the maqui belt of Israel. Ph. D. thesis. Jerusalem: Hebrew University 1960. (In Hebrew with English summary.)

NAVEH, Z.: Mediterranean ecosystems and vegetation types in California and Israel. Ecology **48**, 345–359 (1967).

NAVEH, Z.: Comparative dynamics of mediterranean Oak Savannas in Israel and California. Proc. XI Int. Bot. Cong. Seattle, 157 (1969).

NAVEH, Z.: Effect of integrated ecosystem management on productivity of a degraded mediterranean hill pasture in Israel. Proc. XI Int. Grassld. Cong., Queensland, 59–63 (1970).

NAVEH, Z.: The conservation of ecological diversity of mediterranean ecosystems through ecological management. 605–622. In: The scientific management of animal and plant communities for conservation (DUFFEY, E. and WATT, A. S., Eds). Oxford: Blackwell 1971.

NAVEH, Z., RON, B.: Agro-ecological management of mediterranean ecosystems – the basis for intensive pastoral hill-land use in Israel. Proc. X Int. Grassld. Cong., Helsinki 4/15, 871–874 (1966).

NIR, D.: Geomorphology of Israel. Jerusalem: Akdamon 1970. (In Hebrew).

OAKLEY, K. P.: On man's use of fire, with comments on toolmaking and hunting. Viking Fd Publ. Anthrop. **31,** 176–193 (1961).

REIFENBERG, A.: The struggle between the desert and the sown. Rise and fall of the Levant. Jerusalem: Jewish Agency Publication Department 1955.

RICKLI, M.: Das Pfanzenkleid der Mittelmeerländer. Berlin: Hans Huber 1943.

RON, Z.: Agricultural terraces in the Judean mountains, Israel. Israel Expl. J. **16,** 33–49, 111–122 (1966).

SAUER, C. O.: The agency of man on earth. 49–64. In: Man's role in changing the face of earth. (THOMAS, W. L., Ed.). Chicago: Chicago University Press 1957.

SELIGMAN, N., ROSENSAFT, Z., TADMOR, N., KATZNELSON, J., NAVEH, Z.: Natural pastures in Israel, vegetation, carrying capacity and improvement. Tel Aviv: Sifriat Hapoalim 1959. (In Hebrew with English summary).

SEMPLE, E. C.: The geography of the mediterranean region. New York: Henry Holt 1931.

STEWART, O. C.: Fire as the first great force employed by man. 115–133. In: Man's role in changing the face of earth (THOMAS, W. L., Ed.). Chicago: Chicago University Press 1957.

TAYLOR, F. H.: The destruction of the soil in Palestine. Bull. Soil Conserv. Bd. Palest. 2 (1946).

TISDALE, E. W.: A study of dry-land conditions and problems in portions of Southwest Asia, North Africa and the Eastern Mediterranean. Report 3. Dry-land Res. Inst., University Cal. Riverside 1967. (Mimeo).

UNESCO-FAO: Bioclimatic map of the mediterranean zone. Explanatory notes. Arid Zone Research 21. Paris: UNESCO 1963.

VITA-FINZI, C.: The mediterranean valleys. Geological changes in historical times. Cambridge: Cambridge Press 1969.

WALTER, H.: Die Vegetation der Erde in öko-physiologischer Betrachtung. 64–78. Band II. Die gemässigten und arktischen Zonen. Jena: Gustav Fischer Verlag 1968.

WHYTE, R. O.: Evolution of land use in South-Western Asia. 57–118. In: A history of land use in arid regions (STAMP, D., Ed). Paris: UNESCO 1961.

ZOHARY, M.: Plant life in Palestine. New York: Ronald Press 1962.

Subject Index

Ecological Studies

In production:

Vol. 5: Arid Zone Irrigation. Edited by B. YARON, E. DANFORS and Y. VAADIA
Vol. 6: K. STERN and L. ROCHE: Genetics of Forest Ecosystems
Vol. 8: H. LIETH: Phenology and Seasonality Modeling

In preparation:

A. BAUMGARTNER: Forest Energy Budgets
F. GOLLEY: Tropical Ecology
A. HASLER: Land-Water Interactions
O. W. HEAL and D. F. PERKINS: The Ecology of British Moors and Mountain Grasslands
T. KIRA et al.: Plants, Population and Logistic Growth
M. KLUGE: Crassulacean Acid Metabolism
J. KRANZ: Mathematical Analysis and Modeling of Epidemics
B. OSMOND, O. BJÖRKMAN and D. ANDERSEN: Ecology and Ecophysiology of Atriplex
A. POLJAKOFF-MAYBER: Plants in Saline Environments
B. SLAVIK: Methods of Studying Water Plant Relations
F. WIELGOLASKI: Scandinavian Tundra Biome